# RELIEF, DRAINAGE AND URBAN AREAS

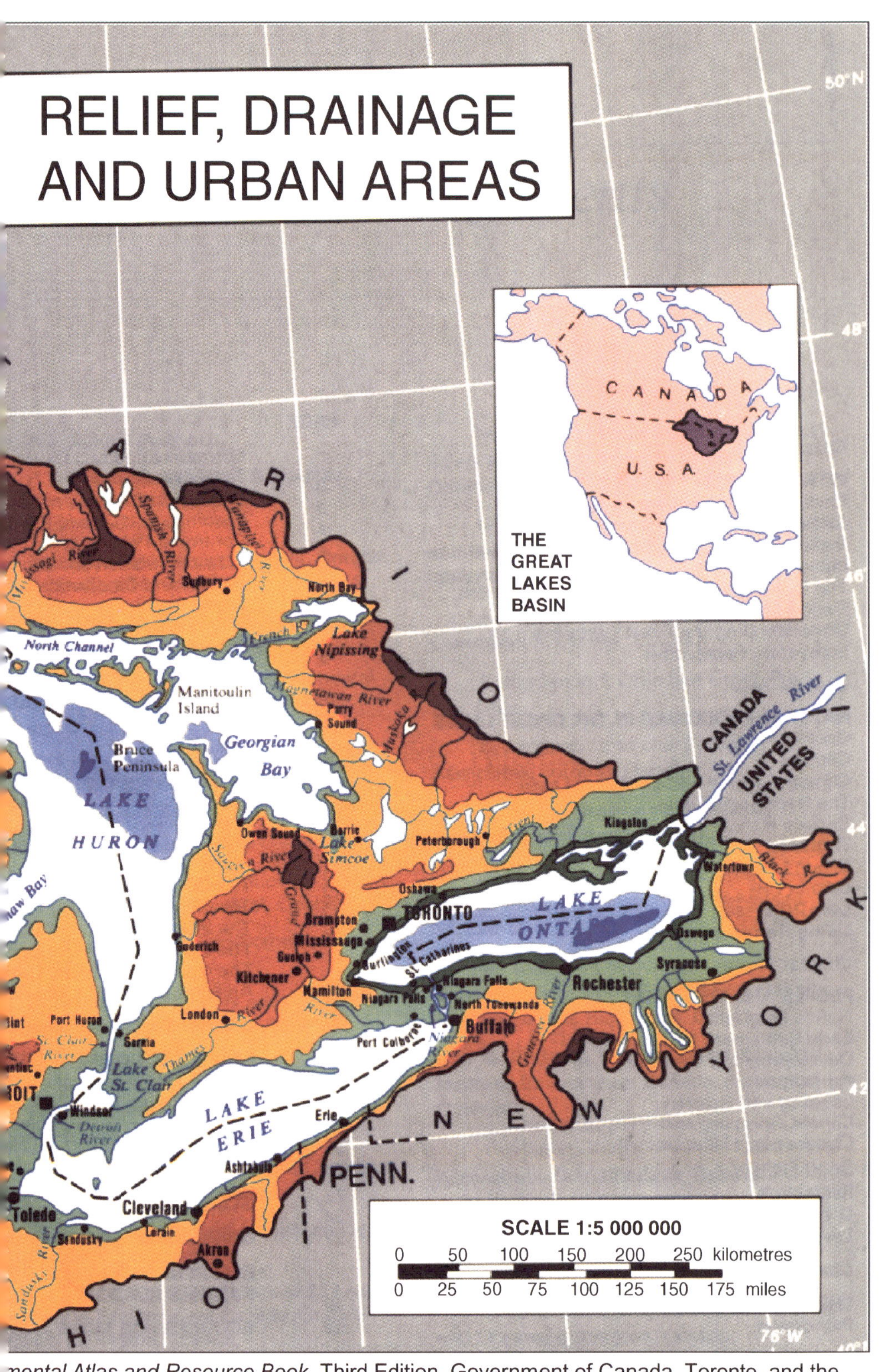



*...mental Atlas and Resource Book*, Third Edition, Government of Canada, Toronto, and the
...95)

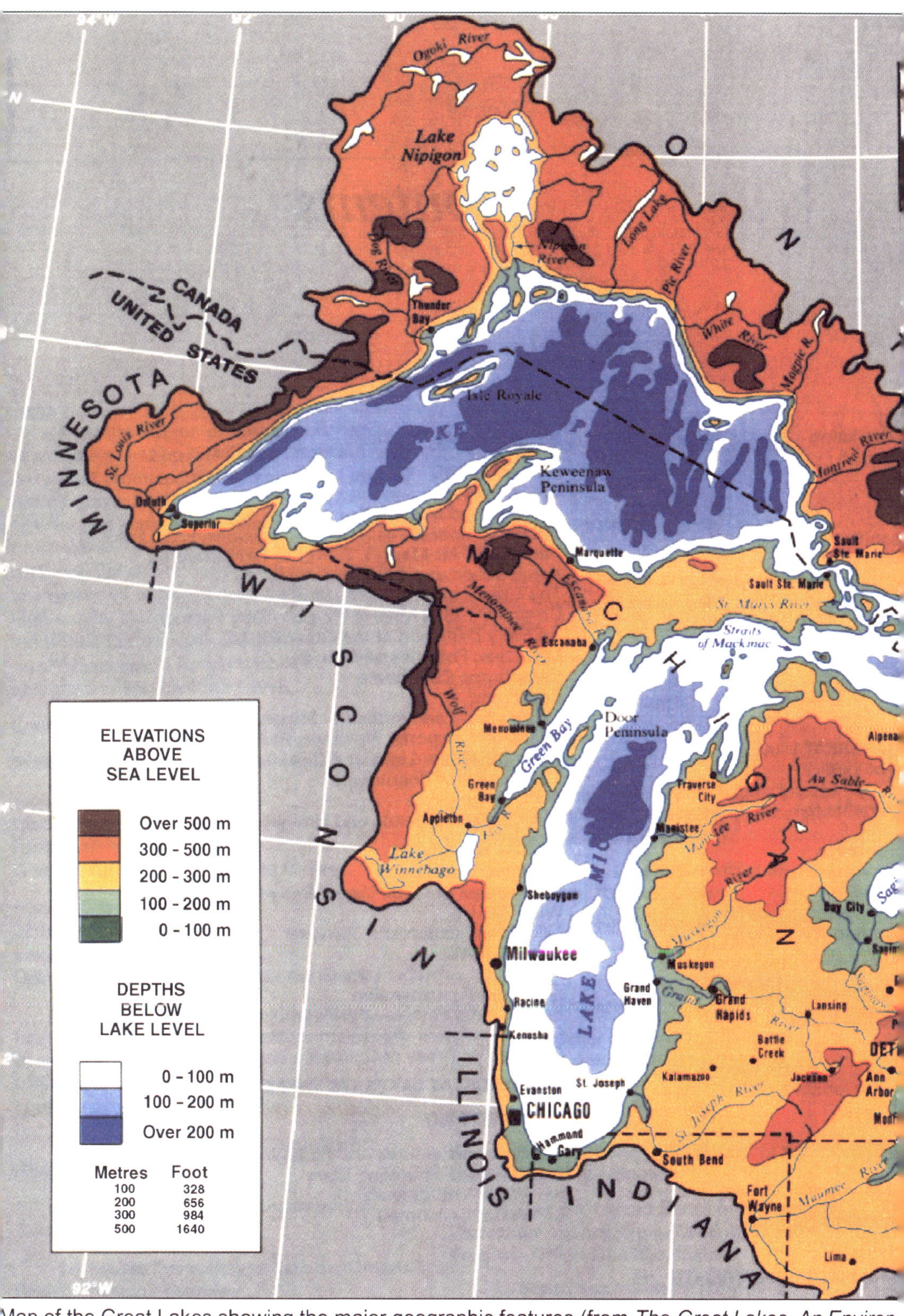


Map of the Great Lakes showing the major geographic features (from *The Great Lakes, An Environ*
United States Environmental Protection Agency, Great Lakes National Program Office, Chicago, 19

# The Handbook of Environmental Chemistry

Editor-in-Chief: O. Hutzinger

Volume 5  Water Pollution
Part N



Advisory Board:

# The Handbook of Environmental Chemistry
## Recently Published and Forthcoming Volumes

# Persistent Organic Pollutants in the Great Lakes

Volume Editor:
Ronald A. Hites

With contributions by

J. E. Baker · T. F. Bidleman · D. L. Carlson · K. Coady
S. J. Eisenreich · J. P. Giesy · P. A. Helm · R. A. Hites
K. C. Hornbuckle · L. M. Jantunen · P. D. Jones · K. Kannan
S. A. Mabury · J. W. Martin · D. C. G. Muir · J. L. Newsted
R. J. Norstrom · J. H. Offenberg · J. Ridal · M. F. Simcik
J. Struger · D. L. Swackhamer

 Springer

Environmental chemistry is a rather young and interdisciplinary field of science. Its aim is a complete description of the environment and of transformations occurring on a local or global scale. Environmental chemistry also gives an account of the impact of man's activities on the natural environment by describing observed changes.

The Handbook of Environmental Chemistry provides the compilation of today's knowledge. Contributions are written by leading experts with practical experience in their fields. The Handbook will grow with the increase in our scientific understanding and should provide a valuable source not only for scientists, but also for environmental managers and decision-makers.

The Handbook of Environmental Chemistry is published in a series of five volumes:

Volume 1: The Natural Environment and the Biogeochemical Cycles
Volume 2: Reactions and Processes
Volume 3: Anthropogenic Compounds
Volume 4: Air Pollution
Volume 5: Water Pollution

The series Volume 1 The Natural Environment and the Biogeochemical Cycles describes the natural environment and gives an account of the global cycles for elements and classes of natural compounds. The series Volume 2 Reactions and Processes is an account of physical transport, and chemical and biological transformations of chemicals in the environment.

The series Volume 3 Anthropogenic Compounds describes synthetic compounds, and compound classes as well as elements and naturally occurring chemical entities which are mobilized by man's activities.

The series Volume 4 Air Pollution and Volume 5 Water Pollution deal with the description of civilization's effects on the atmosphere and hydrosphere.

Within the individual series articles do not appear in a predetermined sequence. Instead, we invite contributors as our knowledge matures enough to warrant a handbook article.

Suggestions for new topics from the scientific community to members of the Advisory Board or to the Publisher are very welcome.

Library of Congress Control Number: 2005938926

ISSN 1433-6863
ISBN-10 3-540-29168-7 Springer Berlin Heidelberg New York
ISBN-13 978-3-540-29168-8 Springer Berlin Heidelberg New York
DOI 10.1007/b13133



**Springer is a part of Springer Science+Business Media**

springer.com



Cover design: E. Kirchner, Springer-Verlag
Typesetting and Production: LE-TEX Jelonek, Schmidt & Vöckler GbR, Leipzig

Printed on acid-free paper    02/3141 YL – 5 4 3 2 1 0

## Prof. Dr. D. Mackay

Department of Chemical Engineering
and Applied Chemistry
University of Toronto
Toronto, ON, M5S 1A4, Canada

## Prof. Dr. A. H. Neilson

Swedish Environmental Research Institute
P.O. Box 21060
10031 Stockholm, Sweden
*ahsdair@ivl.se*

## Prof. Dr. J. Paasivirta

Department of Chemistry
University of Jyväskylä
Survontie 9
P.O. Box 35
40351 Jyväskylä, Finland

## Prof. Dr. Dr. H. Parlar

Institut für Lebensmitteltechnologie
und Analytische Chemie
Technische Universität München
85350 Freising-Weihenstephan, Germany

## Prof. Dr. S. H. Safe

Department of Veterinary
Physiology and Pharmacology
College of Veterinary Medicine
Texas A & M University
College Station, TX 77843-4466, USA
*ssafe@cvm.tamu.edu*

## Prof. P. J. Wangersky

University of Victoria
Centre for Earth and Ocean Research
P.O. Box 1700
Victoria, BC, V8W 3P6, Canada
*wangers@telus. net*

# The Handbook of Environmental Chemistry
## Also Available Electronically

For all customers who have a standing order to The Handbook of Environmental Chemistry, we offer the electronic version via SpringerLink free of charge. Please contact your librarian who can receive a password or free access to the full articles by registering at:

springerlink.com

If you do not have a subscription, you can still view the tables of contents of the volumes and the abstract of each article by going to the SpringerLink Homepage, clicking on "Browse by Online Libraries", then "Chemical Sciences", and finally choose The Handbook of Environmental Chemistry.

You will find information about the

- Editorial Board
- Aims and Scope
- Instructions for Authors
- Sample Contribution

at springer.com using the search function.

# Preface

Environmental Chemistry is a relatively young science. Interest in this subject, however, is growing very rapidly and, although no agreement has been reached as yet about the exact content and limits of this interdisciplinary discipline, there appears to be increasing interest in seeing environmental topics which are based on chemistry embodied in this subject. One of the first objectives of Environmental Chemistry must be the study of the environment and of natural chemical processes which occur in the environment. A major purpose of this series on Environmental Chemistry, therefore, is to present a reasonably uniform view of various aspects of the chemistry of the environment and chemical reactions occurring in the environment.

The industrial activities of man have given a new dimension to Environmental Chemistry. We have now synthesized and described over five million chemical compounds and chemical industry produces about hundred and fifty million tons of synthetic chemicals annually. We ship billions of tons of oil per year and through mining operations and other geophysical modifications, large quantities of inorganic and organic materials are released from their natural deposits. Cities and metropolitan areas of up to 15 million inhabitants produce large quantities of waste in relatively small and confined areas. Much of the chemical products and waste products of modern society are released into the environment either during production, storage, transport, use or ultimate disposal. These released materials participate in natural cycles and reactions and frequently lead to interference and disturbance of natural systems.

Environmental Chemistry is concerned with reactions in the environment. It is about distribution and equilibria between environmental compartments. It is about reactions, pathways, thermodynamics and kinetics. An important purpose of this Handbook, is to aid understanding of the basic distribution and chemical reaction processes which occur in the environment.

Laws regulating toxic substances in various countries are designed to assess and control risk of chemicals to man and his environment. Science can contribute in two areas to this assessment; firstly in the area of toxicology and secondly in the area of chemical exposure. The available concentration ("environmental exposure concentration") depends on the fate of chemical compounds in the environment and thus their distribution and reaction behaviour in the environment. One very important contribution of Environmental Chemistry to the above mentioned toxic substances laws is to develop laboratory test methods, or mathematical correlations and models that predict the environmental fate of new chemical compounds. The third purpose of this Handbook is to help in the basic understanding and development of such test methods and models.

The last explicit purpose of the Handbook is to present, in concise form, the most important properties relating to environmental chemistry and hazard assessment for the most important series of chemical compounds.

At the moment three volumes of the Handbook are planned. Volume 1 deals with the natural environment and the biogeochemical cycles therein, including some background information such as energetics and ecology. Volume 2 is concerned with reactions and processes in the environment and deals with physical factors such as transport and adsorption, and chemical, photochemical and biochemical reactions in the environment, as well as some aspects of pharmacokinetics and metabolism within organisms. Volume 3 deals with anthropogenic compounds, their chemical backgrounds, production methods and information about their use, their environmental behaviour, analytical methodology and some important aspects of their toxic effects. The material for volume 1, 2 and 3 was each more than could easily be fitted into a single volume, and for this reason, as well as for the purpose of rapid publication of available manuscripts, all three volumes were divided in the parts A and B. Part A of all three volumes is now being published and the second part of each of these volumes should appear about six months thereafter. Publisher and editor hope to keep materials of the volumes one to three up to date and to extend coverage in the subject areas by publishing further parts in the future. Plans also exist for volumes dealing with different subject matter such as analysis, chemical technology and toxicology, and readers are encouraged to offer suggestions and advice as to future editions of "The Handbook of Environmental Chemistry".

Most chapters in the Handbook are written to a fairly advanced level and should be of interest to the graduate student and practising scientist. I also hope that the subject matter treated will be of interest to people outside chemistry and to scientists in industry as well as government and regulatory bodies. It would be very satisfying for me to see the books used as a basis for developing graduate courses in Environmental Chemistry.

Due to the breadth of the subject matter, it was not easy to edit this Handbook. Specialists had to be found in quite different areas of science who were willing to contribute a chapter within the prescribed schedule. It is with great satisfaction that I thank all 52 authors from 8 countries for their understanding and for devoting their time to this effort. Special thanks are due to Dr. F. Boschke of Springer for his advice and discussions throughout all stages of preparation of the Handbook. Mrs. A. Heinrich of Springer has significantly contributed to the technical development of the book through her conscientious and efficient work. Finally I like to thank my family, students and colleagues for being so patient with me during several critical phases of preparation for the Handbook, and to some colleagues and the secretaries for technical help.

I consider it a privilege to see my chosen subject grow. My interest in Environmental Chemistry dates back to my early college days in Vienna. I received significant impulses during my postdoctoral period at the University of California and my interest slowly developed during my time with the National Research Council of Canada, before I could devote my full time of Environmental Chemistry, here in Amsterdam. I hope this Handbook may help deepen the interest of other scientists in this subject.

Amsterdam, May 1980                                                               *O. Hutzinger*

Twenty-one years have now passed since the appearance of the first volumes of the Handbook. Although the basic concept has remained the same changes and adjustments were necessary.

Some years ago publishers and editors agreed to expand the Handbook by two new open-end volume series: Air Pollution and Water Pollution. These broad topics could not be fitted easily into the headings of the first three volumes. All five volume series are integrated through the choice of topics and by a system of cross referencing.

The outline of the Handbook is thus as follows:

1. The Natural Environment and the Biochemical Cycles,
2. Reaction and Processes,
3. Anthropogenic Compounds,
4. Air Pollution,
5. Water Pollution.

Rapid developments in Environmental Chemistry and the increasing breadth of the subject matter covered made it necessary to establish volume-editors. Each subject is now supervised by specialists in their respective fields.

A recent development is the accessibility of all new volumes of the Handbook from 1990 onwards, available via the Springer Homepage springeronline.com or springerlink.com.

During the last 5 to 10 years there was a growing tendency to include subject matters of societal relevance into a broad view of Environmental Chemistry. Topics include LCA (Life Cycle Analysis), Environmental Management, Sustainable Development and others. Whilst these topics are of great importance for the development and acceptance of Environmental Chemistry Publishers and Editors have decided to keep the Handbook essentially a source of information on "hard sciences".

With books in press and in preparation we have now well over 40 volumes available. Authors, volume-editors and editor-in-chief are rewarded by the broad acceptance of the "Handbook" in the scientific community.

Bayreuth, July 2001                                                            *Otto Hutzinger*

# Contents

# Foreword

I grew up in Detroit, Michigan. One day when I was a little boy my grandparents took me to a park on the shore of what I was told was the Detroit River. I remember being amazed to see very large ships passing by. It wasnt long before I learned that the Detroit River was part of the Laurentian Great Lakes system and that these big ships moved among all five of the lakes. In our social studies class (which was mostly a class in maps), I learned that the land I was seeing across the Detroit River was a foreign country called Canada and that the United States-Canadian international border ran through four of the Great Lakes. We also learned that the Great Lakes were closely tied to the cultural heritage of both countries and that the lakes had served as a highway leading to the exploration, settlement, and eventual industrialization of this region of North America.

The Great Lakes are big – the total area of the Great Lakes' drainage basin is $\sim$ 40% larger than France – and the Great Lakes contain the largest reservoir of fresh, surface water on earth; only the polar ice caps contain more fresh water. The lakes are an especially valuable natural resource to the United States, to Canada, and to the world. A brief study of a map shows that the Great Lakes basin is highly populated in the south and includes such major cities as Chicago, Illinois; Detroit, Michigan; Cleveland, Ohio; and Toronto, Ontario. Large concentrations of industry and agriculture are also located in the Great Lakes region. All of this anthropogenic activity has had its environmental costs. Agricultural runoff, urban waste, industrial discharge, landfill leachate, and atmospheric deposition all contribute to the pollution of the lakes.

In the late 1960s, growing public concern about the deteriorating environmental quality of the Great Lakes stimulated research on the inputs and behavior of pollutants in the lakes. Governments in both the United States and Canada responded to the concern by regulating pollutant discharges and initiating research on the sources, fates, and effects of pollutants in the lakes. These efforts were formalized in the first Great Lakes Water Quality Agreement between Canada and the United States in 1972.

Because of this increased attention, major reductions were made in pollutant inputs, and by the 1970s some results were clear. Oil slicks disappeared; dissolved oxygen levels improved; odor problems were eliminated; beaches reopened; and algal mats disappeared. Buoyed by these successes, attention shifted to persistent toxic contaminants in the 1980s, as reflected in the 1987 Amended Great Lakes Water Quality Agreement.

The book you are holding brings together what is known about the major classes of these persistent organic pollutants in the Great Lakes. Each chapter reviews our knowledge of the extent of contamination of the various parts of the Great Lakes ecosystem (air, water, sediment, fishes, birds, etc.), what is known about the trends over time of this contamination, and knowledge about the mechanisms by which these pollutants are mobilized in the lakes. Detailed information is presented on polychlorinated biphenyls (PCBs), polychlorinated dibenzo-*p*-dioxins and dibenzofurans (dioxins), pesticides, toxaphene, polychlorinated naphthalenes (PCNs), polycyclic aromatic hydrocarbons (PAHs), brominated flame retardants (including polybrominated biphenyls and diphenyl ethers), and perfluoroalkyl acids (PFAs). As my wife points out, I have not done much; I have simply asked my friends and colleagues to write chapters and added my name to the book's cover. While this is an oversimplification of an editor's job, she is not too far from the mark. Thus, I want to sincerely thank the authors of the various chapters for their more or less prompt attention to researching and writing each chapter. Without these authors, there would be no book and no cover to bear my name. We all hope that you, the reader, will find this book to be a useful resource for research, decision making, and teaching.

Bloomington, Indiana, August 2005                                        Ronald A. Hites

Hdb Env Chem Vol. 5, Part N (2006): 1–12
DOI 10.1007/698_5_038
© Springer-Verlag Berlin Heidelberg 2005
Published online: 23 November 2005

# Persistent Organic Pollutants in the Great Lakes: An Overview

Ronald A. Hites

School of Public and Environmental Affairs, Indiana University, Bloomington, IN 47405,
USA
*HitesR@Indiana.edu*



**Abstract**   This chapter presents background information on the Great Lakes and summarizes the content of each chapter of this book.

# 1
# Introduction: The Great Lakes [1]

The Laurentian Great Lakes (not to be confused with the Great Lakes in Africa) are located near the middle of the North American continent. There are five Great Lakes, and they are called Lakes Superior, Michigan, Huron, Erie, and Ontario; see Fig. 1. Lake St. Clair is, strictly speaking, not one of the Great Lakes, but it is part of the connection between Lakes Huron and Erie. The United States-Canadian international border runs through four of the lakes, and collectively the Great Lakes are closely tied to the cultural heritage of both the United States and Canada. Early on, the lakes were the highway that led to the exploration, settlement, and eventual industrialization of this region of North America. The lakes have provided a thriving commercial fishery (now largely vanished) and water for drinking, transportation, power, industry, and recreation.

The Great Lakes are big; see Table 1. From the westernmost corner of Lake Superior to the easternmost shore of Lake Ontario, these lakes cover a distance of > 1200 km. The total water area of the lakes is 244 000 km$^2$—an area

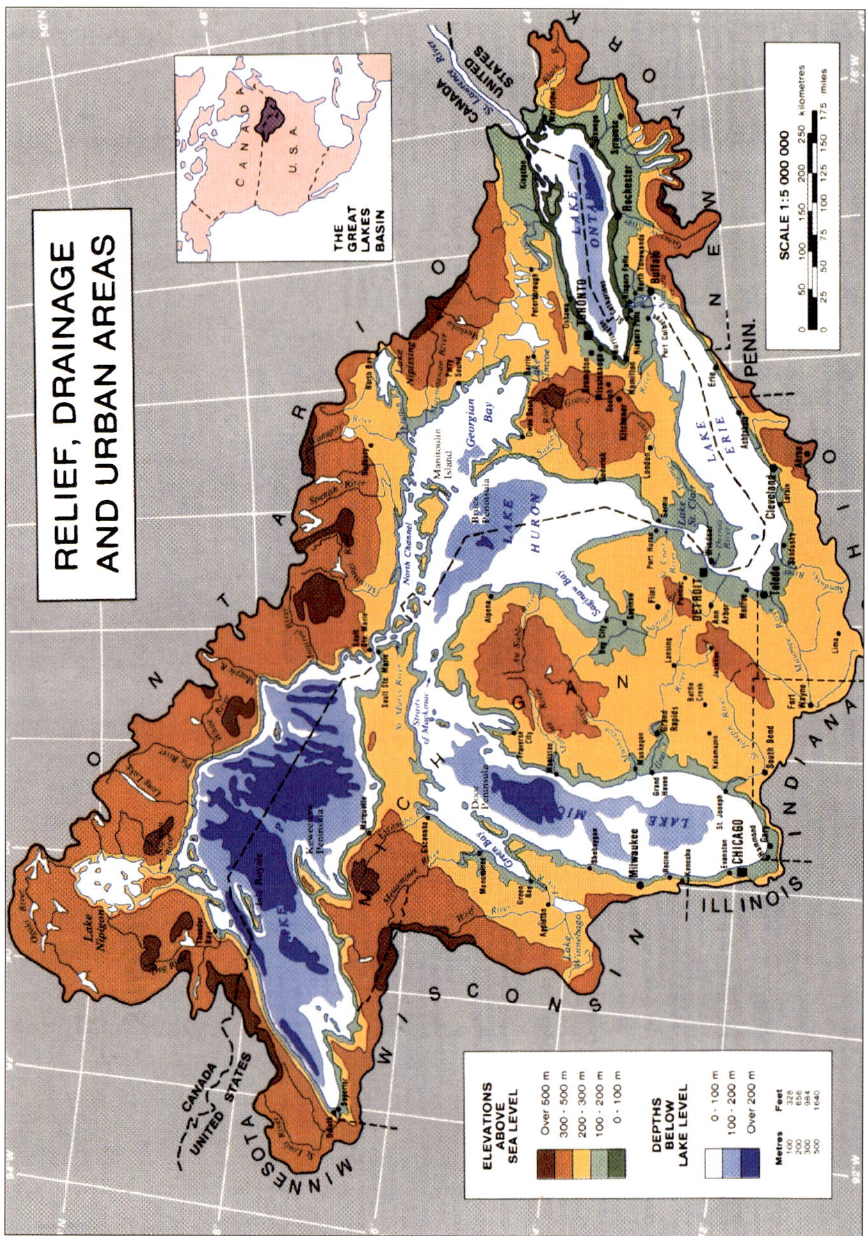


**Fig. 1** Map of the Great Lakes showing the major geographic features (from [1])

about the size of Great Britain. The total area of the Great Lakes' drainage basin is $766\,000\,\mathrm{km}^2$—an area much larger than the combined areas of Germany and Italy and $\sim 40\%$ larger than France. In this drainage basin, the

**Table 1** Physical features and population of the Great Lakes (from [1])

| | Superior | Michigan | Huron | Erie | Ontario | Total |
|---|---|---|---|---|---|---|
| Elevation [a] (in meters) | 183 | 176 | 176 | 173 | 74 | |
| Length (in km) | 563 | 494 | 332 | 388 | 311 | |
| Breadth (in km) | 257 | 190 | 245 | 92 | 85 | |
| Average depth [a] (in meters) | 147 | 85 | 59 | 19 | 86 | |
| Maximum depth [a] (in meters) | 406 | 282 | 229 | 64 | 244 | |
| Volume [a] (in km$^3$) | 12 100 | 4920 | 3540 | 484 | 1640 | 22 700 |
| Water area (in km$^2$) | 82 100 | 57 800 | 59 600 | 25 700 | 19 000 | 244 000 |
| Land drainage area [b] (in km$^2$) | 128 000 | 118 000 | 134 000 | 78 000 | 64 000 | 522 000 |
| Total area (in km$^2$) | 210 000 | 176 000 | 194 000 | 104 000 | 83 000 | 766 000 |
| % Water relative to total area | 39 | 33 | 31 | 25 | 23 | 32 |
| Shoreline length [c] (in km) | 4380 | 2630 | 6160 | 1400 | 1150 | 17 000 [d] |
| Retention time (years) | 191 | 99 | 22 | 2.6 | 6 | |
| Population: U.S. (1990) | 425 500 | 10 057 000 | 1 503 000 | 10 018 000 | 2 704 000 | 24 707 000 |
| Canada (1991) | 181 600 | | 1 191 000 | 1 665 000 | 5 447 000 | 8 484 000 |
| Totals | 607 100 | 10 057 000 | 2 694 000 | 11 683 000 | 8 151 000 | 33 191 000 |

[a] Measured at low water height. [b] Land drainage area for Lake Huron includes the St. Marys River; Lake Erie includes the St. Clair-Detroit system; and Lake Ontario includes the Niagara River. [c] Including islands. [d] These totals are greater than the sum of the shoreline lengths for the lakes because they include the connecting channels (excluding the St. Lawrence River).

lakes themselves cover about 32% of this area. The lakes contain $22\,700\ \mathrm{km}^3$ (or $2.27 \times 10^{16}$ L) of freshwater, much of which is still pure enough to be consumed with minimal treatment. The Great Lakes contain the largest reservoir of fresh, surface water on earth—about 18% of the world's supply; only the polar ice caps contain more fresh water.

The Great Lakes basin is highly populated in the south and includes such major cities as Chicago, Illinois; Detroit, Michigan; Cleveland, Ohio; and Toronto, Ontario. More than 10% of the United States' population and more than 25% of the Canadian population live in the Great Lakes basin. Large concentrations of industry are located in the Great Lakes region; for example, the U.S. and Canadian automobile industry is located there. Agriculture is also an important part of the Great Lakes basin's economy; $\sim 25\%$ of Canadian and $\sim 7\%$ of United States agricultural production is in the Great Lakes basin. Of course, all of this activity is not without its environmental costs. Agricultural runoff, urban waste, industrial discharge, and landfill leachate all contribute to the pollution of the lakes. In addition, the large surface area and slow flushing rates of the lakes makes them vulnerable to the deposition of contaminants from the atmosphere.

Because of the sheer magnitude of the Great Lakes' watershed size, physical characteristics, such as climate and soils, vary across the basin. In the north, the climate is relatively cold, and the terrain is dominated by granite bedrock called the Canadian (or Laurentian) Shield, which consists of Precambrian rocks under a generally thin layer of acidic soils. Conifers dominate the northern forests. In the southern areas of the Great Lakes basin, the climate is warmer, and the soils are deeper with layers of clay, silt, sand, and gravel, which were deposited by glaciers. In the southern basin, the lands can be readily used for agriculture. In several places in the basin, the original landscape has been replaced by sprawling urban development.

The lakes have very different depths, and thus they have very different properties and responses to environmental insult. The depth profile of the lakes is shown in Fig. 2.

**Lake Superior** is by far the largest, deepest, and coldest of the five lakes. This lake has a water retention time of almost 200 years. There is little agriculture in Lake Superior's basin (largely because of the poor soils and cold climate); instead most of the basin is forested. The sparse human population around Lake Superior results in relatively little pollution entering the lake as a result of local human activity, but because of its large surface area and its small drainage area, large amounts of pollutants are delivered by deposition from the atmosphere.

**Lake Michigan** is the second largest of the lakes, and it is the only lake entirely within the United States. The northern part of Lake Michigan is in the colder and less developed upper Great Lakes region, and this region is sparsely populated, except for the Fox River Valley, which drains into Green Bay. In fact, this bay has been contaminated by wastes from a large concen-

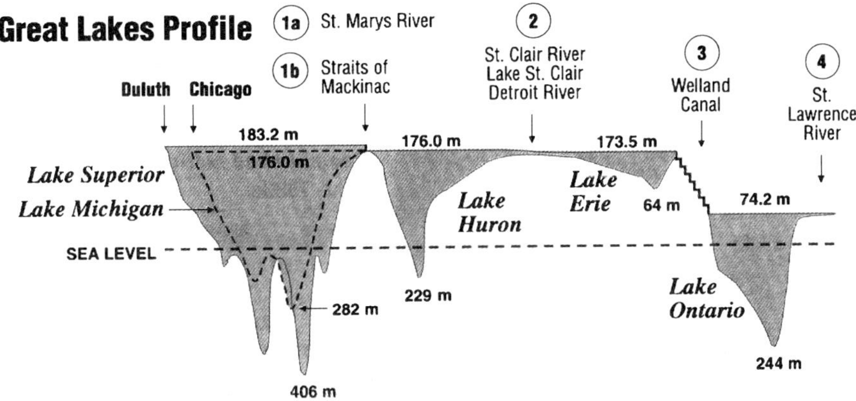


**Fig. 2** Diagram indicating the relative depths (in meters below the lake surface) and surface elevations (in meters above sea level) of the five Great Lakes measured along the long axis of each lake. The vertical exaggeration is 2000 : 1. The inter-lake lock and river systems are numbered. From [1]

tration of pulp and paper mills in this part of Wisconsin. The Southern Lake Michigan basin is among the most urbanized areas in the Great Lakes system, being the home of the greater Milwaukee, Chicago, and Gary metropolitan areas. About 12 million people live in this region.

**Lake Huron** (including Georgian Bay) is the third largest of the lakes. The human impacts in the Lake Huron basin are relatively minor, consisting mostly of recreational uses. For example, many Canadians have cottages on the shallow, sandy beaches of Lake Huron or along the rocky shores of Georgian Bay. On the United States side of the lake, the Saginaw River basin is farmed and contains the Flint, Saginaw, and Bay City metropolitan areas. Saginaw Bay has also received industrial waste from the chemical industry centered in Midland, Michigan.

**Lake Erie** is the smallest and shallowest of the lakes, and it has been heavily impacted by urbanization and agriculture. Lake Erie receives runoff from agricultural areas in southwestern Ontario, northern Ohio, and southern Michigan. In addition, there are numerous metropolitan areas located in the Lake Erie basin; these include Cleveland, Ohio, and Buffalo, New York. The average depth of Lake Erie is only 19 meters, and as a result, it warms rapidly in the spring and summer, and it frequently freezes over in winter. Lake Erie also has the shortest water residence time of the lakes (2.6 years), and thus, the condition of this lake can change rapidly.

**Lake Ontario** is much deeper than its upstream neighbor, having an average depth of 86 meters and a water residence time of $\sim$ 6 years. Major Canadian urban industrial centers, such as Hamilton and Toronto, Ontario, are located on its shore. The southern (U.S.) shore of Lake Ontario is less urbanized and is not intensively farmed, except for a narrow band along the lake itself. Lake Ontario has also received contaminants from the chemical industry operating along the Niagara River in New York.

In the late 1960s, growing public concern about the deterioration of environmental quality of the Great Lakes stimulated research on the inputs and behavior of pollutants in the lakes. Governments in both the United States and Canada responded to the concern by regulating pollutant discharges and initiating research on the sources, fates, and effects of pollutants in the lakes. These efforts were formalized in the first Great Lakes Water Quality Agreement between Canada and the U.S. in 1972.

Because of this increased attention, major reductions were made in pollutant inputs, and by the 1970s, some results were clear. Floating debris and oil slicks disappeared; dissolved oxygen levels improved; odor problems were eliminated; beaches reopened; and algal mats disappeared. Attention shifted to persistent toxic contaminants in the 1980s, as reflected in the 1987 Amended Great Lakes Water Quality Agreement. Given that some of these toxic substances accumulate as they move through the food chain, top predators such as lake trout, fish-eating birds (such as cormorants, ospreys, and herring gulls), and people can receive relatively high doses of these contaminants.

This book brings together what is known about the major classes of these persistent organic pollutants (the so-called POPs). Each chapter reviews our knowledge of the extent of contamination of the various parts of the Great Lakes ecosystem (air, water, sediment, fishes, birds, etc.), what is known about the trends over time of this contamination, and information about the mechanisms by which these compounds are mobilized in the lakes. The following section presents abstracts of the contents of each chapter.

# 2
# Overview of Chapters

## 2.1
## Polychlorinated Biphenyls

Polychlorinated biphenyls (PCBs) were widely used in the Great Lakes region primarily as additives to oils and industrial fluids, such as dielectric fluids in transformers. PCBs are persistent, bioaccumulative, and toxic to animals and humans. The compounds were first reported in the Great Lakes natural environment in the late 1960s. At that time, PCB production and use was near the maximum level in North America. Since then, inputs of PCBs to the Great Lakes have peaked and declined: Sediment profiles and analyses of archived fish indicate that PCB concentrations have decreased markedly in the decades following their phase-out in the 1970s. Unfortunately, PCB concentrations in some fish species remain too high for unrestricted safe consumption. PCB concentrations remain high in fish because of their persistence, tendency to bioaccumulate, and the continuing input of the compounds from uncontrolled sources. PCBs are highly bioaccumulative, and many studies have shown that the complex food webs of the Great Lakes contribute to the focusing of PCBs in fish and fish-eating animals. PCB concentrations in the open waters are in the range of 100–300 pg/L and are near equilibrium with the regional atmosphere. PCBs are hydrophobic yet are found in the dissolved phase of the water column and in the gas phase in the atmosphere, and they continue to enter the Great Lakes environment. The atmosphere, especially near urban-industrial areas, is the major source to the open waters of the lakes. Other sources include contaminated tributaries and in-lake recycling of contaminated sediments. Until these remaining sources are controlled or contained, unsafe levels of PCBs will be found in the Great Lakes environment for decades to come.

## 2.2
## Dioxins

Good information exists on the occurrence, geographical distribution, and temporal trends of dioxins in Great Lakes' air, water, sediments, fish, seabirds, and people. Dioxin congener patterns and concentrations in sediment indicate that atmospheric input dominated in Lake Superior, southern Lake Michigan, and Lake Erie. Inputs from the Saginaw River to Lake Huron and from the Fox River to upper Lake Michigan added some dioxin loading to these lakes above atmospheric deposition. Lake Ontario was heavily impacted by the input of dioxins, particularly 2,3,7,8-tetrachlorodibenzo-p-dioxin, from the Niagara River. Sediment core and bio-monitoring data revealed that dioxin contamination peaked in most lakes in the late 1960s to early 1970s, followed by rapid, order of magnitude declines in the mid to late 1970s. The downward trend stalled in some lakes in the 1980s, but seems to have continued after the late 1990s, probably in response to various remediation efforts and reductions in dioxin emissions to the atmosphere. During the height of contamination, effects attributed in whole or in part to dioxin contamination included reproductive failure in lake trout and herring gulls in Lake Ontario. Aryl hydrocarbon receptor-mediated sub-lethal effects may still be occurring in seabirds and fish, but much of this is thought to be due to dioxin-like PCBs rather than dioxins.

## 2.3
## Pesticides

Spatial distributions and temporal trends of pesticides in sediment, water, and fish indicate good progress in the reduction of persistent organochlorine pesticides in the Great Lakes Basin. Concentrations of many of the organochlorine pesticides have decreased significantly in Great Lakes wildlife, subsequent to restriction of these compounds' usage in the Great Lakes basin in the 1970s and the 1980s. Nevertheless, concentrations of several organochlorine pesticides in the Great Lakes are currently either not declining or are declining only slowly. The current concentrations of several organochlorine pesticides seem to be at a steady state, such that the amount lost to the sediments and exported either to the atmosphere or flowing out through the St. Lawrence River is balanced by input from rivers and the atmosphere. Thus, further reductions in the input of organochlorine pesticides and the continued recovery of wildlife populations are dependent, in part, on the control of new inputs.

The frequency of detection of current-use pesticides in the Great Lakes generally increased in the order Superior < Huron and Georgian Bay < Ontario < Erie. The highest concentrations among current-use pesticides were measured for atrazine, metolachlor, 2,4-D, and diazinon in the western basin of Lake Erie, because of the close proximity to areas where these pesticides are applied in both agricultural and urban settings. Future monitoring activities should focus on the major pesticides that are in current use. At present, programs are in place to analyze for about half of the pesticides used. Additive and synergistic effects of pesticide mixtures must be examined more closely, since existing guidelines have been developed for individual pesticides only.

## 2.4
## Toxaphene

Toxaphene is a major persistent organic contaminant in air, water, and fish in the Great Lakes. The story of toxaphene in the Great Lakes, like that of most other persistent organochlorine compounds, has only become clear after the ban on the use of this pesticide in the mid-1980s. The spatial and temporal trends of toxaphene in the Great Lakes are now reasonably well documented. Highest concentrations in fish and lake water are found in Lake Superior. Concentrations of toxaphene declined in lake trout from Lakes Michigan, Huron, and Ontario during the 1990s (half-lives of 5–8 years) but not in Lake Superior. Recent measurements suggest no declines from the mid-1990s to 2000 in all four lakes. Modeling has demonstrated that the colder temperatures and low sedimentation rates in Lake Superior, and to some extent in Lake Michigan, conspire to maintain high toxaphene concentrations in the water column. Sediment core profiles from Lakes Michigan, Ontario, and Superior all show declining inputs in the past 10–20 years, mirroring reduced emissions following toxaphene deregistration in the United States in 1986. Atmospheric transport of toxaphene from agricultural soils in the southern United States continues and modeling results suggest that 70% of the atmospheric inputs to the Great Lakes are due to long range atmospheric transport and deposition from outside of the basin itself. Some degradation is apparent in the Lake Superior food web based on non-racemic enantiomer fractions for selected chlorobornanes, but for most congeners this process is slow and does not result in negative food web biomagnification in Lake Superior. High proportions of hexa- and heptachlorobornanes have been found in some lake sediments and tributary waters, indicating that slow degradation, mainly via dechlorination, is proceeding within the Great Lakes basin.

Toxaphene concentrations probably did not reach levels that would, by themselves, cause effects on salmonid reproduction, survival, or growth in the Great Lakes. The levels and effects of toxaphene in fish-eating birds and mammals (such as mink) in the Great Lakes have never been thoroughly investigated; however, it seems likely that exposure levels for birds and mammals would have been lower than for PCBs. In some Great Lakes jurisdictions, concerns remain about human exposure to toxaphene via consumption of Lake Superior lake trout. Given the long half-lives in fish and water, elevated toxaphene is likely to remain a contaminant issue in the Great Lakes until the middle of the 21st century.

## 2.5
### Polychlorinated Naphthalenes

Polychlorinated naphthalenes (PCNs) have entered the Great Lakes through the production and use of Halowax technical mixtures, as trace contaminants in Aroclor PCB mixtures, and through industrial processes such as chlor-alkali production and waste incineration. Air concentrations of PCNs were highest in urban areas, and congener profiles indicate that evaporative emissions relating to past uses are the dominant sources, but combustion processes also contribute. Sediment measurements indicate that the highest PCN concentrations are in the Detroit River, and congener profiles indicate Halowax contamination from past inputs. Fish from this area had the highest reported concentrations in the Great Lakes region, followed by lake trout from Lake Ontario. Estimates of dioxin toxic equivalents of PCNs indicate their contributions are as important as the dioxin-like PCBs in some aquatic species, and more important in air and sediments. No time trend information for PCNs in the Great Lakes exists, and further spatial assessment, and toxic equivalent comparisons of PCNs, dioxins, and PCBs in additional fish species should be undertaken.

## 2.6
### Polycyclic Aromatic Hydrocarbons

Polycyclic aromatic hydrocarbons (PAHs) are produced during the incomplete combustion of organic material. PAHs can also be produced through natural, non-combustion processes, and they may be present in uncombusted petroleum. Uncombusted petroleum can be a direct source to the waters of the Great Lakes, but combustion sources discharge PAHs into the coastal atmosphere. Atmospheric deposition of combustion-related PAHs seems to be the dominant source to the Great Lakes, except in nearshore areas where point sources can be significant. Once airborne, PAHs partition in the atmosphere between the gas and particle phases and can undergo long-range transport. During transport, PAHs can be degraded or modified by photochemical reactions. Both the original PAH species and their degradation products can be washed out of the atmosphere by wet and dry deposition, air-water exchange, and air-terrestrial exchange. Once in an aquatic system, PAHs partition between the dissolved and particle phases. In general, PAHs are particle reactive and settle out in sediments. PAH contamination of Great Lakes sediments is higher in the near-shore regions where ports, harbors, and urban/industrial areas are the densest. In the open lake areas, sediment concentrations are uniform, with Lake Superior having slightly less PAHs in its surficial sediments. That portion of the PAHs that does not partition to particles can bioaccumulate in the lipid reserves of organisms.

PAHs accumulated in an organism may be metabolized to more toxic by-products or exert toxicity in their original form. When combined with ultraviolet radiation, this toxicity is greatly enhanced. In coastal areas where concentrations can be quite high, PAHs can be toxic to all forms of aquatic life during at least part of their life cycle. PAHs are expected to remain an ecological threat to the Great Lakes well into the future. This threat may even increase with the increasing combustion needed for increasing population centers and greater transportation needs. Of particular concern is the short-term increase in PAH concentrations that can result from dredging of ports and harbors, where highly contaminated sediments have been buried.

## 2.7
### Brominated Flame Retardants

Brominated flame retardants in the Great Lakes have not been as well studied as many of the polychlorinated pollutants, especially the PCBs, but in the last 5–10 years, there has been some significant progress. The ubiquity of these compounds in the sediment and fishes of the lakes has now been well established, and perhaps more alarming, it is now clear that the concentrations of some of these compounds are actually increasing. This observation is particularly important given that the concentrations of most of the other persistent organic pollutants in the lakes are decreasing. Despite their production cessation in the mid-1970s, polybrominated biphenyls (PBBs) are still present in fishes and sediment from most of the lakes. In general, these PBB concentrations are decreasing slowly, if at all. Polybrominated diphenyl ethers (PBDEs) are present in air, fishes, birds, and sediment from the lakes. In lake trout and in herring gull eggs, the PBDE concentrations have doubled every 3–5 years; in sediment cores, the doubling time is ∼ 15 years. Hexabromocyclododecanes (HBCD) are also present in fishes and sediment from the lakes, but at much lower levels compared to the PBDEs. Two previously unreported flame retardants [1,2-bis(2,4,6-tribromophenoxy)ethane (TBE) and 1,2,3,4,5-pentabromoethylbenzene] have been found in the air and sediment of the lakes. Clearly, it would be good to monitor the concentrations of all of these compounds in (at least) the sediment and fishes of the lakes to determine long-term trends. One might want to focus especially on those BFRs that will continue to be in production; these include deca-BDE, HBCD, and TBE.

## 2.8
## Perfluorinated Compounds

Perfluoroalkyl acids (PFAs) are released into the environment via their manufacturing processes, their use in commercial products, or indirectly via oxidation of precursor molecules containing perfluoroalkyl chains. PFA precursors are diverse and include perfluorinated alcohols and perfluoroalkyl sulfonamide derivatives. Products in which PFAs and their precursors have been used include wetting agents, lubricants, stain resistant treatments, and firefighting foams. The PFAs in the environment comprise two general classes: perfluoroalkyl carboxylates such as $CF_3(CF_2)_xCO_2^-$ and perfluoroalkyl sulfonates, such as $CF_3(CF_2)_xSO_3^-$. The predominant PFA in biota samples from the Great Lakes is perfluorooctane sulfonate (PFOS), but a homologous series of perfluoroalkyl carboxylates, where $x = 6\text{--}13$, is also detected in most samples at lesser concentrations.

The environmental behavior of most PFAs is not well studied, and our knowledge of the physicochemical properties of perfluorooctane sulfonate and carboxylate is limited. Both compound classes are persistent in the environment and are not expected to volatilize into the atmosphere to a significant extent, but they have much greater water solubilities than similar chlorinated compounds. Concentrations of PFOS in surface waters are usually less than those of perfluorooctanoic acid, but PFOS accumulates in aquatic organisms to a greater extent and appears to biomagnify in the food web of the Great Lakes region. PFAs or their precursors have been measured in air, surface waters, sediments, aquatic invertebrates, and in the tissues of fish, fish-eating water birds, mink, otter, and other wildlife from the Great Lakes. Although the sources of PFAs to the Great Lakes are not well understood, fluorotelomer alcohols and perfluorooctylsulfonamides degrade to perfluoroalkyl carboxylates and PFOS, respectively, in laboratory studies. On the basis of preliminary and incomplete information, current concentrations of PFOS in the Great Lakes environment do not seem to be sufficient to pose a significant risk to most aquatic organisms including fish. However, the margins of safety are less for mammals such as mink and birds, and when the concentrations of all PFAs are considered together, current concentrations may pose some risk to some sensitive species.

## References

1. This section was generously paraphrased from: The Great Lakes, An Environmental Atlas and Resource Book. Third Edition, Government of Canada, Toronto, and the United States Environmental Protection Agency, Great Lakes National Program Office, Chicago, 1995

Hdb Env Chem Vol. 5, Part N (2006): 13–70
DOI 10.1007/698_5_039
© Springer-Verlag Berlin Heidelberg 2005
Published online: 20 December 2005

# Polychlorinated Biphenyls in the Great Lakes

Keri C. Hornbuckle[1] (✉) · Daniel L. Carlson[2] · Deborah L. Swackhamer[3] ·
Joel E. Baker[4] · Steven J. Eisenreich[5]

[1]Department of Civil and Environmental Engineering and IIHR-Hydroscience
and Engineering, University of Iowa, Iowa City, IA 52242, USA
*keri-hornbuckle@uiowa.edu*

[2]Division of Environmental Health Sciences, University of Minnesota,
Minneapolis, MN 55455, USA
*carl2446@umn.edu*

[3]Water Resources Center, University of Minnesota, St. Paul, MN 55108, USA
*dswack@umn.edu*

[4]University of Maryland Center for Environmental Science,
Chesapeake Biological Laboratory, 1 Williams Street, PO Box 38, Solomons, MD 20688,
USA
*baker@cbl.umces.edu*

[5]European Commission, Joint Research Centre Ispra European Chemicals Bureau,
I-21020 Ispra, Italy
*steven.eisenreich@jrc.it*

**Abstract** This chapter reviews the scientific understanding of the concentrations, trends, and cycling of polychlorinated biphenyls (PCBs) in the Great Lakes. PCBs were widely used in the Great Lakes region primarily as additives to oils and industrial fluids, such as dielectric fluids in transformers. PCBs are persistent, bioaccumulative, and toxic to animals and humans. The compounds were first reported in the Great Lakes natural environment in the late 1960s. At that time, PCB production and use was near the maximum level in North America. Since then, inputs of PCBs to the Great Lakes have peaked and declined: sediment profiles and analyses of archived fish indicate that PCB concentrations have decreased markedly in the decades following the phase-out in the 1970s. Unfortunately, concentrations in some fish species remain too high for unrestricted safe consumption. PCB concentrations remain high in fish because of their persistence, tendency to bioaccumulate, and the continuing input of the compounds from uncontrolled sources. PCBs are highly bioaccumulative and many studies have shown that the complex food webs of the Great Lakes contribute to the focusing of PCBs in fish and fish-eating animals. PCB concentrations in the open waters are in the range of $100–300\,\mathrm{pg\,L^{-1}}$, and are near equilibrium with the regional atmosphere. PCBs are hydrophobic yet are found in the dissolved phase of the water column and in the gas phase in the atmosphere, and they continue to enter the Great Lakes environment. The atmosphere, especially near urban-industrial areas, is the major source to the open waters of the lakes. Other sources include contaminated tributaries and in-lake recycling of contaminated sediments. Until these remaining sources are controlled or contained, unsafe levels of PCBs will be found in the Great Lakes environment for decades to come.

**Keywords** Bioaccumulation · Mass balance · PCBs · Toxicity


**Abbreviations**

| | |
|---|---|
| PCBs | Polychlorinated biphenyls |
| $\mathrm{Sol_{aq}}$ | Water solubility |
| VP | Vapor pressure |
| $K_{\mathrm{H}}$ | Henry's Law constant |
| $K_{\mathrm{OW}}$ | Octanol–water partition coefficient |
| $K_{\mathrm{OA}}$ | Octanol–air partition coefficient |
| AhR | Aryl hydrocarbon receptor |
| AHH | Aryl hydrocarbon hydroxylase |
| PCDD/F | Polychlorinated dibenzo-$p$-dioxins and dibenzofurans |
| FDA | Food and Drug Administration |
| ELS | Early life stage |
| BNL | Benthic nepheloid layer |
| GLFMP | Great Lakes Fish Monitoring Program |

TEQ     Toxicity equivalence
BAF     Bioaccumulation factor
BCF     Bioconcentration factor
BSAF    Biota-sediment accumulation factor
BMF     Biomagnification factor
LMMB    Lake Michigan Mass Balance

# 1
# Introduction

## 1.1
## History: Discovery in the Great Lakes

Polychlorinated biphenyls (PCBs) were first identified in environmental samples from the Great Lakes region by Gilman Veith, who found PCBs in lake trout and bloater chubs from Lake Michigan in 1968 [1]. Jensen [2] had just demonstrated the existence of PCBs in the Swedish environment, and the Michigan Department of Environmental Quality was interested in whether some unusual peaks in the gas chromatograms of fish extracts might also be PCBs. Veith confirmed the identity of PCBs in these extracts by mass spectrometry, and the newly established Federal Research Laboratory in Duluth, MN (of which Veith eventually served as director, now the US EPA Mid-Continent Division of the National Health Effects Research Laboratory) began investigating this contamination.

The intense research on PCBs in the Great Lakes over the intervening decades is important in three ways. First, the region is a major economic resource and population center for the USA and Canada, and as such the human and environmental health of the region is of great significance. Second, the ecosystem is unique and valuable in its own right. And third, as a heavily studied region with well-defined but diverse characteristics, studying the behavior of PCBs in the Great Lakes has served to greatly enhance our knowledge of the fate and transport of PCBs and other contaminants in the environment in general.

PCBs contaminated the environment of the Great Lakes as a result of historical use and discharge of PCBs from many different industries within the Great Lakes basin, including packing plants, paper mills, breweries, tanneries, machine shops, and foundries. These industries prospered in the Great Lakes throughout the 20th century, with major growth concurrent with the growth in production and sales of PCBs between 1929 and the mid-1970s. At this time, PCBs were common additives to oil to prevent breakdown of the oil and to maintain specific viscous properties when heated. PCBs were commonly used by the power industry in electrical transformers, capacitors, hydraulic equipment, and as lubricants. They were used as plasticizers in rubber and synthetic resins. These useful compounds were also added to many products used directly by the public or by light industry, including carbonless copy paper, adhesives, waxes, inks, pesticide extenders, sealants, and caulks [3]. New uses of PCBs were discontinued in the mid-1970s and all production ceased by 1979. Since then, PCBs have been removed from active use although they are still found in high concentrations in waste disposal sites and in reservoirs of old stockpiles throughout the Great Lakes basin.

## 1.2
## Formulations

PCBs were sold by Monsanto in North America as technical mixtures of PCB congeners. A PCB congener is a chlorinated biphenyl (one to ten chlorines) with a unique chemical structure. There are 209 possible congeners, although less than $\sim 150$ were produced for industrial use. Twelve different technical mixtures were produced by reacting biphenyls with anhydrous chlorine in the presence of an iron catalyst. The various mixtures were produced by controlling the residence time of the chlorine in the reactor. The technical mixtures were generally named according to the percentage chlorine in the PCB mixture. For example, Aroclor 1242 was 42% chlorine by mass while Aroclor 1254 was 54% chlorine by mass. The most common mixtures used were Aroclors 1242, 1248, 1254, and 1260 [3]. The exact mass percentage for the production and sales of the individual mixtures is unknown, although more than half of PCBs produced in the USA were in the form of Aroclor 1242, used for capacitors and transformers [4]. Aroclor 1242 was also used in the emulsion applied to carbonless copy paper and has been identified as a dominant mixture released from paper mills on the Fox River, Wisconsin [5]. A mixture of 1242 and 1248 was used in the hydraulic fluid of a die-casting industry in Waukegan, Illinois [6]. In both systems, the original mixture of congeners is still evident in sediments. The total mass of Aroclors produced in North America is estimated to be $570 \times 10^6$ kg [7] but the mass used in the Great Lakes basin is unknown.

## 1.3
## Widespread Contamination

There are at least 31 locations in the Great Lakes Basin identified by the US-Canada Water Quality Agreement as Areas of Concern due to impairment to beneficial uses [8]. Of the 14 identified impairments, PCBs are implicated in seven: restrictions on fish and wildlife consumption; degradation of fish and wildlife populations; fish tumors or other deformities; bird or animal deformities or reproduction problems; degradation of benthos (bottom dwelling organisms); restrictions on dredging activities; and added costs to agriculture or industry. Most of the sites that are contaminated with PCBs were originally exposed as a result of direct use and disposal. These sites continue to contribute to PCBs in the Great Lakes system though tributary flow, volatilization, erosion, and uptake by local fish and other biota. The majority of sites have not been fully characterized and the extent of contamination is unknown. Some of the sites, including the Fox River, Green Bay, and Waukegan Harbor (Lake Michigan), have associated PCB inventories (Table 1). The mass of PCBs in these areas is very large. In some cases, the total amount of PCBs in the Area of Concern is larger than that estimated for all the sediment in the

**Table 1** Published sediment inventories of PCBs in the Great Lakes

| System | Inventory (kg) | Refs. |
| --- | --- | --- |
| Lake Michigan (open lake) | 75 000 | [10] |
| Waukegan Harbor (Lake Michigan) | 900[a] | [9] |
| Green Bay (Lake Michigan) | 14 565 | [11, 12] |
| Lower Fox River (Lake Michigan) | 28 602 | [13] |
| Lake Ontario (open lake) | 130 000 | [14] |
| Lake Superior (open lake) | 4900 | [10] |

[a] More than $100 \times 10^3$ kg have been removed from navigational channels in Waukegan Harbor [9]

**Fig. 1** Sources and cycling of PCBs in the Great Lakes [15]

open waters of the Great Lakes. For example, Waukegan Harbor was burdened with as much as $136 \times 10^3$ kg of PCBs before dredging began in 1990 [9]. That is more than the entire PCB inventory buried in the sediments of Lake Michigan's open waters [10].

PCBs can travel from their sources via the air or water. Volatilization and deposition (gas absorption) are known major routes for PCB input and loss to the Great Lakes. Direct input via tributaries is also still significant in some places. In the lakes, partitioning to water column particles leads to settling and eventual burial of PCBs, but sediment may be frequently resuspended, effectively serving as an additional source. Uptake by phytoplankton and subsequent bioaccumulation through the food chain results in continuing contamination of Great Lakes predators with substantial amounts of PCBs. Indeed, consumption of fish from the Great Lakes is the major exposure pathway for humans in the region. PCBs are now distributed throughout all compartments of the Great Lakes, and several programs routinely monitor concentrations in fish, birds, and air. Figure 1 describes the pathways (except degradation pathways) that contribute to PCB cycling in the lakes and uptake into the food chain. Degradation (photolysis, metabolism, and other forms of chemical decay) is not effective in decreasing total PCB concentrations in dilute systems.

## 2
## Physical-Chemical Properties

The 209 different PCB congeners vary widely in their properties because of differing degrees and patterns of chlorination. Determining values of relevant physical-chemical constants ($Sol_{aq}$, VP, $K_H$, $K_{OW}$, and $K_{OA}$) is difficult both because of the large number of congeners and the difficulty of working with very hydrophobic compounds. Often, there is a large amount of disagreement between published literature values. Two studies, published in 2000 [16] and 2003 [17], reviewed all available data to determine suggested values for a suite of properties, but only for a limited number of congeners.

Nonetheless, there is widespread agreement in the literature on the overall trends and relative magnitudes of PCB properties. Figure 2 shows data for PCB vapor pressure, octanol–water ($K_{OW}$) and octanol–air ($K_{OA}$) partitioning coefficients, and Henry's Law constant [18–21]. The sources were chosen because they represented self-consistent values for large numbers of congeners, and they are widely considered the best values available. In each case, values were determined experimentally for some congeners, and these data were used to predict the values of the remaining congeners.

The values of solubility, vapor pressure, $K_{OW}$, and $K_{OA}$ vary over nearly six orders of magnitude for PCB congeners. Such a large range means the behavior in the environment should also vary greatly among congeners. The exception is Henry's Law constant, which is relatively invariant for all congeners, although the spread in values increases with increasing chlorination. Note that other published values for the Henry's Law constant, while different, are still within two orders of magnitude of those shown in Fig. 2 [16, 17, 22] and thus the range is still small relative to other physical and chemical properties.

Ranges are given in Table 2 for the properties plotted in Fig. 2 and for solubility [16, 17]. In most cases, a variation of one to two orders of magnitude is observed among PCBs with the same number of chlorines. For vapor pres-

**Table 2** Ranges of physical-chemical properties for PCBs. All values for 25 °C except log $K_{OA}$ (20 °C)

| Number of chlorines | log VP (Pa) | log $K_H$ (Pa m$^3$ mol$^{-1}$) | log $K_{OA}$ | log $K_{OW}$ | log Sol$_{aq}$ (mol m$^{-3}$) |
|---|---|---|---|---|---|
| 1 | 0 | 1 | 7 | 4 to 5 | − 2 |
| 2 | − 1 to − 0 | 1 | 7 to 8 | 5 | − 4 to − 2 |
| 3 | − 2 to − 1 | 1 to 2 | 8 to 9 | 5 to 6 | − 4 to − 3 |
| 4 | − 3 to − 1 | 1 to 2 | 8 to 10 | 5 to 6 | − 6 to − 4 |
| 5 | − 3 to − 2 | 1 to 2 | 9 to 11 | 6 to 7 | − 5 |
| 6 | − 4 to − 2 | 1 to 2 | 9 to 11 | 6 to 7 | − 6 to − 5 |
| 7 | − 4 to − 3 | 1 to 2 | 10 to 12 | 7 to 8 | − 6 |
| 8 | − 5 to − 3 | 1 to 2 | 10 to 12 | 7 to 8 | − 6 |
| 9 | − 5 to − 4 | 1 to 2 | 11 to 12 | 8 | − 7 |
| 10 | − 5 | 2 | 12 | 8 | − 7 |

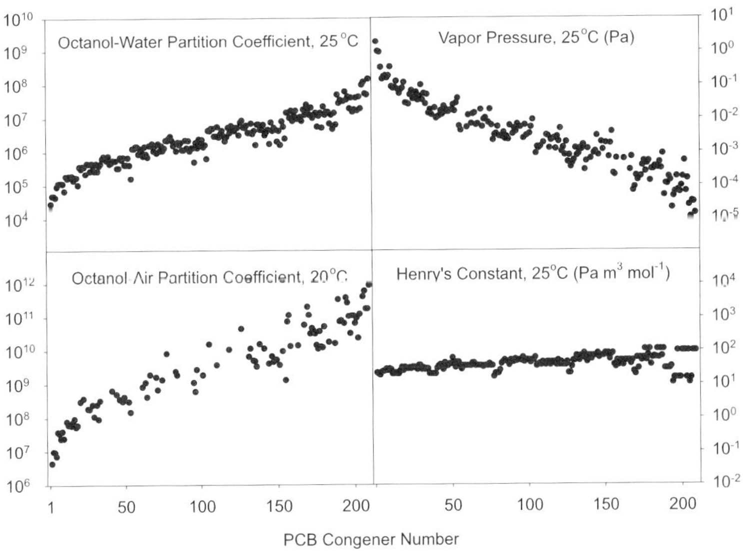


**Fig. 2** Properties of PCB congeners [18–21]

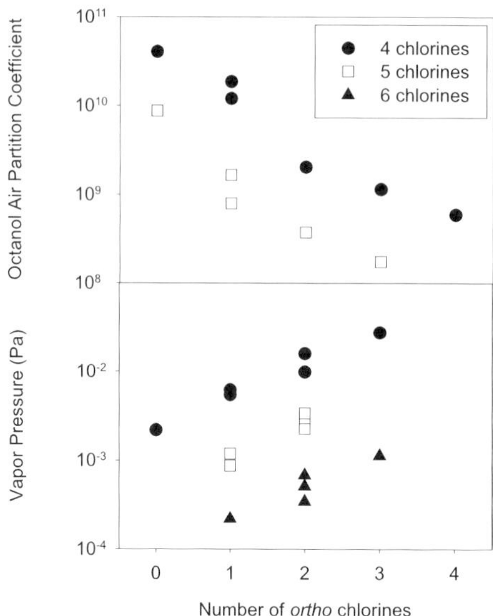



**Fig. 3** Variation of the octanol–air partition coefficient and vapor pressure within homolog groups as a function of the number of *ortho* chlorine substituents [18, 23]

sure and $K_{OA}$, we can see that this variation is partly explained by chlorine position (Fig. 3). Measured values of the properties vary with the number of chlorines in the *ortho* position [18, 23]. The vapor pressure increases, and $K_{OA}$ decreases, exponentially with more *ortho* chlorines. Indeed, for these two parameters, the variation *within* a homolog group is greater than the variation *between* homolog groups (for congeners with the same number of *ortho* chlorines).

Based on the observed properties of PCBs, we would expect to see a large range of behavior in environmental transport and partitioning. In particular, differences in bioconcentration, vapor–particle partitioning, and water–particle partitioning should be apparent both between and within homolog groups. We would not expect one congener to be representative of an entire homolog group.

## 2.1
## Analytical Challenges

The analysis of samples on a congener basis was not possible in the early years of investigation of PCBs in the Great Lakes region. The commercial availability of high resolution capillary gas chromatographic columns and the concurrent synthesis, identification, and production of individual PCB congener standards [24] in the mid-1980s was a significant breakthrough. Through a transition period before the congener composition of individual chromatographic peaks was known, capillary GC analysis of Great Lakes water improved "Aroclor-level" quantification by reducing interferences and by improving matches to Aroclor standards through multiple linear regressions routines like COMSTAR [25] and PCBQ [26]. The pioneering work of Mullin and coworkers [24] and the determination of PCB congener physicochemical properties $K_{OW}$ by Hawker and Connell [21], and of vapor pressures and Henry's Law constants by Burkhard et al. [27, 28]) catalyzed congener-specific geochemical studies in the Great Lakes in the 1980s and 1990s. More recently, analytical methods for PCBs have been further improved through the use of high resolution mass spectrometry, which provides substantially better signal-to-noise ratios and, therefore, instrumental sensitivity [29]. Due largely to high per-sample costs, high resolution mass spectrometry has not been widely used to date to measure PCB congeners in Great Lakes waters.

Congener-specific analysis of PCBs is now the norm for determination in natural matrices. Determination by Aroclor mixture is ineffective in the open waters and in biological samples. Congener distributions in natural matrices are often very different compared to any of the original Aroclor mixtures. This is primarily a result of weathering that removes or enriches some congeners over others. Weathering of PCB mixtures via volatilization, sedimentation, bioaccumulation, metabolism, and other natural processes has tremendous impact on the overall half-life of PCBs in the environment, their ultimate fate, and their toxicological impact on humans and organisms. The mechanisms that control these processes are of major concern to Great Lakes scientists.

# 3
# Human and Wildlife Toxicological Concerns

Many of the properties described above, such as degree of chlorination, low water solubility, and hydrophobicity, lead to concerns about PCB toxicological effects. They are resistant to breakdown or metabolism and are bioaccumulative, and possess the ability to pass through cell membranes and bind to a variety of receptors that elicit their toxicity. There have been decades of research demonstrating that exposure to PCBs results in the risk of adverse impacts on wildlife populations as well as human populations. These impacts vary by species, and the degree and level to which they are exposed, thus resulting in a wide variety of toxicological endpoints (e.g., reproductive impairment, cancer, chloracne). Furthermore, the number and placement of the chlorine atoms on the biphenyl ring greatly affects toxicity as well as physical properties. PCBs with no *ortho*-substituted chlorines exhibit dioxin-like effects associated with the induction of numerous enzymes, most notably that of microsomal cytochrome P4501A1 and its associated monooxygenase activity, aryl hydrocarbon hydroxylase (AHH) [30]. PCBs with *ortho*-substituted chlorines exhibit a different mode of toxicity [31, 32] that is linked to adverse neurological effects. Several studies have implicated low molecular weight PCBs and their metabolites as potential tumor-initiating compounds that may induce oxidative DNA damage or formation of DNA adducts [33, 34].

Much of our current understanding of these impacts arose from research on wildlife and human populations within the Great Lakes basin. Despite being banned in North America more than 25 years ago, the toxicological risks from PCB exposure are still a present-day concern. For example, current PCB concentrations in fish in the Great Lakes are sufficiently large as to drive the need for fish consumption advisories for sport fish and to restrict commercial fisheries.

Humans, fish, and wildlife are exposed to PCBs from a number of different exposure routes, including respiration, water consumption, food consumption, and dermal contact with contaminated water, soil, etc. However, the dominant exposure route is food consumption, and especially the consumption of fish. This is because PCBs bioaccumulate effectively in fats and lipids [35]. As a result, most PCB congeners (including the most toxic ones) biomagnify in the food web, which means that the higher levels of the food web will have greater concentrations than lower ones (see full discussion in the section on food web dynamics). As a result, adverse effects have been observed in fish-eating animals and birds, and in humans that have consumed fish (occupationally exposed workers are not considered in this discussion).

The literature on PCB toxicological effects is vast, and only those studies that relate more directly to the Great Lakes will be discussed here. There have been numerous workshop publications as well as reviews on this topic over time [3, 4, 36–48].

## 3.1
## Effects on Fish and Other Animals

While much attention is paid to the consumers of fish, there are also concerns that PCBs have impacts on fish themselves. The native species of lake trout (*Salvalinus namaycush*) was extirpated in the Great Lakes in the mid-1900s (with the exception of a few isolated self-reproducing populations in Lake Superior), likely due to a combination of stressors including overfishing, the invasion of the sea lamprey, and synthetic toxic organic compounds such as PCBs and PCDD/Fs. Efforts to restore the lake trout to the Great Lakes began in the 1950s under the direction of the Great Lakes Fisheries Commission, and while stocked fry resulted in adult stocks, no natural reproduction was achieved until recently [49]. Given that PCBs are the dominant organic contaminant across the lakes, much of the research has focused on the impacts of PCBs on fish reproduction.

The results of numerous laboratory studies on fish and a number of other aquatic species (daphnids, algae, minnows) were used to establish water quality criteria for regulatory purposes, using the LD$_{50}$ of the most sensitive species to protect aquatic life in the Great Lakes. PCB-126 has also been shown to impact two species of frogs (green, *Rana clamitans*, and leopard, *Rana pipiens*) at high concentrations in laboratory studies [50]. Several laboratory studies have implicated PCBs, particularly the AHH PCBs, in early life stage mortality (ELS) of fish [51–56].

Congener 126 is the most potent of the PCBs in binding to the AHH receptor of fish [57]. A number of studies demonstrated the correlation between organic contaminant levels and fish reproductive effects, but specific mechanisms were not identified [58–60]; it was assumed that PCBs and PCDD/Fs were the causative contaminants. Wilson et al. [61] demonstrated that fish eggs injected with Lake Michigan lake trout extract exhibited embryonic toxicity.

A definitive study by Guiney at al. [55] demonstrated the link between AHH-active compound exposure to lake trout eggs and blue-sac syndrome, an edema condition that develops in exposed fry and results in mortality. This established a mechanism and dose-response that was hypothesized to explain some of the natural reproduction impairment of lake trout in the Great Lakes. Cook and colleagues [56] tested this hypothesis for Lake Ontario by estimating the dose of AHH compounds to lake trout over time from lake sediment cores, and concluded that the lack of natural reproduction in Lake Ontario was consistent with their contaminant exposure. PCBs typically comprise the majority of the AHH-active compounds in Great Lakes fish [62]. However, other factors alone and in combination with organic contaminants may play a role in the low natural reproduction of lake trout. For example, over-exploitation, lack of genetic diversity, sea lamprey predation, and thiamine deficiency in eggs have all been suggested as possible stressors on lake trout recruitment. ELS and swim-up syndrome can both be caused by thiamine deficiency, and are not associated with exposure to AHH-active contaminants [63].

More recently, PCB-126 has been implicated in disrupting thyroid function in fish (see references in [64]). Brown et al. [64] demonstrated in laboratory studies that lake trout exposed to PCB-126 at $40\,\mu g\,kg^{-1}$ in food had increased levels of the growth regulator T4, but unaffected T3 levels, and thus the thyroid system could compensate. However, PCB exposure could result in significant impacts when T4 needs are greatest, such as during temperature changes, periods of rapid growth, or metamorphosis.

One of the most sensitive mammals to dioxin-like PCB and dioxin exposure is the mink. Effects have been documented in both farmed mink [65, 66] and wild mink [67–72]. Extensive work was done in the 1970s and 1980s by Auerlich, Ringer, and colleagues [65, 66, 73–76] that documented the impact of feeding Great Lakes fish to mink. These impacts included reproductive failure and mortality. A recent study [77] showed the presence of squamous epithelial lesions on the mandibles and maxillae of wild mink from a heavily PCB-contaminated area along the Kalamazoo River in Michigan. While seen previously in laboratory studies [78], this is the first report of such an effect in wild mink, which can lead to significant tooth loss.

Fish-eating and colonial nesting birds have also been adversely impacted by contaminant exposures, with effects ranging from mortality and chick deformities to reproductive effects and thyroid toxicity to alterations in nesting behavior, female–female pairing, and laying of super-clutches [42, 79–90]. While most studies have examined effects in wild populations, the effects are correlated with the exposures to dioxin and dioxin-like PCBs, mostly through eating contaminated Great Lakes fish. Populations that have been affected include terns, herring gulls, eagles, kestrels, and cormorants from throughout the Great Lakes, concentrated in areas with greater concentrations of contaminants.

## 3.2
## Effects on Humans

There is a growing body of literature on PCB toxicity to humans, with current research examining more subtle effects associated with chronic exposures, and the associations of those effects with specific congeners. One of the first epidemiological studies to demonstrate effects from exposure to the contaminants in Great Lakes fish was done by Joseph and Sandra Jacobson and colleagues [91–93]. This study examined development and behavior outcomes in children born to mothers who consumed varying amounts of Lake Michigan fish that were highly contaminated with PCBs. They found deficits in a number of measures, including head circumference, birth weight, and gestational age, which related to fish consumption by the mothers in a dose-dependent fashion, after controlling for a number of confounding variables. This cohort of children has been followed for several years, and additional deficits have been associated with their exposure to PCBs in utero as measured by cord blood concentrations, including effects on cognitive function, memory, motor activity, and intellectual development [94–99]. Caution must be exercised in the interpretation of the epidemiology studies, as results correlated with fish consumption or PCB measurements may covary with other contaminants. Thus PCBs may be a tracer of, or acting in concert with, other toxic contaminants that produce some of these effects [100]. Nevertheless, the weight of evidence is strong in various human studies that have isolated the impacts of PCBs [101]. Many other similar findings corroborating the Michigan study. The multigenerational effects of PCBs on infants have been corroborated by other human epidemiological studies conducted in New York [102], The Netherlands [103], Germany [104], and the Faroe Islands [105]. The agreement of many animal studies with primates [44, 106–108] and other animals (see references in [48]) has lent significant support to the conclusions of the human studies.

Neurological effects in humans have been a particular focus of recent research, and appear to be related to exposures of the *ortho*-substituted congeners as well as the co-planar configurations [43, 44, 107, 109–112]. Impacts on immune function [113], some aspects of thyroid function [114], gender ratios, and endocrine disruption [40, 115] have also been documented.

Consumption advisories in the Great Lakes are designed to protect people from the harmful effects of PCBs and other contaminants in fish. There are several benchmarks that are used to assess PCB levels in fish. The Food and Drug Administration (FDA) regulates allowable contaminants (including PCBs) in food as part of interstate commerce, which includes the commercial fisheries. The FDA Action Level is $2\,\mu g\,g^{-1}$ wet weight (ww) in the edible portion. The FDA Action Levels are based on evaluating economic impacts on the regulated industry in light of concerns for public health. The FDA Action Limit for PCBs for fish was lowered from $5\,\mu g\,g^{-1}$ (ww) to $2\,\mu g\,g^{-1}$ (ww) as the levels in most of the commercial species of interest dropped below the $2\,\mu g\,g^{-1}$ (ww) benchmark in the 1980s. This is not considered protective of public health, nor was it derived by any risk assessment approach.

The individual states issue Fish Consumption Advisories for fish in their inland waters, which are aimed at providing anglers health-based guidance on the consumption of sport fish. Each state approaches the development of their guidance differently. Because the Great Lakes are shared across eight US states and two Canadian provinces, a common protocol for an advisory for PCBs was developed so that conflicting advice was not given on fish common to several jurisdictions [116]. PCBs were chosen since they are the contaminant that drives the Fish Consumption Advisories in every state and province (with one exception by the Ministry of Environment of Ontario which issues an advisory for toxaphene in Lake Superior). Thus each state and province issues separate guidance on fish consumption relative to PCBs in Great Lakes fish compared to their fish from inland waters.

The continued elevated concentrations of PCBs in sport fish in the Great Lakes necessitates that fish consumption advisories be issued by all of the US Great Lakes states and the Canadian provinces of Ontario and Quebec. Unlimited consumption is not advised until fish tissue concentrations decline below $0.05\,\mu g\,g^{-1}$ (ww), a level not expected to be reached for many decades (see Sect. 8). Thus PCBs will be of significant concern well into the future.

## 4
## Atmospheric Processes

Concentrations of PCBs in air vary as a function of temperature. The temperature effect has been observed in samples collected in urban and remote areas and has been recognized as a major factor in observed concentrations since the late 1980s and early 1990s [117, 118]. The first convincing report of the seasonal trend came in 1992 when Hoff and coworkers published the results of an annual study of PCB concentrations measured at Egbert, Ontario [119, 120]. This pair of papers first showed that gas-phase PCBs and many other persistent organic compounds exhibited strong seasonal signals, with summertime concentrations exceeding wintertime concentrations by almost an order of magnitude. Their papers also showed that there was a good fit between the logarithm of the concentrations and inverse temperature. This relationship is predicted by the Clausius–Clapeyron equation if volatilization and sorption to surfaces is governing the atmospheric concentrations. Since then, others have confirmed that the temperature effect can be observed in the Great Lakes region on both a diurnal and seasonal basis [121–125]. At Chicago, there is enough data to demonstrate that the temperature effect is generally more pronounced for more chlorinated congeners [126], with significant variation within homolog groups, as with other physical properties.

As part of a mass balance study of PCBs entering Lake Michigan, a stochastic model was developed to describe the measured variation in air concentrations at more than a dozen sites on and around Lake Michigan [127–130]. Figure 4 illustrates the results of this effort for three of the sites. Airborne PCBs in Chicago were predicted to vary on a daily and seasonal basis by more than an order of magnitude with concentrations reaching as high as 15 ng m$^{-3}$ once or twice a year. South Haven had much lower concentrations, and the remote Sleeping Bear Dunes site had concentrations nearly two orders of magnitude less than Chicago. Concentrations have since declined somewhat (Hites RA, personal communication), but otherwise the model remains consistent with observations.

Total airborne PCBs are dominated by the gas phase. Less than 10% of PCBs in ambient air are associated with aerosol particles, even in urban areas. For example, the average PCB gas-phase concentration (sum of $\sim$ 120 congeners) in Milwaukee was $1.9 \pm 0.78$ ng m$^{-3}$ (standard deviation), while the average and standard deviation for the particulate-associated PCBs was $0.05 \pm 0.02$ ng m$^{-3}$. Particulate phase PCBs therefore accounted for less than 5% of the total atmospheric concentration [131]. This has been observed by many researchers and has been successfully predicted as a function of PCB physical-chemical properties [132–137]. For example, Pankow [138] and Harner and colleagues [23, 139, 140] have shown that the octanol–air partition coefficient ($K_{OA}$) can be used to accurately predict gas–particle partitioning of PCBs and other semivolatile organic compounds. This correlation has important implications about the nature of gas–particle partitioning of PCBs. The reason why PCBs are found in the gas phase is because of the low organic carbon content and total surface area (and volume) provided by atmospheric particles.

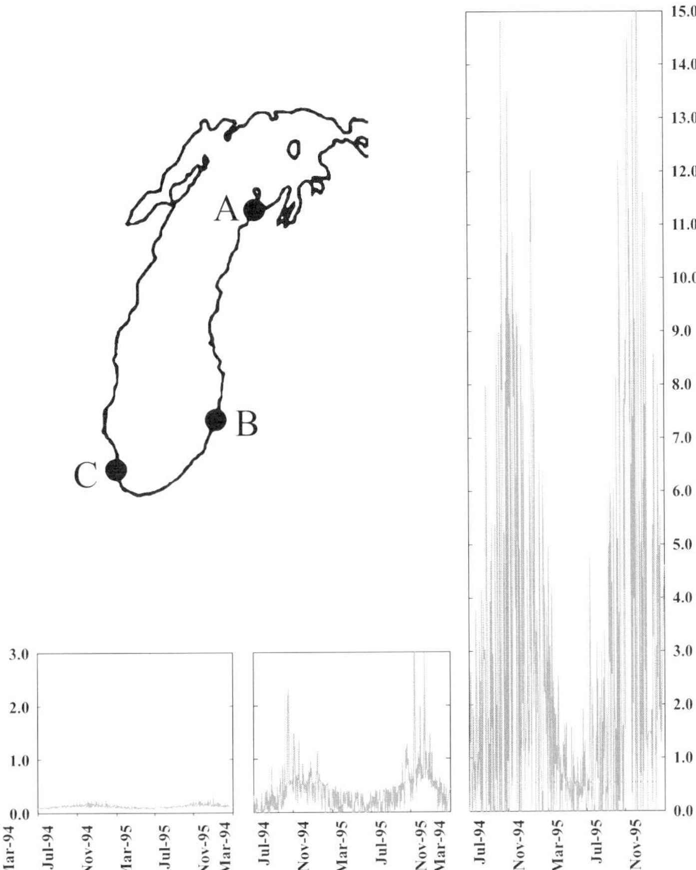

**Fig. 4** Concentrations of airborne PCBs as a function of season at Sleeping Bear Dunes (*left plot and point A*), South Haven (*center plot and point B*), and Chicago (*right plot and point C*) around Lake Michigan. Concentrations shown are daily predictions based on fits to data

PCBs in air are enriched in the less chlorinated congeners although the higher molecular weight compounds are regularly observed (albeit at very low concentrations). Figure 5 shows the mean congener distributions found in air from Milwaukee [131] and Chicago (Hites RA, personal communication). The congener profile is fairly similar (qualitatively) in both cities. The atmospheric PCB profiles share their most prominent congeners with one of the two technical mixtures that were commonly used in the Great Lakes region, although significant differences are observed as a result of environmental processing and degradation. The preponderance of less chlorinated PCBs in the atmospheric samples may simply be a result of the greater volatility of these congeners, and does not necessarily indicate the relative emissions of Aroclor 1242 versus Aroclor 1254 in the region.

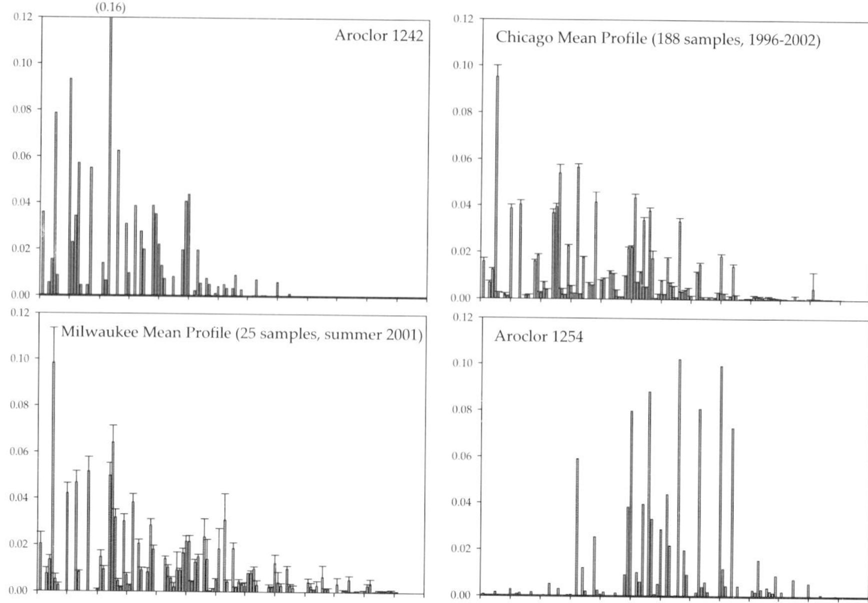

**Fig. 5** Mean PCB congener profiles for two technical mixtures (Aroclor 1242 and Aroclor 1254) and at two urban locations (Milwaukee and Chicago). Data from Milwaukee is from [131] and data from Chicago is from the IADN network (Hites RA, 2005, Personal communication). Congener distributions for Aroclor 1242 and 1254 are from [141]

## 4.1
## Atmospheric Sources

Present day PCB sources for the atmosphere are not well defined [142–144]. Without much exaggeration, the extent of understanding about atmospheric PCB sources is that they are larger in heavily populated cities and industrial areas and they are dominated by volatilization sources. Much more is known about non-atmospheric sources of PCBs. Direct discharges from industry have decreased since the phase-out. Disposal and handling of PCB contaminated soils and sediments are also controlled, at great expense and effort. But, current atmospheric sources are unregulated and uncontrolled because their locations are unknown and their relative contribution to human and ecosystem exposure is not well understood.

The major sources of PCBs, globally, are found in northern latitudes. Breivik et al. [144, 145] published a complete assessment of global historical usage of PCBs and, using a mass balance approach, predicted emissions of PCBs to the air as a function of their use and climate factors (primarily temperature). They identified major sources of PCBs as directly contaminated soils, fires, open use, use in capacitors, use in closed systems, disposal to landfills, waste incineration, and PCB destruction. They found that use of PCBs was focused in the USA, Europe, the former Soviet Union, and Japan. The consumption of PCBs was strongly linked to population, and Breivik et al. used population densities to illustrate the distribution of PCB sources.

PCB concentrations are high in cities because of historical use. Cities in the Great Lakes basin have a long history of PCB use in many industries such as steel mills, aluminum processing, paper mills, electrical generation and distribution, packing plants, breweries, tanneries, machine shops, and foundries [146]. These industries developed through the middle part of the 20th century and served as the economic foundation for many lakeside cities. During this time, PCBs were common additives to cutting fluids, electrical transformers, light ballasts, and lubricants [147]. After 1977, all new uses of PCBs were eliminated and industries began to remove PCBs from active use.

Atmospheric sources are presumed to be on the decline, although there are few measurements from before the 1990s. In the 1990s, several major measurement initiatives began that have given scientists a much better picture of the trends in airborne PCBs. For example, in 1990 the Integrated Atmospheric Deposition Network (IADN) installed five master stations to measure PCBs and other compounds in background air [148]. In 1995, the first urban monitoring site was installed by IADN in Chicago. Data from all the sites indicate declines in concentrations of airborne PCBs, despite strong interannual variability (Fig. 6). Most, but not all, of the variability is due to temperature. Hites

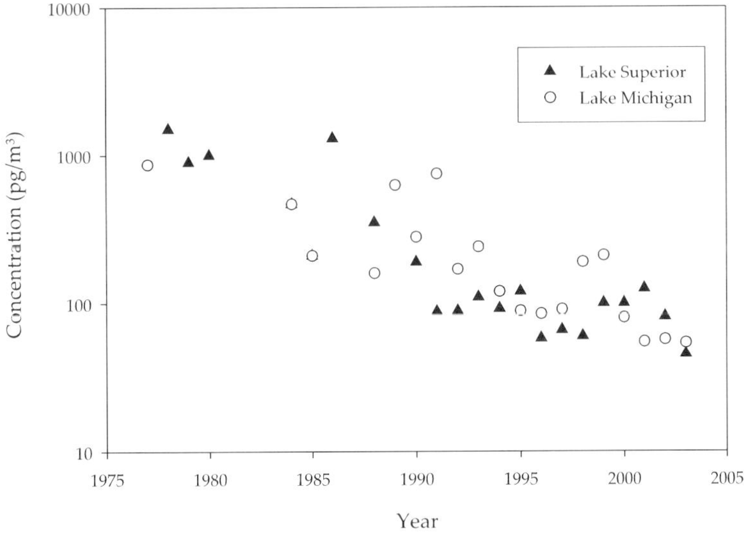


**Fig. 6** Annual average gas-phase PCB concentrations in non-urban air near Lake Michigan and Lake Superior, as compiled in [124] including additional data from IADN (Hites RA, 2005, Personal communication)

and coworkers have provided a full statistical review of the relatively slow decline ($\sim 3\%$ per year since 1990) in airborne PCBs [124, 149–151].

Data from long-term monitoring stations confirm major atmospheric emissions in the cities of Chicago, Cleveland, and the New York City area; these sites are either part of IADN or the New Jersey Atmospheric Deposition Network [152–160]. Extensive analysis based on air trajectories for sites in IADN indicate that Chicago is an important source area for the western Great Lakes region, while the heavily populated East Coast from Boston to Washington, DC is a source for the eastern portion of the Great Lakes [161].

Despite the clear evidence of PCB sources in urban air, the magnitude of these sources is unknown. Emission inventories that have proved useful for other pollutants (mercury, dioxins, and furans) are inadequate for PCBs. Emission inventories for PCBs maintained by the US EPA show a total release of about 15 kg year$^{-1}$ released to the atmosphere in the entire Great Lakes region [162]. This value grossly underestimates what is actually released. For demonstration purposes, we can compare this emission value to observed concentrations. For example, consider the air over a city as a well-mixed box. Let us assume that the bulk of the PCB emissions from the urban landscapes of Chicago, Milwaukee, Detroit, Cleveland, Toronto, or Buffalo into the local atmosphere are each focused in a $5 \times 5$ km area with an atmospheric mixing height of 1 km. Then the surface-normalized burden for atmospheric $\Sigma$PCBs at 1 ng m$^{-3}$ average annual concentration is 1 $\mu$g m$^{-2}$ or 25 g in the 25 km$^3$ of atmosphere. Occasionally, PCB concentrations are as high as 10 ng m$^{-3}$, in which case the surface-normalized burden is 10 $\mu$g m$^{-2}$ or 250 g. Emission rates of around 1 kg per *day* in *each* city are necessary to achieve the observed atmospheric concentrations, assuming a slow wind speed and residence time of the atmosphere of 30 min. Thus the vast majority of PCB atmospheric sources are unaccounted for.

People living near potential sources of atmospheric PCBs are very concerned about the magnitude of emissions and their exposure. In East Chicago, IN, for example, citizens are concerned about the effects of dredging on airborne concentrations of PCBs. As a result, environmental and community advocacy groups have released public documents citing their concerns and some have actively evaluated the planned activities by the Army Corp of Engineers [163, 164]. This concern appears to have motivated the US EPA and the US ACE to issue several documents evaluating the major dredging methods, including hydraulic dredging and clamshell dredging [165–167]. The activity of the community also seems to have encouraged a more open dialog about the relationship between dredging activities and atmospheric concentrations of PCBs. The public concern is reasonable. PCB exchange between air and water is dynamic. Human activities in heavily contaminated water are likely to have some impact on atmospheric emissions of PCBs.

## 4.2
## Calculating Air–Water Exchange

Air–water exchange of PCBs is often determined using coordinated measurements of PCBs in air and water and the modified two-film theory. The two film theory of mass transfer was first described in 1923 by Whitman [168] and its modifications have been used many times for predicting air–water exchange of PCBs in the Great Lakes, smaller lakes, rivers, harbors, and bays [131, 169–177]. Its basic assumption is that molecular diffusion across the air–water interface is the rate-limiting step. Turbulence in air and water is very fast and so the total flux can be calculated as a function of molecular diffusion and film width only. This is a straightforward application of Fick's law, which states that the flux of mass across a boundary is a function of the concentration gradient and a diffusion or mass transfer coefficient. Examples of Fick's law in the environment include uptake of oxygen into surface waters, absorption of carbon dioxide by the earth's oceans, release of hydrogen sulfide from aerated ground water, and emission of methane gas from swamps. Examples in the human body include transfer of oxygen, carbon dioxide, nicotine and other gases between inhaled air and blood circulating in the lung. Equation 1 is the general form for predicting the direction and rate of exchange:

$$F_{\text{gas,net}} = k_{\text{ol}} \left( C_{\text{w}} - \frac{C_{\text{g}}RT}{H} \right) \tag{1}$$

where $F_{\text{gas,net}}$ is the net mass flux (mass area$^{-2}$ time$^{-1}$), $C_{\text{w}}$ is the concentration dissolved in water, $C_{\text{g}}$ is the concentration in the gas phase, $R$ is the universal gas constant, $T$ is temperature, and $H$ is Henry's Law constant. The terms of Eq. 1 that are in parentheses define the chemical concentration gradient. The concentration gradient is measured from samples collected near the air–water surface. The mass transfer coefficient (length time$^{-1}$) is a rate constant. The value of the mass transfer coefficient is determined by the nature of the interface and the molecular diffusivity of the chemical. In the case of transport across a physical membrane, the mass transfer coefficient could be approximated by the molecular diffusivity of the compound divided by the width of the membrane. Liss and Slater defined the model for air–water exchange in the environment [178]. This analogy makes the mathematical derivation of the mass transfer coefficient easier. The overall mass transfer coefficient, $k_{\text{ol}}$, is described as two resistances in series:

$$\frac{1}{k_{\text{ol}}} = \left( \frac{1}{k_{\text{w}}} + \frac{RT}{Hk_{\text{g}}} \right) \tag{2}$$

where $k_{\text{w}}$ is the mass transfer coefficient through a stagnant water layer and $k_{\text{g}}$ is the mass transfer coefficient through a stagnant air layer. The other terms are the same as before. If the stagnant layers were real (i.e., constant and measurable) then the mass transfer coefficients could be calculated as a function of the thickness of the layers. If an oil film is present, the mass transfer equation will include a third term for the mass transfer coefficient for each congener in oil. However, one cannot measure these widths, so the mass transfer coefficients are found through experiment. Many laboratory and field studies have been conducted to determine the values of $k_w$ and $k_g$ as a function of measurable factors such as molecular diffusivity, wind speed, surface roughness, and wave height. Classic experiments include studies of oxygen and carbon dioxide uptake in wind tunnels and flumes. Recently studies in open oceans, rivers, and sheltered lakes have been performed using volatile tracers like sulfurhexafluoride and helium isotopes [179–183]. The design of the tracer experiments is to predict the mass transfer coefficients as a function of measured wind speed and molecular diffusivities (expressed as Schmidt numbers). These findings permit determination of mass transfer coefficients for compounds with other diffusivities and under other meteorological conditions.

Input via wet and dry particle deposition is not as important for the Great Lakes as for other bodies of water because of their large surface area relative to the size of their drainage basins. This results in air–water gas exchange dominating the atmosphere–lake interaction. However, direct deposition is not negligible. Deposition occurs through two processes: deposition of atmospheric particles, and precipitation. The magnitude of these fluxes depends on concentrations in the atmosphere, which vary themselves, as discussed above. Particulate deposition can be measured directly [152, 158, 184], but it is difficult to produce an artifact-free collection surface that accurately reflects the conditions of the Great Lakes water surface. Therefore, PCB particle deposition is often modeled as a function of the size of the aerosol particles and the concentration of PCBs on those particles. Particle size distributions have been measured over and around the Great Lakes but there are no reports of PCB congener concentrations as a function of particle size. Current analytical methods are not sensitive enough to collect such data and so most studies of PCB deposition have assumed that PCB concentrations are not a function of particle size [129, 131, 185]. Deposition fluxes are then modeled as a function of deposition velocity to water surfaces [186–188].

Wet deposition may be an important source of PCBs to the Great Lakes. Unfortunately, like dry particle deposition, it is difficult to measure directly because of the low PCB concentrations, the problem of sampling artifacts, and the distribution of rain and PCB concentrations over space and time. Some trends are evident, however [189]. Snow has been shown to be much more effective than rain in scavenging PCBs from both vapor and particulate phases at a site in Minnesota [190, 191]. This leads to a seasonal effect, in that precipitation concentrations of PCBs are highest in the winter in Chicago (Hites, personal communication), as with the concentrations of many other persistent organic pollutants [192]. Possible reasons include sorption to an aqueous or organic film on the snowflake, a filtering effect of the snowflake in scavenging particles from the atmosphere, and lower temperatures associated with snow events [190, 193]. The environmental conditions and hydrometer properties can also have a large effect on scavenging efficiency. Research in the southern Lake Michigan area has shown that scavenging ratios for PCBs vary significantly (over more than two orders of magnitude for total PCBs) from storm to storm, and even over the course of the same storm, in the case of rain [194].

Air–water exchange fluxes for the Great Lakes have been calculated in many cases [195], and they have been found to be highly variable even within a lake [129], as might be expected with the large variability of the region. Proximity to population centers is one important variable that affects the net direction of flux [131, 159, 196]. A discussion of the magnitude of atmospheric exchange processes can be found in Sect. 9.

# 5
# Tributaries

Industrial emissions of PCBs to the lakes via rivers, tributaries, and connecting channels have been a significant source in many places. Local areas with very high PCB concentrations have been identified throughout the basin (see Table 3 and Fig. 7 for examples of sediment and stream concentrations). In some cases, local contamination with PCBs alone or in combination with other contaminants has been linked to adverse effects in wildlife [77, 197–199]. For example, mink living near a contaminated reach of the Kalamazoo River in Michigan displayed lesions of the jaw known to be caused by PCB-126 and 2,3,7,8-tetrachlorodibenzo-$p$-dioxin [77], as mentioned earlier.

Long after industrial emissions have been reduced, contaminated sediments in affected reaches can still serve as a significant PCB source for the Great Lakes. The most-studied Great Lakes tributary is the Lower Fox River in Wisconsin, which drains to Lake Michigan via Green Bay. Paper mills in the area first starting using PCBs to make carbonless copy paper in 1954, and it

**Table 3** Sediment PCB concentrations in Great Lakes tributaries (ng g$^{-1}$ dry weight)

| Location | Min | Max | Average | Date | Refs. |
|---|---|---|---|---|---|
| Ashtabula, OH | 500 | 7000 | 2000 | 1998 | [201] |
| Fox, WI | 132 | 223 000 | 24 300 | 1987–1994 | [202] |
| Detroit, MI/ON | 8 | 25 000 | 4800 | 1998 | [203] |
| Rouge, MI | 470 | 10 900 | 2500 | 1998 | [203] |
| Manitowoc/Pine Creek, WI | < 50 | 1 900 000 | 44 000 | 1993–1995 | [204] |
| Milwaukee, WI | < 50 | 870 000 | | 1993–1995 | [204] |

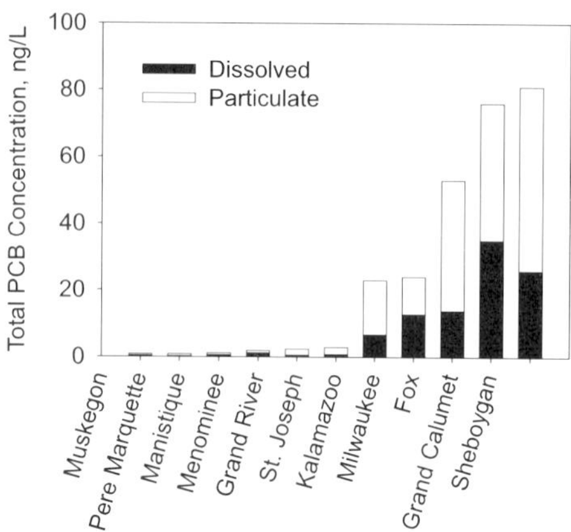


**Fig. 7** PCB concentrations in Lake Michigan streams [200]

is estimated that between $120 \times 10^3$ and $400 \times 10^3$ kg of PCBs were released to the Fox River since that time from the manufacture and recycling of carbonless copy paper [13]. An estimated 95% of PCBs in Green Bay originated in the Fox River; as much as $70 \times 10^3$ kg of PCBs may have been transported to Green Bay sediments in this way [13]. Prevailing counterclockwise currents led to the accumulation of PCBs in the eastern part of the bay, where concentrations were measured as high as $1302 \, \mathrm{ng \, g^{-1}}$ [13]. In comparison, the maximum observed concentration of PCBs in the sediment of northern Lake Michigan was $91.2 \, \mathrm{ng \, g^{-1}}$, and the maximum for the entire lake was $220 \, \mathrm{ng \, g^{-1}}$, with averages of $40–70 \, \mathrm{ng \, g^{-1}}$ for regions of the lake [200].

# 6
# Sediments

## 6.1
## Accumulation

PCBs accumulate in bottom sediments and are recalcitrant to decay. As a result, sediments harbor some of the most concentrated levels of PCBs near regions of intense PCB usage. For example, measurements as high as $3.6 \, \mathrm{g \, kg^{-1}}$ dry weight (0.36%) have been found in the sediments of the Waukegan harbor [205]. These local areas had severe problems, driving regulatory initiatives. Researchers then expanded their studies to explore contamination lakewide. One large study collected more than 1700 surficial sediment samples in a grid across the Great Lakes Basin between 1969 and 1975 [206], at the time of peak PCB concentrations. Average concentrations were lowest in Lake Superior ($3.3 \, \mathrm{ng \, g^{-1}}$) and highest in Lakes Ontario ($57.5 \, \mathrm{ng \, g^{-1}}$) and Erie

(94.6 ng g$^{-1}$). Concentrations were highly variable spatially in all lakes with much higher concentrations typically found in depositional areas [206]. The most recent lakewide surveys show significant decreases in concentrations since the 1970s and 1980s [207, 208].

The history of environmental exposure to PCBs is well recorded in Great Lakes sediments. Sediment cores show increases in concentrations and accumulation rates that begin in the early part of the 20th century and peak in the late 1960s to mid-1970s [209, 210]. Since the 1970s, most sediment cores show significant declines in PCB accumulations [211]. The most recently dated samples (Fig. 8) show that PCB accumulation rates in the sediment of Grand Traverse Bay have decreased to a level similar to that recorded for 1960, a time of widespread PCB use in the region [210]. Accumulation of PCBs in sediments exhibit slow declines and reflect the continuing exposure to these compounds in today's environment.

Accumulation in sediments from isolated lakes has sometimes been used to determine atmospheric inputs [212]. For example, rates of accumulation in Lake Superior [213] were found to be similar to rates in lakes with atmospheric sources only [205, 212], confirming that the atmosphere is the largest source of PCBs.

PCB accumulation in the sediment is not a straightforward function of downward settling of falling particles. Several field and laboratory studies have shown that PCBs can be recycled back to the water column via biological

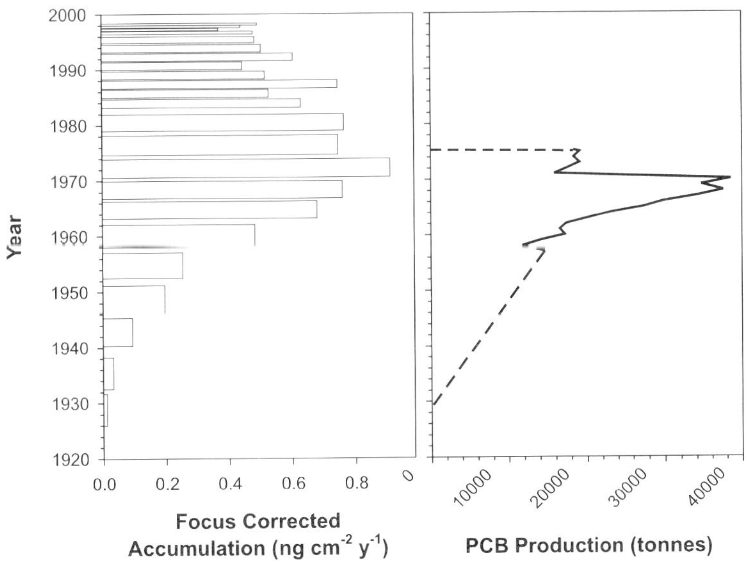


**Fig. 8** PCBs in a sediment core collected in Grand Traverse Bay, Lake Michigan (modified from [210] and PCB production data [3])

and physical processes occurring at the sediment–water interface. Baker and Eisenreich used a submersible to sample the benthic nepheloid layer (a region of particle-rich lake water extending a few meters above the lake floor), the sediment boundary layer, and sediment cores [214]. They found a dynamic system: the nature of particles changes with the season, and cycling within the lake occurs [214]. There is rapid settling of organic-rich particles to the benthic environment, where organic material is utilized by biota [214]. Further research revealed that the accumulation rate of PCBs was much less than the flux of PCBs associated with settling particles [213, 215]. Calculations show that 50% of PCBs in the Lake Superior water column are transported to the benthic environment each year via settling particles. This represents a total flux approximately 17 times that of the estimated atmospheric deposition. However, of the PCBs associated with settling particles, only 1–10% end up in the sediments [213, 216].

The sorptive properties of PCBs are largely controlled by their hydrophobicity and by the particle organic carbon content, which is generally associated with the clay or fine-sized particles. As a result, PCBs are carried by fine particles, deposited to sediments, and focused into the more quiescent depositional basins. Once delivered to the bottom sediments, PCB burial can be slowed by sediment resuspension and bioturbation. PCBs may also partition into porewater or bind to colloidal organic matter and migrate within and from the sediments via diffusion or bioirrigation. The potential effect of these processes is to alter the depositional history of the contaminant as recorded in the sediments and to increase the residence time of the contaminant in the ecosystem. PCBs may also be subject to biotic or abiotic transformation in sediments, which can further alter the sediment input history. Studies on the accumulation of PCBs have found that their depositional history is often preserved in the sediment bed, as local and non-local redistribution processes are not sufficiently intense over the time scale of HOC inputs to alter the profiles significantly. However, much of the work was done in the 1980s, and there have been only a few more cores collected since then [210]. Sediment profiles of PCBs are needed to determine the recovery of aquatic ecosystems impacted by PCB deposition, the effectiveness of chemical bans, and the data needed to model chemical behavior. The diagenesis of PCBs within sediments over time is important to evaluate the extent that redistribution and transformation processes have altered the recent historical record. In addition, previous estimates of the total sediment burden of PCBs, which are necessary to construct a mass balance, have not reported on sufficient cores to form a reliable estimate. The work of Eisenreich et al. [209] and Wong et al. [14] represents a unique field study of PCB accumulation and diagenesis in sediments. The investigators collected sediment cores from the same location in Lake Ontario in 1981 and in 1990. The results are illustrated in Fig. 9. Between the two cores, there is a shift in the depth of peak PCB concentration with the peak moving downcore at the rate of sedimentation. In both cores, it is clear

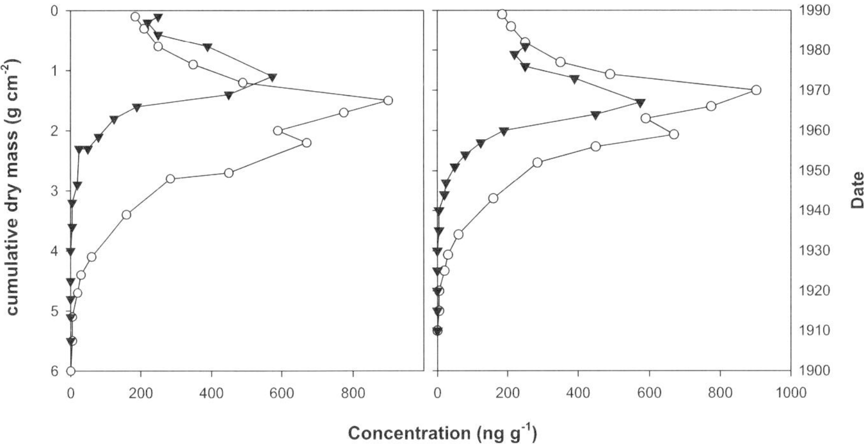

**Fig. 9** PCB accumulation in Lake Ontario sediment, modified from [14]. The *left plot* shows the downward shift in the PCB peak as cleaner sediment accumulates at the sediment water interface and older sediment is buried. The *right plot* shows that the peak of the PCB accumulation rate occurred in $\sim$ 1970 and that peak is retained over time. The core collected in 1981 (*filled triangles*) was analyzed for 25 congeners while the core collected in 1990 (*open circles*) was analyzed for 85 congeners. When corrected for the same congeners, there is no evidence of PCB decay or loss from the sediments over the 9-year period

that the depositional history shows a peak at about 1970, similar to most other Great Lakes cores and the record of PCB production. Moreover, focusing-corrected inventories of the two cores are similar at about $320-340\,\mathrm{ng\,cm^{-2}}$, with inventories statistically the same.

## 6.2
## Resuspension

PCBs that have historically accumulated in the sediments may return to the water through resuspension–desorption and through continuous diffusion from sediment porewaters.

Resuspension moves a large mass of PCBs into the water column. As part of the Episodic Events: Great Lakes Experiment (EEGLE) study in southern Lake Michigan, Hornbuckle and colleagues showed that hundreds of kilograms of PCBs are resuspended during major wintertime storms [217–219]. The origin of the resuspended sediments includes coastal sediments and sediments from deep regions of the lake, where PCB concentrations are highest. This episodic input of PCBs into the lake would not be considered in the annual loadings determined by mass balance studies.

It is possible that resuspension of PCBs during major storms may be as large a source as the tributaries. However, the fate of the resuspended contaminants is unclear. The EEGLE study found that most of the resuspended

PCBs appear to remain sorbed to the sediment particles [220]. Concentrations of dissolved PCBs did not increase after a large storm, despite evidence of resuspended PCBs in the particle phase [219]. In fact, Bogdan et al. reported *decreased* concentrations of dissolved PCBs after a major storm in southern Lake Michigan [218]. The authors suggested that the resuspended sediments actually cleansed dissolved PCBs from the water. In either case, it appears that resuspended PCBs may simply return to the lake bottom after the storm subsides. This hypothesis is well supported in the experimental literature. For example, in a well-constructed laboratory study, Gong and Depinto measured and modeled the rate of desorption from suspended particles pre-equilibrated with PCB congeners 52 and 153 [221, 222]. They show that desorption exhibits two-stage behavior: a fast rate followed by a much slower rate. The fraction of PCBs lost during the first stage depends on how long the particles have been equilibrated with the PCBs. When PCBs have been in contact with the sediment for a very long time ($\sim$3 years), only about 10% of the PCB is desorbed from the particles in 24 h. Even after a year of suspension, these particles only desorb about 50% of the total exchangeable PCB. A review article by Pignetello and Xing shows that these findings are consistent with many other hydrophobic organic compounds [223]. Short-term resuspension of PCB contaminated sediments appears to be a small or even negligible source of PCBs in the Great Lakes.

A study of resuspension and downward settling of sediments in Grand Traverse Bay supports the contention that resuspension events do not change the net flux of PCBs in the system. Schneider et al. deployed sequencing sediment traps at two locations in the western arm of Grand Traverse Bay, Lake Michigan [224]. The traps collected integrated samples of settling particles every two weeks from May 1997 to September 1999. Storm-driven episodic events, which occurred only 20% of the time, accounted for 65% of both the mass flux and total PAH flux. The annual PCB flux was not influenced by these episodic events: only 18% of the total PCB flux occurred during these events. PAHs and PCBs appear to be tracing different types of particles in the water column. Several large mass flux events characteristic of seiches were observed simultaneously in the benthic nepheloid layer (BNL) at both the northern and southern sites. The particles settling as a result of these resuspension events had lower PCB and PAH concentrations than particles settling at other times, suggesting that material settling into the traps on the high mass flux days is a mixture of the less contaminated resuspended sediment and the "regular" contaminant-rich particles settling into the BNL.

Diffusion of PCBs from highly contaminated surficial sediments in areas of concern may be an important net source to the Great Lakes. In fact, PCB diffusion from the porewaters of heavily contaminated coastal and riverine sediments may justify the removal of those sediments. Porewater diffusion in open lake sediments is not a major source. Diffusion of PCBs from porewater is governed by an effective diffusion or mass transfer coefficient and the sediment–water concentration gradient. The concentration gradient is controlled by the fraction of organic carbon in the sediment and the sediment–water equilibrium distribution coefficient [225]. Ortiz et al. reported the equilibrium distribution constant, $K_{OC}$, and the mass transfer coefficients for PCBs in sediment–water systems similar to those found in the Great Lakes [225]. While the rate of diffusion from sediment porewaters is slow, the constant release may result in a significant source of PCBs to waters exposed to contaminated sediments.

# 7
# Water Column Processes

From a geochemical perspective, the inventories of chemicals in the water column play a central role in the aquatic cycling of pollutants. The large quantity of water in the Great Lakes is a significant reservoir for anthropogenic chemicals, moderating the long-term movement of chemicals from their sources to their ultimate environmental fates. Although a large mass of PCBs have been historically stored in the Great Lakes water column, PCB concentrations dissolved in the water are generally quite low (femtogram to nanogram per liter range for individual congeners), which presents significant analytical challenges. Nonetheless, the Great Lakes region enjoys a nearly 30-year record of high-quality measurements of water column PCBs, especially in Lakes Superior and Michigan.

The Great Lakes water columns are cold, deep, and generally contain only modest amounts of suspended solids and dissolved organic carbon. PCB congeners distribute between truly dissolved, colloidal, and particulate phases according to their individual hydrophobicities and the amount of suspended solids and dissolved organic matter in the water column. The dissolved–particulate partitioning of PCBs controls the relative magnitude of fate processes, the exposure to and accumulation within the food web, and the ecosystem-level response to altered loadings. While the two-phase partitioning model is the standard description of PCB behavior in surface waters, field measurements from the Great Lakes consistently show deviations from this model [216, 226]. Specifically, the PCB congener partitioning is substantially more variable than can be attributed to sampling or analytical variance, and the relationship between observed PCB partition coefficients and their octanol–water partitioning coefficients is "flatter" that predicted by empirical models. Field-measured partition coefficients for higher molecular weight PCB congeners are often less than $K_{OW}$, perhaps due to either slow sorption kinetics or the presence of "third phase" colloidal material, which biases the operationally-defined dissolved PCB measurements. Interestingly, the lower chlorinated PCB congeners in the Great Lakes often have partition coefficients greater than predicted from models. The mechanism(s) for this enrichment of these less hydrophobic congeners on Great Lakes particles is unclear, but may involve differential partitioning into plankton [227], and/or magnification through microbial loops [228].

## 7.1
## Concentrations

PCB congeners and other synthetic organic chemicals are present in the waters of the Great Lakes at concentrations generally ranging from femtograms per liter to nanograms per liter, with the majority of PCB congeners in the "dissolved" or "non-filterable" phase (Table 4). Measuring these low concentrations requires quantitative concentration of the analytes from large volumes of water. Historically, sampling techniques for PCBs in the Great Lakes involve a particle separation step (usually filtration through flat glass fiber filters with about 1 $\mu$m nominal size cutoffs, and, less commonly, continuous-flow centrifugation) followed by isolation of PCBs from the filtrate by adsorption to a pre-cleaned resin (usually XAD-2) or extraction via liquid–liquid partitioning. Resin isolation can efficiently extract PCBs from large volumes of Great Lakes water, but requires careful cleaning and conditioning prior to use and attention to not exceed optimal flowrates and resin capacity. Liquid–liquid partitioning has been employed either in batch mode (extracting a few liters with dichloromethane in sealed glass bottles aboard ship or in standard glassware in the laboratory) or using the Goulden counter-current liquid–liquid extraction system [229]. Due to the large volume of solvent required to extract sufficient amounts of Great Lakes water, liquid–liquid extraction techniques are susceptible to elevated blanks. These techniques have routinely been used to process tens to hundreds of liters of Great Lakes water, resulting in methods sufficiently sensitive to reliably measure total PCBs greater than about 1 ng L$^{-1}$.

It is recognized that filtration is operational, that colloidal-bound PCB congeners are not retained by the filter, and that operational "dissolved" measurements may be biased positively by colloidal material. Techniques to measure truly dissolved PCBs include gas sparging, differential diffusion into membrane-bound lipids (e.g., semipermeable membrane devices, [230]), and selective adsorption (e.g. non-equilibrium solid phase microextraction [231, 232]). Unfortunately, none of these techniques has sufficient sensitivity to reliably and unambiguously measure truly dissolved PCB congeners at the levels present in the Great Lakes.

The first published measurements of PCBs in the Great Lakes surface waters are those of Veith et al. [233], who reported total PCB concentrations in western Lake Superior (at the Duluth EPA laboratory intake) of 0.8 ng L$^{-1}$ in 1972. An initial attempt to survey PCB levels in the Great Lakes was that of Glooschenko et al. [234], who sampled Lakes Superior and Huron in late July and early August, 1974. Two liters of surface water (1 m depth) was

**Table 4** Concentrations of PCBs in the water column of the Great Lakes. All values are the sum of PCB congeners measured in both the dissolved and particle phases

| Lake | Location | Sampling year | Sampling month | mean t-PCB ng/L | $t$-PCB SD | min $t$-PCB ng/L | max $t$-PCB ng/L | N | Method | Refs. |
|---|---|---|---|---|---|---|---|---|---|---|
| Erie | | 1986 | April/May | 1.378 | – | 0.341 | 3.513 | 21 | Capilliary | [239] |
| Erie | | 1993 | April/May | 0.70 | 0.67 | 0.2 | 1.6 | 6 | Capilliary | [240] |
| Huron | Open Lake | 1986 | May | 0.631 | – | 0.186 | 2.342 | 17 | Capilliary | [239] |
| Huron | Georgian Bay | 1986 | May | 0.688 | – | 0.279 | 1.434 | 7 | Capilliary | [239] |
| Huron | | 1993 | April/May | 0.13 | 0.03 | 0.088 | 0.16 | 5 | Capilliary | [240] |
| Huron | | 1994–1995 | | 0.17 | – | – | – | 13 | Capilliary | [241] |
| Michigan | Chicago | 1976 | – | 41 | – | – | – | 2 | Packed | [242] |
| Michigan | Beaver Island | 1976 | – | 30 | – | – | – | 1 | Packed | [242] |
| Michigan | | 1979 | August | 2.88 | 3.37 | – | – | 8 | Packed | [237] |
| Michigan | Green Bay | 1980 | September | 3.5 | 2.7 | 1.1 | 20 | 9 | Capilliary | [243] |
| Michigan | | 1980 | September | 1.8 | 1.8 | 0.4 | 7.9 | 19 | Capilliary | [243] |
| Michigan | | 1980 | April | 5.66 | 1.12 | 4.9 | 7.1 | 4 | Packed | [237] |
| Michigan | | 1980 | July | 6.36 | 1.3 | 4.8 | 7.9 | 7 | Packed | [237] |
| Michigan | | 1991 | September | 0.47 | 0.06 | – | – | 9 | Capilliary | [244] |
| Michigan | | 1992 | Spring | 0.424 | 0.058 | – | – | | Capilliary | [245] |
| Michigan | | 1993 | April/May | 0.21 | 0.03 | 0.17 | 0.27 | 5 | Capilliary | [240] |
| Michigan | | 1994 | July | 0.12 | 0.05 | 0.07 | 0.20 | 14 | Capilliary | [156] |

**Table 4** (continued)

| Lake | Location | Sampling year | Sampling month | mean t-PCB ng/L | t-PCB SD | min t-PCB ng/L | max t-PCB ng/L | N | Method | Refs. |
|---|---|---|---|---|---|---|---|---|---|---|
| Michigan | Green Bay | 1994–1995 | | 0.75 | 0.65 | 0.25 | 1.67 | 27 | Capilliary | [241] |
| Michigan | Open Lake | 1994–1995 | | 0.26 | 0.10 | 0.13 | 0.67 | 307 | Capilliary | [246] |
| Michigan | | 1994 | May | 0.21 | 0.02 | 0.19 | 0.24 | 4 | Capilliary | [156] |
| Michigan | | 1995 | January | 0.31 | 0.03 | 0.27 | 0.36 | 5 | Capilliary | [156] |
| Michigan | Grand Traverse Bay | 1997 | | 0.31 | 0.56 | 0.09 | 2.34 | | Capilliary | [247, 248] |
| Michigan | Grand Traverse Bay | 1998 | | 0.35 | 0.46 | 0.08 | 1.45 | | Capilliary | [247, 248] |
| Michigan | Grand Traverse Bay | 1999 | | 0.17 | 0.04 | 0.12 | 0.22 | | Capilliary | [247, 248] |
| Michigan | | 1998 | | 0.35 | 0.10 | 0.16 | 0.60 | 11 | Capilliary | [218] |
| Michigan | | 1999 | | 0.27 | 0.052 | 0.12 | 0.56 | 23 | Capilliary | [219] |
| Michigan | | 2000 | | 0.16 | 0.029 | 0.071 | 0.28 | 19 | Capilliary | [219] |
| Michigan | | 2003 | | 0.149 | 0.042 | 0.092 | 0.199 | 5 | Capilliary | [249] |
| Ontario | | 1986 | April | 1.41 | – | 0.484 | 2.614 | 31 | Capilliary | [239] |
| Ontario | | 1993 | April/May | 0.22 | 0.02 | 0.19 | 0.25 | 5 | Capilliary | [240] |
| Superior | | 1971–1975 | | 400 | – | – | 2000 | 84 | Packed | [250] |
| Superior | Western Arm ERL Lab Intake | 1972 | | 0.8 | – | – | – | | Packed | [233] |

**Table 4** (continued)

| Lake | Location | Sampling year | Sampling month | mean t-PCB ng/L | t-PCB SD | min t-PCB ng/L | max t-PCB ng/L | N | Method | Refs. |
|---|---|---|---|---|---|---|---|---|---|---|
| Superior | | 1972–1975 | | 110 | – | – | 1200 | 160 | Packed | [250] |
| Superior | | 1974–1975 | | 50 | – | – | 300 | 16 | Packed | [250] |
| Superior | | 1974 | November | 6 | – | 5 | 7 | 2 | Packed | [235] |
| Superior | | 1976 | December | 5 | – | – | – | 1 | Packed | [235] |
| Superior | Western arm/ Duluth Harbor | 1976 | January | – | – | 10 | 20 | 2 | Packed | [235] |
| Superior | Isle Royale | 1976 | January | – | – | 50 | 157 | 3 | Packed | [235] |
| Superior | | 1978 | July | 1.3 | 1.3 | 0.4 | 7.4 | 28 | Packed | [236, 251, 252] |
| Superior | | 1979 | June | 3.8 | 1.9 | 0.3 | 8.4 | 35 | Capilliary | [236, 251, 252] |
| Superior | | 1980 | August | 0.9 | 0.4 | 0.3 | 2.1 | 56 | Capilliary | [236, 251, 252] |
| Superior | | 1983 | June– October | 0.8 | 0.07 | 0.3 | 0.8 | 28 | Capilliary | [252, 253] |
| Superior | | 1986 | August | 0.55 | 0.37 | – | – | 5 | Capilliary | [169] |
| Superior | | 1986 | May | 0.337 | – | 0.193 | 0.578 | 19 | Capilliary | [239] |
| Superior | | 1988 | July | 0.33 | 0.04 | – | – | 5 | Capilliary | [254] |
| Superior | | 1990 | August | 0.32 | 0.03 | – | – | 6 | Capilliary | [254] |
| Superior | | 1992 | May | 0.18 | 0.02 | – | – | 5 | Capilliary | [254] |
| Superior | | 1993 | April/May | 0.08 | 0.01 | 0.07 | 0.1 | 3 | Capilliary | [240] |

Dashes indicated data that was not reported.

extracted from 18 stations in Lake Huron and 16 sites in Lake Superior. Extracts were purified with Florisil liquid-solid chromatography and analyzed with a packed column gas chromatograph with an electron-capture detector. PCBs were not detected in any of the water samples at a quantification limit of 0.1 ppb (0.1 $\mu$g L$^{-1}$ or 100 ng L$^{-1}$). Swain [235] reported total PCB concentrations in Lake Superior of 6 ng L$^{-1}$ in November, 1974 and 5 ng L$^{-1}$ in December, 1976. Corresponding concentrations in the western arm of Lake Superior and on Isle Royale ranged from 10–20 and 50–157 ng L$^{-1}$, respectively, in January 1976 [235]. Eisenreich et al. [236] measured total PCB concentrations in Lake Superior in July 1978 that averaged $1.3 \pm 1.3$ ng L$^{-1}$ (range 0.4–7.4 ng L$^{-1}$, N = 28) and Rice et al. [237] report levels of $2.9 \pm 3.4$ ng L$^{-1}$ (N = 8) in Lake Michigan in August 1979. Each of these studies in the 1970s used packed column chromatography with electron capture detection. The results may be biased high due to coelution of PCB and toxaphene components. Toxaphene is enriched in Lake Superior relative to the other Great Lakes [238].

## 7.2
### Recycling in the Water Column

PCB congener dynamics in the Great Lakes water column is controlled by partitioning between dissolved chemical and the atmosphere, suspended particles (including plankton), and settling particles. Unlike shallower, more productive lakes with significant net sedimentation, burial of PCB congeners in Great Lakes sediments is inefficient, with considerable recycling within the water column [215, 217, 224]. Since the lakes have been net sources of PCBs to the atmosphere via volatilization since at least the late 1980s [169, 254–256], the waters of the Great Lakes served as a decadal-scale PCB capacitor – accumulating contaminants derived from the atmosphere, river loads, and local sources during periods of increasing PCB environmental burdens, then slowly releasing the chemicals into the regional atmosphere. Although sufficient PCBs accumulate in Great Lakes sediments to preserve historical loadings, long-term burial removes only a small fraction of the PCBs that cycle through the water column.

Water column recycling of PCBs is driven by efficient partitioning into rapidly-settling, organic-rich particles that move PCBs from surface waters to near the lake floor on time scales of days to months [213, 215, 217, 224]. This rapid communication between the atmosphere and near-bottom environments shortens lakewide contaminant residence times and exposes epibenthic organisms to atmospheric contaminants. Once delivered to the lake floor, more than 90% of the particle mass is degraded and the associated PCBs released into bottom waters, where they are remixed throughout the water column during turnover. Early evidence for this seasonal cycle were the observations of Eisenreich et al. [236] that total PCB concentrations in western Lake Superior surface waters were significantly higher in the spring prior to stratification (3.8 ng L$^{-1}$ in June, 1979) than later in the summer (1.3 ng L$^{-1}$ in July, 1978 and 0.9 ng L$^{-1}$ in August, 1980).

## 7.3
## Trends

PCBs were measured in the surface waters of Lakes Superior and Michigan using consistent congener-based quantification between about 1980 and the mid-1990s. Average total PCB concentrations reported for the two lakes are plotted versus the year of collection in Fig. 10. As noted earlier by Jeremiason et al. for Lake Superior [254], and by Offenberg and Baker [156] and Schneider et al. [210] for Lake Michigan, total PCB concentrations decreased exponentially in each lake during that period. In Lake Superior, the first-order rate of decline is 0.08 year$^{-1}$ over that period. At this rate of decline, the current PCB concentration in Lake Superior surface waters should be less than 0.01 ng L$^{-1}$ (10 pg L$^{-1}$). This extrapolation assumes that the rate of decline between 1980 and 1993 continued unabated in Lake Superior.

The temporal history of PCBs in Lake Michigan is apparently more complex (Fig. 10). There appears to be little if any change in total PCB concentrations between the spring 1993 measurements of Anderson et al. [240], those of Baker and colleagues in northern Lake Michigan in 1997–99 [247], and the EEGLE Program in southern Lake Michigan in 1998–2000 [217, 218], although the most recent measurement in 2003 was 149 pg L$^{-1}$ [249]. Earlier, Pearson et al. [244] reported a first-order loss rate of $0.078 \pm 0.018$ year$^{-1}$ for the period 1980–1991. Offenberg and Baker [156] extended this analysis, resulting in a loss rate between 1980 and 1994 of $0.17 \pm 0.03$ year$^{-1}$. These more recent measurements suggest that the rate of decrease in Lake Michigan has slowed (Fig. 10). This is consistent with the slow declines for PCBs in lake trout (discussed in Sect. 8). Whether this is "real", reflects the inherent difficulty of accurately characterizing an exponentially declining inventory, or is due to some unknown bias cannot be determined. Clearly, as PCB concentrations decline in the Great Lakes, the analytical challenge increases, and the potential for positive bias due to field and laboratory contamination increases. We note that there are no PCB data available for Lake Superior after 1993, so it is not possible to determine whether a similar apparent leveling off of the decline is occurring in Lake Superior.

Based on temporal extrapolations, it is likely that the PCB concentrations in the open waters of the lakes are less than 0.1 ng L$^{-1}$, requiring sampling and analytical methods than can reliably measure PCB congeners with method quantification limits in the pg L$^{-1}$ range. This will require a reinvestment in method development, including in situ isolation samplers, ship-board clean rooms, and dedicated laboratory facilities. The application of high resolution mass spectrometry (EPA Method 1668A) greatly improves instrumental sen-

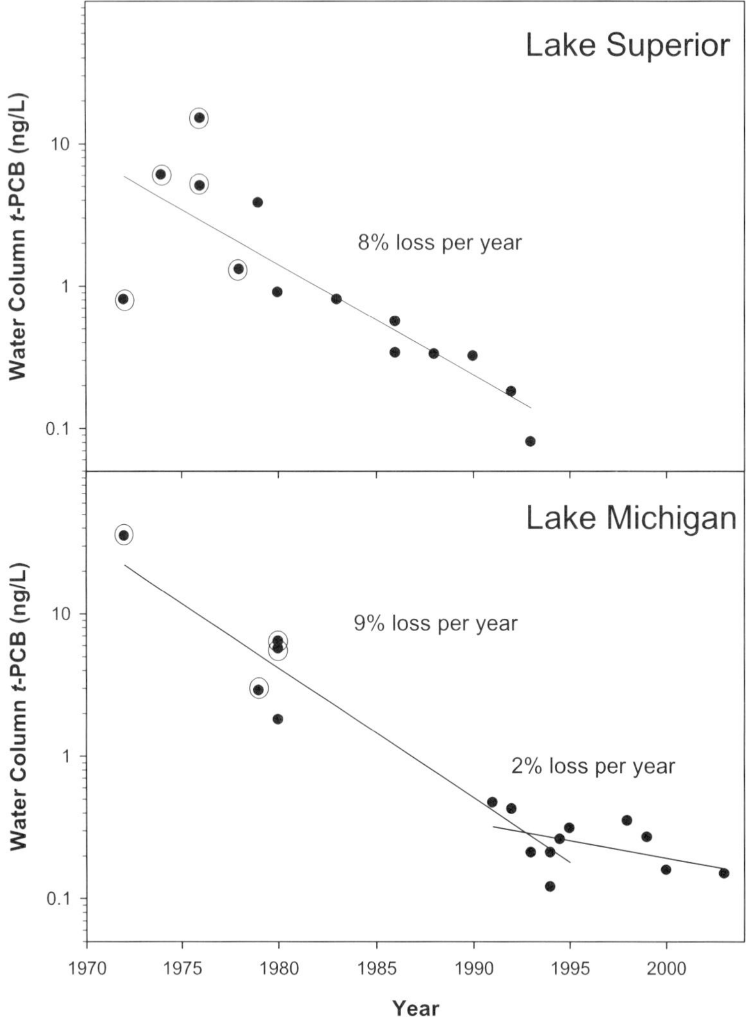


**Fig. 10** Concentrations of PCBs in Lake Superior and Lake Michigan. References are listed in Table 4

sitivity but does not relieve the requirement of sufficiently clean field and laboratory blanks for the PCB congeners.

Prior to the late 1970s, PCBs were present in the Great Lakes water columns at relatively high concentrations but were not detectable with the available sampling and analytical tools. Currently, PCBs are present at much lower levels, and probably are not detectable using the "classic" methods used between 1980 and 1999. During the past two decades, concentrations of PCBs have declined in the Great Lakes water columns. Unfortunately, even very low concentrations of PCBs in water can cause unacceptably high concentrations in fish. For this reason, risk-based water quality guidelines for PCBs will require ever improving methods to continue the excellent long-term record of PCBs in the Great Lakes water column.


# 8
# Food Web Dynamics

PCBs are often the prime example of a persistent, bioaccumulative, toxic (PBT) chemical. To fully understand the dynamics of PCB bioaccumulation, it is first necessary to define several terms. *Bioconcentration* refers to the accumulation of a chemical in an organism resulting from an equilibrium distribution of the chemical between the organism's tissue and its environment (usually referring to water). *Biomagnification* is the accumulation of a chemical by an organism from water and food exposure that results in a concentration that is *greater* than would have resulted from water exposure only and thus *greater* than expected from equilibrium. Compounds that biomagnify have greater concentrations in higher trophic levels of food webs. *Bioaccumulation* is a generic term that refers to either process, and describes what is observed in the field. These concepts are often described quantitatively, using ratios of what is found in the organism or trophic level compared to water, sediment, or another trophic level:

- Bioconcentration factor:
  $BCF = C_{org}/C_{water}$ at equilibrium (expressed in equivalent units)
- Bioaccumulation factor:
  $BAF = C_{org}/C_{water}$ observed in the environment (expressed in equivalent units)
- Biomagnification factor:
  $BMF = BAF/BCF$ (unitless)
- Biota-sediment accumulation factor:
  $BSAF = C_{org,lipid}/C_{sed,oc}$ (expressed in equivalent units)

Note that the BSAF is an empirical ratio and does not necessarily indicate a mechanistic relationship or route of exposure.

There are many factors that affect the bioaccumulation of PCBs and other chemicals. The degree to which a chemical bioaccumulates is related primarily to its lipophilicity (or hydrophobicity), expressed as the octanol–water partition coefficient, $K_{OW}$ [257, 258]. Because PCBs have $K_{OW}$s that span many orders of magnitude, PCB congeners have a wide range of bioaccumulation potential. The factors that influence bioaccumulation via $K_{OW}$ include the degree of chlorination and the substitution pattern of the chlorination. These two factors also determine the extent to which the molecule is metabolized; the more readily the molecule is metabolized, the less it bioaccumulates. The molecular configuration, a direct function of degree and pattern of chlorination, is thought to control the ease of passage of the molecule across cell membranes, such that compounds having a cross-sectional diameter greater than 9.5 Å are limited in their ability to bioaccumulate [259].

## 8.1
### Factors Affecting Fish Concentrations of PCBs

Properties of the organism that affect the degree of bioaccumulation include the amount and type of lipids (fatty organisms accumulate more PCBs), age (longer exposures lead to higher bioaccumulation of PCBs), metabolic systems (different species metabolize PCBs differently), and the diet of the organism (higher trophic levels have greater concentrations of PCBs). For example, the greater PCB concentrations in lake trout compared to rainbow trout are thought to be mostly due to age differences [260].

Thus, overall, the bioaccumulation of PCBs is a function of the uptake of PCBs from water and food, and the losses due to metabolism, excretion, and growth dilution; all are a function of time. A general model for estimating the concentration of PCBs ($C_{fish}$) in fish is:

$$dC_{fish}/dt = \text{water uptake} + \text{food uptake} - \text{losses} \tag{3}$$

or

$$dC_{fish}/dt = (k_u \times C_{water}) + (\alpha \times F \times C_{prey}) - [(k_x + k_g + k_m) \times C_{fish}] \tag{4}$$

where $k_u$ is the uptake rate constant from water, $\alpha$ is the assimilation efficiency in the gut, $F$ is the feeding rate for a given food preference, $C_{prey}$ is the concentration of prey in the diet, $k_x$ is the first-order excretion and egestion rate constant, $k_g$ is the first-order rate constant of dilution due to growth, and $k_m$ is the first order metabolism rate constant [261–265]. In top predators, as much as 98% of the PCB burden is due to uptake from food rather than water [262]. The biomagnification process is a result of the very slow clearance rates (excretion and egestion) relative to the uptake rates from water and food – the differences can be on the order of $10^6$ [264, 266]. It should also be noted that fish and mammalian enzyme systems will hydroxylate certain PCB congeners, and these hydroxylated PCBs have also been shown to bioaccumulate [267].

The length of the food chain has the largest effect on the observed accumulation of PCBs in top predators [268]. However, one of the more sensitive aspects of trophic transfer in food webs is the initial uptake of PCBs into the primary trophic level, phytoplankton [263]. The BCF from water to phytoplankton is approximately $10^5$–$10^6$, and subsequent BMFs to higher trophic levels are on the order of two to five. Due to the life cycle of most phytoplankton, their rate of growth is on the same order as their rate of uptake of PCBs, thus making their uptake very time-dependent [227, 269–271]. Bioaccumulation into top predators will be greater if phytoplankton have a chance to reach equilibrium (low growth conditions; oligotrophic conditions) or reduced if the phytoplankton grow quickly and sediment to the bottom before being consumed (high growth conditions; eutrophic conditions) [271].

Top predator concentrations can also be affected by changes in their food web. Madenjian et al. have showed that the variability seen in lake trout concentrations of PCBs and other organochlorine contaminants is driven by changes in prey concentration [260, 272–274]. The changes in food web structure [275] and lipid content [276] over time are likely responsible for some of the changes seen in the concentrations of PCBs in lake trout in the 1980s to 1990s [277]. The introduction of the invasive zebra mussel into the Great Lakes ecosystem in the 1980s and 1990s, for example, has been calculated to reduce PCB concentrations by nearly 50% in some forage fish [278].

Because of the concern over PCB exposure from eating contaminated fish, there has been widespread attention to documenting and studying the mechanism of PCB accumulation in the food webs of the Great Lakes. It is well-established in laboratory experiments that PCBs bioconcentrate in fish and other aquatic organisms [270, 279], and that they biomagnify in the environment [280], including the Great Lakes [281, 282].

PCBs have been documented to occur in nearly all trophic levels throughout the Great Lakes ecosystem [282], including bacteria (Hudson, unpub-

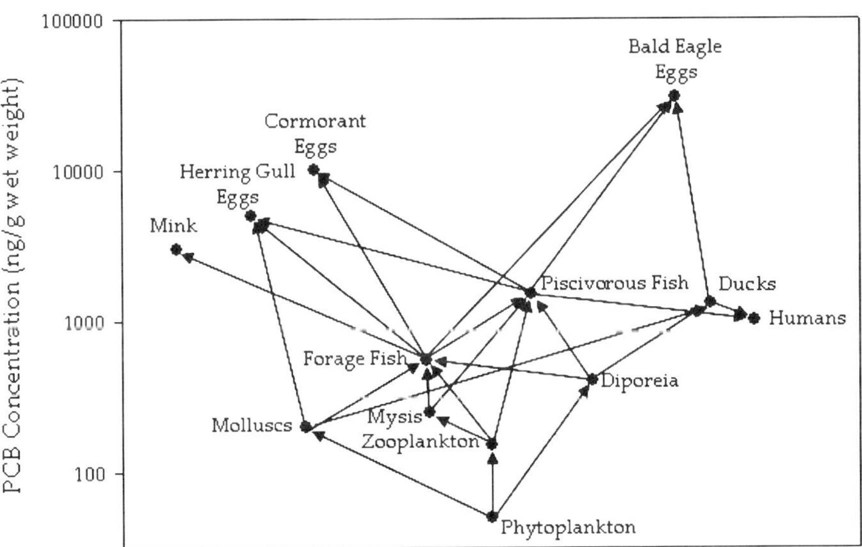


**Fig. 11** Great Lakes food web, with approximate PCB concentrations. Note that PCB concentrations vary significantly among species in different locations, and among individuals of the same species. Concentrations in humans are estimated from the blood plasma concentrations in humans that eat Lake Michigan fish [314] and the distribution of PCBs in the body of rats [35]. See text for additional references

lished data, 2005); phytoplankton [271, 283]; zooplankton [281, 283, 284]; benthic invertebrates [283, 285–287]; zebra mussels [283, 288–291]; sea lamprey [292]; snapping turtles [293]; fish [277, 294–304]; ducks [305]; fish-eating birds such as cormorants [86, 306], herring gulls [85, 306–308], and eagles [307, 309–313]; and fish-eating wildlife such as mink [46, 67–69, 72, 77]. These data all support the conclusion that PCBs bioaccumulate and that the more hydrophobic congeners (those with five or more chlorines) biomagnify in food webs. A simplified diagram of the Great Lakes food web, with approximate concentrations of total PCBs, is shown in Fig. 11.

In the Great Lakes basin, many studies have also documented that PCBs accumulate in humans, and that concentrations are greater in those who consume greater amounts of Great Lakes fish [314–322]. PCB exposure is also high in at-risk populations [323], such as charter boat anglers [320] and subsistence populations such as Native Americans [324–327]. For example, PCB blood plasma concentrations in fish-eaters in Michigan were approximately 14 ppb on average, compared to 5 ppb for the general population [314].

## 8.2
### Trends in Food Web Components

Both the USA and Canada have ongoing monitoring programs to assess the change in PCBs and other contaminants over time. In Canada, the Great Lakes Herring Gull Monitoring Program has been in operation since 1974 [328]. For fish, the US EPA, in cooperation with the US GS and Great Lakes States, has monitored PCBs in composites of lake trout (Lakes Superior, Michigan, Huron, and Ontario) and walleye (Lake Erie only) of constant size range (600–700 mm lake trout, 400–500 mm walleye) collected biennially from master sites within each lake since 1970, and the Department of Oceans and Fisheries Canada has a long-standing program of measuring PCBs in individual whole lake trout that are 4+ years in age. These programs were developed to provide exposure data for assessing human and wildlife effects, and provide some of the most extensive databases of PCBs in the world. The US program also analyzes coho and chinook salmon fillets from selected tributaries around the Great Lakes, and the Canadian program includes the analysis of smelt composites to assess trends in the dominant forage in Lake Ontario.

The current trend data for PCBs in lake trout collected for the US program are shown in Fig. 12 (Swackhamer, unpublished data, 2005). These data indicate that PCBs declined rapidly following their ban in both the USA and Canada in the early to mid-1970s. This decline was consistent with a first-order decay, and has been described in that manner previously [277, 304]. The rate of decline in PCBs in Lakes Superior, Huron, and Michigan clearly changed in the mid-1980s [277]. Since then, the PCB concentrations in the upper three lakes have declined at a much slower rate, if at all. The half-lives in the 1970s were on the order of 4–6 years, while the half-lives since the

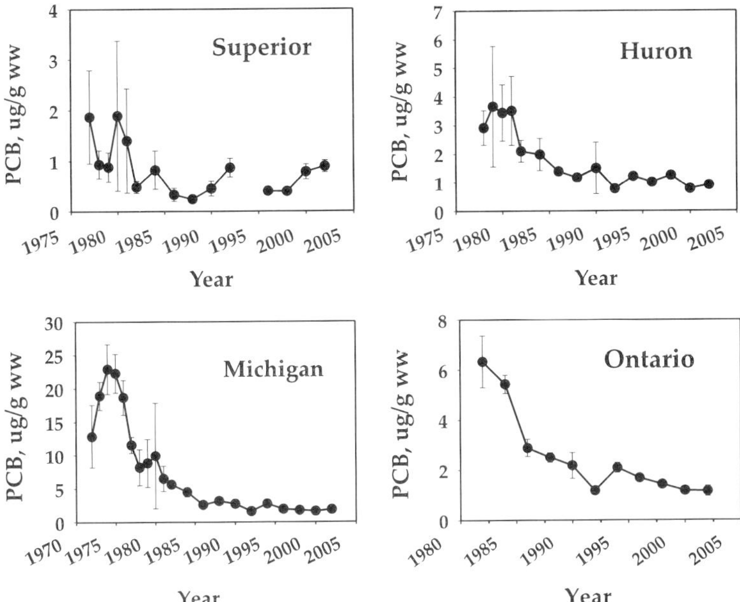

**Fig. 12** Trends in PCBs in lake trout from Lakes Superior, Michigan, Huron, and Ontario. Data are means of approximately 10 composites of 5 whole fish, each 600–700 mm in length. All fish from a given lake are from the same site

mid-1980s are around 10–20 years (depending on the time range used for calculation) for Lakes Michigan and Huron, and concentrations have actually increased in Lake Superior. The rate of decline of PCBs in Lake Erie walleye and Lake Ontario lake trout has not changed significantly over time, with half-lives of 18 and 9 years, respectively.

## 8.3
## Trophic Transfer and Biomagnification Studies

In the bioaccumulation process, not every congener biomagnifies to the same extent [329]. This is because $K_{OW}$, configuration, uptake and excretion rate constants, and metabolism vary by congener. This can be clearly seen in Fig. 13, which shows the PCB chlorine number for selected compartments of the Lake Michigan food web. Chlorine number is the average PCB chlorination level in the sample: low chlorine numbers are found in samples enriched in the less chlorinated, more volatile congeners while high chlorine numbers are found in samples enriched in the heavier congeners. The congener distribution in top predators is quite different than the congener distributions observed in the original technical mixtures and even in the current sources. In Lake Michigan, the congener distributions measured in

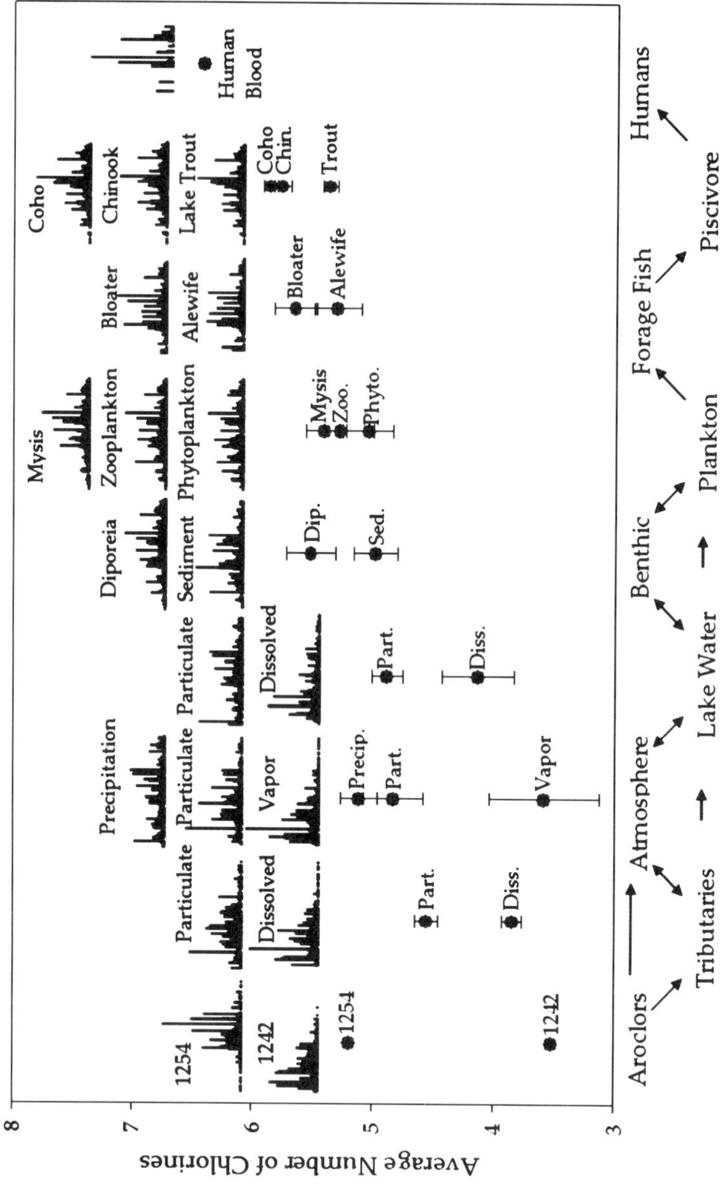


**Fig. 13** Degree of chlorination and congener profiles of two Aroclor formulations [330] and PCBs in various media in Southern Lake Michigan. Dates of most samples vary from 1994 to 2000. Tributary data is from the Kalamazoo River and represents the mean and standard deviation in 27 dissolved and 26 particulate phase water samples [241]. Atmospheric samples are from Chicago (Hites RA, 2005, Personal communication) and represent 48 samples. Lake water [219, 241], sediment [331], and biota values [241] are from near the Saugatuck region in southeast Lake Michigan (GLFMP). Human blood samples were taken from frequent consumers of Lake Michigan fish in 1992 [314]

the most commonly used Aroclor mixtures (1242, 1248, and 1254) bear little resemblance to biotic samples, including those at the bottom of the food chain. Weathering due to volatilization, metabolism, and bioaccumulation has a major impact on the congener distributions. Figure 13 illustrates that there is a gradual shift of the congener composition from a dominance of the tri- and tetra-chlorinated congeners in the free forms (dissolved and gas phases), to penta-chlorinated congeners in sorbed phases (particles, sediments, phytoplankton), to a dominance of the hexa-chlorinated congeners in top predators. There is also a gradual increase in the percentages of the penta- through octa-chlorinated homologues with increasing trophic level.

Of particular interest is the bioaccumulation of the AHH PCB congeners. These congeners, as mentioned above, are thought to contribute to toxic effects in a number of species. The AHH congeners have been shown to preferentially biomagnify in the Lake Michigan lower food web relative to total PCBs [284]. In other words, the percentage contribution of AHH to total PCBs increases with increasing trophic level. This may be due to a greater degree of metabolism of the less-chlorinated congeners compared to the relatively non-reactive AHH congeners, which have four chlorines or more. Smith et al. [62] reported a similar finding in a study of fish and piscivorous birds in Lake Michigan. The potency, expressed as dioxin TEQs, increased with trophic transfer in the upper food web. Metcalfe and Metcalfe found that AHH congeners 77, 126, and 151 were not accumulated as well as other congeners with similar $K_{OW}$s between forage and predator fish in Lake Ontario, and attributed this to a greater propensity for metabolism [332]. Thus, preferential biomagnification is observed in the lower and upper food web, but not between fish tropic levels. The AHH PCBs make up more than 95% of the total dioxin-like toxicity in the Great Lakes food web [41].

There have been a few large-scale field studies of PCBs in the food webs of the lakes. These include two EPA mass balance studies, the Green Bay Mass Balance Study and the Lake Michigan Mass Balance Study. The field data collected for these studies was used to calibrate the complex contaminant models, which were developed as the primary goal of the studies. The ultimate objective of these mass balance studies was to predict concentrations of PBTs in top predator fish from only knowing the external loadings of the PBTs. Thus the models linked food web models to fate and transport, hydrologic, and nutrient models. To calibrate the food web models, an extensive collection of all major trophic levels over both space and time was done and analyzed for PCBs and other selected analytes.

The most extensive field collection effort to date on the Great Lakes, the Lake Michigan Mass Balance Study, collected net phytoplankton (< 100 μm), two sizes of zooplankton (> 100 μm and > 500 μm), *Mysis relicta* by handpicking from net tows, *Diporeia* sp. by handpicking from sled tows, five species of prey fish consumed by lake trout [alewife (*Alosa pseudoharengus*), rainbow smelt (*Osmerus mordax*), bloater (*Coregonus hoyi*), slimy sculpin

(*Cottus cognatus*), deepwater sculpin (*Myoxocephalus thompsoni*)], and lake trout. Lower food web components were collected from 11 locations in the lake on seven different cruises, including all seasons over a 2 year period. The fish were collected from three sites over the same 2 year period. The concentrations of PCBs in each of these food web components are shown in Fig. 14. The data are lipid-normalized to allow for better comparison across trophic levels. Note that concentrations increase with increasing trophic level, even when normalized to lipids. These data demonstrate that PCBs biomagnify in Great Lakes food webs, and are the most comprehensive and self-consistent data in the world for PCBs in a food web of a large aquatic system.

In the lower levels of the food web, there was significant seasonal and spatial variability in the concentrations and corresponding BAFs [333]. The phytoplankton and zooplankton concentrations were greater in southern and nearshore sites compared to northern and open lake sites. The spatial variability in BAFs, and the lack of a strong correlation of BAF with $K_{OW}$ for a given sample collection, indicated that the PCBs were not in equilibrium with phytoplankton, net zooplankton, or Mysis. The collections where PCBs were closest to equilibrium were in winter and late summer, two periods when phytoplankton growth is lowest and when there is be more time for PCBs to reach equilibrium with the cells [227, 271]. The variability in BAF in zooplankton and *Mysis* tracked the variability in phytoplankton, and also reflected the rapid seasonal changes in lipid content in zooplankton. Thus the growth of phytoplankton, and the change in lipid content in zooplank-

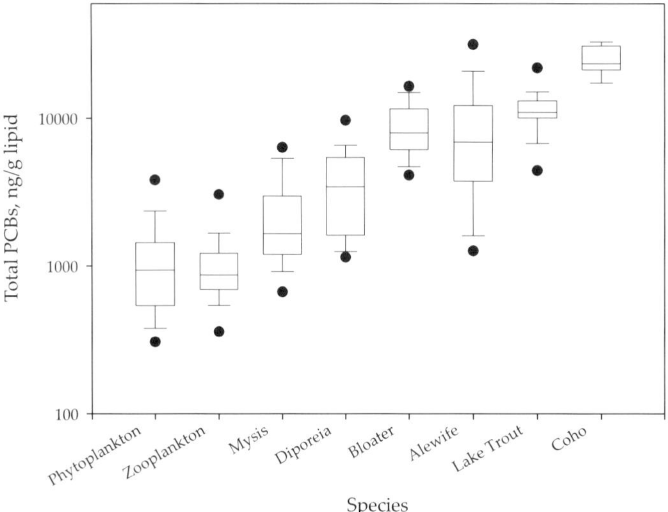


**Fig. 14** Lipid-normalized PCB concentrations in the Lake Michigan food web, showing 95th, 90th, 75th, 50th, 25th, 10th, and 5th percentiles (data from the Lake Michigan Mass Balance study)

ton, occur over relatively short time frames such that PCBs do not easily reach equilibrium in these lower trophic levels. Conversely, the *Diporeia* BAFs showed little change in space or time, reflecting their stable environment and less dynamic life cycle. These observations of PCB seasonal dynamics in lower trophic levels were also observed in a study of Grand Traverse Bay, in northern Lake Michigan [247, 334, 335]. Stapleton and colleagues [247] also found that growth, lipid, and diet changes explained much of the observed seasonal variability in the Grand Traverse Bay food web. Furthermore, stable isotope analyses combined with the PCB data indicated that the majority of PCBs reaching the food web were atmospheric in origin, as opposed to coming from the sediment [334].

The LMMB also provided further insight into the food web dynamics of top predators. For example, Madenjian et al. [336] compared the forage fish and lake trout PCB concentrations at the three different collection sites, and found the forage fish concentrations did not vary with space, but lake trout concentrations were greater in the Sturgeon Bay site compared to the Saugatuck site. Using dietary information obtained from detailed gut analyses, they were able to conclude that dietary differences led to a more PCB-enriched diet in the fish from the Sturgeon Bay site. Additional data were collected by these researchers, and they were able to demonstrate that coho salmon retain approximately 50% of the PCBs in their prey [337], while lake trout retain approximately 80% of the PCBs in their prey [338].

# 9
## Mass Balance for PCBs in the Great Lakes: Lake Michigan

With the accumulation of research into the various transport processes and partitioning behaviors of PCBs discussed above, a clearer picture of the overall fate of PCBs in the Great Lakes environment began to emerge. Models could be constructed based on this knowledge and physical/chemical properties. The earliest such models were published for Lake Ontario [339, 340]. However, more data were needed. Surprisingly little was known about the relative magnitudes of PCB sources to natural waters. Over the last decade, there have been several major research studies to quantify all the PCB sources to Lake Michigan [128–130, 217, 218, 244] and Green Bay [170, 341]. Although there have been studies in Lake Huron's Saginaw Bay [342], Lake Erie [343], and Lake Ontario [344–348], only Lake Michigan has benefited from a coordinated field study to assess all of the possible sources. The purpose of the Lake Michigan Mass Balance (LMMB) study was to develop a lakewide model that required inputs of atmospheric and tributary loadings of PCBs into a complex suite of linked models including a hydraulics model, nutrient dynamics and carbon model, toxics model, and a food web model. The model included all inputs, losses, and internal cycling of PCBs. The overall objective was to be able to predict concentrations of PCBs in top predator fish with an uncertainty factor of two. This is the most complex model for a toxic compound ever developed, and it required an enormous set of environmental and PCB measurements to support it.

The LMMB study included the collection of over 25 000 air, water, sediment, and biota samples at more than 200 locations in and around the lake. All samples were analyzed for 110 individual PCB congeners [241]. The Lake Michigan Mass Balance program reported a preliminary mass budget of PCBs (Fig. 15).

The inputs include the atmosphere, resuspension, and tributary flows. All samples were analyzed by congener-specific methods. Only the sum of the congeners is shown here. The losses include volatilization, export to Lake Huron, downward settling of sediments, and permanent burial. The inputs include gas absorption, wet and dry deposition, flow from tributaries, and resuspension. Non-tributary discharges were assumed to contribute a negligible PCB mass and were not surveyed. The budget is based on data collected in 1994 and 1995. However, more current measurements indicate that the budget remains a good measure of the state of PCBs through the early 2000s. Although other mass budgets have been developed for Lake Michigan [159, 243, 244], the analysis reported by the EPA is the first that used a consistent PCB data set.

Construction and calibration of the model has informed us a great deal about the overall dynamics of PCBs in the Great Lakes. The atmosphere is the largest source of PCBs to Lake Michigan. More than 2000 kg of atmospheric

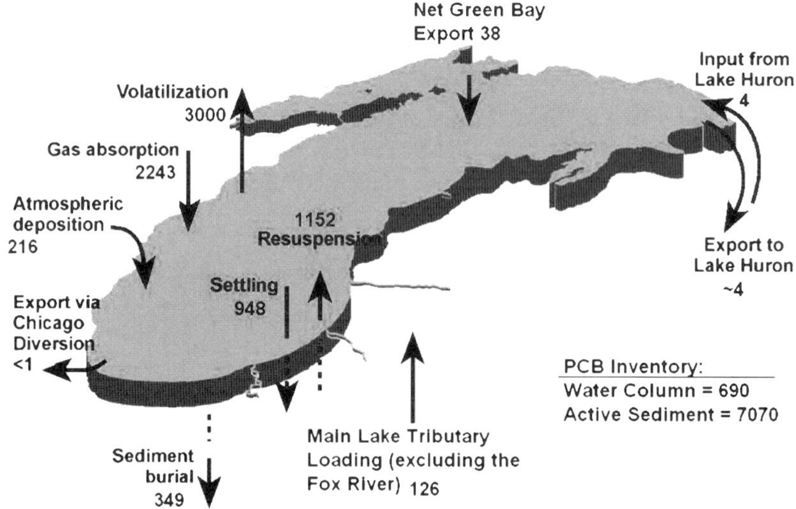


**Fig. 15** The magnitude and direction of various processes (kg year$^{-1}$) that move PCBs into and out of Lake Michigan each year, modified from [241]

PCBs enter Lake Michigan each year. The majority of that input is through gas exchange. Deposition of PCBs associated with aerosol particles accounts for 100 kg each year [129]. Deposition of PCBs in rain contributes 90 kg each year. Atmospheric deposition (gas, dry particle, and wet deposition) is larger than inputs from resuspension of contaminated sediments and larger than inputs from direct discharge and contaminated tributaries. Resuspension of contaminated sediments is an important source of PCBs to the water, but still not as large as gross atmospheric deposition. Although not considered in the LMMB effort, a field study of resuspension after major storms concluded that annual wintertime storms contribute as much as 400 kg of PCBs per event [217]. There may be more than one major event each year, but the total amount of PCBs resuspended is still only about 1200 kg, and a large fraction of that mass may immediately return to the lake floor. The tributary load is only about 380 kg and that includes the Fox River discharge to Green Bay. A significant portion of the PCBs discharged by the Fox River are retained or volatilized from Green Bay and never reach Lake Michigan [241]. Most direct discharges from industrial waste, urban runoff, and wastewater treatment plants are included in the estimate of tributary loads.

The importance of the atmospheric pathway is easily misjudged. For example, sediment export from contaminated tributaries as a source of PCBs to Lake Michigan is often cited as a justification for dredging [163]. However, atmospheric releases from highly contaminated tributaries and harbors may be a larger net source of PCBs to the open lake than the corresponding direct discharge. In Milwaukee, atmospheric sources contribute about 150 kg each year to the open waters of Lake Michigan. The Milwaukee River, on the other hand, contributes only about 12 kg annually [131, 241]. The mass balance effort and other quantitative studies of PCB sources shows that the atmosphere is the primary source of PCBs to the lake. This implies that the atmosphere is also a major source of PCBs ultimately accumulated in fish.

The importance of atmospheric exchange processes is one of the most striking findings of the mass budget analysis. Atmospheric PCB losses from Lake Michigan are also large. In the dilute water of Lake Michigan, hydrolysis, photolysis, and microbial decay are considered negligible loss mechanisms for PCBs. Volatilization is larger than net burial of PCBs into the sediments and much larger than export to Lake Huron.

Predicting long-term trends in whole lake mean PCB concentrations using the mass balance model approach is problematic. PCBs are clearly declining in Lake Michigan as shown from the long-term monitoring results discussed above. Therefore, the PCB inputs must be less than the losses. This is difficult to observe in the full mass balance model because of the uncertainties in each of the input and loss calculations. For example, atmospheric inputs and losses are both large but the difference between them is small. Propagation of error shows that the net whole-lake annual exchange of PCBs across the air–water interface is not significantly different from zero. While this may seem disappointing, the analyses of uncertainty simply shows how dynamic the system is with respect to PCB exchange. On a smaller temporal scale, such as a day or a month, the difference between inputs and outputs can be much larger, especially near major sources like Chicago. For example, net gas exchange of total PCBs in southern Lake Michigan is illustrated for four consecutive days in Fig. 16. The figure shows that small changes in wind direction, temperature, and atmospheric PCB concentrations can change the direction of air–water exchange from net volatilization to net deposition. PCB gas exchange in Lake Michigan is highly dynamic over time and space.

The results of the Lake Michigan Mass Balance project are consistent with an analysis of PCB trends in Lake Superior, which, like Lake Michigan, has a long retention time among other similar characteristics. Jeremiason et al. examined the PCB congener profiles in buried sediment and long-term PCB trends in Lake Superior water and sediments [254]. The authors concluded, via a mass balance modeling exercise, that volatilization explains most of the post-1978 decline in PCB water concentrations. The importance of atmospheric exchange in Lake Michigan and Lake Superior can be explained by their physical characteristics, which are similar to many large lakes, including most of the Laurentian Great Lakes. They have a long retention time (inflows and outflows are small relative to lake volume), shallow depth relative to lateral dimensions (allowing for efficient settling of particles), and a small drainage basin relative to lake area [15]. Smaller lakes with large rela-

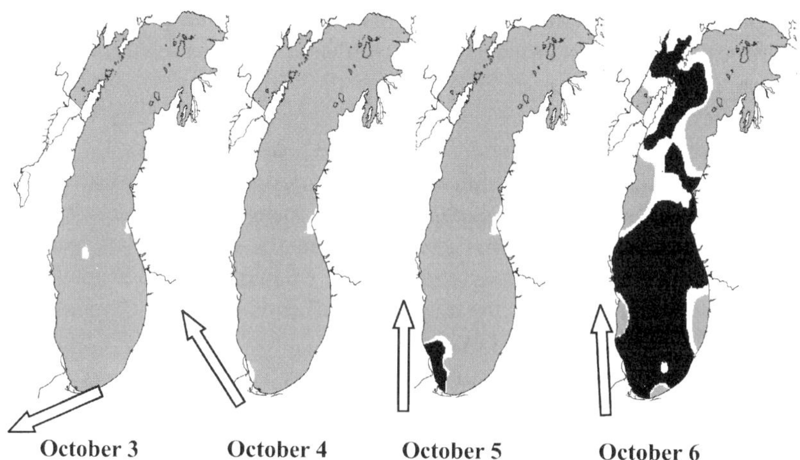


**Fig. 16** Net gas exchange of total PCBs to Lake Michigan on four consecutive days, starting with 3 Oct 1994 on the *left*. The *darkest shade* represents net deposition (most evident on 6 Oct). *White* represents zero net exchange and the *light shade* represents net volatilization fluxes. *Arrows* indicate the average wind direction on that date. The average daily temperatures for 3–6 Oct were: 14.1 °C, 13.6 °C, 13.2 °C, and 17.9 °C, respectively. Reproduced from [130]

tive catchment areas, greater relative depth, and short residence times would be expected to have a greater influence from riverine inputs in the mass balance equations. As an example, in a small Arctic lake almost 80% of PCB loss was attributed to PCBs in the outflowing water [349]. Lake Erie has intermediate characteristics: it has a short retention time, but an average depth that is much less than Lake Superior or Lake Michigan [15]. In this case, the atmosphere would be expected to have an even greater influence on the mass balance than in Lake Michigan, but there could also be substantial input and export terms from connecting channels. Clearly, mass balance studies cannot be easily generalized, and characteristics of individual lakes must be considered carefully in formulating hypotheses regarding PCB fate.

## 10
## Conclusions

PCBs are clearly declining in the Great Lakes environment. The long-term data from sediments, archived fish, and long term monitoring studies illustrate the results of reducing PCB sources to the lakes. Without legislation to stop production of these compounds and without regulation to remove residual industrial sources, such a reduction would not have been achieved. The enormous decrease in PCB concentrations is truly a success story.

The research behind this success story has also provided an exhaustive and detailed understanding of the environmental processes affecting the fate and transport of PCBs. The story of PCBs provides guidance for predicting the behavior of similar chemicals for which there is much less information available. This, too, is a success story.

PCBs are still a great concern in the Great Lakes ecosystem. Research continues to show that these chemicals are hazardous to animals and humans. The latest findings show particularly alarming effects on neurological function in humans and animals. Unfortunately, exposure to these compounds continues despite decades of effort and millions of dollars spent. The largest current sources are atmospherically driven. These are difficult to identify or control, and are nearly immune from the traditional tactics of point source regulation and enforcement. Atmospheric sources do not derive from a pipe that can be shut off or from an industrial process that can be retooled. Atmospheric sources are not, however, completely elusive. Scientists have shown that the most widely observed characteristics of atmospheric PCBs are also the clues to their origin. Atmospheric PCBs come from volatilization processes and they are at the highest concentration in the industrial regions where they were used. This means that surface contamination near sites where they were used, stored, or disposed should be monitored and remediated. Protection of future generations of humans and other animals depends on our creativity and determination to halt this continuing degradation of our environment.

 

## References


1. Veith G (1968) Environmental chemistry of the chlorobiphenyls in the Milwaukee River. PhD Dissertation, Water Chemistry Program, University of Wisconsin, Madison, Wisconsin
2. Jensen S (1966) New Sci 32:612
3. ATSDR (2000) Toxicological profile for polychlorinated biphenyls (PCBs). US Department of Health and Human Services, Agency for Toxic Substances and Disease Registry
4. NRC (1979) Polychlorinated biphenyls. National Research Council, National Academy of Sciences, Washington, DC
5. Imamoglu I, Christensen ER (2002) Water Res 36:3449
6. Swackhamer DL, Armstrong DE (1988) J Great Lakes Res 14:277
7. Brown JF, Wagner RE, Feng H, Bedard DL, Brennan MJ, Carnahan JC, May RJ (1987) Environ Toxicol Chem 6:579
8. International Joint Commission (2003) Status of restoration activities in Great Lakes areas of concern: a special report
9. Zarull MA, Hartig JH, Maynard L (1999) Ecological benefits of contaminated sediment remediation in the Great Lakes basin. Sediment Priority Action Committee, Great Lakes Water Quality Board, International Joint Commission
10. Golden KA, Wong CS, Jeremiason JD, Eisenreich SJ, Sanders G, Hallgren J, Swackhamer DL, Engstrom DR, Long DT (1993) Water Sci Technol 28:19
11. WDNR (2003) White paper no 19 – Estimates of PCB mass, sediment volume, and surface sediment concentrations in Operable Unit 5, Green Bay, using an alternative approach. Wisconsin Department of Natural Resources
12. WDNR (2003) White paper no 18 – Evaluation of an alternative approach of calculating mass, sediment volume, and surface concentrations in Operable Unit 5, Green Bay. Wisconsin Department of Natural Resources
13. Kovatch E (2002) Remedial investigation report, Lower Fox River and Green Bay, Wisconsin. Prepared by the RETEC Group for Wisconsin Department of Natural Resources
14. Wong CS, Sanders G, Engstrom DR, Long DT, Swackhamer DL, Eisenreich SJ (1995) Environ Sci Technol 29:2661
15. Government of Canada and US EPA (1995) The Great Lakes: An environmental atlas and resource book, 3rd edn.
16. Shiu W-Y, Ma K-C (2000) J Phys Chem Ref Data 29:387
17. Li N, Wania F, Lei YD, Daly GL (2003) J Phys Chem Ref Data 32:1545
18. Falconer RL, Bidleman TF (1994) Atmos Environ 28:547
19. Bamford HA, Poster DL, Huie RE, Baker JE (2002) Environ Sci Technol 36:4395
20. Zhang X, Schramm K-W, Henkelmann B, Klimm C, Kaune A, Kettrup A, Lu P (1999) Anal Chem 71:3834
21. Hawker DW, Connell DW (1988) Environ Sci Technol 22:382
22. Goss K-U, Wania F, McLachlan MS, Mackay D, Schwarzenbach RP (2004) Environ Sci Technol 38:1626
23. Harner T, Bidleman TF (1996) J Chem Eng Data 41:895
24. Mullin MD, Pochini CM, McCrindle S, Romkes M, Safe SH, Safe LM (1984) Environ Sci Technol 18:468
25. Burkhard LP, Weininger D (1987) Anal Chem 59:1187
26. Capel P, Rapaport R, Eisenreich S, Looney B (1985) Chemosphere 14:439
27. Burkhard LP, Armstrong DE, Andren AW (1985) Environ Sci Technol 19:590

28. Burkhard LP, Andren AW, Armstrong DE (1985) Environ Sci Technol 19:500
29. US EPA (1999) Method 1668, revision A: Chlorinated biphenyl congeners in water, soil, sediment, and tissue. HRGC/HRMS US EPA, Office of Water
30. Van den Berg M, Birnbaum L, Bosveld ATC, Brunstrom B, Cook P, Feeley M, Giesy JP, Hanberg A, Hasegawa R, Kennedy SW, Kubiak T, Larsen JC, van Leeuwen FXR, Liem AKD, Nolt C, Peterson RE, Poellinger L, Safe S, Schrenk D, Tillitt D, Tysklind M, Younes M, Waern F, Zacharewski T (1998) Environ Health Perspect 106:775
31. Hansen LG (1998) Environ Health Perspect 106:171
32. Robertson LW, Hansen LG (2001) Recent advances in the environmental toxicology and health effects of PCBs. University Press of Kentucky, Lexington, KY
33. Espandiari P, Glauert HP, Lehmler HJ, Lee EY, Srinivasan C, Robertson LW (2003) Toxicol Appl Pharmacol 186:55
34. Oakley GG, Devanaboyina US, Robertson LW, Gupta RC (1996) Chem Res Toxicol 9:1285
35. Kania-Korwel I, Hornbuckle KC, Peck A, Ludewig G, Robertson LW, Sulkowski WW, Espandiari P, Gairola CG, Lehmler H-J (2005) Environ Sci Technol 39:3513
36. Eisler R (1986) Polychlorinated biphenyl hazards to fish, wildlife and invertebrates: a synoptic review. US Fish and Wildlife Service
37. Safe S (1991) Environ Carcin Ecotoxicol Rev c9:261
38. Safe SH (1994) CRC Crit Rev Toxicol 24:87
39. US EPA (1994) Health assessment document for 2,3,7,8-tetrachloro-$p$-dioxin (TCDD) and related compounds. US EPA
40. Foster WG (1995) Environ Health Perspect Suppl 103 Suppl. 9:63
41. Metcalfe CD, Haffner GD (1995) Environ Rev (Ottawa) 3:171
42. Grasman KA, Scanlon PF, Fox GA (1998) Environ Monit Assess 53:117
43. Faroon O, Jones D, De Rosa C (2000) Toxicol Ind Health 16:305
44. Rice DC (2001) Human Ecol Risk Assess 7:1059
45. van der Oost R, Beyer J, Vermeulen NPE (2003) Environ Toxicol Pharmacol 13:57
46. Giesy JP, Kannan K (2001) Organohalogen Cmpd 52:71
47. D'Itri FM, Kamrin MA (eds) (1983) PCBs: Human and environmental hazards. Butterworth, Boston, MA
48. Robertson LW, Hansen LG (eds) (2001) PCBs: Recent advances in environmental toxicology and health effects. Proceedings of the PCB workshop, Lexington, Kentucky, 9-12 April 2000. University Press of Kentucky, Lexington, KY
49. Selgeby JH (1995) J Great Lakes Res 21 (Suppl. 1):1
50. Rosenshield ML, Jofre MB, Karasov WH (1999) Environ Toxicol Chem 18:2478
51. Walker MK, Peterson RE (1991) Aquat Toxicol 21:219
52. Zabel EW, Walker MK, Hornung MW, Clayton MK, Peterson RE (1995) Toxicol Appl Pharmacol 134:204
53. Zabel EW, Cook PM, Peterson RE (1995) Environ Toxicol Chem 14:2175
54. Walker MK, Cook PM, Butterworth BC, Zabel EW, Peterson RE (1996) Fund Appl Toxicol 30:178
55. Guiney PD, Cook PM, Casselman JM, Fitzsimons JD, Simonin HA, Zabel EW, Peterson RE (1996) Can J Fish Aquat Sci 53:2080
56. Cook PM, Robbins JA, Endicott DD, Lodge KB, Guiney PD, Walker MK, Zabel EW, Peterson RE (2003) Environ Sci Technol 37:3864
57. Van den Berg M, Birnbaum L, Bosveld ATC, Brunstrom B, Cook P, Feeley M, Biesy JP, Hanberg A, Hasegawa R, Kennedy SW, Kubiak T, Larsen JC, van Leeuwen FXR, Djien Liem AK, Nolt C, Peterson RE, Poellinger L, Safe S, Schrenk D, Tillitt D, Tysklind M, Younes M, Waern F, Zacharewski T (1998) Environ Health Perspect 106:775

58. Mac MJ, Edsall CC (1991) J Toxicol Environ Health 33:375
59. Mac MJ, Schwartz TR (1992) Chemosphere 25:189
60. Mac MJ, Schwartz TR, Edsall CC, Frank AM (1993) J Great Lakes Res 19:752
61. Wilson PJ, Tillitt DE (1996) Mar Environ Res 42:129
62. Smith LM, Schwartz TR, Feltz K, Kubiak TJ (1990) Chemosphere 21:1063
63. Fitzsimons JD (1995) J Great Lakes Res 21:267
64. Brown SB, Evans RE, Vandenbyllardt L, Finnson KW, Palace VP, Kane AS, Yarechewski AY, Muir DCG (2004) Aquat Toxicol 67:75
65. Aulerich RJ, Ringer RK, Iwamoto S (1973) J Reprod Fertil Suppl 19:365
66. Aulerich R, Ringer R, Saronoff J (1986) Arch Environ Cont Toxicol 15:393
67. Proulx G, Weseloh DV, Elliott JE, Teeple SM, Anghern PAM, Mineau P (1987) Bull Environ Contam Toxicol 39:939
68. Foley RE, Jackling SJ, Sloan RJ, Brown MK (1988) Environ Toxicol Chem 7:363
69. Giesy JP, Verbrugge DA, Othout RA, Bowerman WW, Mora MA, Jones PD, Newsted JL, Vandervoort C, Heaton SM, Aulerich RJ, Bursian SJ, Ludwig JP, Dawson GA, Kubiak TJ, Best DA, Tillitt DE (1994) Arch Environ Cont Toxicol 27:213
70. Heaton SN, Bursian SJ, Giesy JP, Tillitt DE, Render JA, Jones PD, Verbrugge DA, Kubiak TJ, Aulerich RJ (1995) Arch Environ Cont Toxicol 29:411
71. Heaton SN, Bursian SJ, Giesy JP, Tillitt DE, Render JA, Jones PD, Verbruge DA, Kubiak TJ, Aulerich RJ (1995) Arch Environ Cont Toxicol 28:334
72. Tillitt DE, Gale RW, Meadows JC, Zajicek JL, Peterman PH, Heaton SN, Jones PD, Bursian SJ, Kubiak TJ, Giesy JP, Aulerich RJ (1996) Environ Sci Technol 30:283
73. Aulerich RJ, Ringer RK (1977) Arch Environ Contam Toxicol 6:279
74. Bleavins MR, Aulerich RJ, Ringer RK (1980) Arch Environ Cont Toxicol 9:627
75. Hornshaw TC, Aulerich RJ, Johnson HE (1983) J Toxicol Environ Health 11:933
76. Shipp EB, Restum JC, Bursian SJ, Aulerich RJ, Helferich WG (1998) J Toxicol Environ Health, Part A 54:403
77. Beckett KJ, Millsap SD, Blankenship AL, Zwiernik MJ, Giesy JP, Bursian SJ (2005) Environ Toxicol Chem 24:674
78. Render JA, Bursian SJ, Rosenstein DS, Aulerich RJ (2001) Vet Human Toxicol 43:22
79. Fry DM (1996) Comments Toxicol 5:401
80. Fry DM (1995) Environ Health Perspect 103:165
81. McNabb FMA, Fox GA (2003) Evol Dev 5:76
82. Fisher S, Bortolotti G, Fernie K, Smits J, Marchant T, Drouillard K, Bird D (2001) Arch Environ Cont Toxicol 41:215
83. Larson JM, Karasov WH, Sileo L, Stromborg KL, Hanbidge BA, Giesy JP, Jones PD, Tillitt DE, Verbrugge DA (1996) Environ Toxicol Chem 15:553
84. Feyk LA, Giesy JP (1998) In: Principles and processes for evaluating endocrine disruption in wildlife. Proceedings conference on endocrine disruption in wildlife, Kiawah Island, SC, Mar 1996, p 121
85. Fox GA, Trudeau S, Won H, Grasman KA (1998) Environ Monit Assess 53:147
86. Ryckman DP, Weseloh DV, Hamr P, Fox GA, Collins B, Ewins PJ, Norstrom RJ (1998) Environ Monit Assess 53:169
87. Lorenzen A, Moon TW, Kennedy SW, Fox GA (1999) Environ Health Perspect 107:179
88. Tillitt DE, Ankley GT, Giesy JP, Ludwig JP, Kurita-Matsuba H, Weseloh DV, Ross PS, Bishop CA, Sileo L et al. (1992) Environ Toxicol Chem 11:1281
89. Leatherland JF (1998) Toxicol Ind Health 14:41
90. Hoffman DJ, Melancon MJ, Klein PN, Eisemann JD, Spann JW (1998) Environ Toxicol Chem 17:747

91. Fein GG, Jacobson JL, Jacobson SW, Schwartz PW, Dowler JK (1984) J Pediatr 105:315
92. Jacobson SW, Fein GG, Jacobson JL, Schwartz PM, Dowler JK (1985) Child Dev 56:853
93. Jacobson JL, Humphrey HEB, Jacobson SW, Schantz SL, Mullin MD, Welch R (1989) Am J Public Health 79:1401
94. Jacobson JL, Jacobson SW (1993) J Great Lakes Res 19:776
95. Jacobson JL, Jacobson SW (1996) N Engl J Med 335:783
96. Jacobson JL, Jacobson SW (2001) In: Robertson LW, Hansen LG (eds) PCBs, recent advances in environmental toxicology and health effects. University Press of Kentucky, Lexington, KY
97. Jacobson JL, Jacobson SW, Humphrey HEB (1990) J Pediatr 116:38
98. Jacobson JL, Jacobson SW, Humphrey HEB (1990) Neurotoxicol Teratol 12:319
99. Jacobson JL, Jacobson SW, Padgett RJ, Brumitt GA, Billings RL (1992) Dev Psych 28:297
100. Seegal RF (1999) Environ Res 80:S38
101. Schantz SL, Gardiner JC, Gasior DM, McCafferey RJ, Sweeney AM, Humphrey HEB (2004) Psychol Schools 41:669
102. Stewart PW, Reihman J, Lonky EI, Darvill TJ, Pagano J (2003) Neurotoxicol Teratol 25:11
103. Patandin S, Lanting CI, Mulder PGH, Boersma ER, Sauer PJJ, Weisglas-Kuperus N (1999) J Pediatr 134:33
104. Walkowiak J, Wiener J-A, Fastabend A, Heinzow B, Kramer U, Schmidt E, Steingruber H-J, Wundram S, Winneke G (2001) The Lancet 358:1602
105. Grandjean P, Weihe P, Burse VW, Needham LL, Storr-Hansen E, Heinzow B, Debes F, Murata K, Simonsen H, Ellefsen P, Budtz-Jorgensen E, Keiding N, White RF (2001) Neurotoxicol Teratol 23:305
106. Tilson H, Jacobson J, Rogan W (1990) Neurotoxicol Teratol 12:239
107. Rice DC (1995) Environ Health Perspect Suppl 103 Suppl. 9:71
108. Swain WR (1991) J Toxicol Environ Health 33:587
109. Schantz SL, Gasior DM, Polverejan E, McCaffrey RJ, Sweeney AM, Humphrey HEB, Gardiner JC (2001) Environ Health Perspect 109:605
110. Seegal RF, Bush B, Shain W (1990) Toxicol Appl Pharmacol 106:136
111. Shain W, Bush B, Seegal R (1991) Toxicol Appl Pharmacol 111:33
112. Seegal RF, Bemis JC, Okoniewski RJ (1998) Organohalogen Cmpd 37:1
113. Tryphonas H (1995) Environ Health Perspect Suppl 103 Suppl. 9:35
114. Persky V, Turyk M, Anderson HA, Hanrahan LP, Falk C, Steenport DN, Chatterton R Jr, Freels S, Boddy J, Budd M, Burkett M, Fiore B, Humphrey HEB, Johnson R, Lee G, Monaghan S, Reed D, Shelley T, Sonzogni W, Steele G, Wright D (2001) Environ Health Perspect 109:1275
115. Karmaus W, Huang S, Cameron L (2002) J Occup Environ Med 44:8
116. Anderson HA, Amrhein JF, Shubat P, Hesse J (1993) Protocol for a uniform Great Lakes fish advisory. Great Lakes Sport Fish Advisory Task Force
117. Manchester-Neesvig JB, Andren AW (1989) Environ Sci Technol 23:1138
118. Hermanson MH, Hites RA (1989) Environ Sci Technol 23:1253
119. Hoff RM, Muir DCG, Grift NP (1992) Environ Sci Technol 26:266
120. Hoff RM, Muir DCG, Grift NP (1992) Environ Sci Technol 26:276
121. Hornbuckle KC, Eisenreich SJ (1996) Atmos Environ 30:3935
122. Panshin SY, Hites RA (1994) Environ Sci Technol 28:2008
123. Simcik MF, Basu I, Sweet CW, Hites RA (1999) Environ Sci Technol 33:1991
124. Buehler SS, Basu I, Hites RA (2002) Environ Sci Technol 36:5051
125. Lee RGM, Hung H, Mackay D, Jones KC (1998) Environ Sci Technol 32:2172

126. Carlson DL, Hites RA (2005) Environ Sci Technol 39:740
127. Green ML, Depinto JV, Sweet C, Hornbuckle KC (2000) Environ Sci Technol 34:1833
128. Miller SM, Sweet CW, Depinto JV, Hornbuckle KC (2000) Environ Sci Technol 34:55
129. Miller SM, Green ML, Depinto JV, Hornbuckle KC (2001) Environ Sci Technol 35:278
130. Hornbuckle KC, Green ML (2003) Ambio 32:406
131. Wethington DM, Hornbuckle KC (2005) Environ Sci Technol 39:57
132. Foreman W, Bidleman T (1987) Environ Sci Technol 21:869
133. Pankow JF, Bidleman TF (1992) Atmos Environ Part A 26:1071
134. Finizio A, Mackay D, Bidleman T, Harner T (1997) Atmos Environ 31:2289
135. Simcik MF, Franz TP, Zhang HX, Eisenreich SJ (1998) Environ Sci Technol 32:251
136. Kaupp H, McLachlan MS (1999) Chemosphere 38:3411
137. Tasdemir Y, Vardar N, Odabasi M, Holsen TM (2004) Environ Pollut 131:35
138. Pankow JF (1998) Atmos Environ 32:1493
139. Harner T, Bidleman TF (1998) Environ Sci Technol 32:1494
140. Harner T, Bidleman TF (1998) J Chem Eng Data 43:40
141. Frame GM, Wagner RE, Carnahan JC, Brown JF, May RJ, Smullen LA, Bedard DL (1996) Chemosphere 33:603
142. Pacyna JM, Breivik K, Munch J, Fudala J (2003) Atmos Environ 37:S119
143. Alcock RE, Gemmill R, Jones KC (1999) Chemosphere 38:759
144. Breivik K, Sweetman A, Pacyna JM, Jones KC (2002) Sci Total Environ 290:199
145. Breivik K, Sweetman A, Pacyna JM, Jones KC (2002) Sci Total Environ 290:181
146. Council NR (2001) A risk management strategy for PCB contaminated sediments. National Academy, Washington, DC
147. Gurda J (1999) The making of Milwaukee. Milwaukee County Historical Society. Burton & Mayer, Brookfield, Wisconsin
148. Gatz DF, Sweet CW, Basu I, Vermette SJ, Harlin K, Bauer S (1994) Great Lakes atmospheric deposition network (IADN): Data report 1990–1992. Illinois State Water Survey
149. Hillery BR, Simcik MF, Basu I, Hoff RM, Strachan WMJ, Burniston D, Chan CH, Brice KA, Sweet CW, Hites RA (1998) Environ Sci Technol 32:2216
150. Buehler SS, Hites RA (2002) Environ Sci Technol:354A
151. Buehler SS, Basu I, Hites RA (2004) Environ Sci Technol 38:414
152. Holsen TM (1991) Environ Sci Technol 25:1075
153. Cotham WE, Bidleman TF (1995) Environ Sci Technol 29:2782
154. Simcik MF, Zhang HX, Eisenreich SJ, Franz TP (1997) Environ Sci Technol 31:2141
155. Hsu Y-K, Holsen TM (2000) In: Using models to develop air toxics reduction strategies: Lake Michigan as a test case. International Joint Commission International Air Quality Advisory Board, IJC, Milwaukee, WI
156. Offenberg JH, Baker JE (2000) J Great Lakes Res 26:196
157. Hsu YK, Holsen TM, Hopke PK (2003) Environ Sci Technol 37:681
158. Tasdemir Y, Odabasi M, Vardar N, Sofuoglu A, Murphy TJ, Holsen TM (2004) Atmos Environ 38:2447
159. Offenberg J, Simcik M, Baker J, Eisenreich SJ (2005) Aquat Sci 67:79
160. Totten LA, Gigliotti CL, VanRy DA, Offenberg JH, Nelson ED, Dachs J, Reinfelder JR, Eisenreich SJ (2004) Environ Sci Technol 38:2568
161. Hafner WD, Hites RA (2003) Environ Sci Technol 37:3764
162. Wu CY, Nelson C, Cabrera-Rivera O, Quan J, Asselmeier D, Bates J, Moore S, Baker G, McGeen D, McDonough R, Sliwinski R, Altenburg R, Velalis T, Judson H, Wong P, Wagemakers J, Moy D (2001) In: 10th international emission inventory conference, one atmosphere, one inventory, many challenges. US EPA, Denver, CO

163. Reible DD, Riley K (2004) TOSC review of Indiana harbor and canal dredging and disposal alternatives analysis. Conducted at the request of the Grand Calumet Task Force and Citizens for a Clean Environment Technical Outreach Services for Communities (TOSC) Program, Michigan State University
164. Short Elliott Hendrickson Inc. (2004) Review and analysis of risk, proposed confined disposal facility. Prepared for the East Chicago Waterway Management District
165. Army Corps of Engineers Chicago District (2004) Indiana harbor and canal dredging and disposal activities
166. Estes TJ, Schroeder PR, Loikets WR, Taylor ER, Agrawal V, Caine C, Gallas R (2003) Indiana harbor and canal dredging and disposal alternatives analysis: Evaluation of relative disposal requirements, emissions and costs for mechanical and hydraulic dredging alternatives. US Army Engineer Research and Development Center
167. Gumm GJ (2001) Indiana harbor and canal maintenance dredging and disposal activities: Project status report. Army Corps of Engineers, Chicago District
168. Whitman WG (1923) Chem Metall Eng 29:146
169. Baker JE, Eisenreich SJ (1990) Environ Sci Technol 24:342
170. Achman DR, Hornbuckle KC, Eisenreich SJ (1993) Environ Sci Technol 27:75
171. Bidleman TF, McConnell LL (1995) Sci Total Environ 159:101
172. Dickhut RM, Gustafson KE (1995) Mar Pollut Bull 30:385
173. Pirrone N, Keeler GJ, Holsen TM (1995) Environ Sci Technol 29:2123
174. McConnell LL, Kucklick JR, Bidleman TF, Ivanov GP, Chernyak SM (1996) Environ Sci Technol 30:2975
175. Offenberg JH, Baker JE (1999) J Air Waste Manage Assoc 49:959
176. Park JS, Wade TL, Sweet S (2001) Atmos Environ 35:3315
177. Clark JF, Schlosser P, Stute M, Simpson HJ (1996) Environ Sci Technol 30:1527
178. Liss PW, Slater PG (1974) Nature 247:181
179. Matthews CJD, St Louis VL, Hesslein RH (2003) Environ Sci Technol 37:772
180. Wanninkhof RH, Asher W, Weppernig R, Chen H, Schlosser P, Langdon C, Sambrotto R (1993) J Geophys Res 98:20237
181. Upstill-Goddard R, Watson A, Liss P, Liddicoat M (1990) Tellus 42B:364
182. Watson AJ, Ledwell JR (2000) J Geophys Res-Oceans 105:14325
183. Ho DT, Schlosser P, Caplow T (2002) Environ Sci Technol 36:3234
184. Shahir UM, Li YH, Holsen TM (1999) Aerosol Sci Technol 31:446
185. Franz TP, Eisenreich SJ, Holsen TM (1998) Environ Sci Technol 32:3681
186. Williams RM (1982) Atmos Environ 16:1933
187. Pryor SC, Barthelmie RJ, Geernaert LLS, Ellermann T, Perry KD (1999) Atmos Environ 33:2045
188. Zufall MJ, Dai WP, Davidson CI, Etyemezian V (1999) Atmos Environ 33:4273
189. Simcik MF, Hoff RM, Strachan WMJ, Sweet CW, Basu I, Hites RA (2000) Environ Sci Technol 34:361
190. Franz TP, Eisenreich SJ (1998) Environ Sci Technol 32:1771
191. Wania F, Mackay D, Hoff JT (1999) Environ Sci Technol 33:195
192. Carlson DL, Basu I, Hites RA (2004) Environ Sci Technol 38:5290
193. Lei YD, Wania F (2004) Atmos Environ 38:3557
194. Offenberg JH, Baker JE (2002) Environ Sci Technol 36:3763
195. Hoff RM, Strachan WMJ, Sweet CW, Chan CH, Shackleton M, Bidleman TF, Burniston DA, Cussion S, Gatz DF, Harlin K, Schroeder WH (1996) Atmos Environ 30:3505
196. Zhang HX, Eisenreich SJ, Franz TR, Baker JE, Offenberg JH (1999) Environ Sci Technol 33:2129

197. Anderson MJ, Cacela D, Beltman D, Teh SJ, Okihiro MS, Hinton DE, Denslow N, Zelikoff JT (2003) Biomarkers 8:371
198. de Solla SR, Bishop CA, Brooks RJ (2002) Environ Toxicol Chem 21:922
199. Schrank CS, Cormier SM, Blazer VS (1997) J Great Lakes Res 23:119
200. McCarty HB, Schofield J, Miller K, Brent RN, Van Hoof P, Eadie B (2004) Results of the Lake Michigan mass balance study: polychlorinated biphenyls and trans-nonachlor data report. US EPA
201. Imamoglu I, Li K, Christensen ER (2002) Environ Toxicol Chem 21:2283
202. Cacela D, Beltman DJ, Lipton J (2002) Environ Toxicol Chem 21:1591
203. Kannan K, Kober JL, Kang Y-S, Masunaga S, Nakanishi J, Ostaszewski A, Giesy JP (2001) Environ Toxicol Chem 20:1878
204. Steuer JS, Hall DW, Fitzgerald SA (1999) Distribution and transport of polychlorinated biphenyls and associated particulates in the Hayton Millpond, South Branch Manitowoc River, 1993–95. US Geological Survey Water-Resources Investigations Report 99-4101
205. Swackhamer DL, McVeety BM, Hites RA (1988) Environ Sci Technol 22:664
206. Thomas RL, Frank R (1983) Physical behavior of PCBs in the Great Lakes. In: Mackay D, Paterson S, Eisenreich SJ, Semmens MS (eds) PCBs in sediment and fluvial suspended solids in the Great Lakes. Ann Arbor Science, Ann Arbor, MI, p 245
207. Marvin CH, Charlton MN, Stern GA, Braekevelt E, Reiner EJ, Painter S (2003) J Great Lakes Res 29:317
208. Painter S, Marvin C, Rosa F, Reynoldson TB, Charlton MN, Fox M, Thiessen PAL, Estenik JF (2001) J Great Lakes Res 27:434
209. Eisenreich SJ, Capel PD, Robbins JA, Bourbonniere R (1989) Environ Sci Technol 23:1116
210. Schneider AR, Stapleton HM, Cornwell J, Baker JE (2001) Environ Sci Technol 35:3809
211. Marvin CH, Painter S, Charlton MN, Fox ME, Thiessen PAL (2004) Chemosphere 54:33
212. Swackhamer DL, Armstrong DE (1986) Environ Sci Technol 20:879
213. Jeremiason JD, Eisenreich SJ, Baker JE, Eadie BJ (1998) Environ Sci Technol 32:3249
214. Baker JE, Eisenreich SJ (1989) J Great Lakes Res 15:84
215. Baker JE, Eisenreich SJ, Eadie BJ (1991) Environ Sci Technol 25:500
216. Baker JE, Eisenreich SJ, Swackhamer DL (1991) Organic substances and sediments in water. In: Baker RA (ed) Field-measured associations between polychlorinated biphenyls and suspended solids in natural water: an evaluation of the partitioning paradigm, vol 2. Lewis, Chelsea, MI, p 79
217. Hornbuckle KC, Smith GL, Miller SM, Eadie BJ, Lansing MB (2004) J Geophys Res-Oceans 109:Art. no C05017 MAY 18
218. Bogdan JJ, Budd JW, Eadie BJ, Hornbuckle KC (2002) J Great Lakes Res 28:338
219. Miller SM (2003) The effects of large-scale episodic sediment resuspension on persistent organic pollutants in Southern Lake Michigan. PhD Dissertation, Civil and Environmental Engineering, University of Iowa, Iowa City, IA
220. Digiano FA, Miller CT, Yoon JY (1993) J Environ Eng-ASCE 119:72
221. Gong YY, Depinto JV, Rhee GY, Liu X (1998) Water Res 32:2507
222. Gong YY, Depinto JV (1998) Water Res 32:2518
223. Pignatello JJ, Xing BS (1996) Environ Sci Technol 30:1
224. Schneider AR, Eadie AS, Baker JE (2002) Environ Sci Technol 36:1181
225. Ortiz E, Luthy RG, Dzombak DA, Smith JR (2004) J Environ Eng-ASCE 130:126
226. Baker JE, Capel PD, Eisenreich SJ (1986) Environ Sci Technol 20:1136

227. Skoglund RS, Stange K, Swackhamer DL (1996) Environ Sci Technol 30:2113
228. Wallberg P, Andersson A (2000) Environ Toxicol Chem 19:827
229. Foster GD, Gates PM, Foreman WT, Mckenzie SW, Rinella FA (1993) Environ Sci Technol 27:1911
230. Huckins JN, Tubergen MW, Manuweera GK (1990) Chemosphere 20:533
231. Langenfeld JJ, Hawthorne SB, Miller DJ (1996) Anal Chem 68:144
232. Potter DW, Pawliszyn J (1994) Environ Sci Technol 28:298
233. Veith GD, Khehl DW, Puglisi FA, Glass GE, Eaton JG (1977) Arch Environ Cont Toxicol 5:487
234. Glooschenko W, Strachan WMJ, Sampson RCJ (1976) Pest Monit J 10:61
235. Swain WR (1978) J Great Lakes Res 4:398
236. Eisenreich SJ, Capel PD, Looney BB (1983) Physical behaviors of PCBs in the Great Lakes. In: Mackay D, Paterson S, Eisenreich SJ, Simmons MS (eds) PCB dynamics in Lake Superior water. Ann Arbor Science, Ann Arbor, MI p 141
237. Rice CP, Meyers PA, Brown GS (1983) Physical behavior of PCBs in the Great Lakes. In: Mackay D, Paterson S, Eisenreich SJ, Semmens MS (eds) Role of surface microlayers in the air-water exchange of PCBs. Ann Arbor Science, Ann Arbor, MI p 157
238. Swackhamer DL, Pearson RF, Schottler S (1999) Environ Sci Technol 33:3864
239. Stevens R, Neilson M (1989) J Great Lakes Res 15:377
240. Anderson DJ, Bloem TB, Blankenbaker RK, Stanko TA (1999) J Great Lakes Res 25:160
241. US EPA (2005) Lake Michigan Mass Balance Study loadings report
242. Murphy TJ, Rzeszutko CP (1977) J Great Lakes Res 3:305
243. Swackhamer DL, Armstrong DE (1987) J Great Lakes Res 13:24
244. Pearson RF, Hornbuckle KC, Eisenreich SJ, Swackhamer DL (1996) Environ Sci Technol 30:1429
245. Bicksler JL (1996) Polychlorinated biphenyls in the spring-time water column of the Great Lakes. PhD Dissertation, University of Minnesota, Minneapolis, MN
246. US EPA (2005) Data request to the Great Lakes Protection Program, LMMB dataset
247. Stapleton HM, Skubinna J, Baker JE (2002) J Great Lakes Res 28:52
248. Stapleton HM (2000) Accumulation of atmospheric and sedimentary PCBs and toxaphene in a Lake Michigan food web. MS Thesis, University of Maryland, College Park, MD
249. Henderson S, Swackhamer DL, Simcik MF (2005), in preparation
250. Strachan WMJ, Glass GE (1978) J Great Lakes Res 4:389
251. Capel PD, Eisenreich SJ (1985) J Great Lakes Res 11:447
252. Eisenreich S (1987) In: Hites RA, Eisenreich SJ (eds) The chemical limnology of nonpolar organic contaminants: polychlorinated biphenyls in Lake Superior, sources and fates of aquaticpollutants. American Chemical Society, Washington, DC, p 393
253. Baker JE, Eisenreich SJ, Johnson TC, Halfman BM (1985) Environ Sci Technol 19:854
254. Jeremiason JD, Hornbuckle KC, Eisenreich SJ (1994) Environ Sci Technol 28:903
255. Hornbuckle KC, Eisenreich SJ (1995) Environ Sci Technol 29:848
256. Hornbuckle KC, Jeremiason JD, Sweet CW, Eisenreich SJ (1994) Environ Sci Technol 28:1491
257. Mackay D (1982) Environ Sci Technol 16:274
258. Mackay D, Frasar A (2000) Environ Pollut 110:372
259. Opperhuizen A, Van der Velde EW, Gobas FAPC, Liem DAK, Van der Steen JMD (1985) Chemosphere 14:1871
260. Madenjian CP, Carpenter SR, Rand PS (1994) Can J Fish Aquat Sci 51:800
261. Thomann RV (1989) Environ Sci Technol 23:699

262. Thomann RV, Connolly JP (1984) Environ Sci Technol 18:65
263. Thomann RV, Connolly JP, Parkerton TF (1992) Environ Toxicol Chem 11:615
264. Connolly JP, Pedersen CJ (1988) Environ Sci Technol 22:99
265. Gobas FAPC, Pasternak JP, Lien K, Duncan RK (1998) Environ Sci Technol 32:2442
266. Gobas FAPC, Zhang X, Wells R (1993) Environ Sci Technol 27:2855
267. Campbell Linda M, Muir Derek CG, Whittle DM, Backus S, Norstrom RJ, Fisk AT (2003) Environ Sci Technol 37:1720
268. Rassmussen JB, Rowan DJ, Lean DRS, Carey JH (1990) Can J Fish Aquat Sci 47:2030
269. Skoglund RS, Swackhamer DL (1994) Environmental chemistry of lakes and reservoirs. In: Baker LA (Ed) Processes affecting the uptake and fate of hydrophobic organic contaminants by phytoplankton. American Chemical Society, Washington, DC, p 559
270. Swackhamer DL, Skoglund RS (1991) The role of phytoplankton in the partitioning of hydrophobic organic contaminants in water. In: Baker RA (ed) Organic substances and sediments in water, vol II. Lewis, Chelsea, MI, p 91
271. Swackhamer DL, Skoglund RS (1993) Environ Toxicol Chem 12:831
272. Madenjian CP, Carpenter SR, Eck GW, Miller MA (1993) Can J Fish Aquat Sci 50:97
273. Madenjian CP, Carpenter SR, Noguchi GE (1993) Arch Environ Cont Toxicol 24:78
274. Madenjian CP, Whittle DM, Elrod JH, O'Gorman R, Owens RW (1995) Environ Sci Technol 29:2610
275. Madenjian CP, Fahnenstiel GL, Johengen TH, Nalepa TF, Vanderploeg HA, Fleischer GW, Schneeberger PJ, Benjamin DM, Smith EB, Bence JR, Rutherford ES, Lavis DS, Robertson DM, Jude DJ, Ebener MP (2002) Can J Fish Aquat Sci 59:736
276. Madenjian CP, Elliott RF, DeSorcie TJ, Stedman RM, O'Connor DV, Rottiers DV (2000) J Great Lakes Res 26:427
277. De Vault DS, Hesselberg R, Rodgers PW, Feist TJ (1996) J Great Lakes Res 22:884
278. Morrison HA, Gobas FAPC, Lazar R, Whittle DM, Haffner GD (1998) Environ Sci Technol 32:3862
279. Veith GD, Lefoe DD, Bergstedt BV (1979) J Fish Res Board Can 36:1040
280. Connell DW (1990) Bioaccumulation of xenobiotic compounds. CRC, Boca Raton
281. Oliver BG, Niimi AJ (1988) Environ Sci Technol 22:388
282. Swackhamer DL (1996) Issues Environ Sci Technol 6:137
283. Hanari N, Kannan K, Horii Y, Taniyasu S, Yamashita N, Jude DJ, Berg MB (2004) Arch Environ Cont Toxicol 47:84
284. Trowbridge AG, Swackhamer DL (2002) Environ Toxicol Chem 21:334
285. Landrum PF, Robbins JA (1990) Sediments: chemistry and toxicity of in-place pollutants. In: Baudo R, Giesy JP, Muntau H (ed) Bioavailability of sediment-associated contaminants to benthic invertebrates. Lewis, Ann Arbor, MI, p 237
286. Landrum PF (1988) Aquat Toxicol 12:245
287. Jackson LJ, Carpenter SR, Manchester-Neesvig JB, Stow CA (1998) J Great Lakes Res 24:808
288. Marvin CH, Howell ET, Kolic TM, Reiner EJ (2002) Environ Toxicol Chem 21:1908
289. Comba ME, Metcalfe-Smith JL, Kaiser KLE (1996) Water Qual Res J Can 31:411
290. Bruner KA, Fisher SW, Landrum PF (1994) J Great Lakes Res 20:725
291. Bruner KA, Fisher SW, Landrum PF (1994) J Great Lakes Res 20:735
292. MacEachen DC, Russell RW, Whittle DM (2000) J Great Lakes Res 26:112
293. Pagano JJ, Rosenbaum PA, Roberts RN, Sumner GM, Williamson LV (1999) J Great Lakes Res 25:950
294. Kannan K, Yamashita N, Imagawa T, Decoen W, Khim JS, Day RM, Summer CL, Giesy JP (2000) Environ Sci Technol 35:566
295. De Vault DS, Clark JM, Lavhis G, Weishaar J (1988) J Great Lakes Res 14:23

296. Baumann PC, Whittle DM (1988) Aquat Toxicol 11:241
297. Borgmann U, Whittle DM (1991) J Great Lakes Res 17:368
298. Wong CS, Mabury SA, Whittle DM, Backus SM, Teixeira C, DeVault DS, Bronte CR, Muir DCG (2004) Environ Sci Technol 38:84
299. Newsome WH, Andrews P (1993) J AOAC Int 76:707
300. Miller MA, Kassulke NM, Walkowski MD (1993) Arch Environ Cont Toxicol 25:212
301. Gerstenberger SL, Gallinat MP, Dellinger JA (1997) Environ Toxicol Chem 16:2222
302. Gerstenberger SL, Dellinger JA (2002) Environ Toxicol 17:513
303. De Vault DS (1985) Arch Environ Cont Toxicol 14:587
304. De Vault DS, Willford WA, Hesselberg RJ, Nortrupt DA, Rundberg EGS, Alwan AK, Bautista C (1986) Arch Environ Cont Toxicol 15:349
305. Custer CM, Custer TW (2000) Environ Toxicol Chem 19:2821
306. Kannan K, Hilscherova K, Imagawa T, Yamashita N, Williams LL, Giesy JP (2001) Environ Sci Technol 35:441
307. Weseloh DV, Hughes KD, Ewins PJ, Best D, Kubiak T, Shieldcastle MC (2002) Environ Toxicol Chem 21:1015
308. Grasman KA, Fox GA, Scanlon PF, Ludwig JP (1996) Environ Health Perspect 104:829
309. Bowerman WW, Best DA, Evans ED, Postupalsky S, Martel MS, Kozie KD, Welch RL, Scheel RH, Durling KF et al. (1990) Organohalogen Cmpd 4:203
310. Bowerman WW, Best DA, Grubb TG, Zimmerman GM, Giesy JP (1998) Environ Monit Assess 53:197
311. Kozie KD, Anderson RK (1991) Arch Environ Cont Toxicol 20:41
312. Giesy JP, Bowerman WW, Mora MA, Verbrugge DA, Othoudr RA, Newsted JL, Summer CL, Aulerich RJ, Bursian SJ, Ludwig JP, Dawson GA, Kubiak TJ, Best DA, Tillitt DE (1995) Arch Environ Cont Toxicol 29:309
313. Dykstra CR, Meyer MW, Warnke DK, Karasov WH, Andersen DE, Bowerman WWIV, Giesy JP (1998) J Great Lakes Res 24:32
314. Humphrey HEB, Gardiner JC, Pandya JR, Sweeney AM, Gasior DM, McCaffrey RJ, Schantz SL (2000) Environ Health Perspect 108:167
315. He J-P, Stein AD, Humphrey HEB, Paneth N, Courval JM (2001) Environ Sci Technol 35:435
316. Anderson HA, Falk C, Hanrahan L, Olson J, Burse VW, Needham L, Paschal D, Patterson D Jr, Hill RH Jr (1998) Environ Health Perspect 106:279
317. Hovinga ME, Sowers M, Humphrey HEB (1992) Arch Environ Cont Toxicol 22:362
318. Hanrahan LP, Falk C, Anderson HA, Draheim L, Kanarek MS, Olson J (1999) Environ Res 80:S26
319. Cole DC, Sheeshka J, Murkin EJ, Kearney J, Scott F, Ferron LA, Weber J-P (2002) Arch Environ Health 57:496
320. Falk C, Hanrahan L, Anderson HA, Kanarek MS, Draheim L, Needham L, Patterson D Jr (1999) Environ Res 80:S19
321. Cole DC, Kearney J, Ryan JJ, Gilman AP (1997) Chemosphere 34:1401
322. Humphrey HE, Budd ML (1996) Toxicol Ind Health 12:499
323. Hicks HE, Ashizawa A, Cibulas W, De Rosa CT (2002) Organohalogen Cmpd 59:139
324. Gerstenberger SL, Tavris DR, Hansen LK, Pratt-Shelley J, Dellinger JA (1997) J Toxicol Clin Toxicol 35:377
325. Gerstenberger SL, Dellinger JA, Hansen LG (2000) J Toxicol Clin Toxicol 38:729
326. Fitzgerald EF, Brix KA, Deres DA, Hwang SA, Bush B, Lambert G, Tarbell A (1996) Toxicol Ind Health 12:361
327. Fitzgerald EF, Deres DA, Hwang SA, Bush B, Yang BZ, Tarbell A, Jacobs A (1999) Environ Res 80:S97

328. Hebert CE, Norstrom RJ, Weseloh DVC (1999) Environ Rev (Ottawa) 7:147
329. Froese KL, Verbrugge DA, Ankley GT, Niemi GJ, Larsen CP, Giesy JP (1998) Environ Toxicol Chem 17:484
330. Frame GM, Cochran JW, Boewadt SS (1996) J High Resolut Chromatogr 19:657
331. Burkhard LP, Cook PM, Lukasewycz MT (2004) Environ Sci Technol 38:5297
332. Metcalfe TL, Metcalfe CD (1997) Sci Total Environ 201:245
333. Trowbridge AG (2001) Polychlorinated biphenyls in the lower trophic levels of the Lake Michigan food web. PhD Dissertation, Environmental Health, University of Minnesota, Minneapolis, MN
334. Stapleton HM, Masterson C, Skubinna J, Ostrom P, Ostrom NE, Baker JE (2001) Environ Sci Technol 35:3287
335. Stapleton HM, Baker JE (2003) Arch Environ Cont Toxicol 45:227
336. Madenjian CP, DeSorcie TJ, Stedman RM, Brown EHJ, Eck GW, Schmidt LJ, Hesselberg RJ, Chernyak SM, Passino-Reader DR (1999) J Great Lakes Res 25:149
337. Madenjian CP, Elliott RF, Schmidt LJ, Desorcie TJ, Hesselberg RJ, Quintal RT, Begnoche LJ, Bouchard PM, Holey ME (1998) Environ Sci Technol 32:3063
338. Madenjian CP, Hesselberg RJ, Desorcie TJ, Schmidt LJ, Stedman RM, Quintal RT, Begnoche LJ, Passino-Reader DR (1998) Environ Sci Technol 32:886
339. Mackay D, Diamond M (1989) Chemosphere 18:1343
340. Mackay D (1989) J Great Lakes Res 15:283
341. Hornbuckle KC, Achman DR, Eisenreich SJ (1993) Environ Sci Technol 27:87
342. Richardson WL, Smith VE, Wethington R (1983) In: Mackay D, Peterson S, Eisenreich SJ, Simmons MS (eds) Dynamic mass balance of PCB and suspended solids in Saginaw Bay – A case study in physical behavior of PCBs in the Great Lakes.
343. Froese KL, Verbrugge DA, Snyder SA, Tilton F, Tuchman M, Ostaszewski A, Giesy JP (1997) J Great Lakes Res 23:440
344. Ling H, Diamond M, Mackay D (1993) J Great Lakes Res 19:582
345. Mackay D, Sang S, Vlahos P, Diamond M, Gobas F, Dolan D (1994) J Great Lakes Res 20:625
346. Diamond ML, Poulton DJ, Mackay D, Stride FA (1994) J Great Lakes Res 20:643
347. Halfon E, Allan RJ (1995) Environ Int 21:557
348. Williams DJ, Kuntz KW, Sverko E (2003) J Great Lakes Res 29:594
349. Helm PA, Diamond ML, Semkin R, Strachan WMJ, Teixeira C, Gregor D (2002) Environ Sci Technol 36:996

Hdb Env Chem Vol. 5, Part N (2006): 71–150
DOI 10.1007/698_5_040
© Springer-Verlag Berlin Heidelberg 2005
Published online: 2 December 2005

# Polychlorinated Dibenzo-*p*-dioxins and Dibenzofurans in the Great Lakes

Ross J. Norstrom

Centre for Analytical and Environmental Chemistry, Department of Chemistry,
Carleton University, Ottawa, ON K1S 5B6, Canada
*ross.norstrom@sympatico.ca*

**Abstract** The history of "dioxin", PCDD/F, contamination in the Great Lakes is reviewed. Occurrence, geographical distribution, and temporal trends in air, water, sediments, fish, seabirds, snapping turtles, and humans are presented, and eco/human toxicological implications reviewed. Patterns and concentrations in sediment indicate that atmospheric input dominated in Lake Superior, lower Lake Michigan, and Lake Erie. Inputs from the Saginaw River to Lake Huron and Fox River to upper Lake Michigan added some PCDD/F loading to these lakes above atmospheric deposition. Lake Ontario was heavily impacted by input of PCDD/Fs, particularly 2378-TeCDD, from the Niagara River. Sediment core and biomonitoring data revealed that PCDD/F contamination peaked in most lakes in the late 1960s to early 1970s, followed by rapid, order of magnitude declines in the mid- to late 1970s. The downward trend stalled in some lakes in the 1980s, but seems to have continued after the late 1990s, probably in response to various remediation efforts and reductions in PCDD/F emissions to the atmosphere. During the height of contamination, effects attributed in whole or in part to PCDD/F contamination included reproductive failure in lake trout and herring gulls in Lake Ontario. AHR-mediated sublethal effects may still be occurring in seabirds and fish, but much of this is thought to be due to dioxin-like PCBs rather than PCDD/Fs.

**Keywords** Dioxins · Effects · Furans · Levels · Sources · Trends

**Abbreviations**

| | |
|---|---|
| AHR | Aryl hydrocarbon receptor |
| AHR-congeners | PCDD/Fs with chlorine at the 2,3,7,8-positions, PCBs with chlorine at the $3,3'4,4'$-positions, not more than one chlorine at $2,2',6,6'$-positions |
| BMF | Biomagnification factor |
| BSAF | Biota-sediment bioaccumulation factor |
| Congener | Any member of a compound class, e.g., PCDDs, PCDFs or PCBs |
| EROD | Ethoxyresorufin-*o*-deethylase |
| Homolog | Group of isomers with the same carbon skeleton and number of chlorines, e.g., TeCDDs |
| H4IIE | Rat hepatoma cell line |
| PCBs | Polychlorinated biphenyls |
| PCDD | Polychlorinated dibenzo-*p*-dioxin |
| PCDF | Polychlorinated dibenzofuran |
| PCDD/F | Polychlorinated dibenzo-*p*-dioxin and -furan |
| TeCDD (F) | Tetrachlorodibenzo-*p*-dioxin (-furan) |
| PnCDD (F) | Pentachlorodibenzo-*p*-dioxin (-furan) |
| HxCDD (F) | Hexachlorodibenzo-*p*-dioxin (-furan) |
| HpCDD (F) | Heptachlorodibenzo-*p*-dioxin (-furan) |

| | |
|---|---|
| OCDD (F) | Octachlorodibenzo-*p*-dioxin (-furan) |
| 2378-TeCDD (and similar) | 2,3,7,8-Tetrachlorodibenzo-*p*-dioxin (commas are left out of all formulae for brevity) |
| SPMD | Semi-permeable membrane device |
| $\sum$- | Total concentration of congeners in the group |
| 2,4,5-T | 2,4,5-Trichlorophenoxyacetic acid |
| TEQ | 2,3,7,8-TeCDD toxic equivalent concentration |
| TEF | 2,3,7,8-TeCDD toxic equivalent factor (relative toxicity/potency to 2,3,7,8-TeCDD) |
| WHO | World Health Organization |

# 1
# Introduction

The history of "dioxins", polychlorinated dibenzo-*p*-dioxins (PCDDs), and polychlorinated dibenzofurans (PCDFs) in the Great Lakes really begins in 1978 when 2378-TeCDD was reported to be in fish from the Tittabawassee River in Michigan downstream from a large DOW chemical complex in Midland [1], and in Lake Ontario fish. The Michigan findings were subsequently confirmed by Harless and Lewis [2, 3]. PCDD/Fs had been widely recognized as an important class of environmental contaminants when this news broke. Early concern about PCDD/Fs had been primarily about industrial accidents, chemical manufacturing waste, the fungicide pentachlorophenol and the herbicide, 2,4,5-T. This herbicide, and its production wastes, were known to contain 2378-TeCDD as a byproduct from condensation of 2,4,5-trichlorophenol during its production [4]. TeCDD was a subject of considerable interest in the USA because of exposure of servicemen and women to Agent Orange, a defoliant used in the Viet Nam war, which contained 2,4,5-T as one of the main ingredients. It was also known that a wide range of PCDD/Fs were emitted by municipal waste incinerators and other combustion processes when a source of chlorine was present [5], and that commercial PCB mixtures contained a variety of PCDFs [6]. It was not until much later that bleached kraft pulp mills were identified as a specific source of 2378-TeCDD and 2378-TeCDF [7]. The finding of 2378-TeCDD in Great Lakes fish caused the scientific community to take notice. The Love Canal waste dumpsite issue was gaining wide coverage in the press about the same time [8], so stories about "the most toxic chemical known to man" created immediate consternation in the Great Lakes community, which put pressure on government agencies in both the USA and Canada to address the issue. The result was a flurry of activities in the mid- to late-1980s to survey concentrations of 2378-TeCDD in a variety of fish, birds, and sediments from around the Great Lakes, and to scope the implications to health of fish, wildlife, and humans.

Among these was a preliminary investigation in 1980 of 2378-TeCDD in herring gull eggs, which found concentrations in eggs from Saginaw Bay,

Lake Huron and Lake Ontario to be four to six times higher than in Lakes Michigan, Huron (main body), and Erie [9]. Although the data were semi-quantitative, they provided early evidence that Saginaw Bay and Lake Ontario were the areas of most concern. Herring gull eggs from Lake Ontario, 1981, were reanalyzed using improved methods and found to have $132\,ng\,kg^{-1}$ of 2378-TeCDD [10]. This study also provided the first evidence of the presence of 12378-PnCDD and HxCDDs in the Great Lakes. Immediate suspicion fell on effluent and waste disposal from the large number of chlorine-based chemical industries in the Niagara Falls, NY area, especially Love Canal and other dumpsites along the Niagara River.

Herring gull eggs collected in 1971 and archived in the Canadian Wildlife Service Specimen Bank were also analyzed and found to be contaminated with $1225\,ng\,kg^{-1}$ of 2378-TeCDD [11]. This concentration was well above the $LD_{50}$ of 2378-TeCDD in chicken embryos, $250\,ng\,kg^{-1}$ [12]. At the time, there were no data on the toxicity of 2378-TeCDD in wild birds. When these early 1970s Lake Ontario herring gull egg concentrations were lined up against the complete failure of herring gull eggs to hatch in the same period, due primarily to early death of embryos [13, 14], it was assumed that the chemical culprit had been apprehended. The story turned out to be more complicated than that, as usual. We now know that the herring gull is about 50 times less sensitive than the chicken to 2378-TeCDD toxicity [15]. However, these early surveys provided considerable ammunition to begin comprehensive studies on PCDD/F contamination in the Great Lakes.

In the 25 years since dioxin concerns began in the Great Lakes, an enormous amount of information has been generated on sources, deposition, concentrations in sediments and biota, and temporal trends of PCDD/Fs in the Great Lakes environment, which are the subject of this review. We also have a much better understanding of their toxicity to fish and wildlife. The dioxin issue is a mature one in the Great Lakes. That is to say, considerable efforts to eliminate sources of PCDD/Fs resulted in environmental concentrations decreasing one to two orders of magnitude since the peak contamination in the 1970s, to the point where currently there is less concern than for other contaminants. PCBs are generally considered to be a significantly greater problem to fish and wildlife in the Great Lakes than PCDD/Fs at present, although there may be some exceptions, e.g., sublethal effects in lake trout in Lake Ontario [16]. Concerns about human exposure to PCDD/Fs related to heavy consumption of sport fish remain [17, 18].

## 2
## Occurrence and Geographical Distribution

### 2.1
### Air

### 2.1.1
### Concentrations in Air

Because of the extremely low concentrations of PCDD/Fs in ambient air, there are very few published studies that provide direct measurement of concentrations in air from the Great Lakes region. All of these are for the 1986–1988 period. While there are several reports of homolog concentrations in air, there is a lack of data on concentrations of the 2378-substituted congeners. This is unfortunate, since it is only these congeners which bioaccumulate and are of toxicological interest in fish, wildlife, and humans.

Most of the information on atmospheric loading of PCDD/Fs to the Great Lakes comes from estimates of deposition fluxes through analysis of sediment cores rather than direct measurement of air concentrations. The contribution of PCDD/F atmospheric loading to the Great Lakes relative to point sources (subregional atmospheric deposition, or water-born) was estimated to be 100% for Lake Superior, 80% in southern Lake Michigan, 40% in northern Lake Michigan, and 10% in Lake Ontario in the mid-1990s [19].

No estimates of the relative contribution of atmospheric and non-atmospheric sources have been made for Lake Huron and Lake Erie. Lake Huron has a relatively small urban population and industrial base compared to Lakes Michigan, Erie, and Ontario, but has a potentially significant input from the chlorine chemical industry at the head of Saginaw Bay. This appears to have resulted in higher relative HxCDF and HpCDF contributions to the PCDD/F profile in southern Lake Huron sediments in 1981 [20]. Lake Huron is connected by a wide channel to upper Lake Michigan, facilitating water exchange. These two lakes have nearly the same level and are often considered to be one lake hydrologically. Consequently, the relative importance of atmospheric loading of PCDD/Fs in Lake Huron proper may be similar to northern Lake Michigan. Surface sediment PCDD/F congener profiles in Lake Erie are consistent with a primarily atmospheric loading for this lake [20].

Eitzer and Hites [21, 22] reported PCDD/F homolog concentrations in ambient air from regions near the Great Lakes in 1987–1988. These studies were conducted in the Bloomington and Indianapolis, IA area, south of Lake Michigan, and a rural site in Wisconsin south of Lake Superior. There were distinct differences in concentrations and profiles in the urban and rural areas. Total PCDD concentrations ranged from 0.24 pg m$^{-3}$ at the rural site to 2.5 pg m$^{-3}$ in the cities. The corresponding range in total PCDF concentrations was similar, 0.18–2.6 pg m$^{-3}$. These concentrations were similar to those reported in

European air. The profile in urban air (Indianapolis) was most consistent with a combustion source, having a relatively high concentration of PCDFs in which HxCDFs were the major contributor. The suburban (Bloomington) and rural (Trout Lake) profiles were more similar, having a lower relative contribution from PnCDFs and HxCDFs. Average PCDD/F homolog concentrations in air (vapor and particles) and rain (dissolved and particles) in Bloomington, IN, sampled monthly between August 1985 and July 1988 are shown in Fig. 1 [22]. The distribution favored particles over vapor and dissolved phases as the number of chlorines increases. Thus, most of the TeCDFs were in the vapor and dissolved phases, while most of the HpCDD/Fs and OCDD/Fs were adsorbed to particles. It was postulated that photodegradation and washout of PCDFs and lower chlorinated PCDDs during transport from urban/industrial sources resulted in the rural profile. However, the relatively high concentrations of OCDD and HpCDDs in rain may be due to photochemical formation from pentachlorophenol [23]. The PCDD/F homolog profile in Great Lakes sediments was similar to that in "average rain", indicating that atmospheric transport was an important source to the Great

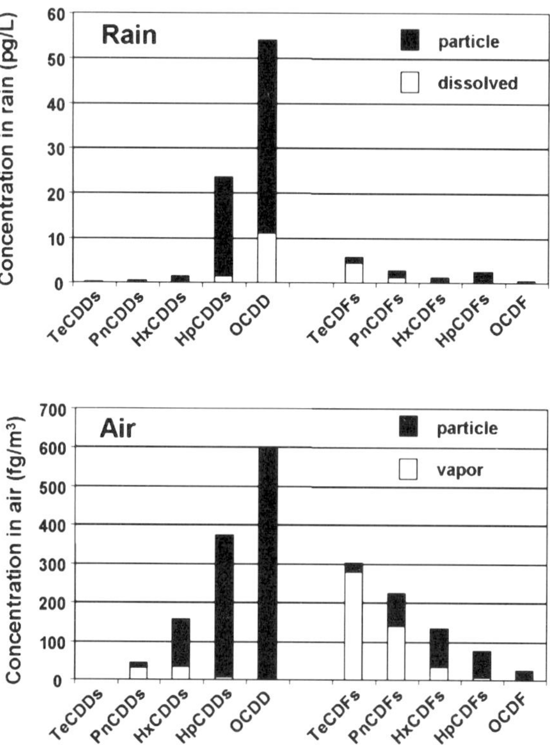


**Fig. 1** Average concentrations (pg L$^{-1}$) of PCDD/F homologs in air and rain in Bloomington, IN, sampled monthly between Aug 1985 and July 1988 (from Eitzer and Hites [22])

Lakes. Flux calculations based only on wet and dry particle deposition were close to measured sediment fluxes. PCA analysis confirmed that wet and dry particle deposition was much more important than dry vapor deposition, based on homolog patterns.

Steer et al. [24] determined PCDD/F homolog concentrations in ambient air in Ontario in 1988. Two of the sites were highly urban/industrial areas, Windsor/Detroit and Toronto, and one was a rural site in central Ontario east of Georgian Bay, Lake Huron. OCDD and OCDF blank concentrations were high, so reliable concentrations of these compounds could not be determined. However, based on the results of Eitzer and Hites [21, 22], the concentrations of OCDD and OCDF were in the expected range. Average ambient concentrations of the PCDD/F homologs in air were in the same range ($0.1-1\ \mathrm{pg\,m^{-3}}$) as those found by Eitzer and Hites [21, 22]. Similar to these studies, TeCDFs were a lower proportion of total PCDFs at the rural site than the two urban sites, indicating selective loss during transport. TeCDDs were anomalously higher relative to the PnCDD, HxCDD, and HpCDD homologs than any other study of ambient air in the Great Lakes region. TeCDDs and PnCDDs concentrations were two to three times higher in Windsor than Toronto, but there was not much difference in the HxCDD and HpCDD concentrations among any of the sites. PCDF concentrations also tended to be higher in Windsor.

PCDD/F homolog concentrations were also determined in air at Windsor, ON and Walpole Island, ON, in 1987–1988, preliminary to establishment of an large municipal incinerator in Detroit [25]. The detection limits were not high enough to detect TeCDDs ($0.06\ \mathrm{pg\,m^{-3}}$) or PnCDD ($0.16\ \mathrm{pg\,m^{-3}}$) at either site, or any PCDFs at Walpole Island. PCDFs concentrations were lower than those of PCDD at Windsor, $0.14-0.56\ \mathrm{pg\,m^{-3}}$. Concentrations of PCDDs in Windsor air were highest when the wind was from the north ($3.86\ \mathrm{pg\,m^{-3}}$), lowest when from the west ($1.22\ \mathrm{pg\,m^{-3}}$), and when from south and east ($1.9\ \mathrm{pg\,m^{-3}}$). This agreed with a Detroit urban source contributing from the north. Concentrations of PCDD/Fs in this study were comparable to those measured by Steer et al. [24] in air from Windsor, except that lower TeCDD concentrations were found, assuming that concentrations $> 0.1\ \mathrm{pg\,m^{-3}}$ would have been detected. The profile was dominated by OCDD $>$ HpCDD $>$ HxCDD. The vapor–particle distribution was highly dependent on the number of chlorines, about 35% of HxCDDs and 18% of HpCDDs. These percentages are somewhat higher than found by Eitzer and Hites [22], who measured 23% of HxCDDs and 1.5% of HpCDDs in the vapor phase. The directional dependence of concentrations was much more dramatic at Walpole Island. $\Sigma$-PCDDs were undetectable or barely detectable, $0.11\ \mathrm{pg\,m^{-3}}$, when winds were from the east or north, but $0.53-0.59\ \mathrm{pg\,m^{-3}}$ when winds were from the south or west, that is, from the Detroit urban area.

Edgerton et al. [26] determined atmospheric concentrations of PCDD/F homologs in ambient air from several sites in Ohio, south of Lake Erie, in 1987. Two sites were near municipal waste and sewage sludge incinerators. Total

PCDD/F concentrations ranged from 1.9 to 6.4 pg m$^{-3}$. The profile was dominated TeCDFs > OCDD ≈ HpCDDs > HxCDDs ≈ HxCDFs ≈ PnCDFs. Source apportionment based on principle components analysis indicated that 72–81% of the source in the urban areas (biased by being near incinerators) was municipal waste incineration, 28% was "urban background". The rural site had a profile surprisingly similar to municipal incinerators. It was suggested to be representative of Great Lakes regional background concentrations.

Smith et al. [27] studied atmospheric concentrations of PCDD/F homologs at two sites in the Niagara Falls, NY area, two to three times per month during the period December 1986 to April 1988. The sites were close to a chlorine-based chemical industry complex and several dumpsites. Concentrations of total PCCD/Fs ranged from 0.5 to 22 pg m$^{-3}$. Concentrations and profiles were highly dependent on wind direction. Concentrations of PCDD/Fs in air not passing over urban/industrial areas were similar to those found in other studies of ambient air. Unusually high concentrations (> 10 pg m$^{-3}$) at both sites appeared to be associated with an energy-from-waste facility in which TeCDFs and HxCDFs were most important.

## 2.1.2
### Air Deposition Models

A thorough analysis of atmospheric transport and deposition to the Great Lakes has been carried out using the HYSPLIT model developed by the US National Atmospheric and Oceanic Administration (NOAA) [28, 29]. An emissions inventory of PCDD/Fs for North America in 1996 was used as input to the model. Factors considered in the fate and distribution were meteorological data, vapor–particle partitioning, aerosol characteristics, reaction with hydroxyl radicals, photolysis, and dry and wet deposition. The model was generally satisfactory at estimating fluxes, except for HpCDD and OCDD, which appeared to be underestimated by about a factor of four. The model output was summarized as 2378-TeCDD toxic equivalent concentrations (TEQs) based on the WHO mammalian 2378-TeCDD toxic equivalent factors (TEFs) [30]. Since HpCDD and OCDD were estimated to contribute only 2% of TEQs, the model was considered to be valid for the purpose intended.

A possible explanation for the under-prediction of HpCDD and OCDD atmospheric deposition fluxes was given by Baker and Hites [31]. They found a gross discrepancy in the global mass balance of emission versus deposition of OCDD and, to a lesser extent, HpCDD. Deposition was about 40 times higher than emissions for OCDD and six times higher for HpCDD. Given the exhaustive work on developing inventories of PCDD/F emissions, Baker and Hites believed that there must be an unrecognized source of OCDD and HpCDD to the atmosphere. They hypothesized that photolytic condensation of pentachlorophenol in water droplets was the source of the excess OCDD and HpCDD. Experimental evidence was provided that this conversion occurred in water at wavelengths great than 290 nm, with or without the presence of hydrogen peroxide as an OH radical donor. There is a kinetic problem with this theory of OCDD/HpCDD formation from pentachlorophenol in water droplets in the environment. The condensation reaction is bimolecular, and the rate is therefore proportional to the square of the pentachlorophenol concentration. The experiments were conducted at pentachlorophenol concentrations of $100–1000\ \mu g\,L^{-1}$, while average concentrations in rain were estimated to be $0.02\ \mu g\,L^{-1}$. On the basis of concentration alone, the rate of conversion in rain would be $2.5 \times 10^7$ times slower than the $100\ \mu g\,L^{-1}$ photolysis experiment.

Another contributor to the deposition-inventory deficit of OCDD and OCDF may be underestimation of the contribution from forest fires. Gullet and Touati [32] found that previous estimates were nearly an order of magnitude too low. Forest fires gain in importance as other sources decline, and contribute an increasingly significant proportion of PCDD/F loading to the atmosphere, perhaps comparable to domestic and industrial sources.

The major anthropogenic sources of PCDD/F emissions to the North American atmosphere that contribute to TEQ deposition in the Great Lakes have been estimated [28, 29]. The fraction of total estimated atmospheric TEQ deposition to Lake Superior in 1996 from various sources is shown in Fig. 2.

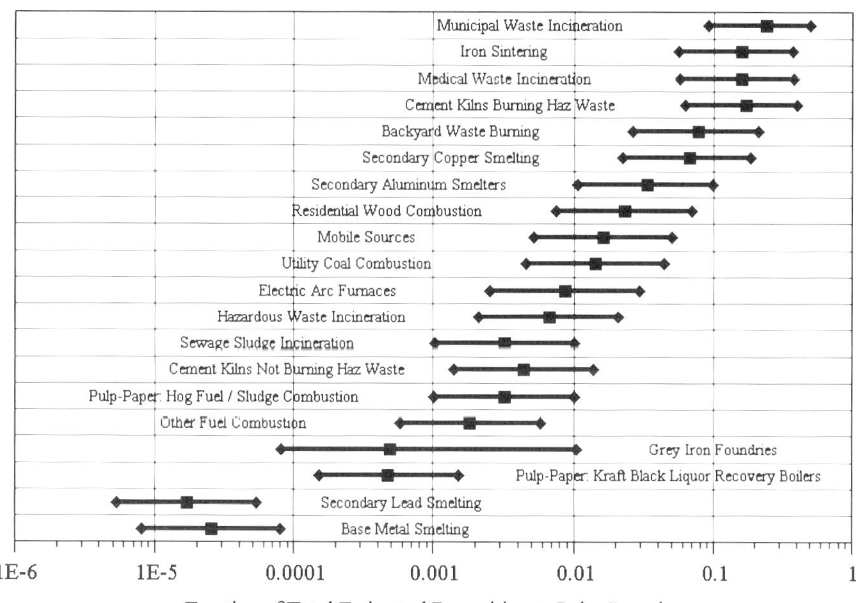


**Fig. 2** Major sources of PCDD/F emissions to the North American atmosphere that contributed to TEQ deposition in the Great Lakes in 2001. Reproduced with permission from Cohen [28, 29]

The proportional contribution of these sources varied among the lakes. Split into three broad source sectors, incineration contributed the most TEQ deposition in all of the Great Lakes. The metals sector contributed the majority of the remaining TEQs, followed by a small contribution from fuel combustion. The lakewide average contribution to total TEQ deposition for the three sectors was about 70% for incineration, 27% metals, and 3% fuel combustion. On a per capita basis, the basin-wide contribution from incineration was sixfold lower in Canada than the USA, about equal from metals, and twofold higher from fuel combustion.

An estimate of the air emissions and distribution of atmospheric deposition of PCDD/F-derived TEQs to each lake (g year$^{-1}$) from inside and outside the Great Lakes watershed based on the HYSPLIT model is shown in Fig. 3. Lakes Michigan, Superior, and Huron have the highest total deposition, in large part because of their bigger surface area. The proportion of atmospheric sources from inside the Great Lakes watershed contributing to total atmospheric deposition (1996 estimates) ranged from a high of 41% in Lake Michigan to a low of 28% in Lake Superior and Lake Erie. Inside watershed atmospheric sources were intermediate in Lake Huron (37%) and Lake Ontario (34%). The distribution of the contribution of all North American sources of PCDD/Fs to atmospheric deposition in Lake Superior is shown in Fig. 4. Atmospheric sources of PCDD/Fs (primarily incinerators) from all over eastern North America as far away as northeast Texas and southern Florida were identified as significant contributors of PCDD/F-related TEQs in the Great Lakes.

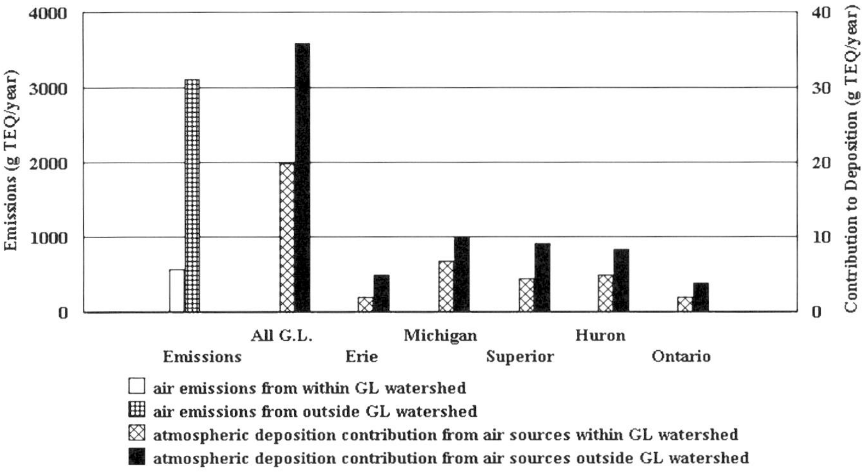


**Fig. 3** Estimate of the air emissions and distribution of atmospheric deposition of PCDD/F-derived TEQs to the Great Lakes (g TEQ year$^{-1}$) from inside and outside the Great Lakes watershed, based on the NOAA HYSPLIT model. Reproduced with permission from Cohen [28, 29]

Predicted individual congener contribution to atmospheric deposition of TEQs in Lake Superior in 1996 is given in Fig. 5. Overall, PCDFs, especially 23478-PnCDF, had a larger contribution than PCDDs. The most toxic congener, 2378-TeCDD, provided about 5% of total TEQs in all of the lakes. Most

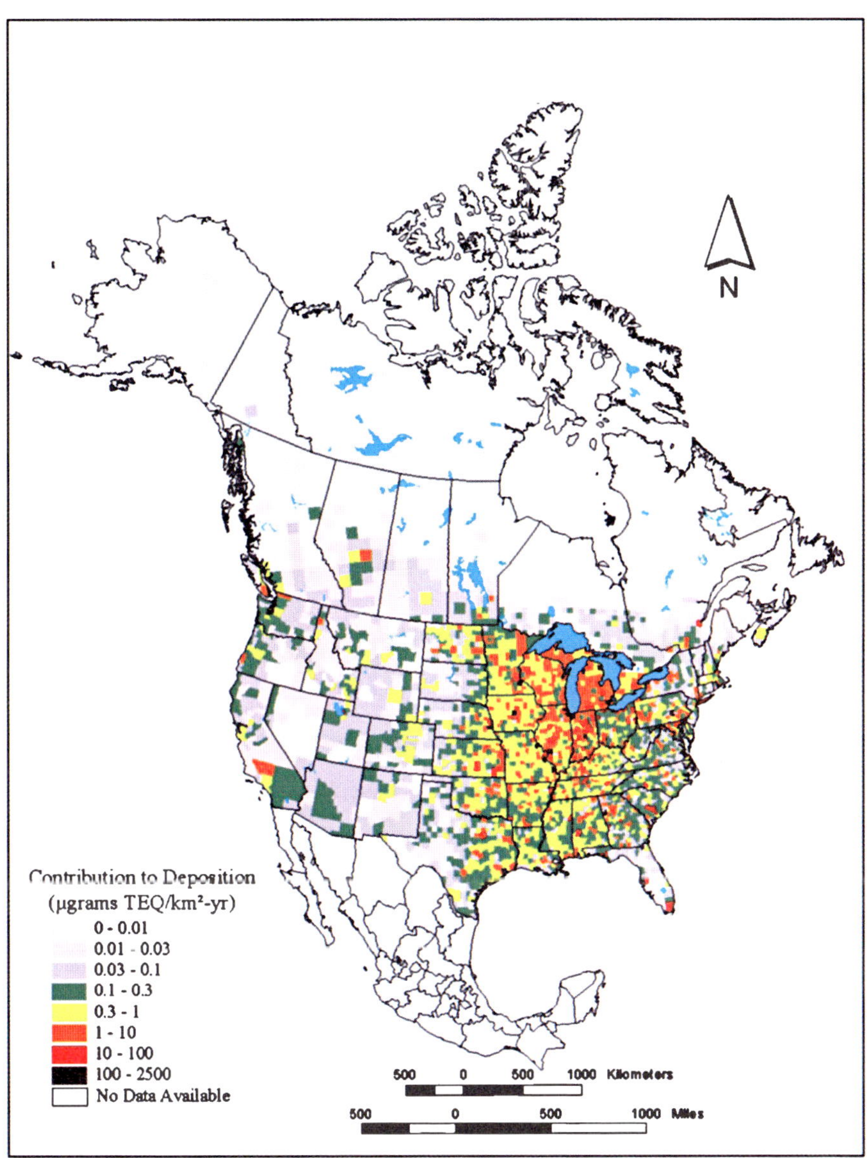


**Fig. 4** Contribution of all North American sources of PCDD/Fs ($\mu$g TEQ km$^{-2}$ year$^{-1}$) to atmospheric deposition in Lake Superior in 1996 based on the NOAA HYSPLIT model. Reproduced with permission from Cohen [28]

of the deposition was wet, rather than dry (particle and vapor) deposition. It must be kept in mind that the major contributor to total TEQ exposure of fish, wildlife, and humans in most areas of the Great Lakes is from a few PCB congeners that exhibit dioxin-like toxicity. The significance of TEQs are discussed in a later section.

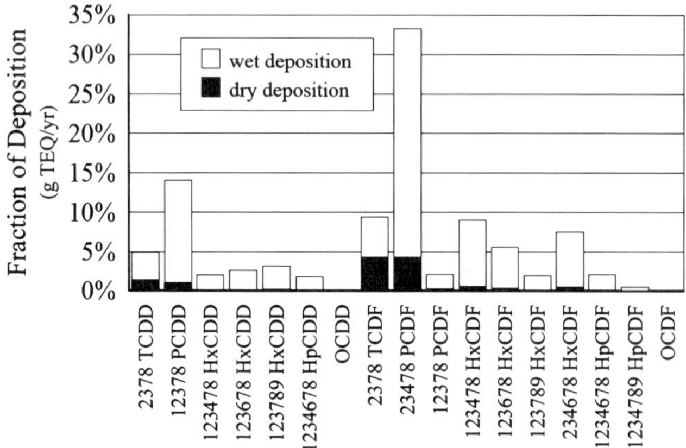

**Fig. 5** Fraction of individual AHR congener contribution to total atmospheric deposition of TEQs in Lake Superior in 1996, based on the NOAA HYSPLIT model (from Cohen et al. [28])

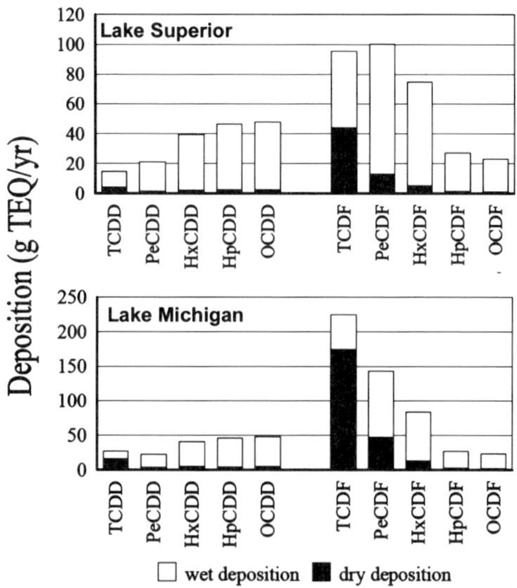

**Fig. 6** Flux of PCDD/F homologs (g TEQ lake$^{-1}$ year$^{-1}$) estimated by the NOAA HYSPLIT model for Lake Superior and Lake Michigan in 1996. Reproduced with permission from Cohen et al. [28]

The flux of PCDD/F homologs (g year$^{-1}$ for each lake) estimated by the HYSPLIT model for Lake Superior and Lake Michigan in 1996 is shown in Fig. 6 [28]. The major differences between the two lakes were the higher dry deposition loadings of TeCDF and PnCDF in Lake Michigan, which is due to its proximity to major urban/industrial areas. The pattern of PCDD/F deposition in the Lake Erie was similar to Lake Michigan, while Lake Huron and Lake Ontario depositional patterns were more similar to Lake Superior.

## 2.2
## Water

There are even fewer measurements of ambient concentrations of PCDD/Fs in Great Lake water than in air. Concentrations of 2378-substituted PCDD/Fs in water, which may result in bioaccumulation of toxicologically significant sub-ng g$^{-1}$ concentrations in fish and seabirds, are at the tens of femtograms per liter ($10^{-14}$ g L$^{-1}$) range, which is beyond the sensitivity of most analytical methods. There are no measurements of PCDD/F concentrations in water of any of the Great Lakes proper. Only two published reports were found giving PCDD/F concentrations in water from Great Lakes connecting channels. Both were drinking water surveys employing 10 L samples.

## 2.2.1
## Saginaw River

Indirect measurement of PCDD/F concentrations in the Saginaw River system flowing into Saginaw Bay, Lake Huron, was carried out by the deployment of semi-permeable membrane devices (SPMDs) in the mid-1990s [33]. The Saginaw River drains an area having a high concentration of chlorine chemical industry in the Midland, MI area, into Saginaw Bay, Lake Huron. Congener-specific concentrations in water, estimated from concentrations in SPMDs and a linear kinetic model, are given in Table 1. Measurable, relatively uniform concentrations of 2378-TeCDD were found in SPMD contents at all sites, translating into a concentration in water from 0.03 to 0.1 pg L$^{-1}$. The major contaminant was 2378-TeCDF at predicted concentrations ranging from 1 to 6 pg L$^{-1}$. Predicted OCDD concentrations ranged from undetectable to 1 pg L$^{-1}$. There were approximately fivefold increases in 2378-TeCDF, 12378-PnCDF, 23478-PnCDF, and 123478-HxCDF concentrations just downstream from the Tittabawassee River entrance. These same four PCDF congeners furnished most of the TEQs found in sediments and flood plain soils along the Tittabawasee River, while PCDDs played a minor role [34]. The Tittabawassee River did not contribute significant additional loading of PCDDs at the time of sampling. At the mouth in Saginaw Bay, predicted 2378-TeCDF and 12378-PnCDF concentrations were higher (two- to threefold) than headwater concentrations of the Saginaw River, while the HxCDF concentra-

**Table 1** Estimated concentrations (pg L$^{-1}$) of PCDD/Fs in Saginaw River water at various sites along the Saginaw River in the mid-1990s

|  | Shiawas-see | Titta-bawassee | Zilwaukee | Middle-ground | Bay City | Saginaw Bay |
|---|---|---|---|---|---|---|
| 2378-TeCDD | 0.03 | 0.10 | 0.06 | 0.08 | 0.10 | 0.08 |
| 12378-PnCDD | 0.01 | 0.02 | 0.01 | 0.01 | nd | 0.08 |
| OCDD | 1.00 | 0.20 | 0.80 | nd | nd | 0.20 |
| Σ-PCDD | 1.04 | 0.32 | 0.87 | 0.09 | 0.10 | 0.36 |
| 2378-TeCDF | 1 | 6 | 3 | 3 | 3 | 2 |
| 12378-PnCDF | 0.08 | 0.60 | 0.20 | 0.20 | 0.20 | 0.20 |
| 23478-PnCDF | 0.07 | 0.40 | 0.10 | 0.20 | 0.10 | 0.10 |
| 123478-HxCDF | 0.02 | 0.10 | 0.03 | 0.04 | 0.03 | 0.04 |
| 123678-HxCDF | 0.08 | 0.10 | 0.07 | 0.07 | 0.06 | 0.08 |
| Σ-PCDF | 1.25 | 7.20 | 3.40 | 3.51 | 3.39 | 2.42 |

Data are based on concentrations in SPMDs containing about 8 g of triolein suspended in the river for 28 days, and a kinetic model [33]. The Shiawassee River, Tittabawassee River, and the Cass River join to form the Saginaw River, which flows to Saginaw Bay, Lake Huron at Bay City. The major chlorine chemical industry is in Midland MI, on the Tittabawassee River

tions were more similar to background. PCDD/F concentrations calculated for the Saginaw River in the mid-1990s were much lower than those in Niagara River water in 1986 [35], but may have been higher historically. Episodes of PCDF input from the Tittabawasee River associated with flooding and erosion of contaminated flood plain soils or resuspension of contaminated sediments, such as occurred in 1994, may result in much higher concentrations in Saginaw River water and increased loading to Saginaw Bay [34].

## 2.2.2
## Detroit River

Jobb et al. [36] surveyed PCDD/F concentrations in 399 raw and treated drinking water supplies throughout Ontario in 1989. Positive results were found only for 33 raw and four treated water samples in the St. Clair–Detroit River corridor. The only PCDD/F found was OCDD at concentrations of 20–175 pg L$^{-1}$. Although blank values were not given, it can be assumed that the concentrations were not false positives, based on lack of detection in over 300 samples from outside this area. The treated water samples from Wallaceburg, Walpole Island, and Windsor had OCDD concentrations of 20–46 pg L$^{-1}$, in the same range as the raw water samples. There does not seem to be any obvious reason why water in the St. Clair–Detroit River corridor was abnormally contaminated with OCDD. In 1983 this system, including Lake St. Clair, was found to be contaminated by a number of relatively volatile chlorinated compounds (pentachlorobenzene, hexachlorobenzene, hexachlorobutadiene, octachlorostyrene) associated with the production of perchloroethylene and carbon tetrachloride and/or production of chlorine in the Sarnia, Ontario area [37]. Although OCDD has not been specifically associated with manufacture of these chemicals, there may be some connection with this industry.

### 2.2.3
### Niagara River

Meyer et al. [35] analyzed drinking water supplies from various places in New York State in 1986. They found $2.1\,pg\,L^{-1}$ of TeCDFs ($1.2\,pg\,L^{-1}$ 2378-TeCDF) in the soluble fraction of finished water from one of 20 community water systems, Lockport, NY, which receives its water from the Niagara River. OCDD ($5\,pg\,L^{-1}$) was also found, but the distilled water blank concentration of OCDD was even higher ($6.5\,pg\,L^{-1}$). The Lockport raw water supply from the Niagara River had significant PCDF contamination: TeCDFs ($18\,pg\,L^{-1}$), PnCDFs ($27\,pg\,L^{-1}$), HxCDFs ($85\,pg\,L^{-1}$), 1234678-HpCDF ($210\,pg\,L^{-1}$), and OCDF ($230\,pg\,L^{-1}$). It is not clear how much of this contamination was associated with the soluble fraction. No 2378-TeCDF ($< 0.7\,pg\,L^{-1}$) was found. Apart from HpCDF and OCDF, the major 2378-substituted congeners were 12378-PnCDF ($2\,pg\,L^{-1}$), 123478 HxCDF ($39\,ng\,L^{-1}$) and 123678-HxCDF ($9\,pg\,L^{-1}$). This pattern was well reflected in sediment and sludge from the system, and typical of Niagara River sediments [38].

Hallett and Brooksbank [39] quote an internal Environment Canada report in which water and suspended sediment samples from the Niagara River were sometimes found to contain measurable concentrations of various PCDD/F homologs. PCDD concentrations were $< 1.4\,pg\,L^{-1}$ in one or two out of six samples, however, five or six samples contained OCDD concentrations up to $3.6\,pg\,L^{-1}$. TeCDF concentrations were very high in one sample, $156\,pg\,L^{-1}$; four out of six samples had PnCDF concentrations up to $317\,pg\,L^{-1}$. PCDD/Fs were detected in suspended sediments more frequently, up to $2530\,pg\,L^{-1}$ of OCDF. TeCDDs were not detected in any sample at unknown detection limits.

In the Niagara River Mussel Biomonitoring Program, caged mussels were suspended at various sites along the river for 21 days as an indirect measure of the concentration of organic contaminants in water. Mussels accumulate contaminants directly from the water and from the particulate matter they ingest. The last survey was in 2000 [38]. The most frequently detected PCDD/Fs in mussels were TeCDDs. At the Bloody Run Creek site all of the TeCDD was 2378-TeCDD ($23\,pg\,g^{-1}$ wet weight). Bloody Run Creek historically was contaminated by drainage from the Hyde Park dump site, which contained 2,4,5-trichlorophenol manufacturing wastes. It is not stated which TeCDD congeners were present at the other sites, but based on the calculated TEQs, they were mostly not 2378-TeCDD. Mussels caged at the Pettit Flume site accumulated the highest concentrations of PCDD/Fs. Concentrations dropped off 20- to 30-fold downstream from this site.

From sediment analyses, only 10% of PCDD/F loading in Lake Ontario in the mid-1990s was from the Great Lakes region atmosphere, the remainder was local atmospheric or non-atmospheric sources [19]. Lake Erie outflow water has relatively low PCDD/F contamination [38], and therefore contributes little to Lake Ontario loading. Since Niagara River water upstream and downstream from Niagara Falls on the US side was highly contaminated with PCDD/Fs at least until the mid-1980s, the Niagara River is most likely the primary source of PCDD/F loading to Lake Ontario. This is especially true for 2378-TeCDD, HxCDFs, HpCDFs, and OCDF, which are a higher proportion of total PCDD/Fs in Lake Ontario sediment than in the other Great Lakes [20]. Biomonitoring and sediment core data, which are discussed below, revealed that PCDD/F loading to Lake Ontario had already decreased substantially due to changes in industrial activity and processes before the "dioxin" problem was discovered in the 1980s. It is therefore likely that historical concentrations of PCDD/Fs in Niagara River water were as much as an order of magnitude higher than those found in the analyses conducted in the mid-1980s. Considerable effort was expended to limit movement of PCDD/Fs into the Niagara River from historical waste dumps in the 1980s and 1990s, which probably had some additional influence on PCDD/F loading to Lake Ontario. Despite recent sediment remediation efforts at sites such as Gill Creek and Pettit Flume in the upper Niagara River, there was evidence of bioavailable PCDD/F contamination in water and sediments in 2000 [38]. Thus, PCDD/F contamination is still entering the Niagara River and being flushed into Lake Ontario.

## 2.3
## Lake Sediments

Because sediments build up over time, eventually sealing off lower layers, sedimentation functions as a main removal mechanism (primarily in the oceans) of highly hydrophobic contaminants like PCDD/Fs from the biosphere. Unlike in water and air, concentrations of PCDD/Fs in Great Lakes sediment are high enough that analysis of individual 2378-substituted PCDD/F congeners is feasible even in areas receiving their PCDD/F loading only from atmospheric deposition. Depth profiling of PCDD/F concentrations in dated sediment cores also provides us with the only means of determining historical trends in loading, since biomonitoring programs only go back as far as the early 1970s, while much of the growth in the use of chlorinated organic compounds occurred in the 1950s and 1960s.

Czuczwa and coworkers documented the PCDD/F homolog concentrations in sediment cores from several areas in and around the Great Lakes. Profiles of PCDD/F relative concentrations in Saginaw Bay and southern Lake

Huron, 1981, were compared to various sources [40, 41]. The homolog profiles were similar among areas, including a surface grab sample from the mouth of the Saginaw River. It should be noted that the relative contribution of TeCDFs and PnCDFs found in these studies was lower than in more recent analyses, as pointed out by Pearson et al. [42]. Underestimation of these homologs may be due to large variations in sensitivity among isomers in the electron capture negative ion (ECNI) MS technique used by Czuczwa and coworkers in the earlier publications. OCDD and OCDF were the dominant congener/homologs. The isomer makeup was similar to combustion. Depth profiles in one core from southern Lake Huron offered fairly high resolution. The horizon of appearance of OCDD, HpCDD, and HpCDF was around 1945, and concentrations increased steadily between 1950 and 1965 and remained relatively steady till 1981. This profile matched quite closely the US production of chloro-aromatic compounds. It was concluded that most of the PCDD/Fs in these sediments originated from combustion of chlorinated organic compounds in various wastes. A similar conclusion was based on concentrations of PCDD/Fs in three dated layers (1935, 1953, and 1982) of a sediment core from Siskiwit Lake on an uninhabited island in northern Lake Superior, which experiences deposition only from the atmosphere [43]. There was about a fivefold increase in deposition between 1935 and 1953.

PCDD/F homologs were determined by Czuczwa and Hites [20] in urban air particulates and surface sediments sampled from the middle of Lake Michigan (1982), lower Lake Huron (1975, 1981), eastern Lake Erie (1981, 1983), and western Lake Ontario (1983) near the Niagara River mouth. Homolog profiles from each of the lakes were compared to those of various sources (Fig. 7). The similarity of the sediment profiles and urban air particulates is striking, e.g., the dominance of OCDD, followed by HpCDDs and HpCDFs. Lake Ontario was the exception. In Lake Ontario, OCDF was the second most abundant PCDD/F, followed by HpCDFs. The authors proposed that this may have been due to chemical waste disposal in the Niagara River drainage from HCB or PCP manufacture in this area, both which have been shown to produce OCDF byproduct. Principal components analysis showed that Lake Ontario sediment profiles classified somewhat nearer to PCP than sediments from the other lakes, which classified closely with air particulates. Disposal of electrolytic sludge from the use of graphite electrodes for chlorine production at four sites along the US side of the Niagara River until about 1970 may also be a source. Rappe et al. [44] showed that pitch-impregnated graphite electrode sludges contained high concentrations of 2378-substituted TecDF, PnCDFs, HxCDFs, and OCDF. The only PCDD detected was OCDD. Excavation soil from a chlor-alkali plant employing graphite electrodes had a PCDF profile similar to the Lake Ontario sediments [45].

Czuczwa and Hites [20] also calculated fluxes of PCDD/Fs to Siskiwit Lake on Isle Royale, Lake Superior, 1920–1984, and for Lake Erie, 1950–1984. The Siskiwit profile showed a rapid increase in the PCDD/F flux in the 1950s and

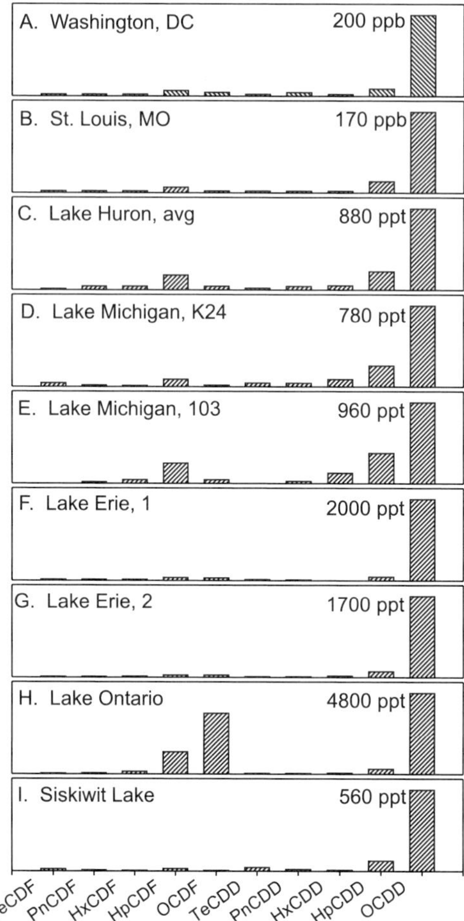



**Fig. 7** PCDD/F homolog profiles in urban air particulates and surface sediments sampled from mid-Lake Michigan (1982), lower Lake Huron (1975, 1981), eastern Lake Erie (1981, 1983), and western Lake Ontario (1983) near the Niagara River mouth. Reproduced with permission from Czuczwa and Hites [20]. Concentrations of OCDD are given in the *upper right hand corner* (ppt = ng kg$^{-1}$ dw)

1960s in line with the increased manufacture and use of chlorinated aromatic compounds. The flux of PCDD/Fs to Lake Erie maximized in the 1970s, and appeared to be decreasing in the early 1980s. This was attributed to reduction in emissions due to passing of the Clean Air Act in the USA in 1970.

Pearson et al. [19, 42] determined the accumulation of PCDD/F homologs in sediment cores from two control lakes near Lake Superior, Lake Superior, Lake Michigan, and Lake Ontario in 1994 (Fig. 8). There were no clear maxima for PCDDs accumulation rates in Lake Superior cores. However, there were maxima around 1960–1970 for PCDFs in the depositional zone of

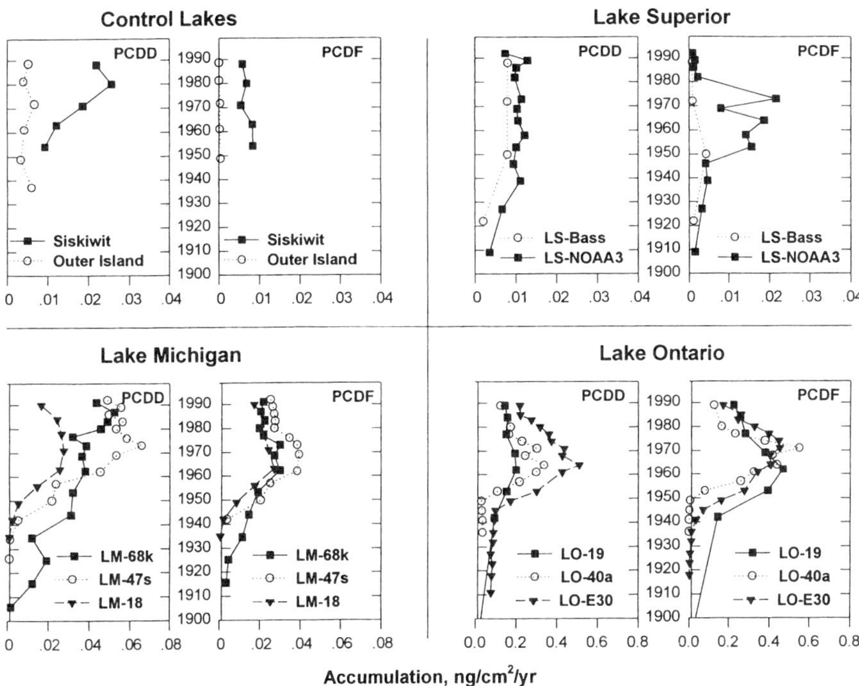


**Fig. 8** Accumulation rate of PCDD/F homologs (ng cm$^{-2}$ year$^{-1}$) in sediment cores from two control lakes near Lake Superior, Lake Superior, Lake Michigan, and Lake Ontario in 1994. Reproduced with permission from Pearson et al. [19, 42]

Lake Superior. Historical accumulation rates, homolog compositions (from PCA) and atmospheric deposition rates all indicated a primarily atmospheric source to Lake Superior and the control lakes, with an additional TeCDF and PnCDF contribution to Lake Superior in the 1950s and 1960s. The authors were unable to conclude confidently that paper mills were the source of elevated PCDF, although it was stated that the homolog composition was consistent with bleached kraft mill discharge. The PCDF source in Lake Superior decreased after 1980.

There was continued increase in PCDD accumulation rates in the two northern Lake Michigan cores in the 1960–1980 period, suggesting a nonatmospheric source during this period [42]. Subtracting estimates of atmospheric input, about two-thirds of the PCDD in northern Lake Michigan was thought to be non-atmospheric. Classification with potential sources by PCA indicated that pentachlorophenol, effluent from papers mills using recycled stock, and sewage effluent could have produced this PCDD signal. All three sources were present, but the authors were unable to conclude which of these sources might be responsible for the non-atmospheric loading. PCDDs in the southern basin and PCDFs in all Lake Michigan cores were more consistent with an atmospheric source. However, it was estimated that 65–95% of PCDF entered the southern portion of the lake from a regional source rich in TeCDF and PnCDF and was distributed around the lake. This is consistent with modeled atmospheric depositional flux to the lake, which was predicted to have higher dry deposition of TeCDFs and PnCDFs than other lakes (Fig. 6). Both PCDD and PCDF deposition rates peaked around 1960–1970, and decreased in the 1970s and 1980s.

In Lake Ontario, PCDF accumulation rates were similar in all three depositional basins, but PCDD accumulation was higher in the eastern sediments [42]. The latter was inconsistent with a Niagara River source, indicating a source of PCDD contamination in Eastern Lake Ontario. The accumulation rate of PCDDs and PCDFs was seven- to 14-fold higher in Lake Ontario than Lake Michigan. By comparison to Lake Michigan, which has a similar population density, it was concluded that $> 65$–95% of the loading of both PCDDs and $> 95$% of PCDFs to Lake Ontario was non-atmospheric, probably via the Niagara River. PCA analysis of homolog profiles including potential sources from the 1950s to the present showed Lake Ontario sediments to have a homolog profile between that of pentachlorophenol and electrolytic sludge, which is highly enriched in PCDFs [44, 45]. Sediment inventories to 1994 of $\Sigma$-PCDD/F were estimated to be $870 \pm 330$ kg in Lake Superior, $1700 \pm 710$ kg in Lake Michigan, and $5800 \pm 800$ kg in Lake Ontario (Pearson et al. 1997).

Marvin et al. [46] found the lakewide average concentration of PCDD/F-related TEQs in surficial sediments in Lake Ontario 1997–1998 to be 101 pg g$^{-1}$ dry weight (dw), five times higher than in Lake Erie, 18.8 pg g$^{-1}$dw. Average concentration of $\Sigma$-PCDD/Fs was 2.81 ng kg$^{-1}$dw (range 0.38 – 14.2 ng kg$^{-1}$). Despite the differences in average TEQ concentration between lakes, the number of sites exceeding the Canadian Probable Effect Level (PEL) of 21.5 ng kg$^{-1}$ TEQs was only about 1.5-fold lower in Lake Erie than in Lake Ontario, 40% versus 58%. The reason for this is that most of the PCDD/Fs in Lake Ontario are in the three depositional basins. TEQ concentrations in these areas exceeded 200 ng kg$^{-1}$. The authors conclude that industries along the Niagara River were the primary source of PCDD/F contamination, in agreement with other studies. The depth profile of TEQ changes in the Mississauga basin (Fig. 6) showed a peak of about 300 pg g$^{-1}$ TEQs between 1940 and 1970, with a steady decline until about 1980 and little change thereafter to 1998, although still elevated at 100 pg g$^{-1}$ TEQs. This finding is not in good agreement with Pearson et al. [19, 42], who found decreases of PCDD/F concentrations (presumably also TEQs derived from them) between 1980 and 1990 in Lake Ontario cores. The concentration versus time profile in Marvin et al. [46] is also much broader than that obtained by Pearson et al. [19, 42] for Lake Ontario, although the peak is approximately the same. It is not clear if the cores in Marvin et al. [46] were focus-corrected, which may be the reason for the disagreement. In a more detailed survey of PCDD/F concentrations in Lake Ontario sediment, Marvin et al. [47] reported an average concentration of PCDD-related TEQs of 111 ng kg$^{-1}$ dry weight, similar to previous findings [20, 46]. The authors note that the basin sediments were enriched in higher chlorinated PCDFs, 2378-TeCDF, and 123478-HxCDF, which was also noted by Czuczwa and Hites [20], and consistent with the pattern of contamination found in Niagara River water in 1986 [35].

## 2.4
## River and Bay Sediments

### 2.4.1
### Fox River/Green Bay Sediments

Smith et al. [48] surveyed concentrations of dioxin-like PCBs and 2378-TeCDD/F in sediment from Green Bay, Lake Michigan and various tributaries to the Great Lakes (1983–1984). Sediment concentrations of 2378-TeCDF were remarkably similar in sediments from Green Bay, the main tributaries to Green Bay (Menominee River and Fox River), Cuyahoga River, OH (southern Lake Erie), Raisin River, MI (western Lake Erie), and the Saginaw River, MI (Saginaw Bay, Lake Huron), ranging from 26 to 97 ng kg$^{-1}$. Sediments from Lake Pepin on the Mississippi River had much lower concentrations of 2378-TeCDF, < 1 ng kg$^{-1}$. Concentrations of 2378-TeCDD were much more variable, < 10 ng kg$^{-1}$ in all areas except sediments from the Fox River (14 ng kg$^{-1}$) and the Saginaw River (15 ng kg$^{-1}$). The Fox River is impacted by upstream paper and pulp mill operations. There has been little characterization of PCDD/F contamination of Green Bay and Lake Huron from the Fox River, because the main issue in this area was PCB contamination from processing recycled paper. However, the findings of elevated 2378-TeCDD and 2378-TeCDF concentrations in Green Bay sediment and the excess inventory of PCDD/Fs in northern Lake Michigan, which could not be explained by atmospheric deposition [19, 42], suggest there may be some PCDD/F input from the Fox River. There was a chlor-alkali plant on the Fox River system.

### 2.4.2
### Lake Superior Bay Sediments

PCDD/Fs were determined in sediments from a harbor near a wood-preserving plan in Thunder Bay, Lake Superior, 1988. The plant used pentachlorophenol [49]. TeCDDs and PnCDDs were below detection, 0.02–0.9 ng g$^{-1}$ dw. Most important PCDD/Fs were OCDD (< 0.4–980 ng g$^{-1}$), HpCDDs (< 0.4–320 ng g$^{-1}$), and OCDF (< 1–400 ng g$^{-1}$). Lesser concentrations of HxDDs, HxCDFs, and HpCDFs were frequently present (< 0.03–36 ng g$^{-1}$). The contamination was highly associated with oil/grease and PAH concentrations, suggesting a common source. There is no indication from Lake Superior sediment core records that contamination from this or similar wood-treatment plants contributed significantly to the PCDD/F profile seen in Lake Superior.

Sherman et al. [50] surveyed PCDD/F concentrations (1988) in Jackfish Bay, Lake Superior, which receives effluent from the only bleached kraft mill on the lake, and is the subject of a Great Lakes Remedial Action Plan (RAP). Whole effluent concentrations of TeCDFs ranged from 0.3 to 1.3 ng L$^{-1}$. Suspended solids contained TeCDFs and "traces" of TeCDD, PeCDF, OCDF, and OCDD. OCDD distribution was uniform in sediments from the western arm of Jackfish Bay at concentrations similar to those in sediments from Siskiwit Lake, Isle Royale [43], indicating that the source was atmospheric. There was a strong gradient of TeCDF (two isomers) concentrations from the mouth of Blackbird Creek, which carried the effluent, to the outer reaches of the western arm, in the order of 4 km. Sites in Jackfish Bay proper had much lower concentrations of TeCDFs. Core profiles showed an abrupt appearance of TeCDF between 1973 and 1975, indicating a change in the mill process stream. A change from "cold" to "hot" bleaching and/or the use of oil-based defoamers containing dibenzofuran was apparently the cause. Pearson et al. [19, 42] suggested the presence of a non-atmospheric source of TeCDF and PnCDF to Lake Superior in the 1950s and 1960s. Based on sediment core analysis, it appears unlikely that the Jackfish Bay pulp mill was responsible for significant lakewide contamination by PCDFs during this period, but the possibility still remains, e.g., through disposal of graphite electrode wastes.

### 2.4.3
### Saginaw River/Saginaw Bay Sediments

The Saginaw River, which flows into the head of Saginaw Bay, Lake Huron, is the tributary garnering the most interest as a source of PCDD/F contamination to the Great Lakes. Saginaw River and Bay have been designated as an Area of Concern (AOC) by US EPA. There are 87 industrial facilities and 127 wastewater treatment plants in the watershed. However it is the main tributary of the Saginaw River, the Tittabawassee River, which is of primary interest as a source of PCDD/Fs [34]. The Dow Chemical plant at Midland, MI, manufactured a wide variety of organochlorine chemicals that may have produced PCDD/Fs as a byproduct, including mustard gas, Agent Orange (2,4-D/2,4,5-T mixture), 2,4,5-trichlorophenol, 2,4-D, chlorpyrifos, vinyl chloride monomer, and ethylene dichloride.

PCDD/Fs concentrations in Saginaw River sediments upstream and downstream of the Tittabawassee River confluence sampled in the early to mid-1990s (sampling dates were not given, but presumably within the 5 years prior to publication) were determined by Gale et al. [33]. Concentrations of the PCDD/F congeners fully substituted by chlorine at the 2378-positions in sediments are presented in Table 2. These will be referred to as AHR congeners, since toxicity occurs as a result of (or at least is correlated to) binding to the

Aryl Hydrocarbon Receptor, and is the basis of the 2378-TeCDD TEQ additive toxicity model. Concentrations of AHR PCDD/Fs increased an order of magnitude downstream from the confluence of the Tittabawassee River, and increased a further order of magnitude in the lower reaches of the river, presumably because these were depositional zones.

Gale et al. [33] is the only study of sediments in the Great Lakes which allows comparison of the relative contribution of AHR and non-toxic congeners to the homolog patterns. From Table 2, about 85% of $\Sigma$-PCDDs downstream of the Tittabawassee River confluence were AHR congeners. This was a result of the overwhelming contribution of OCDD and HpCDDs. The situation was very different for the lower chlorinated homologs. Non-AHR congeners were found to make up the bulk of TeCDDs, PnCDDs, and HxCDDs, although 2378-TeCDD was about 8% of TeCDDs, which is a higher percentage than expected from atmospheric sources. About half of total TeCDFs were 2378-TeCDF, and half of PnCDFs were 12378-PnCDF and 23478-PnCDF. About one-third of

**Table 2** Concentrations of PCDD/Fs in sediments (ng kg$^{-1}$ dw) at various sites along the Saginaw River in the mid-1990s

|  | Shiawassee River | Tittabawassee R. confluence | Zilwaukee Bridge | Middleground Island | Bay City |
| --- | --- | --- | --- | --- | --- |
| 2378-TeCDD | nd | 1 | nd | 8 | 11 |
| 12378-PnCDD | nd | nd | nd | 5 | 8 |
| 123478-HxCDD | nd | nd | nd | 3 | 5 |
| 123678-HxCDD | nd | 2 | nd | 14 | 27 |
| 123789-HxCDD | nd | 1 | nd | 8 | 14 |
| 1234678-HpCDD | 3 | 33 | 9.7 | 331 | 470 |
| OCDD | 25 | 281 | 82 | 4840 | 3690 |
| $\Sigma$-AHR-PCDDs | 28 | 318 | 92 | 5209 | 4225 |
| $\Sigma$-PCDDs | 37 | 372 | 111 | 5720 | 5000 |
| 2378-TeCDF | 4 | 55 | 17 | 221 | 453 |
| 12378-PnCDF | 1 | 23 | 7 | 77 | 237 |
| 23478-PnCDF | nd | 15 | 6 | 58 | 160 |
| 123478-HxCDF | nd | 26 | 6 | 96 | 233 |
| 123678-HxCDF | nd | 6 | nd | 27 | 61 |
| 234678-HxCDF | nd | 3 | nd | 12 | 25 |
| 1234678-HpCDF | 5 | 62 | 19 | 541 | 743 |
| 1234789-HpCDF | nd | 5 | nd | 25 | 40 |
| OCDF | 10 | 90 | 28 | 909 | 1180 |
| $\Sigma$-AHR-PCDFs | 20 | 285 | 83 | 1966 | 3132 |
| **$\Sigma$-PCDFs** | **35** | **514** | **161** | **3450** | **4100** |

Data are from Gale et al. [33]. The Shiawassee River, Tittabawassee River, and the Cass River join to form the Saginaw River, which flows with only minor tributary creeks to Saginaw Bay, Lake Huron at Bay City. The major chlorine chemical industry is in Midland MI, on the Tittabawassee River

HxCDFs were AHR congeners, primarily 123478-HxCDF. AHR PCDFs made up 52–57% of Σ-PCDFs at the first four sites, 76% at Bay City.

Hilscherova et al. [34] surveyed PCDD/F concentrations in sediments and flood plain soils along the Tittabawasee River. Downstream of the Dow Chemical plant at Midland, MI, PCDD/F concentrations in sediments were one to two orders of magnitude higher than upstream, and similar to those in the lower reaches of the Saginaw River [33]. Tittabawassee River composite sediment PCDFs were composed primarily of 2378-TeCDF > 12378-PnCDF > 23478-PnCDF > 123478-HxCDF $\approx$ 1234678-HpCDF $\approx$ OCDF. PCDDs were primarily OCDD. Flood plain soils downstream of Midland had even higher concentrations – an astonishing 14.8 $\mu$g kg$^{-1}$ of Σ-PCDDs and 10.6 $\mu$g kg$^{-1}$ of Σ-PCDFs, compared to 0.32 $\mu$g kg$^{-1}$ Σ-PCDDs and 0.05 $\mu$g kg$^{-1}$ Σ-PCDFs in flood plain soils upstream of Midland.

## 2.4.4
### Detroit River Sediments

The Detroit River drains all of the upper lakes into Lake Erie. Upstream from Detroit, there is a major chemical industry complex on the Canadian side of the St. Clair River, which drains Lake Huron into Lake St. Clair. The Detroit/Windsor urban area and major industrial complexes associated with automobile manufacturing may all be potential PCDD/F sources.

Concentrations of PCDD/Fs, along with other OCs, were measured in suspended sediments at nine sites along the Detroit River in 1999–2000 [51]. Primarily PCDFs were detected. Concentrations of 2378-TeCDF ranged from ND to 25 ng kg$^{-1}$ dw, median 1.3 ng kg$^{-1}$. One site along the Trenton Channel, Monguagon Creek, west of Grosse Ile near the mouth of the river, had the highest concentrations of all PCDD/Fs. The contamination profile at this site was dominated by 2378-TeCDF (25 ng kg$^{-1}$), 12378-PnCDF (19 ng kg$^{-1}$), 123478-HxCDF (110 ng kg$^{-1}$), 23478-PnCDF (65 ng kg$^{-1}$), and 123678-HxCDF (26 ng kg$^{-1}$). TEQ concentrations at this site were about six times lower in 1999 than in 2000, indicating that the contaminant loading was highly variable annually, possibly seasonally. PCDD/Fs contributed the bulk of the TEQs at this site in 2000, but PCBs were also an important contributor to TEQs in 1999. At two sites downstream from Monguagan Creek the proportion of 23478-PnCDF in Σ-PCDFs increased. PCBs were also highest at these two sites, and may have been the source of 23478-PnCDF. Monguagon Creek was historically contaminated by steel and chlor-alkali industries. It was concluded that chlor-alkali effluents were the main source of high PCDF contamination. This a reasonable conclusion, given the preponderance of 1234-substituted AHR congeners, a pattern very similar to that found in the Tittabawassee River. At all stations, CB126 was > 50% of PCB TEQs, average 73%. PCNs were also high at the Trenton Channel sites, and may contribute to TEQs.

## 2.4.5
## Niagara River Sediments

The Niagara River has one of the largest concentrations of chlorine chemical industries in North America. It was attractive historically because of the availability of inexpensive electricity for making chlorine by electrolysis of brine, and the abundance of process water. Four chlorine-producing facilities in the Niagara River area were identified by Mumma and Lawless [52]. EPA has developed fact sheets on five sites which may have contributed PCDD/Fs contamination to the Niagara River. These are discussed in Sect. 4.4 on sources in the Niagara River.

PCDD/F concentrations in sediments along the Niagara River in 2000 were determined by Richman [38] as part of the Niagara River Mussel Biomonitoring Program conducted in support of the Niagara River Toxic Management Plan. Results are given in Table 3. Sediments in the Chippawa Channel west of Grand Island and at Fort Erie at the head of the Niagara River had low PCDD/F contamination (12–165 ng g$^{-1}$ dw) with a homolog pattern simi-

**Table 3** Concentrations (ng kg$^{-1}$ dw) of PCDD/Fs in sediments near various sources of contamination along the US side of the Niagara River in 2000

| | Two Mile Creek | Pettit Flume up-stream | at | down-stream | Gill Creek | Bloody Run Creek up-stream | down-stream |
|---|---|---|---|---|---|---|---|
| 2378-TeCDD | 3 | 1 | 640 | nd | 4 | 32 | 2800 |
| 12378-PnCDD | 5 | nd | 1200 | 17 | 6 | nd | 59 |
| 123478-HxCDD | 4 | nd | 1200 | 19 | 13 | nd | 190 |
| 123678-HxCDD | 18 | nd | 2700 | 36 | 69 | 11 | 2400 |
| 123789-HxCDD | 12 | 4 | 2000 | 25 | 39 | 6 | 1400 |
| 1234678-HpCDD | 22 | 92 | 15000 | 250 | 1100 | 41 | 11000 |
| OCDD | 2300 | 740 | 33000 | 1100 | 12000 | 52 | 8500 |
| Σ-PCDD | 2364 | 837 | 55740 | 1447 | 13231 | 142 | 26349 |
| 2378-TeCDF | 5 | 12 | 8400 | 100 | 10 | 7 | 95 |
| 12378-PnCDF | 6 | nd | 3700 | 47 | 15 | nd | 47 |
| 23478-PnCDF | 13 | 7 | 10000 | 160 | 17 | 6 | 180 |
| 123478-HxCDF | 41 | 48 | 140000 | 2500 | 300 | 54 | 1500 |
| 123678-HxCDF | 14 | 11 | 22000 | 380 | 48 | 9 | 280 |
| 234678-HxCDF | 10 | 5 | 61000 | 91 | 7 | 3 | 61 |
| 1234678-HpCDF | 140 | 180 | 450000 | 7800 | 420 | 110 | 910 |
| 1234789-HpCDF | 12 | 5 | 15000 | 270 | 150 | 5 | 290 |
| OCDF | 300 | 250 | 1100000 | 14000 | 1600 | 140 | 3800 |
| Σ-PCDF | 561 | 518 | 1810100 | 25348 | 2597 | 334 | 7163 |

Data are from Richman et al. [38]. Two Mile Creek and Pettit Flume are 10–15 km from the Lake Erie inlet of the Niagara River; Gill Creek is 3 km upstream of Niagara Falls; Bloody Run Creek is 10 km downstream of the falls, 13 km from the mouth in Lake Ontario

lar to sediments having a primarily atmospheric loading, i.e., OCDD and HpCDDs were dominant. TEQs were 0.2–2.4 ng kg$^{-1}$ dw. These sites are not influenced by effluent from chlorinated organic chemical industries. Sediments at all sites along the Niagara River on the New York side were much more highly contaminated with PCDD/Fs. Concentrations were highly variable. Pettit Flume and Bloody Run Creek were the most highly contaminated. Pettit Flume sediments contained 12-fold higher concentrations of PCDFs than PCDDs. Among the toxic congeners, there was a dominance of 123478-HxCDF and 1234678-HpCDF. This pattern was also found in Gill Creek sediments. It is interesting to note that concentrations and patterns of PCDD/F contamination in sediments downstream from Pettit Flume were nearly identical to those in the raw water pump station sediment from the Lockport, NY water intake 12 years earlier in 1988 [35]. The intake is in the Niagara River at North Tonawanda, also downstream from Petitt Flume.

The fractional contribution of the major PCDD and PCDF congeners to their respective total concentrations in Niagara River sediments in Table 3 is shown in Fig. 9. Despite large variations in concentration among sites, the PCDF pattern is remarkably similar along the river. The major congeners are OCDF,

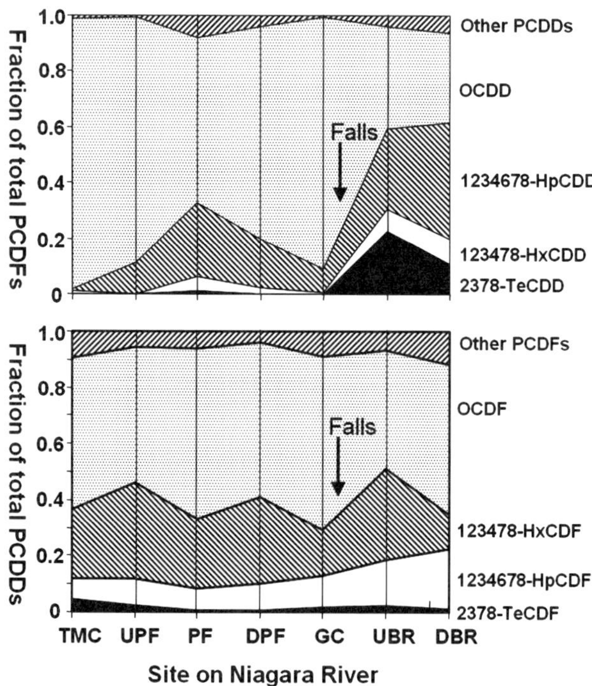



**Fig. 9** Fractional contribution of major PCDD and PCDF congeners to Σ-PCDDs and Σ-PCDFs concentrations in Niagara River sediments in 2000 (Table 3). Adapted from Richman [38]

1234678-HpCDF, and 123478-HxCDF. The proportion of 123478-HxCDF increases at the expense of 1234678-HpCDF downstream from Niagara Falls. The PCDD pattern is much more variable. At Two-Mile Creek, virtually all PCDD contamination was OCDD. At the next four sites, the proportion of 1234678-HpCDD increased, especially at Pettit Flume. Downstream from Niagara Falls, the proportion of 2378-TeCDD jumped to 11–23% of $\Sigma$-PCDDs.

Pearson et al. [42] noted that 1234678-HpCDF constituted a larger proportion of the HpCDF isomers in Lake Ontario sediments than sediments from other Great Lakes, probably originating from dumpsites along the Niagara River. Rappe et al. [44] found high concentrations of 123478-HxCDF and a high concentration of HpCDFs in electrolytic sludge and sediment near a pulp mill in Sweden. Similarly, Kannan et al. [45] found concentrations of PCDD/Fs in soil around a chlor-alkali plant to consist mainly of OCDF > 1234678-HpCDF > 123478-HxCDF $\approx$ OCDD. In marsh and creek sediments, the PCDF contribution decreased relative to PCDDs. Thus, it seems likely that disposal of electrolytic sludge (taffy tar) from graphite electrodes used to produce chlorine from electrolysis of sodium chloride is the source of the highly chlorinated PCDF contamination unique to Lake Ontario.

The high concentrations of 2378-TeCDD found in Bloody Run Creek sediments are a result of runoff from the Hyde Park hazardous waste landfill, which is only 600 m from the Niagara River gorge. Storm sewer effluents from the Love Canal area were also a potential source historically to the Niagara River. Smith et al. [53] found a concentration of 312 $\mu$g kg$^{-1}$ of 2378-TeCDD in a storm sewer sediment at the southern end of the canal a few blocks from the sewer outlet to the Niagara River that had 31 $\mu$g kg$^{-1}$ of 2378-TeCDD in sediments. See Sect. 4.4 for a detailed discussion of sources in the Niagara River.

## 2.5
## Fish

Fish have been in shown in numerous investigations to accumulate primarily AHR congeners of PCDD/Fs, while much of the early analysis of sediments was based on homologs. These sediment data suggested that atmospheric input was the most important source to the Great Lakes proper. Because AHR congeners constitute a relatively small proportion of each PCDD/F homolog from combustion sources, but may be a relatively high proportion in some chemical industry processes (e.g., PCDFs from graphite electrode production of chlorine, 2378-TeCDD from 2,4,5-trichlorophenol production), it is not immediately obvious from much of the sediment data how to factor the relative importance of sources of the PCDD/Fs accumulating in the food web. However, atmospheric sources are not substantially different among lakes, the possible exception being the more particulate-borne TeCDFs and PnCDFs near large urban areas, while local sources to each of the lakes are very different, i.e., the Fox River to Lake Michigan, Saginaw River to Saginaw Bay and Lake Huron, and the Niagara River to Lake Ontario. As will be demonstrated below, there is good evidence that PCDD/F concentrations and profiles in fish are a good reflection of the sediment contamination in each lake, modified by each congener's bioavailability and propensity to bioaccumulate.

### 2.5.1
### Surveys of 2378-TeCDD and 2378-TeCDF

The first report of "dioxin" contamination in the Great Lakes area was in fish from the Tittabawassee River in 1978 [1]. These findings were confirmed by Harless and Lewis [2]. They found that 26 of 35 samples of fish from the Tittabawassee/Saginaw Rivers contained detectable concentrations of 2378-TeCDD from 4 to 690 ng kg$^{-1}$. The highest concentrations were in catfish and carp. In 1982, a more detailed report was given [3]. Mean 2378-TeCDD concentrations in 1978 were: channel catfish 157 ng kg$^{-1}$, carp 55 ng kg$^{-1}$, yellow perch 14 ng kg$^{-1}$, small mouth bass 8 ng kg$^{-1}$, sucker 11 ng kg$^{-1}$, and lake trout, $< 5$ ng kg$^{-1}$. These results sparked a flurry of activity throughout the 1980s to document 2378-TeCDD concentrations in fish throughout the Great Lakes. A summary of the results from these early studies is presented in Table 4. Occasionally, 2378-TeCDF was also determined, but most laboratories did not have the capability to determine other PCDD/F congeners. Concentrations of 2378-TeCDD were $\leq 10$ ng kg$^{-1}$ in fish from Lake Superior, Lake Michigan, and Lake Erie, with the exception of one rainbow trout sample from Lake Superior, and one coho salmon sample from Lake Michigan. Fish from the Tittabawassee/Saginaw River system, Saginaw Bay, Lake Huron, Niagara River, and Lake Ontario had significantly higher concentrations of 2378-TeCDD, mostly in the 10–100 ng kg$^{-1}$ range. When measured, 2378-TeCDF concentrations were in the same range in all lakes, even in Lake Superior.

Besides the studies summarized in Table 4, Fehringer et al. [58] surveyed 2378-TeCDD concentrations in Great Lake fish and Michigan rivers. Dates were not given, but presumably in the early 1980s, most fish had concentrations below the detection limit, which was also not specified. Fish from Saginaw Bay, Lake Huron, and the Tittabawassee River were most frequently found to contain 2378-TeCDD, 15–102 ng kg$^{-1}$ in about one-third of the carp and catfish sampled. Concentrations in Lake Ontario were: brown trout 9 ng kg$^{-1}$, rainbow trout 21 ng kg$^{-1}$, white perch 25 ng kg$^{-1}$, lake trout 46 ng kg$^{-1}$, and coho salmon 35 ng kg$^{-1}$. Kuehl et al. [59] surveyed 2378-TeCDD concentrations in fish in major watersheds throughout the USA. Dates were not given, but early 1980s can be assumed. The fish are identified only as predators (trout and salmon, most likely), or bottom feeders (sucker and carp, most likely). The majority of Great Lakes samples (79%) had detectable concentrations of 2378-TeCDD. Highest concentrations of TeCDD were found

**Table 4** Early reports of concentrations (ng kg$^{-1}$ whole fish or fillet) of 2378-TeCDD and 2378-TeCDF in fish from the Great lakes

| Species | Area | Lake | Year | Ref. | 2378-TeCDD | 2378-TeCDF |
|---|---|---|---|---|---|---|
| Lake trout | | Superior | 1990 | [62] | 3 | 21 |
| Lake trout | Apostle Island | Superior | 1981 | [60] | < 1 | 19 |
| Lake trout | Siskiwit Island | Superior | 1981 | [60] | NA | 10 |
| Lake trout | Whitefish Point | Superior | 1987–88 | [66] | 4 | 11 |
| Lake trout | Marquette | Superior | 1987–88 | [66] | 3 | 14 |
| Lake trout | Grand Marie | Superior | 1987–88 | [66] | < 3 | 4 |
| Coho salmon | Whitefish Point | Superior | 1987–88 | [66] | 2 | 6 |
| Coho salmon | Marquette | Superior | 1987–88 | [66] | 3 | 10 |
| Coho salmon | Grand Marie | Superior | 1987–88 | [66] | 1 | 2 |
| Chinook salmon | Whitefish Point | Superior | 1987–88 | [66] | < 1 | 3 |
| Chinook salmon | Marquette | Superior | 1987–88 | [66] | 5 | 18 |
| Chinook salmon | Grand Marie | Superior | 1987–88 | [66] | < 1 | 3 |
| Rainbow trout | | Superior | 1978–80 | [54] | 8/80 | NA |
| Bloater chub | Keweenaw | Superior | 1981 | [60] | < 1 | 26 |
| Forage fish | | Superior | 1990 | [62] | < 2 | 3 |
| Lake trout | Saugatuk | Michigan | 1981 | [60] | 5 | 35 |
| Lake trout | Saugatuk | Michigan | 1979 | [48] | 5 | 35 |
| Lake trout | Charlevoix | Michigan | 1987–88 | [66] | 9 | 6 |
| Lake trout | Ludington | Michigan | 1987–88 | [66] | 5 | 9 |
| Lake trout | Muskegon | Michigan | 1987–88 | [66] | 10 | 13 |
| Coho salmon | Charlevoix | Michigan | 1987–88 | [66] | 7 | 19 |
| Coho salmon | Ludington | Michigan | 1987–88 | [66] | 7 | 19 |
| Coho salmon | Muskegon | Michigan | 1987–88 | [66] | 8 | 11 |
| Chinook salmon | Charlevoix | Michigan | 1987–88 | [66] | 8 | 6 |
| Chinook salmon | Ludington | Michigan | 1987–88 | [66] | 2 | 30 |
| Chinook salmon | Muskegon | Michigan | 1987–88 | [66] | 15 | 2 |
| Chinook salmon | | Michigan | 1983 | [48] | < 4–3 | 11 ~ 13 |
| Coho salmon | | Michigan | 1978–80 | [54] | 9/79 | NA |
| Carp | Waukegan Harbor | Michigan | 1980 | [48] | < 1 | 29 |
| Largemouth bass | Waukegan Harbor | Michigan | 1980 | [48] | 2 | 68 |
| Carp | Saginaw R. | Huron | 1983 | [48] | 13 | 18 |
| Carp | Saginaw R. | Huron | 1981 | [55] | 31 | NA |
| Sucker | Saginaw R. | Huron | 1981 | [55] | < 2 | NA |
| Northern pike | Saginaw R. | Huron | 1981 | [55] | 4 | NA |
| Carp | Tittabawassee R. | Huron | 1981 | [55] | 71 | NA |
| Sucker | Tittabawassee R. | Huron | 1981 | [55] | 7 | NA |
| Channel catfish | Tittab./Saginaw R. | Huron | 1978 | [3] | 157 | NA |
| Carp | Tittab./Saginaw R. | Huron | 1978 | [3] | 55 | NA |
| Yellow perch | Tittab./Saginaw R. | Huron | 1978 | [3] | 14 | NA |
| Smallmouth bass | Tittab./Saginaw R. | Huron | 1978 | [3] | 8 | NA |
| Sucker | Tittab./Saginaw R. | Huron | 1978 | [3] | 11 | NA |

**Table 4** (continued)

| Species | Area | Lake | Year | Ref. | 2378-TeCDD | 2378-TeCDF |
|---|---|---|---|---|---|---|
| Lake trout | Tittab./Saginaw R. | Huron | 1978 | [3] | < 5 | NA |
| Lake trout | Saginaw Bay | Huron | 1987–88 | [66] | 23 | 9 |
| Coho salmon | Saginaw Bay | Huron | 1987–88 | [66] | 37 | 19 |
| Chinook salmon | Saginaw Bay | Huron | 1987–88 | [66] | 56 | 19 |
| Carp | Saginaw Bay | Huron | 1981 | [55] | 50 | NA |
| Channel catfish | Saginaw Bay | Huron | 1981 | [55] | 61 | NA |
| YOY walleye | Saginaw R./Bay | Huron | 1990 | [56] | 6 | 37 |
| Yearling walleye | Saginaw R./Bay | Huron | 1990 | [56] | 3 | 22 |
| Small walleye | Saginaw R./Bay | Huron | 1990 | [56] | 4 | 37 |
| Medium walleye | Saginaw R./Bay | Huron | 1990 | [56] | 5 | 73 |
| Large walleye | Saginaw R./Bay | Huron | 1990 | [56] | 15 | 93 |
| Large alewife | Saginaw R./Bay | Huron | 1990 | [56] | 2 | 9 |
| YOY gizzard shad | Saginaw R./Bay | Huron | 1990 | [56] | 4 | 82 |
| Yellow perch | Saginaw R./Bay | Huron | 1990 | [56] | 3 | 29 |
| Carp | Saginaw R./Bay | Huron | 1990 | [56] | 21 | 39 |
| Carp | Bay Port | Huron | 1979 | [48] | 27 | 5 |
| Lake trout | | Huron | 1978–80 | [54] | 21 | NA |
| Lake trout | Cheboygan | Huron | 1987–88 | [66] | 2.3 | 10 |
| Lake trout | Port Huron | Huron | 1987–88 | [66] | 13 | 4 |
| Coho salmon | Cheboygan | Huron | 1987–88 | [66] | 1 | 10 |
| Coho salmon | Port Huron | Huron | 1987–88 | [66] | 7 | 18 |
| Chinook salmon | Cheboygan | Huron | 1987–88 | [66] | 2 | 4 |
| Chinook salmon | Port Huron | Huron | 1987–88 | [66] | 10 | 15 |
| Carp | | Huron | 1978–81 | [54] | 26 | NA |
| Channel catfish | | Huron | 1978–82 | [54] | 20 | NA |
| Sucker | | Huron | 1978–83 | [54] | 3 | NA |
| Yellow perch | | Huron | 1978–84 | [54] | < 1 | NA |
| Walleye | Cedar Point | Erie | 1981 | [60] | | 18 |
| Carp | Port Clinton | Erie | 1981 | [60] | < 1 | 5 |
| Coho salmon | | Erie | 1978–80 | [54] | 1 | NA |
| Walleye | | Erie | 1978–80 | [54] | 3 | NA |
| Smallmouth bass | | Erie | 1978–80 | [54] | 2 | NA |
| Carp | | Erie | 1978–80 | [54] | < 2 | NA |
| Carp | Cayuga Creek | Niagara R. | 1978–80 | [54] | 87 | NA |
| Northern pike | Cayuga Creek | Niagara R. | 1978–80 | [54] | 32 | NA |
| Pumpkinseed | Cayuga Creek | Niagara R. | 1978–80 | [54] | 31 | NA |
| Rock bass | Cayuga Creek | Niagara R. | 1978–80 | [54] | 12 | NA |
| Lake trout | | Ontario | 1980s? | [57] | 120 | NA |
| Lake trout | W | Ontario | 1990 | [62] | 44 | 72 |
| Lake trout | | Ontario | 1978–80 | [54] | 51–107 | NA |
| Lake trout | Oswego | Ontario | 1981 | [60] | NA | 34 |
| Coho salmon | | Ontario | 1978–80 | [54] | 20–26 | NA |
| Rainbow trout | | Ontario | 1979–80 | [54] | 17–32 | NA |
| Brook trout | Roosevelt B. | Ontario | 1981 | [60] | 33 | 19 |

**Table 4** (continued)

| Species | Area | Lake | Year | Ref. | 2378-TeCDD | 2378-TeCDF |
|---|---|---|---|---|---|---|
| Brown trout | | Ontario | 1980s? | [57] | 10 | NA |
| Yellow perch | | Ontario | 1980s? | [57] | < 1 | NA |
| White perch | | Ontario | 1980s? | [57] | 70–150 | NA |
| White perch | | Ontario | 1978–79 | [54] | 17–26 | NA |
| Smallmouth bass | | Ontario | 1980s? | [57] | 2 | NA |
| Smallmouth bass | | Ontario | 1979 | [54] | 6 | NA |
| White sucker | | Ontario | 1979 | [54] | < 3 | NA |
| Brown bullhead | | Ontario | 1979 | [54] | 4 | NA |
| Alewife | W | Ontario | 1985 | [63] | 4 | 2 |
| "Forage fish" | E | Ontario | 1990 | [62] | 10 | 30 |

in Lake Ontario predator fish and Saginaw Bay, Lake Huron bottom-feeding fish: 13–41 ng kg$^{-1}$ ww. The highest concentration was in a predator fish from the Oswego area of Lake Ontario. This was the most contaminated sample out of 395 samples that were analyzed nationwide.

The first attempt at determination of more highly chlorinated PCDFs, PCDDs, and PCDDs in Great Lakes fish was by Stalling et al. [60]. Most of the analyses were not isomer-specific. However, the pattern of contamination in each homolog has been well-established in subsequent studies, and inferences can be made that they were AHR congeners. Apart from 2378-TeCDD, lake trout from Lake Superior, Lake Siskiwit (Isle Royale, Lake Superior), and Lake Michigan had no detectable PCDDs at detection limits in the order of 1–5 ng kg$^{-1}$. Carp from the Saginaw Bay and Tittabawassee River had measurable $\Sigma$-PCDDs of 11–385 ng kg$^{-1}$ and $\Sigma$-PCDFs 29–290 ng kg$^{-1}$ covering the whole range TeCDD/F-OCDD/F. Fish from all areas, including Siskiwit Lake, had detectable concentrations of PCDFs, primarily TeCDF (likely all 2378-TeCDF) in the range 5–34 ng kg$^{-1}$. Lake Michigan and Lake Ontario lake trout had high PnCDF (28–61 ng kg$^{-1}$) and HxCDF (8–29 ng kg$^{-1}$) concentrations. No samples from Lake Ontario were analyzed for PCDDs, hence the relatively high concentration of 2378-TeCDD in this lake was not found. Stalling et al. [61] reported further PCDD/F concentrations in fish around the Great Lakes, but did not identify the sites or species specifically. Saginaw Bay, Lake Huron, and Lake Ontario were again recognized as primary areas of concern. Concentrations were particularly high in a fish sample from the Niagara River area: 2378-TeCDD 327 ng kg$^{-1}$, other PCDDs 417 ng kg$^{-1}$, and PnCDFs-OCDF 1015 ng kg$^{-1}$. Concentrations of 2378-TeCDF in this sample were low.

There are practically no reports of concentrations in invertebrates. Sherman et al. [50] presented results from one analysis of the opossum shrimp

*Mysis relicta* from Jackfish Bay, Lake Superior, 1988: 9 ng kg$^{-1}$ 2378-TeCDD, 48 ng kg$^{-1}$ of TeCDFs (probably mostly 2378-), and 16 ng kg$^{-1}$ of PnCDFs. Jackfish Bay received effluents from a bleached kraft mill, and these residues are consistent with this source. Whittle et al. [62] found *Pontoporeia* (presumably *hoyi*, a macrobenthic amphipod) from Lake Ontario in 1990 to have about 10 ng kg$^{-1}$ of 2378-TeCDD. Surface zooplankton had 2–3 ng kg$^{-1}$ TeCDD and about 18 ng kg$^{-1}$ of OCDD in both Lake Ontario and Lake Superior. These concentrations seem higher than would be predicted from concentrations in fish, assuming that they are on a fresh, and not lipid weight basis. However, there are no comparative data.

There are also very few reports of PCDD/F concentrations in smaller fish species which dominate the biomass in the Great Lakes, primarily alewife and rainbow smelt. These species are the principal prey of the larger predator fish and seabirds in most areas of the Great Lakes. Braune and Norstrom [63] analyzed alewife caught in Lake Ontario in 1985. Concentrations of 2378-TeCDD were the highest (4 ng kg$^{-1}$), followed by 2378-TeCDF (2 ng kg$^{-1}$), 23478-PnCDF (2 ng kg$^{-1}$), 12378-PnCDD (1 ng kg$^{-1}$), and 123678-HxCDD (1 ng kg$^{-1}$). These data are similar to those found in a more detailed investigation of concentrations of PCDD/Fs in various age classes of alewife in 1992 as part of a bioaccumulation study (Norstrom, unpublished data, 2004) in line with the lack of a trend in PCDD/F concentrations in Lake Ontario biota during this period. Whittle et al. [62] detected only 2378-TeCDF (3 ng kg$^{-1}$) in forage fish (no species given) from Lake Superior in 1990 (Table 4). However, Lake Ontario forage fish contained a broader range of detectable PCDD/Fs: 2378-TeCDD (10 ng kg$^{-1}$), 2378-TeCDF (30 ng kg$^{-1}$), 12378-PnCDF (6 ng kg$^{-1}$) 23478-PnCDF (11 ng kg$^{-1}$), and 1234678-HxCDF (8 ng kg$^{-1}$). These concentrations are significantly higher than those in 1985 and 1992 alewife from Lake Ontario. However, it is likely the fish in the Whittle et al. [62] study were smelt, since this laboratory routinely analyzes smelt as part of their Great Lakes Monitoring program. Smelt, especially the larger fish, generally have higher concentrations of organochlorine compounds than alewife (Norstrom, unpublished data, 2004).

From these early inquiries, it could be concluded that there were broad low-level source(s) of 2378-TeCDD, 2378-TeCDF, and possibly higher chlorinated PCDFs, in all of the Great Lakes. Lake Siskiwit lake trout had measurable 2378-TeCDF. Since this lake is impacted only by atmospheric deposition, this result was the first indication that the atmosphere was likely to be an important route of PCDF loading in the Great Lakes. These studies also showed that two of the Great Lakes were the most affected by 2378-TeCDD contamination, Lake Huron and Lake Ontario. Lake Huron was thought to be contaminated by the Saginaw River flowing into Saginaw Bay. The relatively low contamination in Lake Erie fish, and high contamination in Niagara River fish pointed to the Niagara River as the main source to Lake Ontario.

## 2.5.2
## Comprehensive Surveys

De Vault et al. [64] carried out the first comprehensive, isomer-specific determination of PCDD/Fs in predator fish (lake trout and walleye, 1984) which had good geographical coverage among the Great Lakes, and compared the same or similar species in each area. PCDD/F concentrations are given in Table 5. Lake Superior and Lake Erie had the lowest total concentrations of both Σ-PCDDs and Σ-PCDFs, and Lake Ontario had by far the highest concentration of Σ-PCDDs, primarily due to 2378-TeCDD. The overall ranking of concentrations in predator fish for PCDDs was Lake Ontario > Lake Huron ≈ Lake Michigan ≈ Lake St. Clair > Lake Erie ≈ Lake Superior. For PCDFs the ranking was Lake Ontario ≈ Lake Michigan ≈ Lake Huron > Lake St. Clair > Lake Erie > Lake Superior. The fractional contributions of the various congeners to Σ-PCDDs and Σ-PCDFs for the data in Table 5 are shown in Fig. 10. The pattern of PCDFs is quite uniform in all lakes except Lake Ontario, which showed a much larger contribution of 123478-HxCDF (17%)

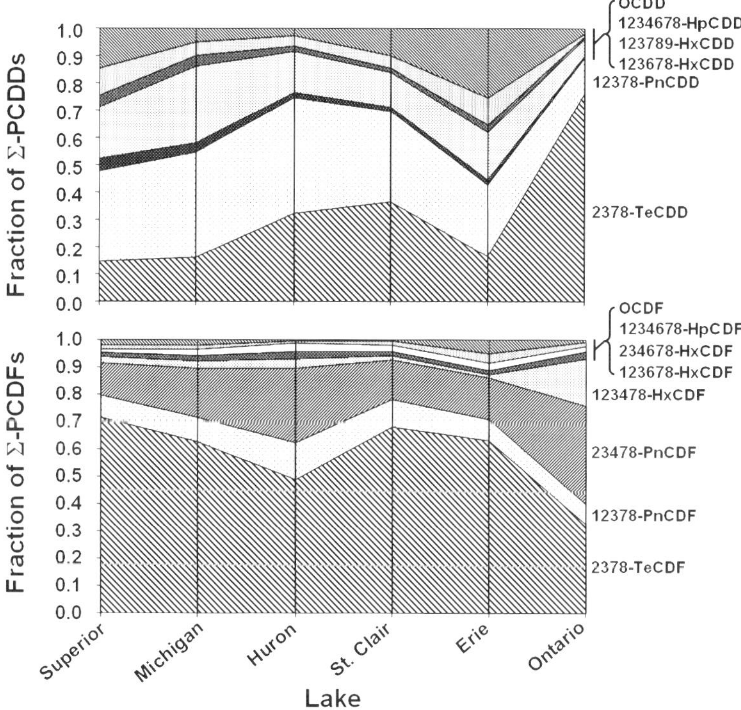


**Fig. 10** Fractional contribution of AHR congeners to Σ-PCDDs and Σ-PCDFs concentrations in lake trout from the Great Lakes in 1984 (Table 5). Adapted from De Vault et al. [64]

**Table 5** Mean concentrations (ng kg$^{-1}$ fresh weight) of PCDD/Fs in lake trout muscle from the Great Lakes; ratio of herring gull egg/lake trout concentrations in Lake Ontario

| | Lake trout | | | Walleye | | Lake trout | Lake trout Lake Ontario, 1977–1993[b] | | | | Gull egg/trout[c] | |
|---|---|---|---|---|---|---|---|---|---|---|---|---|
| | Superior[a] | Michigan[a] | Huron[a] | St. Clair[a] | Erie[a] | Ontario[a] | Mean | SE | Min | Max | Mean | SD |
| 2378-TeCDD | 1.0 | 3.5 | 8.6 | 6.6 | 1.8 | 49 | 37 | 3.9 | 20 | 79 | 3.0 | 1.2 |
| 12378-PnCDD | 2.3 | 8.4 | 11.2 | 5.9 | 2.9 | 8.4 | 6.7 | 0.4 | 4.4 | 11.2 | 1.6 | 0.6 |
| 123478-HxCDD | 0.3 | 0.8 | 0.6 | 0.3 | 0.2 | 0.4 | 0.5 | 0.1 | 0.2 | 1.3 | | |
| 123678-HxCDD | 1.3 | 6.1 | 3.9 | 2.3 | 1.9 | 4.0 | 3.8 | 0.3 | 2.2 | 6.5 | 3.8 | 1.6 |
| 123789-HxCDD | 0.3 | 0.8 | 0.6 | 0.3 | 0.3 | 0.4 | 0.57 | 0.0 | 0.34 | 1.08 | | |
| 1234678-HpCDD | 0.7 | 1.1 | 1.0 | 0.8 | 1.1 | 0.9 | 1.2 | 0.1 | 0.8 | 2.4 | 4.4 | 2.1 |
| OCDD | 1.0 | 1.1 | 0.7 | 1.8 | 2.8 | 1.2 | 9 | 1.5 | 1 | 26 | 1.8 | 1.7 |
| Σ-PCDD | 7.2 | 22 | 27 | 18 | 11 | 65 | 60 | 5.4 | 36 | 113 | | |
| 2378-TeCDF | 15 | 35 | 23 | 25 | 11 | 19 | 32 | 1.8 | 21 | 45 | 0.03 | 0.02 |
| 12378-PnCDF | 1.7 | 4.9 | 6.3 | 3.6 | 1.4 | 4.1 | 7.3 | 0.5 | 4.3 | 10.5 | | |
| 23478-PnCDF | 2.5 | 10.2 | 12.8 | 5.4 | 2.7 | 20 | 28 | 2.3 | 16 | 50 | 0.31 | 0.13 |
| 123478-HxCDF | 0.5 | 1.4 | 1.6 | 0.5 | 0.2 | 9.7 | 11 | 1.4 | 5 | 24 | 0.64 | 0.44 |
| 123678-HxCDF | 0.3 | 1.1 | 1.2 | 0.5 | 0.3 | 1.6 | 2.8 | 0.2 | 1.3 | 4.1 | 2.1 | 2.0 |
| 234678-HxCDF | 0.3 | 1.3 | 1.4 | 0.9 | 0.5 | 1.1 | 1.5 | 0.1 | 1.0 | 2.3 | | |
| 1234678-HpCDF | 0.3 | 0.9 | 0.5 | 0.5 | 0.6 | 0.8 | 1.0 | 0.1 | 0.5 | 1.9 | | |
| OCDF | 0.4 | 1.0 | 0.1 | 0.2 | 0.9 | 0.4 | 1.7 | 0.4 | 0.4 | 6.1 | | |
| Σ-PCDF | 21 | 56 | 47 | 37 | 18 | 56 | 84 | 5.4 | 55 | 130 | | |

[a] From De Vault et al. [64]. Fish were sampled in 1984
[b] Adapted from Huestis et al. [65]. Mean ($N = 15$) of mean ($N = 6$–12) concentrations of PCDD4/Fs, 1977–1993
[c] Mean ratio of pooled herring gull egg to mean lake trout concentrations, 1981–1993 ($N = 7$–12). Calculated from herring gull data in Hebert et al. [71], and lake trout data for the same year [65]

compared to the other lakes (1–3%). The major difference in PCDD patterns among lakes is the much higher relative contribution of 2378-TeCDD in Lake Ontario (76%), but to some extent also in Lake Huron (32%) and Lake St. Clair (37%). Lake Erie had a relatively high proportion of OCDD (26%) compared to the other lakes (2–15%). Partial least squares analysis showed that 12378-PeCDD and 23478-PnCDF concentrations were highly correlated. These two compounds have not been linked as byproducts in any chemical production. DeVault et al. [64] suggested that the high correlation may have been due to a common atmospheric source. They argued that the statistically significant intra- and interlake differences in isomer compositions showed that there were important localized sources, and that the conclusions of Czuczwa and Hites [20] based on homolog compositions that the source of PCDD/Fs to the Great Lakes was primarily atmospheric failed to bring out the differences that show up when isomers are analyzed. However, it should also be noted that sediment isomer compositions (mostly non-AHR congeners) are very different from those in fish (exclusively AHR congeners), therefore the two conclusions may both be correct. Based on the relative uniformity of AHR congener PCDD/F patterns among the Great Lakes (Fig. 10), the DeVault et al. [64] results are consistent with a primarily atmospheric source of AHR PCDD/Fs modified by local inputs of 2378-TeCDD, OCDD, and 23478-PnCDF in all lakes except Lake Ontario. Consistent with findings on PCDD/F contamination in sediments and water from the Niagara River and Lake Ontario discussed previously, the large contributions of 2378-TeCDD, 123478-HxCDF, and 23478-PnCDF to Σ-PCDD/Fs in Lake Ontario predator fish are very likely due to contamination from the Niagara River.

Concentrations of PCDD/F congeners in lake trout in Lake Ontario from the temporal trend study of Huestis et al. [65] are summarized as means, 1977–1993, in Table 5. There was excellent agreement with the PCDD/F results of DeVault et al. [64] for lake trout sampled from Lake Ontario within the same period.

As part of a detailed analysis of lake trout reproductive impairment in Lake Ontario, Cook et al. [16] determined PCDD/F concentrations in lake trout eggs from Lake Ontario, 1978–1991, which are summarized in Table 6. The concentrations are less than one-half of those in lake trout muscle from the same lake and time (Table 5). However, on a lipid weight basis, the egg to female concentration ratio of TeCDD/Fs and PnCDD/Fs was 0.6–0.7, similar to that found in herring gulls [63]. The overall trend in PCDD/F concentrations was similar to that found in lake trout muscle over the same time period, i.e., concentrations of 2378-TeCDD demonstrated the most consistent downward tendency.

Giesy et al. [66] compared PCDD/F concentrations in dorsal muscle and eggs of coho salmon, chinook salmon and lake trout from Lakes Michigan, Superior, and Huron, 1987 and 1988. Data for the three species were presented as concentrations of homologs plus 2378-TeCDD and 2378-TeCDF. The homolog approach renders comparison of higher chlorinated PCDD/F concen-

**Table 6** Concentrations (ng kg$^{-1}$ fresh weight) of PCDD/Fs in lake trout eggs from Lake Ontario, 1978–1991

| | | | Lake Ontario[a] | | | | L. Mich.[b] |
|---|---|---|---|---|---|---|---|
| | 1978 | 1984 | 1987 | 1988 | 1990 | 1991 | 1993 |
| 2378-TeCDD | 26 | 15 | 7.3 | 11 | 6.5 | 6 | 0.7 |
| 12378-PnCDD | 18 | 9.5 | 2.9 | 2.8 | 4.6 | 5.8 | 1.6 |
| 123678-HxCDD | 1.8 | 1.1 | < 1.3 | 0.3 | < 1.3 | < 0.8 | < 0.8 |
| OCDD | na | na | na | 2.2 | 4.9 | 4 | ND |
| $\Sigma$-PCDD | 46 | 26 | 10 | 16 | 16 | 16 | 2 |
| 2378-TeCDF | 18 | 9.5 | 2.9 | 2.8 | 4.6 | 5.8 | 11 |
| 12378-PnCDF | 3.5 | 2.1 | < 2.1 | 1.4 | < 1.3 | 0.7 | 1.4 |
| 23478-PnCDF | 18 | 10 | 4.7 | 4.8 | 7.8 | 4.6 | 3.5 |
| 123478-HxCDF | 5.1 | 4.4 | 1.4 | 1 | 1.8 | 1.6 | < 1 |
| $\Sigma$-PCDF | 45 | 26 | 9 | 10 | 14 | 13 | 16 |

*na* not analyzed, *ND* not detected
[a] From Cook et al. [16]
[b] From Cook et al. [105]

trations with other studies difficult. Therefore 2378-TeCDD and 2378-TeCDF concentrations were included in Table 4. PCDD concentrations increased in the order coho salmon < lake trout < chinook salmon in all lakes. The ranking of PCDF concentrations according to species was different in each lake. Lake Michigan and Lake Huron (except Saginaw Bay) fish dorsal muscle had similar PCDD concentrations, while concentrations in Lake Superior fish were two- to threefold less. Saginaw Bay fish muscle had the highest concentrations in the study: 92–215 ng kg$^{-1}$ ww for $\Sigma$-PCDDs and 250–380 ng kg$^{-1}$ for $\Sigma$-PCDFs. The lipid content in salmon muscle was 4.1–6.7%. Lake trout muscle was significantly leaner in Lake Superior lake trout (6.5–7.1% lipid) than in the other lakes (11.9–14.4% lipid). Therefore, on a lipid weight basis, lake trout had lower PCDD concentrations than the salmon. Concentrations of $\Sigma$-PCDDs and $\Sigma$-PCDFs in eggs were highly correlated with those in muscle, but the relationship was different for the two groups of chemicals: PCDDs concentrations in eggs/muscle were 0.2–0.3, while PCDF concentrations in eggs/muscle were 0.5–0.8 for all three species (based on slopes in Figs. 8, 9, and 10 of Giesy et al. [66]). There does not seem to be any reason why the egg/muscle ratio should be different for PCDDs and PCDFs. On a wet weight basis, the PCDD egg/muscle ratio is in better agreement with Cook et al. [16] than is the PCDF ratio. OCDF and HpCDF concentrations in fish eggs and muscle were anomalously high in all species from all lakes compared to any other study of fish in the Great Lakes (e.g., Table 5). Even air and sediment homolog profiles (Figs. 1 and 7) do not have such a high relatively level of OCDF contribution. OCDF has a very high log $K_{OW}$, and is generally not bioavailable. The BSAF (biota-sediment bioaccumulation factor) for OCDD and OCDF is 0.0001 [16]. Lake trout eggs and muscle in Lake Ontario from the same time period as this study had undetectable OCDF concentrations [16], despite sediment profiles from several studies that clearly indicate that Lake Ontario had the highest OCDF contamination of all the lakes.

## 2.6
## Seabirds and Snapping Turtle Eggs

Because fish-eating seabirds are at the top of the Great Lakes food web, and because biomagnification factors of recalcitrant organochlorine compounds are much higher in birds than in fish of similar size eating a similar diet [67], seabirds have the potential to accumulate relatively high concentrations of PCDD/Fs compared to salmonids. As we have seen, lake trout and various introduced salmon species have the highest concentrations among fish in the Great Lakes. However, it must also be kept in mind that the metabolic capacity of birds and fish is quite different. Both fish and birds accumulate only AHR PCDD/F congeners, that is, those with 2378-positions fully substituted by chlorine [68]. Biomagnification factors of AHR PCDDs in herring gulls are in the same range as moderately recalcitrant organochlorine compounds, hexachlorobenzene, $\beta$-HCH, octachlorostyrene and dieldrin [63]. The most readily bioaccumulated PCDD was 2378-TeCDD ($BMF = 32$), but even this compound had a BMF about threefold lower than highly chlorinated PCBs, mirex and 4,4′-DDE. The BMFs of 2378-TeCDF and 23478-PnCDF in herring gulls were only 1.3 and 6.6, respectively. BMFs are inversely proportional to whole-body clearance rates [67], therefore it appears that PCDFs are metabolized rapidly by the herring gull. This may not be true for other birds; however, an early study showed no or low PCDF accumulation by mallard ducks fed Aroclor 1254 [69]. Furthermore, the rate of PCDD/F metabolism may be tied to enzyme induction and therefore to the degree of exposure. Considering all these factors, we should expect to see quite different patterns of PCDD/F contamination in seabirds compared to salmonids, despite occupying a similar ecological niche in the Great Lakes.

## 2.6.1
## Herring Gull Eggs

The Great Lakes Herring Gull Egg Program was the first biomonitoring program set up to monitor organochlorine contaminant concentration trends and effects on reproduction in a species in the Great Lakes, and is one of the longest-running biomonitoring programs in the world [70]. From 1975 onward, 10–13 fresh eggs were randomly collected from at least two colonies in each of the Great Lakes. Eggs from colonies in the Detroit River, Niagara River, and St. Lawrence River were subsequently added. Herring gulls are a nearly ideal biomonitor species in the lower Great Lakes, since adults are not migratory, and tend to remain in their natal lake all year. Herring gulls are also more piscivorous than the closely related, migratory ringed-bill gull. Recent estimates based on calibration of a organochlorine bioaccumulation model suggested that alewife and smelt constitute 76–83% of the annual diet on an energy intake basis in Lake Ontario herring gulls (Norstrom, unpublished data, 2004). Herring gulls from Lake Superior and Lake Huron, which may freeze over in cold years restricting food availability, tend to move to the lower lakes in winter [70]. Concentrations in herring gull eggs from Lake Superior and to some extent Lake Huron may therefore reflect a combination of exposure from more than one lake, depending on the whole-body half-life of the chemical in the gull.

Samples of herring gull egg homogenates archived in the Canadian Wildlife Service Specimen Bank provided an excellent early means of investigating the extent of 2378-TeCDD contamination among the Great Lakes after the initial results from fish in 1978–1980 indicated that the lakes were contaminated with this compound. Norstrom et al. [9] reported concentrations of 2378-TeCDD of 9–11 $ng\,kg^{-1}$ in herring gull eggs from Lake Superior, Lake Michigan, Lake Huron (main body), and Lake Erie in 1980. Higher concentrations were found in Saginaw Bay, Lake Huron colonies (43–86 $ng\,kg^{-1}$), and Lake Ontario colonies (59 $ng\,kg^{-1}$). These semiquantitative data were obtained using older methodology before $^{13}$C-labeled surrogate standards were available. Recoveries appeared to be concentration-dependent. The concentrations were about twofold lower than those from a subsequent re-analysis of 1980 samples. Nevertheless, the relative contamination among areas established that Saginaw Bay, Lake Huron, and Lake Ontario were the areas of highest concern in the Great Lakes for 2378-TeCDD contamination.

Norstrom and Simon [10], using improved methods, reported concentrations of PCDDs in a pool of three herring gull eggs from Lake Ontario, 1981: 2378-TeCDD (132 $ng\,kg^{-1}$), 12378-PnCDD (36 $ng\,kg^{-1}$), and HxCDDs (82 $ng\,kg^{-1}$). This was the first report of more highly chlorinated PCDDs in Great Lakes biota. Stalling et al. [61] presented a graph showing concentrations of 2378-TeCDD and 12378-PnCDD + 123678-HxCDD in pooled herring gull eggs, 1983, from eight colonies throughout the Great Lakes. Concentrations of 2378-TeCDD ranged from 9–26 $ng\,kg^{-1}$ in eggs from colonies in southern Lake Superior, northern Lake Michigan, southern Lake Huron, Detroit River, Lake Erie, and the Niagara River. Concentrations of 2378-TeCDD were 90 $ng\,kg^{-1}$ in Saginaw Bay, Lake Huron eggs and 141 $ng\,kg^{-1}$ in Lake Ontario eggs. Nearly all concentrations of HpCDDs and OCDD were < 10 $ng\,kg^{-1}$. In general, the sum of the higher chlorinated PCDDs was similar to that of 2378-TeCDD.

Archived samples were used to determine concentrations of 2378-TeCDD, 12378-PnCDD, 123678-HxCDD, 1234678-HpCDD, and OCDD in herring gull eggs 1981–1991 and concentrations of 2378-TeCDF, 23478-PnCDF, 123478-HxDF, and 123678-HxCDF, 1984–1991 [71]. Other congeners usually found in fish were below the detection limit most or all of the time. While there were some obvious decreases in concentrations of the major congeners in some colonies during the 1980s, most of the changes took place in 1981–1983. There were no significant trends between 1984 and 1994, which was the period when there was a complete PCDD/F data set. Therefore mean concentrations for the 1984–1994 period are presented in Table 7. Concentrations of 2378-TeCDD, 1984–1991, were statistically higher (44–87 ng kg$^{-1}$) in Lake Ontario and Saginaw Bay, Lake Huron colonies. Concentrations of 2378-TeCDD ranged from 14–20 ng kg$^{-1}$ in the other colonies. Saginaw Bay herring gull eggs also had two- to threefold higher concentrations of all other PCDD/Fs than the rest of the Great Lakes; 123678-HxCDD and 23478-PnCDF concentrations were statistically higher than in eggs from any other Great Lakes colony. Other than 2378-TeCDD in Lake Ontario eggs and all PCDD/Fs in Saginaw Bay eggs, the concentrations and profiles of PCDD/Fs were relatively constant throughout the Great Lakes, although there was at tendency for 12378-PnCDD and 23478-PnCDF concentrations to be higher in eggs from the upper lakes. The signature Lake Ontario pattern in fish and sediment of high relative 123478-HxCDF concentrations (e.g. lake trout, Fig. 11) compared to other lakes was not very evident, although concentrations of this congener in Lake Ontario herring gull eggs were the next highest after Saginaw Bay eggs among the lakes. Concentrations of all PCDD/Fs in Lake Superior eggs were indistinguishable from those in Lake Michigan eggs, which is probably a reflection of the Lake Superior birds wintering on Lake Michigan [70].

The relative constancy of the herring gull egg and lake trout concentrations during the 1980s afforded an excellent opportunity to compare PCDD/F bioaccumulation in the two species, both of which eat alewife and smelt. The mean ratio of concentrations of the detectable PCDD/Fs in herring gull eggs to those in lake trout from the same years, 1981–1993, is given in Table 5. These gull egg/trout ratios permit conversion of concentrations between the species with variable confidence (relative standard deviations of mean ratios ranged from 40% for 2378-TeCDD to 100% for OCDD and 123678-HxCDF). The maximum ratio, for 2378-TeCDD, 123678-HxCDD, and 1234678-HpCDD, was in the range 3.0–4.8. After correction for the herring gull whole body/egg ratio of 1.2–1.4 [63], this is roughly the ratio of annual food consumption, and therefore influx of contaminant, in a herring gull and lake trout of similar size. For these compounds, it would appear that clearance is slow in both species and therefore food intake rate is the main determinant of concentrations. The gull/trout ratios demonstrate that 2378-TeCDD and 123678-HxCDD have about two- to threefold higher bioaccumulation potential than the other PCDDs and 123678-HxCDF, and five- to tenfold higher than 23478-PnCDF and 123478-HxCDF in herring gull females relative to lake trout. As found in other studies, 2378-TeCDF is the least efficiently bioaccumulated of the AHR PCDD/Fs in the gull.

Assuming that exposure to *patterns* of PCDD/F contamination was similar in herring gulls and lake trout in Lake Ontario, and that all congeners

**Table 7** Mean concentrations (ng kg$^{-1}$ fresh weight) of PCDD/Fs in herring gull eggs from the Great Lakes, 1984–1991. Adapted from Hebert et al. [71]

| | 2378-TeCDD | 12378-PnCDD | 123678-HxCDD | 1234678-HpCDD | OCDD | Σ-PCDD | 2378-TeCDF | 23478-PnCDF | 123478-HxCDF | 123678-HxCDF | Σ-PCDF |
|---|---|---|---|---|---|---|---|---|---|---|---|
| West Lake Superior | 16 | 13 | 14 | 3.6 | 3.6 | 50 | 1.5 | 10.4 | 1.6 | 2.5 | 16 |
| East Lake Superior | 20 | 12 | 12 | 3.9 | 5.5 | 53 | 1.6 | 11.4 | 2 | 2.3 | 17 |
| North Lake Michigan | 14 | 15 | 19 | 2.7 | 7.8 | 59 | 2.6 | 10.5 | 1.5 | 2.3 | 17 |
| Green Bay, LM | 14 | 13 | 14 | 2.5 | 3.8 | 47 | 2.2 | 11.8 | 1.1 | 2.2 | 17 |
| Georgian Bay, LH | 25 | 11 | 11 | 4.0 | 3.8 | 55 | 1.3 | 8.6 | 1.8 | 2.4 | 14 |
| South Lake Huron | 18 | 9 | 9 | 3.6 | 2.6 | 42 | 1.2 | 9.1 | 1.3 | 1.6 | 13 |
| Saginaw Bay, LH | 87 | 23 | 33 | 6.8 | 18.6 | 168 | 3.6 | 23.3 | 5.9 | 7.7 | 41 |
| Detroit River | 17 | 7 | 17 | 5.2 | 9.1 | 55 | 0.9 | 6.4 | 1.1 | 1.5 | 10 |
| West Lake Erie | 16 | 12 | 17 | 3.5 | 7.8 | 56 | 1.6 | 8.1 | 0.7 | 1.4 | 12 |
| East Lake Erie | 17 | 7 | 9 | 4.1 | 5.3 | 42 | 0.7 | 6.1 | 1.6 | 1.9 | 10 |
| Niagara River | 24 | 7 | 9 | 3.5 | 3.4 | 47 | 0.8 | 7.1 | 2.6 | 2.4 | 13 |
| North Lake Ontario | 44 | 7 | 12 | 5.6 | 8.0 | 77 | 0.8 | 5.7 | 3.1 | 3.3 | 13 |
| East Lake Ontario | 69 | 9 | 11 | 3.8 | 6.5 | 99 | 0.8 | 7.0 | 3.5 | 3.6 | 15 |

were accumulated equally efficiently in trout, relative ratios of gull/trout in Table 5 are influenced primarily by relative clearance rates from the gull. On this basis, 2378-TeCDF was cleared 100 times faster than 2378-TeCDD, and 123478-HxCDF was cleared three times faster than 123678-HxCDF in the herring gull. The influence of 123478-HxCDF, which is a congener characteristic of PCDF contamination in Lake Ontario, is therefore muted in herring gulls compared to fish (Table 5). The ranking of congeners from highest to lowest gull/trout ratio is the same as that for herring gull/alewife BMFs given in Braune and Norstrom [63], a further indication that differences in bioaccumulation of PCDD/Fs between herring gulls and lake trout are due to the herring gull's ability to metabolize these compounds. Some differences may be partly a reflection of the more benthic-based food chain of the lake trout, as pointed out by Cook et al. [16].

## 2.6.2
## Double-Crested Cormorant Eggs

Double-crested cormorants are considered a nuisance species by fisherman in the Great Lakes. Cormorants were not present in the Great Lakes historically. Their successful colonization of the Great Lakes is at least partly due to the changes that man's unintentional manipulation of this ecosystem has wrought. The story of the cormorant in the Great Lakes is connected with the story of food web structure changes. Construction and various expansions of the Welland Canal in the 1800s created the first opportunity in modern times for fish to overcome the barrier of Niagara Falls and enter the upper lakes. The alewife, an anadromous, primarily marine species which was previously confined to Lake Ontario, first appeared in Lake Erie in 1931, systematically moved into the upper lakes in the mid-20th century along with lamprey eel, and is now common in Lakes Michigan and Huron, but not in Lake Erie or Lake Superior. Rainbow smelt were introduced into Lake Michigan in 1912. The lamprey eel devastated lake trout populations, which were the main control on abundance of forage fish. It is likely that the cormorant exploited the increased abundance of alewife. The first cormorants in the Great Lakes spread from prairie lake populations to Lake Superior in 1913 and from there to Lake Huron and Georgian Bay in the early 1930s, and finally to Lakes Ontario and Erie in the late 1930s [72]. The Great Lakes population peaked around 1950 and declined thereafter due to 4,4′-DDE-related eggshell thinning. By 1973, the cormorant population in the Great Lakes had declined by 86% and breeding birds had disappeared from Lakes Michigan and Superior [73]. Perversely, it was the increased reproductive success in cormorants caused by decreases in 4,4′-DDE concentrations that was responsible for the large resurgence of cormorant populations in the 1980s.

Three studies have looked at PCDD/F concentrations in eggs of double-crested cormorants in the Great Lakes. These are summarized in Table 8.

**Table 8** Concentrations (ng kg$^{-1}$ fresh weight) of PCDD/Fs in double-crested cormorant eggs from the Great Lakes, 1988–1991

| Colony | Area | Lake | Year | Ref | 2378-TeCDD | 12378-PnCDD | 123478 HxCDD | 123678-HxCDD | 123789 HxCDD | 1234678-HpCDD | OCDD | Σ-PCDD |
|---|---|---|---|---|---|---|---|---|---|---|---|---|
| Little Charity Isl. | Saginaw Bay | Huron | 1998 | [76] | 5 | 10 | 2 | 12 | 2 | 6 | 53 | 89 |
| Scarecrow Island | W, Michigan coast | Huron | 1998 | [76] | 4 | 6 | 1 | 5 | 1 | 4 | 31 | 51 |
| Tahquamenon Island | SE, near outlet | Superior | 1998 | [76] | 4 | 6 | 1 | 7 | 1 | 7 | 45 | 71 |
| | | | | | | | | (Σ-HxCDDs) | | | | |
| Tahquamenon Island | SE, near outlet | Superior | 1998 | [75] | 12 | 3 | | 20 | | 11 | 13 | 59 |
| St. Martin's Shoal | W, near L. Michigan | Huron | 1988 | [75] | 8 | 7 | | 11 | | 7 | 16 | 49 |
| Beaver Islands | N, middle | Michigan | 1988 | [75] | 5 | 4 | | 10 | | 4 | 8 | 31 |
| Green Bay | Mouth of bay | Michigan | 1988 | [75] | 20 | 9 | | 16 | | 8 | 16 | 69 |
| Gravel Island | N shore | Superior | 1989 | [74] | 9 | 14 | 1 | 18 | 5 | 8 | 10 | 65 |
| Cone Island | N shore | Superior | 1989 | [74] | 12 | 21 | 3 | 25 | 8 | 21 | 112 | |
| West Island | N Channel | Huron | 1989 | [74] | 14 | 21 | 2 | 17 | 4 | 13 | 16 | 87 |
| Blackbill Isl. | Georgian Bay | Huron | 1989 | [74] | 18 | 27 | 3 | 21 | 4 | 8 | 4 | 85 |
| East Sister Isl. | W | Erie | 1989 | [74] | 20 | 22 | 2 | 25 | 5 | 8 | 11 | 93 |
| Hamilton Harbour | W | Ontario | 1989 | [74] | 18 | 21 | 2 | 19 | 6 | 11 | 12 | 89 |
| Pigeon Island | E | Ontario | 1989 | [74] | 18 | 17 | 2 | 11 | 3 | 8 | 13 | 72 |
| Pigeon Island | E | Ontario | 1991 | [74] | 36 | 32 | 6 | 37 | 12 | 27 | 25 | 175 |

**Table 8** (continued)

| Colony | Area | Lake | Year | Ref | TeCDF | 2378-PnCDF | 12378 PnCDF | 23478-HxCDF | 123478-HxCDF | 123678-HxCDF | 234678-HxCDF | 123789 HpCDF | 1234678-Σ-PCDF |
|---|---|---|---|---|---|---|---|---|---|---|---|---|---|
| Little Charity Isl. | Saginaw Bay | Huron | 1998 | [76] | 2 | 5 | 12 | 13 | 0.04 | 1.4 | 0.4 | 0.4 | 34 |
| Scarecrow Island | W, Michigan coast | Huron | 1998 | [76] | 2 | 3 | 8 | 8 | 0.82 | 1.3 | 0.8 | 0.5 | 24 |
| Tahquamenon Island | SE, near outlet | Superior | 1998 | [76] | 1 | 1 | 5 | 10 | 0.04 | 0.8 | 0.8 | 0.4 | 19 |
|  |  |  |  |  |  |  |  |  |  | (Σ-HxCDFs) |  |  |  |
| Tahquamenon Island | SE, near outlet | Superior | 1998 | [75] | 1 | 2 | 5 |  |  | 14 |  | 9 | 31 |
| St. Martin's Shoal | W, near L. Michigan | Huron | 1988 | [75] | 2 | 1 | 7 |  |  | < 1 |  | 7 | 17 |
| Beaver Islands | N, middle | Michigan | 1988 | [75] | 3 | 2 | 6 |  |  | < 1 |  | 2 | 13 |
| Green Bay | Mouth of bay | Michigan | 1988 | [75] | 6 | 3 | 2 |  |  | 3 |  | 1 | 15 |
| Gravel Island | N shore | Superior | 1989 | [74] | 1 | ND | 11 | 5 | 3 | 2 | ND | ND | 22 |
| Cone Island | N shore | Superior | 1989 | [74] | 2 | ND | 8 | 4 | 3 | 2 | ND | ND | 19 |
| West Island | N Channel | Huron | 1989 | [74] | < 1 | ND | 12 | 2 | 2 | 2 | ND | ND | 20 |
| Blackbill Isl. | Georgian Bay | Huron | 1989 | [74] | < 1 | ND | 20 | 3 | 2 | 2 | ND | ND | 27 |
| East Sister Isl. | W | Erie | 1989 | [74] | < 1 | ND | 14 | 2 | 1 | 1 | ND | ND | 18 |
| Hamilton Harbour | W | Ontario | 1989 | [74] | 1 | ND | 21 | 7 | 4 | 4 | ND | ND | 37 |
| Pigeon Island | E | Ontario | 1989 | [74] | 1 | ND | 14 | 5 | 2 | 3 | ND | ND | 25 |
| Pigeon Island | E | Ontario | 1991 | [74] | < 1 | ND | 29 | 10 | 4 | 3 | ND | ND | 46 |

Concentrations were in much the same range as those in herring gull eggs. Patterns and concentrations of PCDD/F concentrations in cormorant eggs were remarkably uniform among colonies in Lakes Superior, Huron, Erie, and Ontario in 1988–1989. Concentrations of 2378-TeCDD, 12378-PnCDD, 123678-HxCDD, 1234678-HpCDD, and OCDD were all in the $8–27\,\mathrm{ng\,kg^{-1}}$ range in 1989, regardless of lake [74]. This pattern bears little resemblance to the more resident herring gull, in which 2378-TeCDD concentrations were two- to threefold higher in Lake Ontario and Saginaw Bay eggs. Yamashita et al. [75] found PCDD/F concentrations in cormorant eggs from Lakes Superior, Michigan, and Huron in 1988 to be very similar to those of Ryckman et al. [74], except that concentrations of 12378-PnCDD were three- to fivefold lower. There seems to be a genuine discrepancy in concentrations of this isomer between Ryckman et al. [74] and the other two studies. Hilscherova et al. [76] determined PCDD/Fs in cormorant eggs from Saginaw Bay, Lake Michigan, and Lake Superior in 1998. Compared to the earlier studies, concentrations of 2378-TeCDD appeared to have decreased, concentrations of most other PCDD/Fs were similar, but OCDD concentrations were higher than previous studies. It is not possible to make any definitive statements about trends from these data.

Hilscherova et al. [76] found that PCDD/F concentrations varied as much or more among individuals within a colony as between colonies. This was attributed to variability in pollutant metabolism, but it is more likely due to variability in exposure of individual birds, e.g., time of arrival on colony or different proportion of fish species in the diet. Ryckman et al. [74] proposed that uniformity of PCDD/F concentrations in cormorants may be due to most of the residue being accumulated prior to migrating back to the colony about 1 month before laying eggs. However, they proceeded to rule this out as a factor based on the finding that mirex concentrations were much higher in cormorant eggs from Lake Ontario, which is known to be uniquely contaminated with this chemical, than from the other Great Lakes. However, it may well be that sources of mirex and PCDD/Fs in cormorant eggs should not be compared. Based on relative BMFs, 2378-TeCDD has about one-third the half-life of mirex, photomirex, 4,4′-DDE, and highly chlorinated PCBs in herring gulls [63]. From experimental clearance studies we know the half-life of these recalcitrant compounds in herring gulls is about 1 year or more [67]. Since 2378-TeCDD is the most highly bioaccumulated PCDD/F in herring gulls, the half-life of the other PCDD/Fs is even shorter. If clearance rates of chlorocarbons are similar in herring gulls and cormorants, it is likely that the cormorant will carry over residues of mirex accumulated during its summer and fall residence in Lake Ontario from one year to the next for deposit in eggs, while this may not be the case for residues of PCDD/Fs.

Relative concentrations in diet in wintering areas and on-colony will also affect whether cormorant egg concentrations reflect local concentrations in fish. Exposure to mirex during the winter months is relatively low for all cormorants, while the Lake Ontario birds get a sudden increase in exposure when they return. Note that the same scenario probably occurs for PCBs in Green Bay, Lake Michigan cormorant eggs. That is, slow clearance of PCBs from the bird combined with high exposure on-colony resulted in higher PCB concentrations in Green Bay than in other Great Lakes cormorant eggs [75]. It is probable that differences in exposure of cormorants to PCDD/Fs throughout the year are less dramatic (at least in the past 20 years), which would tend to mask incremental on-colony exposure to these chemicals. Thus, the fidelity of migratory bird egg residues to local fish concentrations is a function of both the half-life of the chemical in the bird and relative exposure during migration and on-colony.

In conclusion, it is likely that a high proportion of the PCDD/Fs seen in cormorant eggs (also true of other migratory fish-eating birds, such as terns and ring-billed gulls) is accumulated outside the Great Lakes basin prior to returning to their colonies in the spring.

## 2.6.3
### Caspian and Forster's Tern Eggs

Like double-crested cormorants, terns are exclusively piscivorous and migratory. Like cormorants, increasing Caspian tern populations have been attributed to the abundance of alewife [77]. Terns were among the first seabird species in the Great Lakes whose reproduction was thought to have been affected by persistent organic chemicals [78]. There are four species of terns nesting in the Great Lakes, however, PCDD/F concentrations have only been measured in Caspian terns and Forster's terns. Caspian terns are island nesters, competing with other seabirds for territory. Forster's terns are marsh nesting, and are far less abundant.

Concentrations of PCDD/Fs in Caspian [75, 79] and Forster's terns [61, 80] is presented in Table 9. Median 2378-TeCDD concentrations were 37 ng kg$^{-1}$ and 8 ng kg$^{-1}$ in Green Bay, Lake Michigan and Lake Poygan (uncontaminated lake) Forster's tern eggs in 1983. The sum of remaining PCDDs was also higher in Green Bay than the uncontaminated control colony in Lake Poygan, 102 ng kg$^{-1}$ versus 25 ng kg$^{-1}$. These concentrations were two- to threefold higher than in Caspian Terns from Green Bay in 1988, and four- to fivefold higher than in Lake Michigan Caspian tern eggs in 1991. Some of the differences between species may be related to time, but also to food chain and proximity of the Forster's tern colony to the mouth of the Fox River, which is the major source of contamination to the bay. There was no 2378-TeCDF found in Forster's tern eggs, however they contained an array of PnCDF–OCDF congeners at a median total concentration of 19 ng kg$^{-1}$ in Green Bay and 9 ng kg$^{-1}$ in Lake Poygan. The lack of 2378-TeCDF is questionable, given concentrations of this compound in Caspian tern eggs.

**Table 9** Concentrations (ng kg$^{-1}$ fresh weight) of PCDD/Fs in tern eggs from the Great Lakes, 1983–1991

| Colony | Area | Lake | Year | Refs. | 2378-TeCDD | 12378-PnCDD | 123678-HxCDD | 1234678-HpCDD | OCDD | Σ-PCDD |
|---|---|---|---|---|---|---|---|---|---|---|
| **Caspian terns** | | | | | | | | | | |
| Gravelly Island | Mouth of Bay | Michigan | 1991 | [79] | 8 | 6 | 10 | ND | ND | 24 |
| | | | 1988 | [75] | 13 | 3 | 9 | 4 | 21 | 50 |
| Hat Island | N, middle | Michigan | 1991 | [79] | 7 | 4 | 6 | ND | ND | 17 |
| | | | 1988 | [75] | 12 | 5 | 10 | 4 | 12 | 43 |
| Ile Aux Galets | N, middle | Michigan | 1991 | [79] | 6 | 5 | 5 | ND | ND | 16 |
| Cousins Island | North Channel | Huron | 1991 | [79] | 22 | 15 | 16 | ND | ND | 53 |
| Halfmoon Island | Georgian Bay | Huron | 1991 | [79] | 22 | 14 | 6 | ND | ND | 42 |
| S. Limestone Island | Georgian Bay | Huron | 1991 | [79] | 8 | 4 | 4 | ND | ND | 16 |
| S. Watcher Island | Georgian Bay | Huron | 1991 | [79] | 26 | 4 | 5 | ND | ND | 35 |
| Channel-Shelter Isl. | Saginaw Bay | Huron | 1991 | [79] | 18 | 9 | 8 | ND | ND | 35 |
| | | | 1988 | [75] | 13 | 5 | 17 | 3 | 10 | 48 |
| Hamilton Harbour | W | Ontario | 1991 | [79] | 31 | 4 | 4 | ND | ND | 39 |
| Pigeon Island | E | Ontario | 1991 | [79] | 27 | 5 | 5 | ND | ND | 37 |
| **Forster's terns** | | | | | | | | | | |
| S. Oconto Marsh | Green Bay | Michigan | 1983 | [80] | 37 | 6 | 22 | 12 | 17 | 94 |
| | | | 1983 | [61] | 47 | NA | NA | NA | NA | 114 |
| Lake Poygan | Near Green Bay | Control | 1983 | [80] | 8 | ND | ND | ND | ND | 8 |
| | | | 1983 | [61] | 9 | NA | NA | NA | NA | 21 |

**Table 9** (continued)

| Colony | Area | Lake | Year | Refs. | 2378-TeCDF | 12378-PnCDF | 123678-HxCDF | 1234678-HpCDF | Σ-PCDF |
|---|---|---|---|---|---|---|---|---|---|
| **Caspian terns** | | | | | | | | | |
| Gravelly Island | Mouth of bay | Michigan | 1991 | [79] | 11 | 4 | ND | ND | 15 |
| | | | 1988 | [75] | 6 | 3 | 6 | 5 | 20 |
| Ile Aux Galets | N, middle | Michigan | 1991 | [79] | 6 | 3 | ND | ND | 9 |
| Hat Island | N, middle | Michigan | 1991 | [79] | 5 | 6 | 11 | 4 | 26 |
| Ile Aux Galets | N, middle | Michigan | 1991 | [79] | 6 | 3 | ND | ND | 9 |
| Cousins Island | North Channel | Huron | 1991 | [79] | 23 | 9 | ND | ND | 32 |
| Halfmoon Island | Georgian Bay | Huron | 1991 | [79] | 7 | 4 | ND | ND | 11 |
| S. Limestone Island | Georgian Bay | Huron | 1991 | [79] | 4 | <1 | ND | ND | 4 |
| S. Watcher Island | Georgian Bay | Huron | 1991 | [79] | 4 | 3 | ND | ND | 7 |
| Channel-Shelter Isl. | Saginaw Bay | Huron | 1991 | [79] | 9 | 5 | ND | ND | 14 |
| | | | 1988 | [75] | 5 | 7 | 13 | 4 | 29 |
| Hamilton Harbour | W | Ontario | 1991 | [79] | 5 | 5 | ND | ND | 10 |
| Pigeon Island | E | Ontario | 1991 | [79] | 4 | 3 | ND | ND | 7 |
| **Forster's terns** | | | | | | | | | |
| S. Oconto Marsh | Green Bay | Michigan | 1983 | [80] | | 3 to 9 | | | 11 to 27 |
| | | | 1983 | [61] | | | | | |
| Lake Poygan | Near Green Bay | Control | 1983 | [80] | | | | | |
| | | | 1983 | [61] | | | | | |

Concentrations of PCDD/Fs in Caspian tern eggs and pattern of contamination was similar to that found in double-crested cormorants. Except for 2378-TeCDF, there was very little geographical variation among lakes. It is notable that Saginaw Bay Caspian tern eggs did not have different patterns or concentrations of PCDD/Fs compared to eggs from northern Lake Michigan. This is in contrast to herring gull eggs from Saginaw Bay, which had significantly higher concentrations of a number of PCDD/Fs than did eggs from other Great Lakes colonies (Table 7). Most of $\Sigma$-PCDDs in Caspian tern eggs was accounted for by 2378-TeCDD in Lake Huron and Lake Ontario. Concentrations of 2378-TeCDD in northern Lake Michigan eggs in 1991, 7 ng kg$^{-1}$, were about half of those in 1988, 12–13 ng kg$^{-1}$, otherwise the PCDD/F concentrations were similar in the two years. Caspian terns had higher and more variable concentrations of 2378-TeCDF than herring gulls and double-crested cormorants, which may be due to the lower ability of the tern to metabolize this compound. Concentrations of 2378-TeCDF were notably high in eggs from Gravelly Island at the mouth of Green Bay, Lake Michigan in 1991 (11 ng kg$^{-1}$), and in the North Channel of Lake Huron (23 ng kg$^{-1}$). The latter value is the highest concentration of 2378-TeCDF ever measured in seabird eggs from the Great Lakes.

Similar to cormorants, it appears that most of the PCDD/Fs found in Caspian tern eggs are accumulated outside the Great Lakes prior to nesting, however, there may be some local influence, especially on 2378-TeCDF concentrations.

### 2.6.4
### Snapping Turtle Eggs

Apart from birds and fish, the only other wildlife that has been analyzed to any extent for PCDD/F content are snapping turtles. Common snapping turtles are omnivorous. They are known to feed on plants, fish, frogs, birds, small mammals, snails, and other slow-moving benthic invertebrates. They inhabit any slow moving shallow, permanent water body, such as swamps, marshes, streams, and ponds. Because of their omnivorous diet, exposure to bioaccumulating contaminants of any kind may be highly variable among individuals, not to speak among areas. However, they are long-lived (up to 24 years), and therefore have the opportunity to accumulate contaminants over a long period of time. Depending on how fast PCDD/Fs are cleared from the turtle (probably mostly by egg production in females), concentrations in tissues and eggs from older turtles may integrate exposure over several years.

PCDD/Fs were first surveyed in tissues from two snapping turtles from the Lake Ontario–St. Lawrence river corridor in 1985 [81]. Concentrations of 2378-TeCDD were 32–107 ng kg$^{-1}$ in liver and 232–474 ng kg$^{-1}$ in fat, while 2378-TeCDF concentrations were much lower, < 2–12 ng kg$^{-1}$. Concentrations of 23478-PnCDF were also relatively high, up to $152\,\mathrm{ng\,kg^{-1}}$ in fat. Much higher concentrations of 2378-TeCDF, PnCDFs, HxCDFs, and HpCDFs were found downstream in the St. Lawrence River below a known source of PCBs near Massena, NY.

Most of the data on concentrations of PCDD/Fs in Great Lakes snapping turtles is for eggs, generated by Bishop and coworkers (Bishop et al. 1991, 1996, 1998). Concentrations of PCDD/Fs in snapping turtle eggs from Bishop et al. [82] are given in Table 10 based on Bishop et al. [82], who summarized PCDD/F concentration data from preceding publications. Congener specific data was obtained for Hamilton Harbour/Cootes Paradise (wetland areas in

**Table 10** Concentrations ($\mathrm{ng\,kg^{-1}}$ fresh weight) of PCDD/Fs in snapping turtle eggs from Lake Erie and Lake Ontario

| | Lake Erie | | Lake Ontario | |
| | BCM, 1989[a] | HH, 1989[b] | HH, 1990[b] | LC, 1991[c] |
| --- | --- | --- | --- | --- |
| 2378-TeCDD | < 0.1 | 12 | 26 | 27 |
| 12378-PnCDD | < 0.1 | 2.6 | 5.4 | 36 |
| 123478-HxCDD | < 0.1 | 0.3 | 0.8 | 2.6 |
| 123678-HxCDD | 0.9 | 2.3 | 5 | 27 |
| 123789-HxCDD | < 0.1 | 0.3 | < 0.1 | 2.8 |
| 1234678-HpCDD | 0.6 | 0.8 | 1.1 | 2.3 |
| 1234679-HpCDD | < 0.1 | < 0.1 | < 0.1 | 4.9 |
| OCDD | 1.6 | 1.7 | 1.8 | 4.4 |
| $\Sigma$-PCDD | 3.1 | 20 | 40.1 | 107 |
| [+1mm] 2378-TeCDF | < 0.1 | < 0.1 | < 0.1 | < 0.1 |
| 12378-PnCDF | < 0.1 | < 0.1 | < 0.1 | < 0.1 |
| 23478-PnCDF | 2.3 | 5.9 | 15 | 24 |
| 12478-PnCDF | < 0.1 | 3.2 | < 0.1 | 31 |
| 23467-PnCDF | < 0.1 | 1.4 | < 0.1 | 7.9 |
| 123478-HxCDF | < 0.1 | 0.3 | 0.4 | 1.2 |
| 123678-HxCDF | < 0.1 | 0.6 | 0.6 | 3.2 |
| 123789-HxCDF | < 0.1 | 0.2 | < 0.1 | 1.8 |
| 234678HxCDF | < 0.1 | 0.3 | < 0.1 | < 0.1 |
| 124678-HxCDF | < 0.1 | 0.3 | < 0.1 | 2.7 |
| 124689-HxCDF | < 0.1 | 0.9 | < 0.1 | 6.2 |
| 1234678-HpCDF | 0.2 | 0.5 | 0.3 | 1.4 |
| 1234789-HpCDF | < 0.1 | 0.1 | < 0.1 | 0.8 |
| OCDF | 0.4 | 0.5 | 0.3 | 1.6 |
| $\Sigma$-PCDF | 2.9 | 14.2 | 16.6 | 81.8 |

Adapted from Bishop et al. [82]
[a] Big Creek Marsh, north shore, eastern Lake Erie, $n = 5$
[b] Hamilton Harbour/Cootes Paradise, western Lake Ontario, $n = 7$ in 1989, $n = 12$ in 1990
[c] Lynde Creek, north shore, mid-Lake Ontario, $n = 8$

a relatively enclosed, industrialized bay on western end of Lake Ontario), Lynde Creek on the north shore of Lake Ontario east of Toronto, and Big Creek Marsh on the north shore of western Lake Erie, 1989–1991. Two other Great Lakes sites, Rondeau Provincial Park, Lake Erie and Cranberry Marsh, Lake Ontario were sampled, but it appeared that turtles from the these areas had little opportunity to feed on fish that had been exposed to lakewide contamination. $\Sigma$-PCDD/Fs in Rondeau turtle eggs were only 4.9 ng kg$^{-1}$, and at Cranberry Marsh were 18.1 ng kg$^{-1}$. The eggs from Hamilton and Lynde creek had comparable concentrations of 2378-TeCDD to liver in lower Lake Ontario/upper St. Lawrence River turtles [81] in 1985. The Lynde Creek turtle eggs had unusually high 12378-PnCDD and 123678-HxCDD concentrations compared to Lake Ontario fish, seabirds, and sediment. This suggests a local source.

Although the AHR PCDD/Fs dominated the pattern of contamination in the snapping turtles as they do in fish and birds, it is interesting to note that measurable concentrations of the non-AHR congeners 12478-PnCDF, 124678-HxCDF, and 124689-HXCDF were identified in both the Lynde Creek and 1989 Hamilton Harbour eggs, but not the 1990 Hamilton Harbour eggs, despite an overall higher $\Sigma$-PCDD/Fs concentration. Concentrations of 12478-PnCDF reached 30 ng kg$^{-1}$ in Lynde Creek turtle eggs, and may be connected to a local source like the high 12378-PnCDD and 123678-HxCDD concentrations. Since fish are not known to accumulate non-AHR congeners, it is likely that the turtles picked them up by eating invertebrates, such as clams and crayfish. It also means that turtles may have a reduced capacity to metabolize non-AHR PCDD/Fs compared to fish and birds. However, the absence of detectable 2378-TeCDF in any sample suggests that snapping turtles are able to metabolize this compound readily, like herring gulls and double-crested cormorants.

## 2.7
### Human Serum

A profile of organochlorine contaminants in serum of Great Lakes sports fish consumers, 1993, was determined in order to identify the contaminants which should be investigated further in a population-based study [17]. Participants were chosen based on consumption of sport fish from Lake Michigan ($n = 10$), Lake Huron ($n = 11$), and Lake Erie ($n = 11$) at least once per week. Median age was 48–56 years, and the sample was roughly half women, half men.

Concentrations of 2378-TeCDD, 123678-HxCDD, 23478-PnCDF, 123478-HxCDF, and 123678-HxCDF were about two- to threefold higher in all Great Lakes sports fish consumers than in the Arkansas comparison group (Table 11). The Lake Huron group of fish consumers stood out as having the highest 2378-TeCDD and 12378-PnCDD concentrations. However, concentrations of the other PCDD/Fs, which constituted the major of $\Sigma$-PCDD/Fs, were similar in Great Lakes and comparison groups. OCDD > 1234678-

**Table 11** Concentrations (ng kg$^{-1}$ lipid weight) of PCDD/Fs in human serum from sport fish consumers in the Great Lakes compared to a control group in Jacksonville, Arkansas. Adapted from Anderson et al. [17]

|  | Lake Michigan | Lake Huron | Lake Erie | Comparison group |
|---|---|---|---|---|
| 2378-TeCDD | 4.7 | 10.5 | 4.3 | 2.8 |
| 12378-PnCDD | 9.8 | 16.0 | 5.8 | 6.6 |
| 123478-HxCDD | 11.4 | 8.4 | 6.6 | 9.0 |
| 123678-HxCDD | 120 | 142 | 115 | 71 |
| 123789-HxCDD | 8.7 | 6.5 | 5.8 | 9.4 |
| 1234678-HpCDD | 144 | 163 | 96 | 124 |
| OCDD | 793 | 919 | 623 | 971 |
| Σ-PCDD | 1087 | 1259 | 844 | 1198 |
| 2378-TeCDF | 2.4 | 2.1 | ND | 3.1 |
| 12378-PnCDF | ND | 1.7 | ND | 1.6 |
| 23478-PnCDF | 20.4 | 22.8 | 10.4 | 5.5 |
| 123478-HxCDF | 11.6 | 16.0 | 10.2 | 8.0 |
| 123678-HxCDF | 9.0 | 10.5 | 7.7 | 5.3 |
| 234678-HxCDF | 6.0 | 4.6 | 5.0 | 3.8 |
| 1234678-HpCDF | 22.1 | 23.0 | 15.0 | 21.3 |
| OCDF | ND | ND | ND | 6.9 |
| Σ-PCDF | 71 | 79 | 49 | 57 |

HpCDD > 123678-HxCDD were > 95% of Σ-PCDD/Fs in all groups. The dominance of these congeners is a very different pattern than found in fish and birds from the Great Lakes (Tables 5–10), a reflection of the broad-based incorporation of atmospheric-sourced PCDD/Fs into the human food chain from meat and dairy products [83]. Bioavailability of the higher chlorinated PCDD/Fs is much less of a factor in modifying source patterns in the atmosphere–plant–livestock route than it is in the atmosphere–water–sediment–fish route.

It was concluded that consumption of Great Lakes sport fish, especially from Lake Huron, increased the PCDD and PCDF contribution to TEQs 1.8-fold and 2.4-fold, respectively, in this group of relatively heavy consumers of Great Lake sport fish. However, the consumption of sport fish contributed a much larger increase in PCB contribution to TEQs (9.6-fold). The fact that Lake Ontario was not included in this survey is cause for concern. Contemporary lake trout and other sport fish from Lake Ontario probably have two to three times higher concentrations of 2378-TeCDD than fish from the Great Lakes included in the Anderson et al. [17] study, based on temporal trends in lake trout and herring gulls (see next section). This surely amplifies the contribution of PCDD/Fs to total TEQs in consumers of sport fish from Lake Ontario.

## 3
## Temporal Trends

### 3.1
### Sediment Cores

Time trends of PCDD/F loading to Great Lakes sediments based on analysis of cores were mostly covered in Sect. 2.3. Pearson et al. [19, 42] provided the most geographically comprehensive picture of depositional trends in Great Lakes sediments to 1990 (Fig. 8). The depositional trends were not always consistent for different cores within each lake. For example, PCDD concentrations in Lake Superior sediment cores changed hardly at all between 1940 and 1990, and actually increased steadily from the early 1900s to 1990 in one Lake Michigan core, but not in others. However, where maxima existed, they were in the 1960–1970 period. Post 1970, decreased deposition of PCDD/Fs to sediments was most pronounced in Lake Ontario, which was the only lake thought to be seriously impacted by non-atmospheric sources. Modeling of PCDD/F deposition from air (Sect. 2.1.2) did not suggest such large differences among lakes in atmospheric deposition profiles. Baker and Hites [31] believed that focus-correction problems may be responsible for some these apparent trend anomalies.

The information on historical PCDD/F atmospheric deposition rates in the Great Lakes was updated by analysis of two sediment cores obtained from Siskiwit Lake, Isle Royale, Lake Superior in 1998 [31]. The trend of $\Sigma$-PCDD and $\Sigma$-PCDF concentrations, 1888–1998, is presented in Fig. 11. There were few notable changes in the pattern of homologs over the whole period. The percent contribution of TeCDFs–HxCDFs increased with increasing $\Sigma$-PCDF concentration, since these homologs accounted for most of the changes. OCDF concentrations remained relatively constant after 1910. There appeared to be no changes in the pattern of PCDD homologs over time. Translated into depositional fluxes, these data indicated a slow increase in PCDD/F deposition between 1888 and 1940. As was observed in previous studies [20, 41, 43], the increases in PCDD/Fs deposition post 1940 tracked the production of chloro-organics in North America. Total PCDD/F depositional flux in Siskiwit Lake peaked at about 9.5 pg cm$^{-2}$ year$^{-1}$ in 1975–1980. Between 1980 and 1998, depositional fluxes decreased about 50%. If this trend has continued since that time, PCDD/F deposition to the Great Lakes from the atmosphere may be approaching the pre-chloroorganic production background level.

US EPA estimated that there were major reductions in emissions from three of the five major PCDD/F sources to the atmosphere: municipal incineration (86%), medical waste incineration (81%), and secondary copper smelting (72%), and a total emission reduction of 77% between 1987 and 1995 [84]. This emission reduction is somewhat optimistic given the 50% de-

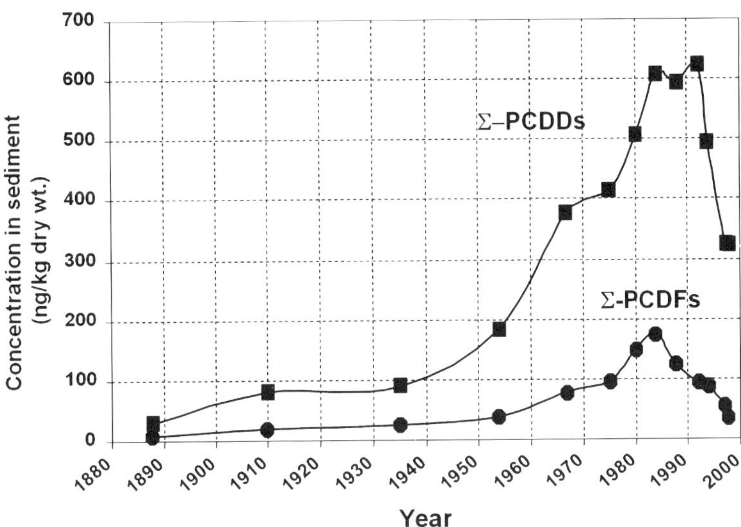



**Fig. 11** Σ-PCDDs and Σ-PCDFs concentrations (ng kg$^{-1}$ dw) in a sediment core from Siskiwit Lake, Isle Royale, Lake Superior, 1888–1998 (from Baker and Hites [31])

crease in Siskiwit depositional fluxes over a similar time period. However, as pointed out previously, approximations of PCDD/F emissions to the atmosphere are lower than actual deposition. This may be due to underestimation of emissions from some some sources, e.g., forest fires [32], vehicle emissions [85], or formation of OCDD and HpCDD from pentachlorophenol in atmospheric water droplets [23]. OCDD and HpCDD constitute 50–70% of Σ-PCDD/F of atmospheric flux to Great Lakes sediments, and are undoubtedly responsible for much of the deficit.

Another possible contribution to the PCDD/F decrease in the 1980s is the phase-out of leaded gasoline, which contained halogenated compounds as lead scavengers. This phase-out began in 1976 in the USA and Canada primarily in response to the introduction of catalytic converters to reduce NOX and SOX emissions, not in response to concerns about lead. US EPA only banned lead in gasoline in 1986. Most of the information about halogenated compounds in gasoline formulations is anecdotal. Thus, several sources note that one of the lead-scavenging compounds used in leaded gasoline was 1,2-dichloroethane, but brominated compounds were also used. Addition of organochlorine compounds to gasoline most likely enhanced formation of PCDD/Fs in vehicle exhaust. Vehicle emissions have been remarkably poorly studied, considering their potential importance. There are no reports of direct emission measurements from North American vehicles in the scientific literature. The most definitive research used to develop emissions inventories was done in traffic tunnels in Europe and the results simply extrapolated based on TEQs/vehicle/km/time. Given the differences in gasoline formulations among countries, not to speak of differences in internal combustion versus diesel engine emissions, it must be concluded that there is large margin for error in PCDD/F emissions inventory estimates for vehicles. A good analysis of present knowledge on the subject of PCDD/F emissions from vehicles can be found in Smit [85].

## 3.2
## Fish

Although programs monitoring organochlorine contaminants in fish were established in the Great Lakes in the 1970s as part of the international Great Lakes Water Quality Agreement, there was not a very systematic effort to archive tissues for subsequent analysis until the 1980s. It was therefore impossible to reconstruct temporal trends in organochlorine concentrations in the 1970s when concentrations were decreasing rapidly in most of the Great Lakes. The greater part of the trend data are also for Lake Ontario.

Temporal trends of PCDD/F concentrations in Lake Ontario lake trout, 1977–1993 were determined by Huestis et al. [65]. Time trends of the mean concentrations of the major congeners are shown in Fig. 12. Very similar trends were found in lake trout eggs from Lake Ontario collected during much the same period, 1978–1991 [16]. There were no strong trends in concentrations of the major PCDF congeners, 2378-TeCDF, and 23478-PnCDF.

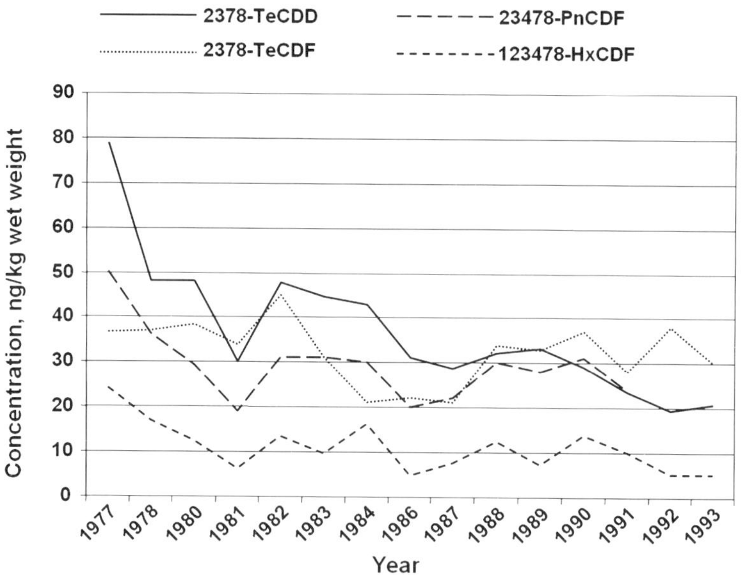


**Fig. 12** Temporal trends of mean concentrations (ng kg$^{-1}$) of the major PCDD/F congeners in Lake Ontario lake trout, 1977–1993 (from Huestis et al. [65])

However, 2378-TeCDD and 123478-HxCDF concentrations declined fairly consistently throughout the 16 year period by about fourfold. The different trends for these two groups of congeners is likely to be linked to different sources. There is no doubt that the majority of 2378-TeCDD and 123478-HxCDF entered Lake Ontario via industrial waste streams and dumps along the Niagara River, which is discussed in more detail later.

## 3.3
## Herring Gull Eggs

As noted in Sect. 2.6.1, the Great Lakes Herring Gull Monitoring Program has proven to be an invaluable resource in tracking long-term temporal trend data for organochlorine compounds in the Great Lakes [70]. Since 1984, shortly after the discovery of PCDD/F contamination of the Great Lakes, composite samples from each of the regular monitor colonies have been monitored annually for PCDD/F concentrations, however, only data for 1984–1991 are published. For 1981–1983 there are PCDD data only; 2378-TeCDD was determined between 1971–1981 in archived eggs from one colony in Lake Ontario [71]. The mean data were discussed in Sect. 2.6.1 and are presented in Table 7.

The trends in concentrations of five major PCDD/Fs found in herring gull eggs at eight colonies (one in each of the Great Lakes, Saginaw Bay, Detroit River, and Niagara River) are presented in Fig. 13. Derivation of sources from these patterns of PCDD/F concentrations is discussed in Sect. 4.1. However, it is obvious from the clumping together of data among lakes and rivers and similarity in annual variations that a common source is responsible for the PCDD/Fs seen in herring gull eggs from most of the Great Lakes. Saginaw Bay eggs stood out as the most contaminated for all five PCDD/Fs. Lake Ontario eggs had similar 2378-TeCDD concentrations as Saginaw Bay eggs. Except for OCDD, there was at least a twofold decrease in concentrations of all PCDD/Fs in Saginaw Bay eggs during the 1980s. Concentrations of 12378-PnCDD and 123678-HxCDD in Saginaw Bay eggs were approaching the average of the other colonies by the end of the 1980s. Concentrations of 2378-TeCDD decreased about fourfold in both Lake Ontario and Saginaw Bay eggs over the 1980s, but were still about four times higher than the average for eggs from the other lakes. Most of the other PCDD congeners showed decreases between 1981 and 1983, but changes thereafter were not significant.

Temporal trends of 2378-TeCDD concentrations in herring gull eggs from colonies in eastern Lake Ontario over a 31 year period, 1971–2002, are shown in Fig. 14. Data for 1971–1982 are for eggs from Scotch Bonnet Island, on the west side of the Prince Edward Peninsula, and those for 1983–1993 are from Snake Island, in the Kingston, ON area [71]. Historically, concentrations of organochlorine compounds have tended to be higher in the Scotch Bonnet

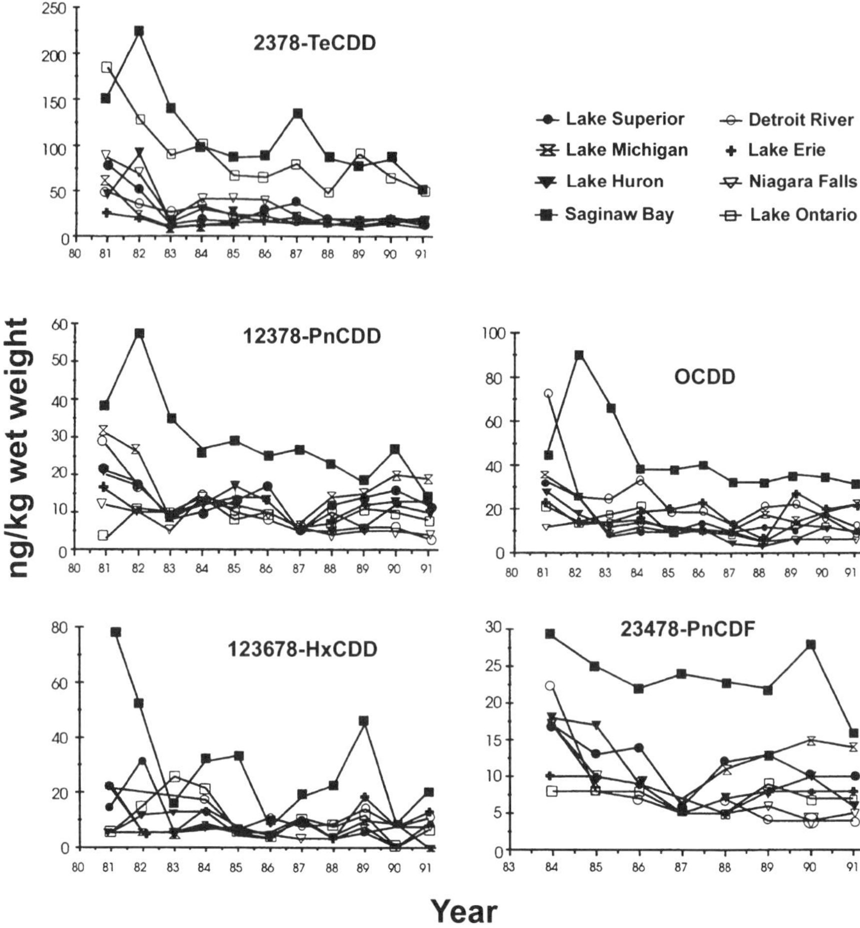



**Fig. 13** Temporal trends of concentrations (ng kg$^{-1}$ wet weight) of the four major PCDDs (1981–1991) and 23478-PnCDF (1984–1991) in pooled herring gull eggs from eight routine monitoring colonies in the Great Lakes. Reproduced with permission from Hebert et al. [71]

colony than other colonies in the area [86]. This is probably due to a higher proportion of fish in the diet of Scotch Bonnet gulls, since the colony is somewhat isolated from human and terrestrial food sources. In 1981 and 1982, the only years that eggs from both colonies were analyzed, 2378-TeCDD was 1.2 and 1.6 times higher in Scotch Bonnet than Snake Island eggs. This represents only a shift up of 0.1 to 0.2 log units in Fig. 13, which does not significantly affect the continuity of data or the apparent trends. Concentrations of 2378-TeCDD have decreased two orders of magnitude in Lake Ontario gull eggs over the last three decades of the 20th century, and appear to be continu-

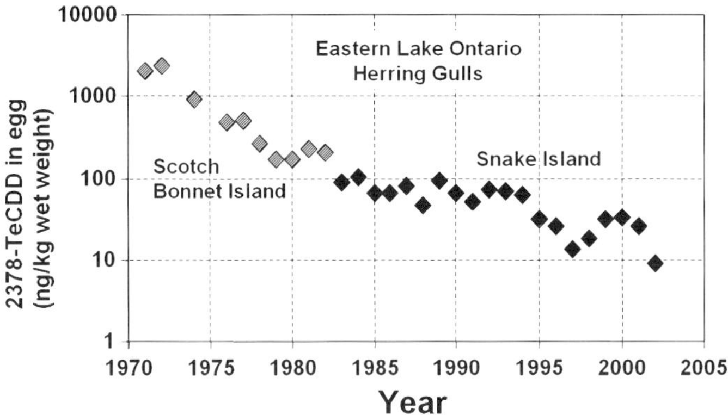



**Fig. 14** Temporal trend of concentrations (ng kg$^{-1}$ wet weight) of 2378-TeCDD in pooled herring gull eggs from two colonies in eastern Lake Ontario, 1971–2002. Data for 1971–1982 are from Hallett and Norstrom [11], for 1984–1991 from Hebert et al. [71], and for 1992–2002 from Weseloh (unpublished)

ing to decrease, although there are very large fluctuations in concentrations in recent years. Linear regression using the whole log-transformed data set (corrected or uncorrected for intercolony differences) is highly significant (P < 0.0001), giving a half-life of about 5.5 years. However, there is a perceptible slowing of the rate of exponential decrease over the long term. Therefore, a bi-exponential or an exponential to an asymptote would be better statistical models of the long term trends.

Extrapolation of the trend in 2378-TeCDD concentrations in herring gull eggs in the 1970s would predict concentrations tenfold *lower* than presently found in any of the Great Lakes. In the mid-1980s to mid-1990s, the trend was relatively flat, probably as a result of continuing input from the Niagara River. Extrapolation of that trend would predict three- to fivefold *higher* concentrations than actually found today, probably because efforts at waste containment in the Niagara River watershed were effective. This illustrates the danger of over-confidence in the prognostic power of environmental trend monitoring data. Unless the system is so stable and well understood that nothing is expected to occur that would disturb the underlying processes (climatic, hydrological, ecological, and source-related) which determine rate of exposure of the species and ecosystem in question, extrapolation of historical temporal trends based on statistical models of any kind is a very uncertain practice.

Thus, regardless of which mathematical model best fits the 2378-TeCDD temporal trends in herring gull eggs, further analysis was not attempted. There is no reason to believe that the environmental dynamics of persistent contaminants in Lake Ontario are necessarily described by processes which are fixed over time. Loadings from point sources in the Niagara River and relative importance of sources are likely to change. Environmental and ecological changes such as temperature fluctuations influencing food availability [87], food web structure changes which affect the trophic level of prey fish and thereby the trophic magnification factors [88], and changes in prey fish abundance [89] are factors influencing contaminant trends in Lake Ontario herring gulls. This results in a lot of ecological "noise". The "waves" with a periodicity of 5–7 years in the long-term trend of 2378-TeCDD in Lake Ontario herring gull eggs have also been noted for PCBs and other chlorocarbons [90]. They were partially explained by cold years affecting fish availability [87].

Concentrations of all PCDD/Fs in herring gull eggs continued to decrease or level off in the 1990s throughout the Great Lakes, including those from Saginaw Bay and Lake Ontario (Weseloh DV, Canadian Wildlife Service, unpublished data, 2004). For example, mean 2378-TeCDD concentrations in herring gull eggs, 1992–2002, relative to concentrations in 1984–1991, averaged over all colonies, was $0.51 \pm 0.12$. This is in excellent agreement with the decrease that would be expected from the diminishing atmospheric loading represented for the Lake Superior region in Fig. 11. The average 2378-TeCDD concentration in Saginaw Bay and Lake Ontario gull eggs in 1992–2002 was still two- to threefold higher ($21-35$ ng kg$^{-1}$) than in eggs from the other areas ($10 \pm 2$ ng kg$^{-1}$). This may have been due to continuing, but decreasing, input from the Saginaw and Niagara Rivers, but may also be partly due to recycling of historical contamination from lake sediments.

# 4
# Sources

## 4.1
## Combustion

Combustion sources contributing via atmospheric deposition to the Great Lakes are discussed in Sect. 2.1 ("Air") and Sect. 2.3 ("Lake Sediments"). Pearson et al. [19, 42] concluded that atmospheric deposition was the primary source of PCDD/Fs to Lake Superior and a major contributor to Lake Michigan. Atmospheric deposition was estimated to contribute only 5–35% of PCDD and 5% of PCDFs to Lake Ontario in 1994. However, these data are for homologs, which may not be valid for the AHR congeners. Czuczwa and Hites [20] concluded that most PCDD/F homologs in Great Lakes sediments came from atmospheric deposition) except in Lake Ontario. Because there are no published reports giving comprehensive congener-specific concentrations of PCDD/Fs in sediment cores from the Great Lakes, we must rely on the biological monitoring data to provide indication of the relative importance of sources of the bioaccumulating AHR congeners.

## 4.2
## Evidence from Herring Gull Eggs and Lake Trout

PCDD/Fs concentrations in herring gull eggs from throughout the Great Lakes, 1984–1991, were evaluated with regard to the contribution of possible sources [70]. Concentration profiles in various sources and environmental compartments of 2378-TeCDD, 12378-PnCDD, 123678-HxCDD, OCDD, and 23478-PnCDF (the major PCDD/Fs found in herring gulls) were modified by the product of a bioavailability index (BSAF) and bioaccumulation index (BMF) relative to 2378-TeCDD to create a transformation index (TI) that accounted for the probable modification of patterns during the course of bioaccumulation in a Great Lakes food web [71]. There were virtually no North American congener-specific data for air, sediments, and water, so European data were used. In this scheme, non-AHR congeners effectively had a TI of 0, since they are not accumulated by fish or herring gulls. The TIs of the other congeners relative to 2378-TeCDD were 12378-PnCDD 0.089, 123678-HxCDD 0.046, OCDD 0.00007, and 23478-PnCDF 0.079. A SIMCA classification analysis of the modified source data was carried out using two classes of normalized, autoscaled herring gull egg PCDD/F concentrations identified by hierarchical cluster analysis [71]. Transformed PCDD/F profiles of air particulates, water particulates, total air, and an incinerator sediment classified within or just outside the 95% probability of matching the patterns found in Lakes Michigan, Superior, Huron (excluding Saginaw Bay), Detroit River, and Lake Erie gull eggs. It was concluded that the PCDD/F contaminant profile seen in herring gulls from these lakes was most consistent with an atmospheric combustion-related source. Transformed sediment PCDD/F profiles from the Passaic River, NJ, which was impacted by 2,4,5,-T as well as an urban atmospheric sources, classified within 95% probability with herring gull eggs from Saginaw Bay, Lake Ontario and the St. Lawrence River, primarily due to the influence of high 2378-TeCDD. These areas were therefore most consistent with an atmospheric source underlying a specific source of 2378-TeCDD. Although there were many uncertainties in this analysis, it provided a statistically meaningful way of reconciling the very large difference in the patterns of concentrations between sources and gull eggs.

Consider the data in Fig. 14 in the context of changes in sources driving the trends in 2378-TeCDD in Lake Ontario over the last three decades of the 20th century. The sediment record clearly shows a peak and rapid decline in 2378-TeCDD in the early 1970s along with many other organochlorine compounds. Hooker Chemicals produced 2,4,5-trichlorophenol for 45 years at its Niagara Falls plant [4]. Global production of 2,4,5-T, one of the major end-uses of 2,4,5-trichlorophenol, peaked between 1960–1968, and declined 20-fold between 1970–1978 due to restrictions on its use [4]. The tenfold decrease in 2378-TeCDD in Lake Ontario herring gull eggs in the 1970s was most likely a reflection of this decline in production. Another important change in industrial practice in the Niagara River area that may have resulted in this decrease was closing of landfill waste disposal sites near the Niagara River. More concerted efforts to contain waste effluents may have also occurred because of the large attention on environmental calamities associated with bioaccumulating organochlorine compounds, principally DDT and PCBs at that time.

It is sufficient to note that present day concentrations of 2378-TeCDD in Lake Ontario herring gulls are low and approach those in Lake Superior, Lake Michigan, Lake Huron (excluding Saginaw Bay), and Lake Erie (4–6 ng kg$^{-1}$), indicating that the influence of point sources in the Niagara River is gradually being supplanted by atmospheric input as the main source to Lake Ontario. As such, future trends in 2378-TeCDD contamination in Lake Ontario and all the other Great Lakes are most likely to be dictated by waste incineration practices in eastern North America and forest fires.

Lake trout data were generally in agreement with the conclusions reached for herring gull eggs, modified by differences in bioaccumulation of the AHR congeners in the two species. Assuming that Lake Superior and Erie patterns of PCDD/F contamination were representative of an atmospheric profile, there were significant non-atmospheric sources of 2378-TeCDD, 12378-PncDD, 123678-HxCDD, and 23478-PnCDF in Lakes Michigan, Huron, and Ontario (especially 2378-TeCDD in Lake Ontario), and 2378-TeCDF in Lake Michigan lake trout (Table 5). Since Lake Ontario herring gull eggs had among the lowest concentrations of 23478-PnCDF (Table 7), there is an inconsistency with the lake trout data. There are potentially multiple sources of 23478-PnCDF to Lake Ontario, which may result in more uneven distribution of this congener in the food web than other PCDD/Fs, although none that appear unique to this lake. The lake trout food web may be more benthic-based (e.g., inclusion of slimy sculpin in the diet) than that of herring gulls [16].

It should be noted that 2,4,5-T was used as a non-agricultural herbicide in the Great Lakes watershed, mostly on roadside ditches and hydro rights-of-way. For example, Frank and Siron [91] estimated that in 1975, 279 kg of 2,4,5-T was used in 11 agricultural watersheds in Ontario, all of them flowing into Lake Huron, Lake Erie, Lake Ontario or connecting channels. This practice was probably common in all of the Great Lakes watershed, and may have been responsible for some 2378-TeCDD input to the Lakes.

## 4.3
## Saginaw River

The Saginaw River is the major source of PCDD/F contamination in Saginaw Bay. In the sections discussing sediments, fish, herring gulls, and humans, we have seen that Saginaw Bay is one of the most contaminated areas of the Great Lakes. However, the influence of this source does not extend to Lake Huron proper to nearly the same extent that the Niagara River influences Lake Ontario. This may be because Saginaw Bay is relatively shallow, long and narrow, and the flow of the Saginaw River is trivial compared to the Niagara River. These factors would encourage sedimentation of much of the PCDD/Fs in Saginaw Bay before they reach the main body of the lake. Nevertheless, Lake Huron samples sometimes have elevated concentrations of PCDD/Fs compared to Lake Michigan, which is probably due to the influence of input from the Saginaw River.

As we can see from Table 2, sediments at the mouth of the river at Bay City, MI had $5\,\mu g\,kg^{-1}$ of $\Sigma$-PCDDs and $4.1\,\mu g\,kg^{-1}$ of $\Sigma$-PCDFs. Flood plain soils had $14.8\,\mu g\,kg^{-1}$ of $\Sigma$-PCDDs and $10.6\,\mu g\,kg^{-1}$ of $\Sigma$-PCDFs [34]. The flood plain soil contamination is important to Lake Huron because the Tittabawassee River and Saginaw River are subject to major flooding, such as the 100-year flood in 1986, in which the Tittabawassee River crested at 3 m over the flood stage [92]. This undoubtedly swept a pulse of PCDD/F-contaminated soil downriver to Saginaw Bay. The 1986 flood was most likely responsible for a near doubling of 2378-TeCDD concentration in eggs of herring gulls nesting at the mouth of the Saginaw River in 1987, although concentrations dropped back to 1985–86 levels in 1988 (Weseloh DV, Canadian Wildlife Service, unpublished data, 2004). Surprisingly, concentrations of other PCDD/Fs in herring gull eggs were not similarly affected. Despite the lack of evidence for elevated 2378-TeCDD or other PCDDs in sediments and flood plain soils, herring gull eggs from the mouth of the Saginaw River consistently have had some of the highest 2378-TeCDD concentrations in the Great Lakes. It is interesting to note that in the early 1980s some parts of the Dow Chemical plant's sewer flow (to a waste treatment facility) had elevated 2378-TeCDD concentrations, with an estimated flow of $8.1\,g\,year^{-1}$ 2378-TeCDD [93]. The source was not from production processes at that time, rather the sewer sump, and therefore historical. A further $6.4\,g\,year^{-1}$ was contributed by waste incinerator effluents. It is not clear how much of this 2378-TeCDD was discharged to the Tittabawassee River.

Principal components analysis including the congener profiles from various potential sources was carried out [34]. It was concluded that pentachlorophenol, Aroclors 1242/1248 and graphite electrode sludge may all have contributed to the contamination in the Tittabawassee River. The prevalence of the six major TeCDF–OCDF AHR congeners is very similar to that found in graphite electrode sludge [44]. Although the relative homolog concentrations vary, the dominance of 123478-HxCDF among HxCDFs, and 1234678-HpCDF among HpCDFs, is also characteristic of Niagara River sediments.

## 4.4
## Niagara River

Lake Ontario is more highly contaminated by PCDD/Fs, primarily 2378-TeCDD, but also some PCDFs, than any other Great Lake. All evidence points to the Niagara River, which has long history of production of chlorinated organic chemicals, as the principal input of PCDD/Fs to Lake Ontario. Pearson et al. (1998) suggested that there was an additional source of PCDDs in eastern Lake Ontario, based on sediment profiles in eastern and western basins. It is worthwhile summarizing information on specific sources from waste sites along the Niagara River and current efforts to contain them. The following are extracts from various EPA Region 2 Site Fact Sheets.

Olin Corporation Niagara Falls Plant just north of the Niagara River manufactured several organic chemicals, including trichlorobenzene, trichlorophenol, and hexachlorocyclohexane (BHC) until 1956. Production of chlorine, caustic, and organic chemicals has ceased. In addition to contaminated soils at the site, a plume of groundwater contamination was identified which contained chlorobenzenes, hexachlorocyclohexane, and chlorinated phenols. The highest concentrations of these chemicals at the Olin Plant occurred in the bedrock between Alundum Road and Gill Creek. There does not appear to have been any attempt to identify PCDD/F contamination from this plant [94].

The DuPont Necco Park site was used for industrial waste disposal from the mid-1930s to 1977, receiving liquid and solid wastes from the nearby DuPont Niagara Plant. Wastes which may have contained PCDD/Fs included: fly ash, sodium sludge waste salts, cell bath, floor sweepings, chlorinolysis wastes, scrap organic mixtures, and off-grade product. Disposal activities were discontinued in 1977. The final design for hydraulic containment was received from DuPont in December 2003. Construction of the remedy is projected to be completed by February 2006 [95].

The Hooker-Hyde Park Site landfill was used to dispose of wastes from 2,4,5-trichlorophenol manufacture, which is a specific source of 2378-TeCDD. It has been estimated that 0.6–1 t of 2378-TeCDD is in the Hyde Park landfill. Cleanup actions at the Hooker-Hyde Park site were completed in June 2003. These consisted of removal of contaminated soils and sediments as well as the leachate control and treatment operations. Remedial construction included the installation of a system of extractions wells, both in the bedrock and overburden. A Leachate Treatment Facility was built on-site. Approximately 400 000 gallons of organic liquids have been extracted and incinerated, and 46 720 t of contaminated sediments were removed from Bloody Run Creek [96]. However, it is unclear whether these remediation efforts are completely effective in stopping seepage through the fractured bedrock into the Niagara River. Sediments sampled in 1979 from Bloody Run Creek, which drains this site, had concentrations of chlorobenzenes, chlorotoluenes, chlorophenols, and dichloro(trifluoromethyl)-substituted aromatic compounds in the 2–90 $mg\,kg^{-1}$ range [97]. As we have seen in Sect. 2.4.5, present-day sediments in the Bloody Run Creek area still have very high concentrations of PCDD/Fs, despite the remediation efforts. The shoreline has not been remediated. It is was thought that a rock slide in 1994 may have responsible for burying some contaminated sediments, but exposing others [38].

The Hooker 102 St. Landfill site is located adjacent to the Niagara River and south of Love Canal. The larger portion of the landfill was primarily in operation from 1943 until 1971. During that time, about 23 500 t of mixed organic solvents, brine sludge, fly ash, electrochemical cell parts, and 300 t of hexachlorocyclohexane process cake, including lindane, were deposited at the site. The smaller portion of the site was in operation from 1948 to about 1970, during which time 20 000 t of mercury brine and brine sludge, more than 1300 t of a mixture of hazardous chemicals, 16 t of mixed concrete boiler ash and fly ash (which may have contained PCDD/Fs) were disposed of at the site. The site has been remediated by construction of a slurry wall around the entire site, the installation of a leachate collection and pumping system, and the installation of a permanent, synthetic/clay cap over the landfill. Monitoring data are collected periodically to verify and confirm that migration of contaminants into the Niagara River has been eliminated. Currently, the EPA is proposing that the site be deleted from the National Priorities List [98]. Historically, this site was probably a major contributor to Niagara River chlorinated organic chemical contamination. Elder et al. [97] found tri- to hexachlorobenzenes, mono- to trichloronaphthalenes and dichlorophenanthrene at concentrations in the $8-200$ mg kg$^{-1}$ range in sediments near the 102 St. landfill.

Occidental Chemical Corporation disposed of approximately 63 000 t of chemical processing wastes into the S-Area landfill from 1947 to 1961. On- and off-site ground water and soil are contaminated with toxic chemicals including organic liquids immiscible with water and chlorinated benzenes. "Dioxin" is also present in ground water at trace levels. Various remedial systems provide physical and hydraulic containment of the 63 000 t of chemical waste buried in the landfill. Approximately 320 000 gallons of contaminated ground water are treated per day, with the treated effluent discharged to the Niagara River via a permitted outfall. Since the startup of the S-Area remedial systems in 1996, over 400 million gallons of contaminated ground water have been treated and 196 000 gallons of organic liquids have been collected for incineration, mostly at a commercial incinerator in Texas [99].

The most notorious of the Niagara Falls area chemical landfill sites is Love Canal, although not necessarily the major contributor to Niagara River contamination. Love Canal is the prototype Superfund site. The landfill was used by Hooker Chemicals and Plastics for the disposal of over 21 000 t of various chemical wastes, including halogenated organics, pesticides, chlorobenzenes, and dioxin starting in 1942. Dumping ceased in 1952. In 1953 the landfill was covered and eventually partially developed, including a school built on the site. Public alarm about exposure to chemicals leaching from the site eventually reopened the issue in 1978. By October 1987, EPA selected a remedy to address the destruction. In 1989, a partial consent decree (PCD) was entered into to address some of the required remedial actions, and Occidental Chemical Corporation was allowed to incinerate wastes off-site. In November 1996, further modification was issued to include off-site EPA-approved incineration and/or land disposal of the stored Love Canal waste materials. After 1998, the sewer and creek sediments and other waste materials were shipped off-site for final disposal. Remedial action was deemed complete in March 2000 [100].

Which of these sites may have contributed to the contamination of the Niagara River and Lake Ontario historically? The rather uniform profile of PCDF contamination in sediments along the river (Fig. 9) supports disposal of graphite electrode wastes as the major source of PCDF contamination presently found in Niagara River sediments [44]. Kaminsky and Hites [101] demonstrated that the chronology of octachlorostyrene contamination in Lake Ontario sediments followed the pattern of use of graphite electrodes for chlorine production in the Great Lakes states. There was a rapid switch to metal anodes beginning in the early 1970s, which presumably would have resulted in a decrease in PCDF- as well as octachlorostyrene-containing wastes. Sediment core chronology indicates that there was a substantial decline in PCDF loading to Lake Ontario in the 1970s [42].

Howdeshell and Hites [102] identified three aromatic fluorinated compounds associated with the production of 4-(trifluoromethyl)-chlorobenzene in sediment core samples from Lake Ontario in 1993. These compounds were specifically associated with the Hyde Park dumpsite near the Niagara River [97], which was also likely to be the main source of 2378-TeCDD (Fig. 9). The average maximum years for concentrations of these fluorinated compounds was 1969–1971 in the Niagara, Mississauga, and Rochester Basins, and 1975 in the Kingston Basin. Pearson et al. [19] noted that the peak PCDF concentration in Lake Ontario sediments corresponded to that of the fluorinated compounds, suggesting that Hyde Park may be the source of the unusual PCDF contamination of Lake Ontario as well as 2378-TeCDD.

Data from the Niagara River Mussel Biomonitoring Program [38] and the Herring Gull Monitoring Program (Table 7, Fig. 14) suggest that the Niagara River continues to be a source of PCDD/F contamination, primarily 2378-TeCDD, to Lake Ontario, although at much lower levels than the 1960s and 1970s. At the risk of over-interpretation, temporal trends of 2378-TeCDD in herring gull eggs (Fig. 14) are consistent with a decreasing input to Lake Ontario after about 1995, subsequent to a decade in which little change in concentration occurred. This renewed onset of 2378-TeCDD decline in the mid-1990s coincides roughly with a number of waste containment efforts along the Niagara River, and may be encouraging evidence of the success of these programs.


# 5
# Effects

Discussion of PCDD/F effects in fish, wildlife and humans is enormously complicated by the fact that a number of PCB congeners also exhibit AHR-related toxicity, some with relative potency approaching that of 2378-TeCDD itself.

PCBs are the main chlorocarbon contaminant in the Great Lakes. Because the mechanism of action is the same or similar for all AHR-active compounds, the effects cannot be attributed to individual compounds or classes of compounds unless their relative potencies as well as concentrations are known. Additivity of toxicity of AHR-active compounds based on appropriately derived TEQs has been established for a number of species, and the TEQ approach is now widely accepted, including use in environmental quality guidelines. There can be large differences in relative potency factors (toxic equivalency factors, TEFs) in fish, birds, and mammals among the various AHR-active compounds, which has led the World Health Organization to recommend different sets of TEFs for these three classes of animals [30]. Thus the same mix of AHR-active PCDD/F and PCB congeners may have different toxicity and different relative contribution of the two classes of compounds to $\Sigma$-TEQs, depending on the animal class. Furthermore, it is clear that there can be large differences in species toxic response to TEQs. For example, Elonen et al. [103] studied the embryotoxicity of 2378-TeCDD in seven species of fish and found a fivefold difference in toxicity (LOEC, $LC_{10}$, $LC_{50}$) from the least sensitive (zebrafish and northern pike) to the most sensitive (fathead minnow). So use of TEQs may help assess *relative* integrated toxicity of a mix of AHR-compounds, but TEQs on their own say nothing about how toxic that concentration is to the animal.

Like homologs, TEQs bury a lot of useful information. If TEQs are reported exclusively, it is very difficult to use the information for any purposes other than the intended one. The tendency in recent years to report TEQs in air and sediment using mammal-based WHO TEFs to satisfy human risk assessments (Sects. 2.1 and 2.3) obviates using these data for risk assessment in fish and birds. It is understandably not feasible to present all the AHR congener-specific data in these reports, however, it would not be all that difficult to generate a complete set of TEQs for fish and birds as well as mammals, and wherever possible, authors should make congener concentrations available. I strongly recommend that this become standard practice.

A comprehensive discussion of TEQ toxicity is beyond the scope of this review, but a brief description of the more important studies will be given.

## 5.1
## Lake Trout

Zacharewski et al. [104] compared in vitro bioassay-derived with chemical analysis-derived TEQs. The study avoided identifying the samples, apart from "fish homogenates", and used five samples from each of the Great Lakes and St. Clair. However, it is likely that the samples were the same lake trout as those in DeVault et al. [64], since the geographical coverage and $\Sigma$-PCDD/Fs concentrations are the same in the two studies. The bioassay-derived TEQs were significantly higher (>twofold) than the chemical analysis-derived TEQs in Lake Erie and Lake Ontario, suggesting that non-PCDD/F AHR-active compounds may have been present. For some reason, the authors did not attempt to include PCBs, which will contribute.

Huestis et al. [65] found that H4IIE-derived bioassay TEQs were significantly correlated to calculated TEQs in lake trout using a variety of different TEF schemes based on EROD induction. Even using H4IIE-TEFs, the bioassay measured about 50% higher TEQs than calculated, similar to the findings of Zacharewski et al. [104]. This probably indicates there were other AHR-active compounds present which were not measured. CB126 and 2378-TeCDD constituted the majority of the calculated TEQs. A trout cell line bioassay had very different TEFs, and consequently different relative contributions of the various PCDD/F and PCB compounds to calculated TEQs. The major contributor was 23478-PnCDF $\approx$ 2378-TeCDD in this scheme. Despite the difference, the significance of correlation and slope of H4IIE-measured and trout TEF-calculated TEQs was very similar to the H4IIE–TEF correlation. Clearly, correlation between chemical concentrations among individuals and years is sufficiently high to give any of the schemes empirical modeling power for bioassay-measured TEQs.

By far the most comprehensive research into AHR-related effects of PCDD/Fs on fish was a retrospective analysis of Lake Ontario lake trout reproductive impairment due to AHR-mediated early life stage mortality [16]. This includes "blue sac disease" as well as sublethal effects, which may increase susceptibility of sac fry and alevins to increased mortality and predation during swim-up. Lake trout are more susceptible to AHR-mediated toxic effects than any other Great Lakes species, with the possible exception of mink. WHO TEFs for fish were used to calculate the 2378-TCDD equivalent ($\text{TEC}_{\text{egg}}$ or TEQ) concentrations in lake trout eggs. The validity of the additive toxicity equivalence model was established through early life stage trout toxicity tests. The WHO fish TEFs are likely to be fairly robust for lake trout, since they were determined primarily from relative potency values for effects in embryos of a related salmonid, rainbow trout, even if the relative sensitivity of the species to 2378-TeCDD toxicity may be different.

BSAFs were determined by analysis of lake trout females and eggs, and the top 1 cm section of a sediment core from the Rochester basin in 1987–1988 from Lake Ontario that was considered representative of lake trout habitat in eastern Lake Ontario. One other core near the mouth of the Niagara River was taken for comparison. Historical concentrations in lake trout eggs were then reconstructed by applying the BSAFs to concentrations of AHR-active compounds in dated sections of both cores. Considerable effort was made to date the cores as accurately and precisely as possible by examination of the $^{210}$Pb and $^{137}$Cs profiles. The core-specific BSAFs were two to three times larger than those calculated using lakewide average sediment concentrations, and should not be construed as the *true* BSAFs, rather a relative value scaled to this particular site in order to take advantage of the higher time resolution offered by a core from a zone of high sedimentation rate. The underlying assumption was that the chronology of concentrations, at least, was the same in lake trout habitat throughout the lake. This was partially justified by the similar chronology of the two sediment cores.

The chronology of estimated TEQs in lake trout from Lake Ontario, 1910–2000, is shown in Fig. 15. The percent contribution of PCB126 to TEQs was about 6% prior to 1960, increased to 14% at the peak TEQs in the late 1960s and early 1970s. Thereafter the percent contribution to TEQs increased to about 22% because 2378-TeCDD concentrations decreased more rapidly than those of PCBs. Approximately 60% of TEQs were attributed to 2378-TeCDD. The importance of 2378-TeCDD to total TEQs is unique to Lake Ontario among the Great Lakes. The contribution of 2378-TeCDD was only 12% of TEQs in lake trout eggs in Lake Michigan [105]. Hebert et al. [71] found the same relative contribution to TEQs in herring gull eggs in these two lakes. The remaining contributors to TEQs in lake trout were mainly 12378-PnCDD and 23478-PnCDF.

It is apparent that the chronology of PCDD/F and PCB loading to Lake Ontario was different. PCB concentrations peaked in the 1969–1972 period, while 2378-TeCDD concentrations appear to have peaked about 1967. Fur-

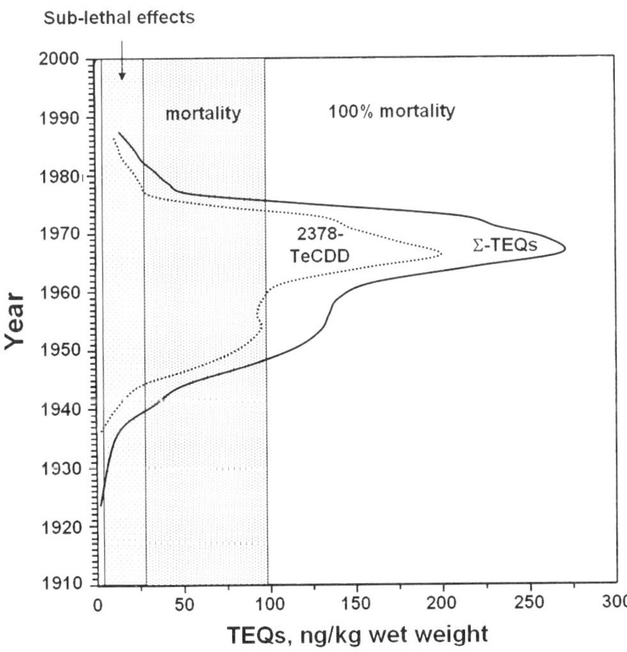


**Fig. 15** Temporal trends of estimated concentrations (ng kg$^{-1}$ wet weight) of 2378-TeCDD and Σ-TEQs in Lake Ontario lake trout eggs, 1910–2000, compared to thresholds for onset of sublethal effects, mortality, and 100% mortality of sac fry (adapted from Cook et al. [16]). Concentrations were estimated from sediment core concentration profiles and BSAFs

thermore, not all of the PCDD/Fs had similar chronology. Between 1940 and the mid-1970s, 12378-PnCDD and 23478-PnCDF loadings were relatively constant, and decreased along with other PCDD/Fs after the mid-1970s. Because 2378-TeCDD, the major contributor to TEQs post-1950, had much more dramatic increases and decreases in loading, the percent contribution of 12378-PnCDD and 23478-PnCDF to TEQs varied over time from about 40% in the 1940s to 10–15% at peak TEQs in the late 1960s and thereafter. There clearly was an early and relatively constant source of these compounds, presumably in the Niagara River, but perhaps other industrialized areas around Lake Ontario or atmospheric deposition. There are no sources of 12378-PnCDD mentioned in the scientific literature which are specific to this congener, so its chronology remains a mystery. Note that 12378-PnCDD was significantly higher in Saginaw Bay herring gull eggs than in eggs from any other colony in the Great Lakes in the 1980s [71]. The trend in 23478-PnCDF loading to sediments was similar to that of CB126, indicating that PCBs may be an important source of 23478-PnCDD.

Application of the TEQ risk assessment model showed that significant mortality of lake trout fry was likely to have occurred in the 1940s, and to have increased to 100% by 1950, where it remained until the mid-1970s (Fig. 15). The model predicted that mortality would have ceased in the mid-1980s after TEQs dropped below the critical threshold of 30 ng kg$^{-1}$. Lake trout population trends in Lake Ontario were estimated from number of fish caught yearly. A thorough analysis of population trends, which included stocking of fry sporadically from the early 1900s to the 1940s, showed that natural reproduction ceased in the 1950s, in congruence with the model. Furthermore, experimental studies showed decreasing mortality from blue sac disease of fry hatched from Lake Ontario lake trout eggs in the 1977–1991 period from about 50% to no incremental mortality, which was very close to the expected improvement based on the TEQ model. The conclusion was that AHR-mediated toxicity resulting in blue sac disease, modeled by TEQs, was primarily responsible for extirpation of lake trout from Lake Ontario post-1940. Most of this effect was due to 2378-TeCDD. However, even in the absence of this contaminant, TEQs from CB126, 23478-PnCDF, and 12378-PnCDD would have been sufficient to cause some mortality in the 1960s and early 1970s.

The good news is that TEQ concentrations have been below the mortality zone for lake trout since the mid-1980s in Lake Ontario. The bad news is that they may still be in the sublethal effects zone, primarily due to 2378-TeCDD concentrations.

Smith et al. [106] investigated embryonic mortality of coho and chinook salmon from the Credit River, Lake Ontario, 1990, correlated to H4IIE-derived TEQs. No correlation was found, but TEQs were correlated with concentrations of PCBs and other chlorocarbons, as expected. The authors pointed out that the relative sensitivity of coho salmon to dioxin-like embryotoxicity is much lower than in lake trout, and that H4IIE is not a good measure of TEQs for fish. However, assuming that concentrations of 2378-TeCDD, the most likely contributor to coho-specific TEQs in Lake Ontario, were correlated to the concentrations of PCBs and other OCs, the H4IIE-derived TEQs are probably still a good relative measure, and the conclusion is sound. Mean H4IIE-derived TEQs were $191 \pm 68$ ng kg$^{-1}$ in chinook salmon eggs and $161 \pm 55$ ng kg$^{-1}$ in coho salmon eggs.

## 5.2
## Herring Gulls

Long before there was any suspicion of effects of PCDD/Fs in fish, it was hypothesized that they may be one of the causes of reproductive failure of herring gulls in Lake Ontario at a time when reproduction was thought to be relatively normal in other Great Lakes [107]. Complete failure of herring gulls to raise chicks in Lake Ontario was noted as early 1972 [14]. Reproduction bounced back from essentially zero to near normal in the space of only 2 years, 1975–1977 [70]. Concentrations of most organochlorine concentrations were decreasing throughout this period. A Spearman Rank Order Correlation was performed between fledging success and egg concentrations of various chemicals, including 2378-TeCDD, $\Sigma$-PCDDs, TEQs (herring gull-based, chicken-based and rat H4IIE-based TEFs), $\Sigma$-PCBs, and hexachlorobenzene. Hexachlorobenzene had the highest correlation ($p = 0.0009$), followed by 2378-TeCDD, $\Sigma$-PCDDs and herring gull-specific TEQs ($p = 0.0072$), but other chemicals also had significant correlations (R.J. Norstrom, Canadian Wildlife Service, Environment Canada, unpublished data, 2004).

Gilbertson et al. [108] had argued that there was a high degree of consistency in the symptoms observed in Lake Ontario herring gull embryos with "Chick Edema Disease", an AHR-mediated toxicity in chickens, and coined the acronym, GLEMEDS, standing for Great Lakes Embryo Mortality, Edema and Deformities Syndrome, The major symptoms of chick edema disease, apart from mortality, were pericardial edema, porphyria, and skeletal deformities. Chick edema disease was first observed in chickens consuming PCDD-contaminated feed [109]. However, the diagnosis of GLEMEDS in herring gulls was not as clear cut as it seemed. Nosek et al. [110] showed that symptoms of chick edema disease were absent in pheasant embryos exposed to 2378-TeCDD at concentrations which were sublethal, and noted that many of the symptoms found in domestic fowl were not exhibited in embryos of wild bird species. Nosek and coauthors suggested that the one toxic response which was common among bird species was embryo mortality. This finding removed one of the major underpinnings of GLEMEDS as the expected suite of AHR-mediated effects in wild birds, leaving embryo mortality.

So, how much of the reproductive failure of herring gulls in Lake Ontario in the early 1970s can be attributed to AHR-mediated effects? A case can be made that relative species sensitivity of birds to embryo mortality is inversely proportional to EROD induction equivalency factors (IEFs) in embryo hepatocytes. At least, the ratio of $LD_{50}$ of 2378-TeCDD in pheasant/chicken is 5–9 [110], and the inverse ratio of IEFs is 10 [15]. The inverse ratio of EROD IEFs for herring gull/chicken is 50 [15]. Therefore the expected $LD_{50\ \text{herring gull}}$ of 2378-TeCDD is in the order of $50 \cdot LD_{50\ \text{chicken}} = 12\,500\ \text{ng\,kg}^{-1}$ [12]. This is sixfold higher than 2378-TeCDD concentrations found in herring gull eggs in the early 1970s ($2000\ \text{ng\,kg}^{-1}$, Fig. 14). However other AHR congeners, including PCBs, were present. When these are added, the total TEQs in early 1970s herring gull eggs was in the order of $9000\ \text{ng\,kg}^{-1}$, or about $0.75\ LD_{50}$ (Norstrom, unpublished data, 2004). This suggests that some of the mortality was due to AHR-mediated toxicity, but much of it associated with PCBs.

Hexachlorobenzene is a well-known porphyrinogen, and may have been the cause of most of the porphyria seen in herring gull embryos. Concentrations of hexachlorobenzene in Lake Ontario herring gull eggs during the period of reproductive failure in the early 1970s were within the crudely-estimated $LD_{50}$ from a herring gull egg-injection study [111], and could account for a large fraction of the reproductive failure.

Herring gull ecotoxicology in the Great Lakes was reviewed by Hebert et al. [70]. AHR-mediated toxicity may be involved in some of the sublethal effects observed, such as reduced hepatic retinoid levels, retinol to retinol palmitate ratio increases in eggs and decreased plasma corticosteroid levels, but no cause-effect linkage could be made because concentrations of all the organochlorine residues were correlated. Since the review [70], Grasman et al. [112] found a correlation between heterophil/lymphocyte ratio and herring gull-specific TEQs (derived from relative potency in *in vitro* embryo hepatocyte EROD induction), but the significance of this finding to immune function was unclear. Lorenzen et al. [113] found basal corticosterone concentrations in herring gulls to be negatively correlated to concentrations of $\Sigma$-PCDD/Fs ($p = 0.011$), $\Sigma$-PCBs ($p = 0.018$), non-*ortho*-PCBs ($p = 0.026$) and herring gull-specific TEQs ($p = 0.048$). Because the correlation was somewhat better with PCDD/Fs than PCBs, there is a higher probability that the effect is genuinely due to PCDD/Fs, but the participation of PCBs cannot be ruled out. Malic enzyme activity was found to be negatively correlated specifically to $\Sigma$-PCDD/Fs ($p = 0.04$), and not to the other measured parameters, including TEQs. This may indicate a PCDD/F-specific (as opposed to an AHR-mediated) effect on fatty acid synthesis, since malic enzyme provides NADPH for this process.

## 5.3
### Other Seabirds and Snapping Turtles

Kubiak et al. [80] studied the reproductive impairment of Forster's terns in Green Bay compared to a relatively uncontaminated site nearby. Increased incubation period, reduced hatchability, lower body weight, increased liver to whole body ratio, and edema were observed to be higher in the Green Bay terns. Only 2378-TeCDD was determined in the eggs. From Table 9, there were likely contributions from higher chlorinated PCDD congeners. However, Green Bay is primarily contaminated by PCBs. Using the best available TEFs at that time, the authors calculated that TEQs from PCBs were one to two orders of magnitude higher than from PCDD/Fs. While the effects were consistent with AHR-mediated toxicity, it was unlikely that PCDD/Fs contributed much to the effects observed.

Sanderson et al. [114] investigated the embryotoxic effects of PCDD/Fs and TEQs in double-crested cormorants from three colonies in British Columbia, one from Saskatchewan, and one from Lake Ontario in 1991. Because there were large differences in relative contamination from PCDD/F and PCB exposure in these sites (high PCDDs in British Columbia, high PCBs in Lake Ontario), factoring cause/effect relationships was more readily achieved. The TEF scheme used to calculate TEQs was H4IIE-based, which grants PCBs a higher contribution than the WHO avian TEFs. Nevertheless, there was a good correlation between EROD induction in embryos and TEQs. These findings lend credence to the use of TEQs as a reasonable integrated determinant of AHR-mediated effects in cormorants.

TEQs in double-crested cormorants and Caspian terns in several Great Lakes colonies in 1988 were calculated using a combination of different TEF schemes. It was concluded that most of the TEQs were due to PCBs, primarily PCB126 and PCB105 [75]. Although the TEF values differed for individual chemicals from the WHO avian TEFs, the conclusion would not change if the WHO TEFs were used. Occurrence of live-deformities (crossed-bill, clubbed-foot) was related to $\Sigma$-TEQs, but was also probably related to $\Sigma$-PCBs concentration. Ludwig et al. [115] found a significant dose-response relationship between TEQs and egg mortality in double-crested cormorants and Caspian terns in the Great Lakes, 1986–1991, and concluded that the evidence was consistent with an AHR-mediated effect.

Ryckman et al. [74] studied the incidence and temporal trends of bill deformities in double-crested cormorants in the Canadian Great Lakes. While the prevalence of bill-deformities was higher than in reference sites (Lake Nipigon and Lake-of-the-Woods), there were no differences in incidence among colonies in the Great Lakes between 1988 and 1996. Given the relatively low (and even) concentrations of PCDD/Fs in cormorant eggs (Table 8), if the cross-bill deformity was AHR-mediated, it was likely to be due to PCBs.

Survival, incidence of abnormalities, and oxidative stress were studied in artificially incubated double-crested cormorant eggs from three colonies in the Great Lakes in 1998 [76]. Survival of the embryos was in the normal range for artificial incubation of cormorant eggs. WHO avian TEFs were used to calculate TEQs. In all cases, TEQs were dominated by non-PCBs. The main focus of the paper was to look at oxidative stress. Although some evidence of oxidative stress was found, it could not be concluded that TEQs contributed to this stress.

In conclusion, historically it appears that there have been several AHR-mediated effects in seabirds in the Great Lakes, which probably contributed to reproductive failure and an increased incidence of live-abnormalities (in cormorants), but most of these were due to the effect of AHR PCB congeners, primarily PCB126. The exceptions may be Lake Ontario and Saginaw Bay, where 2378-TeCDD concentrations and all PCDD/F concentrations, respectively, were very high in the 1970s. Contemporary AHR-mediated effects in Great Lakes seabirds are more likely to be subtle, such as effects on immune system function and fatty acid synthesis, rather than population-level effects such as reduction in reproductive success. Hoffman et al. [116] reviewed PCB and PCDD/F toxicity in birds.

Bishop et al. [82] determined the incidence of abnormalities in artificially incubated snapping turtle eggs from eight sites in Ontario, including three sites in Lake Ontario, two in Lake Erie, and one in the St. Lawrence River, 1989–1991. The St. Lawrence River site is highly contaminated with PCBs. Incidence of total abnormalities (unhatched eggs plus deformities) ranked Lake Ontario > St. Lawrence River > Lake Erie > Algonquin Park. Abnormalities were correlated to EROD activity and CYP1A concentrations among sites. Abnormalities were most significantly correlated to concentrations of mono-*ortho* PCBs and HpCDFs and OCDF, less significantly correlated to PCDDs and lower chlorinated PCBs, and not at all correlated to non-*ortho* PCBs, multi-*ortho* PCBs (i.e., the major congeners) or TEQs calculated using three different TEF schemes. Reanalysis of earlier data showed a significant correlation between abnormal development and PCDD/Fs concentrations. Like seabirds, it was difficult to conclude the role of AHR-mediation in the suite of effects observed, but with EROD enzyme activity and CYP1A findings indicate that a biochemical response was present. There is no basis for determining an appropriate TEF scheme for reptiles, so it is not surprising that the there was no correlation with TEQs.

## 5.4
## Humans

In the autumn of 1993, the Health Departments of five Great Lakes states banded together to study the body burdens of PCDD/Fs and PCBs in heavy consumers of sport fish from the Great Lakes. Serum from 100 subjects was analyzed for PCDD/Fs and non-*ortho* AHR-PCB congeners, and TEQs were calculated. Multiple linear regression was used to find correlations with fish species and number of years consumed, age, body mass index, gender, and lake. Only TEQs were given. Lake Huron subjects had significantly higher PCDD-related TEQs than subjects from Lake Michigan ($p < 0.05$); however, the difference was not large, 10.5 and 8.1 ng kg$^{-1}$ lipid, respectively. Men had higher concentrations of

PCDD-, PCDF- and PCB-related TEQs than women ($p = 0.0001$). There were no significant differences in PCDF-related TEQs among lakes. This was also true for PCB-related TEQs despite a 1.6-fold higher median concentration of PCB-related TEQs in Lake Michigan subjects. $\Sigma$-TEQs were essentially identical in subjects from all three lakes, 21–21.7 ng kg$^{-1}$ serum lipid.

Consumption of lake trout was a significant predictor of concentrations of log ($\Sigma$-PCDFs). Consumption of both lake trout and salmon was a significant predictor of log ($\Sigma$-AHR-PCBs). Fish consumption was not related to $\Sigma$-PCDDs. In the whole data set, the proportional contributions to TEQs were estimated to be 45% for PCDDs, 35% for PCDFs, and 20% for PCBs. This estimate of PCB-related TEQs is very similar to the 25% of average daily intake of TEQs in the USA estimated by Hays and Aylward [83]. If the WHO TEFs had been used, it is likely that the proportions would not have changed significantly, although there would be additional TEQs from mono-*ortho*-AHR-PCBs. Thus, the importance of PCDD/Fs in AHR-mediated effects in humans would appear to be much larger than in wildlife. Paradoxically, the explanation is the relative unimportance of fish in human PCDD/F exposure, which is mainly via meat and dairy products [83]. There is no question that consumption of fish from the Great Lakes increases 2378-TeCDD and PCB exposure however. Anderson et al. [17] found two to seven times higher concentrations of $\Sigma$-PCBs in Great Lakes subjects than in a comparison group. The concentrations ranked Lake Michigan > Lake Huron > Lake Erie. Hays and Aylward [83] note that exposure of the general population to PCDD/Fs has dropped fourfold since 1980. Estimates of 2378-TeCDD exposure in combined data from USA, Canada, Germany, and France revealed about a tenfold decrease in exposure to 2378-TeCDD since 1970.

It is beyond the scope of this review to deal with specific epidemiological studies on effects in humans in the Great Lakes area. Hays and Aylward [83] note that the present daily intake of TEQs in the average US population is estimated to be 1 pg day$^{-1}$. Most agencies around the world have adopted tolerable daily intakes (TDI) of 1–3 pg TEQ day$^{-1}$, while EPA has concluded safe doses two to three orders of magnitude lower. The EPA estimate is so low as to be unachievable and perhaps lower than the baseline pre-industrial exposure, according to Hays and Aylward [83]. They suggested that an acceptable balance between control efforts and negligible human health risks is close to being achieved. However, any incremental increase in exposure from consumption of Great Lakes fish is likely to push exposure of some individuals over the 1–3 pg day$^{-1}$ TDI. Johnson et al. [18] reviewed the public health implications from exposure to persistent toxic substances in the Great Lakes. They concluded that populations, such as heavy Great Lakes fish consumers, are at risk for disruption of reproductive function, and neurobehavioral and developmental deficits in newborns. As is the case for seabirds, most of these effects, if present, are probably due to PCB exposure, rather than PCDD/Fs.

## 6
## Conclusions

All evidence points to Lake Ontario being by far the most seriously impacted by PCDD/F contamination among the Great Lakes, primarily from the chlorine chemical industry along the Niagara River on the US side. This source produced a unique congener profile in sediments in which 2378-TeCDD, 1234678-HxCDD, HpCDFs, and OCDF were more dominant than in other lakes, although OCDD was still the most prevalent congener in all lakes. The current profile of PCDD/F congeners in Niagara River sediments is suggestive of graphite electrode waste from chlorine production, rather than production of specific chemicals, although this may also have occurred. There is no doubt that effluents and wastes from production of 2,4,5-trichlorophenol were responsible for the high relative contribution of 2378-TeCDD to PCDD/F contamination in Lake Ontario. Estimates of sediment inventories to 1994 showed Lake Ontario to have 3.4 times more $\Sigma$-PCDD/Fs than Lake Michigan, and 6.6 times more than Lake Superior. Because bioavailability and bioaccumulation of the highly chlorinated PCDD/Fs are low, it is the most toxic congener, 2378-TeCDD, which is of most interest in Lake Ontario. Sediment core temporal trends showed a rapid increase in 2378-TeCDD between 1940 and the mid-1950s, and an even faster increase beginning around 1960, with doubling times of 5 years to a peak around 1968. Concentrations decreased rapidly after 1972 to 1940s levels by 1980. It is worth speculating that some of the increase of 2378-TeCDD in Lake Ontario in the 1960s was due to increased production of 2,4,5-T for Agent Orange, which was employed in Viet Nam from 1965–1970. Winding down of the Viet Nam War removed much of the market for 2,4,5-T, and this may have contributed to the declines of 2378-TeCDD in the early 1970s. Another factor in the decline was probably increased efforts on the part of chemical industry to contain wastes due to a general awareness of the environmental persistence and biomagnification of organochlorine compounds. Abandonment of graphite electrodes for chlorine production in favor of metal anodes around 1970 undoubtedly reduced the production of PCDF wastes. Virtually all classes of persistent organochlorine compounds experienced similar declines in the 1970s throughout the Great Lakes, but particularly in Lake Ontario.

We should keep in mind that the Hyde Park landfill site, sitting in fractured bedrock only 0.6 km from the Niagara River gorge, is widely reported to contain in the order of 1000 kg of 2378-TeCDD, although an authoritative confirmation of this information was not found. If true, this is a great deal more than has already been flushed into Lake Ontario, considering that 2378-TeCDD is a small percentage of the inventory of 5800 kg total PCDD/Fs in Lake Ontario sediments.

Very strong evidence has been presented, based on sediment core chronology, population data, and experimental toxicology, that lake trout were extirpated from Lake Ontario beginning in the 1940s due to sac fry mortality associated with exposure to 2378-TeCDD and related compounds. From 1950 to 1975, TEQs were above the 100% fry mortality threshold, and exceeded the mortality threshold throughout the whole 40-year period, 1940–1980. It is likely that reproduction of salmon introduced to Lake Ontario was also affected. Other species of fish do not accumulate such high concentrations, and were apparently not affected. Complete lack of herring gull reproduction in Lake Ontario due to death of embryos was first noted in the early 1970s, but probably occurred in the 1960s as well. This reproductive failure was likely to have been partly caused by 2378-TeCDD, although other chemicals, especially hexachlorobenzene and PCBs, were probably equally or more important. Evidence for effects on other species of seabirds is much weaker, but may have been present. Lake Ontario herring gulls are more at risk from Great Lakes contaminants than most seabirds because they remain in the lake all year. Incidence of abnormalities was higher in Lake Ontario snapping turtles than in those from cleaner areas, but no direct cause/effect linkage could be made with PCDD/Fs.

Another Great Lakes area potentially impacted by local PCDD/F contamination is Saginaw Bay, Lake Huron. PCDD/F contamination of the Tittabawassee/Saginaw River system by the Dow facility at Midland, MI has been well-established. Herring gull eggs from a dredge-spoil island at the mouth of the Saginaw River consistently have the highest concentration of PnCDDs–OCDD in the Great Lakes, and are tied with Lake Ontario herring gull eggs for highest 2378-TeCDD concentrations. However, the Saginaw River does not have the same lakewide influence on Lake Huron that the Niagara River does on Lake Ontario. This is probably because flows are very much less in the Saginaw River, and most of the PCDD/F is likely deposited to sediments in Saginaw Bay. There is recent concern for high concentrations of PCDD/Fs in flood plain soils along the Tittabawasee River, which have the potential to contaminate Saginaw Bay during floods. Sediment profiles and concentrations in lake trout and salmon indicate that there is some additional loading of PCDD/Fs in Lake Huron above atmospheric deposition, and this is likely to be from the Saginaw River.

Pulp and paper industrial activity on the Fox river is a major source of PCB contamination to Green Bay, Lake Huron. It is not all that clear whether significant PCDD/F contamination has been associated with this source. One study suggested that northern Lake Michigan sediments received two-thirds of their PCDD from non-atmospheric sources. Concentrations of 23478-PnCDF and 2378-TeCDF tend to be higher in northern Lake Michigan lake trout and herring gull eggs, which is consistent with the an origin in Green Bay, since both these compounds are associated with commercial PCB mixtures. Forster's tern reproduction in Green Bay in the early 1980s was low, but this was most likely due to PCB exposure.

Lake Superior and Lake Erie receive most of their PCDD/F burden from the atmosphere, and have the lowest levels of contamination among the

Great Lakes. This is best demonstrated by comparison of PCDD/F concentrations in Lake Superior lake trout and Lake Erie walleye. The concentrations and profile of contamination are essentially identical in these two samples. Recent modeling efforts have identified a variety of potential atmospheric sources to the Great Lakes, and quantified deposition fluxes separately for each of the lakes. High resolution analysis of a sediment core from an isolated lake on an island in Lake Superior provided the best chronology of atmospheric deposition. PCDD/F depositional fluxes increased slowly between 1888 and 1940, tracking the production of chloro-organics in North America, peaked about 1975–1980, and decreased by about 50% between 1980 and 1998. Total PCDD/F atmospheric emissions in the USA were estimated by EPA to have been reduced 77% between 1987 and 1995. Given the imprecision of the estimate, this is in rather good agreement with the sediment core data.

There is no doubt that substantial progress has been made in reducing both atmospheric and land-based input of PCDD/F contamination in the Great Lakes. In many cases concentrations in fish and seabirds are close to two orders of magnitude lower than historical peaks in the early 1970s. Atmospheric fluxes to the lakes seem to be approaching the pre-organochlorine production background level. Lake Ontario and, to a lesser extent, Lakes Michigan and Huron continue to have additional input from local sources above atmospheric input. However, there is some evidence that even in these lakes, the relative importance of direct input is diminishing.

**Acknowledgements** The author would like to thank Chip Weseloh and the Canadian Wildlife Service for providing both old and new unpublished data on PCDD/F concentrations in herring gull eggs. Mark Cohen is thanked for reviewing the sections of the manuscript on air and generously providing figures on air deposition. Ryan Forsberg is thanked for his efforts in creating the reference data base. Carleton University is thanked for its financial support.

# References


1. The Chlorinated Dioxin Task Force, Michigan Division Dow Chemical USA (1978) The trace chemistries of fire – a source of and routes for the entry of chlorinated dioxins into the environment
2. Harless RL, Lewis RG (1980) 28th Annual Conference on Mass Spectrometry and Allied Topics, p 592
3. Harless RL, Oswald EO, Lewis RG, Dupuy AE Jr, McDaniel DD, Tai H (1982) Chemosphere 11:193
4. Esposito MP, Tiernan TO, Dryden FE (1980) Dioxins. EPA-600/2-80-197
5. Lustenhouwer JWA, Olie K, Hutzinger O (1980) Chemosphere 9:501
6. Bowes GW, Mulvihill M, Simoneit BRT, Buringame AL, Risebrough RW (1975) Nature 256:305
7. Amendola G, Barna D, Blosser R, LaFleur L, McBride A, Thomas F, Tiernan T, Whittemore R (1989) Chemosphere 18:1181

8. Beck EC (1979) The Love Canal tragedy, January 1979. EPA J
9. Norstrom RJ, Hallett DJ, Simon M, Mulvihill MJ (1982) Analysis of Great Lakes herring gull eggs for tetrachlorodibenzo-*p*-dioxins. In: Hutzinger O (ed) Chlorinated dioxins and related compounds: impact on the environment. Pergamon, New York p 173
10. Norstrom RJ, Simon M (1983) Preliminary appraisal of tetra- to octachlorodibenzodioxin contamination in eggs of various species of wildlife in Canada. In: Miyamoto J (ed) IUPAC pesticide chemistry, human welfare and the environment. Pergamon, New York p 165
11. Hallett DJ, Norstrom RJ (1981) 2,3,7,8-TCDD in Great Lakes herring gulls. Environment Canada, Canadian Wildlife Service internal report
12. Allred PM, Strange JR (1977) Arch Environ Contam Toxicol 5:483
13. Gilbertson M, Hale R (1974) Can Field-Natur 88:356
14. Gilbertson M, Hale R (1974) Can Field-Natur 88:354
15. Kennedy SW, Lorenzen A, Jones SP, Hahn ME, Stegeman JJ (1996) Toxicol Appl Pharmacol 141:214
16. Cook PM, Robbins JA, Endicott DD, Lodge KB, Guiney PD, Walker MK, Zabel EW, Peterson RE (2003) Environ Sci Technol 37:3864
17. Anderson HA, Falk C, Hanrahan L, Olson J, Burse VW, Needham L, Paschal D, Patterson D Jr, Hill RH Jr (1998) Environ Health Perspect 106:279
18. Johnson BL, Hicks HE, Jones DE, Cibulas W, Wargo A, De Rosa CT (1998) J Great Lakes Res 24:698
19. Pearson RF, Swackhamer DL, Eisenreich SJ, Long DT (1997) Environ Sci Technol 31:2903
20. Czuczwa JM, Hites RA (1986) Environ Sci Technol 20:195
21. Eitzer B, Hites R (1989) Environ Sci Technol 23:1396
22. Eitzer BD, Hites RA (1989) Environ Sci Technol 23:1389
23. Baker JI, Hites RA (2000) Environ Sci Technol 34:2879
24. Steer P, Tashiro C, Clement R, Lusis M, Reiner E (1990) Chemosphere 20:1431
25. Bobet E, Berard MF, Dann T (1990) Chemosphere 20:1439
26. Edgerton SA, Czuczwa JM, Rench JD (1989) Chemosphere 18:1713
27. Smith RM, O'Keefe PW, Aldous K, Connor S, Lavin P, Wade E (1990) Chemosphere 20:1447
28. Cohen MD, Draxler RR, Artz R, Commoner B, Bartlett P, Cooney P, Couchot K, Dickar A, Eisl H, Quigley J, Rosenthal JE, Niemi D, Ratte D, Deslauriers M, Laurin R, Mathewson-Brake L, MacDonald J (2001) Supporting information for modeling the atmospheric transport and deposition of PCDD/F to the Great Lakes. NOAA, Silver Spring, MD
29. Cohen MD, Draxler RR, Artz R, Commoner B, Bartlett P, Cooney P, Couchot K, Dickar A, Eisl H, Hill C, Quigley J, Rosenthal JE, Niemi D, Ratte D, Deslauriers M, Laurin R, Mathewson-Brake L, McDonald J (2002) Environ Sci Technol 36:4831
30. Van den Berg BM, Birnbaum L, Bosveld AT, Brunstrom B, Cook P, Feeley M, Giesy JP, Hanberg A, Hasegawa R, Kennedy SW, Kubiak T, Larsen JC, van Leeuwen FX, Liem AK, Nolt C, Peterson RE, Poellinger L, Safe S, Schrenk D, Tillitt D, Tysklind M, Younes M, Waern F, Zacharewski T (1998) Environ Health Perspect 106:775
31. Baker JI, Hites RA (2000) Environ Sci Technol 34:2887
32. Gullett BK, Touati A (2003) Atmospheric Environment 37:803
33. Gale RW, Huckins JN, Petty JD, Peterman PH, Williams LL, Morse D, Schwartz TR, Tillitt DE (1997) Environ Sci Technol 31:178
34. Hilscherova K, Kannan K, Nakata H, Hanari N, Yamashita N, Bradley PW, McCabe JM, Taylor AB, Giesy JP (2003) Environ Sci Technol 37:468

35. Meyer C, O'Keefe P, Hilker D, Rafferty L, Wilson L, Connor S, Aldous K (1989) Chemosphere 19:21
36. Jobb B, Uza M, Hunsinger R, Roberts K, Tosine H, Clement R, Bobbie B, LeBel G, Williams D, Lau B (1990) Chemosphere 20:1553
37. Oliver BG, Pugsley CW (1986) Water Pollut Res J Can 21:368
38. Richman LA (2003) Niagara River Mussel Biomonitoring Program, 2000. Ontario Ministry of the Environment
39. Hallett DJ, Brooksbank MG (1986) Chemosphere 15:1405
40. Czuczwa JM, Hites RA (1984) Environ Sci Technol 18:444
41. Czuczwa JM, Hites RA (1985) Historical record of polychlorinated dioxins, furans in Lake Huron sediments. In: Keith LH, Rappe C, Choudhary G (eds) Chlorinated dioxins and dibenzofurans in the total environment II. Butterworth, Toronto p 59
42. Pearson RF, Swackhamer DL, Eisenreich SJ, Long DT (1998) J Great Lakes Res 24:65
43. Czuczwa JM, McVeety BD, Hites RA (1984) Science 226:568
44. Rappe C, Glas B, Kjeller LO, Kulp SE, de Wit C, Melin A (1990) Chemosphere 20:1701
45. Kannan K, Watanabe I, Giesy JP (1998) Toxicol Environ Chem 67:135
46. Marvin CH, Charlton MN, Reiner EJ, Kolic T, MacPherson K, Stern GA, Braekevelt E, Estenik JF, Thiessen L, Painter S (2002) J Great Lakes Res 28:437
47. Marvin CH, Charlton MN, Stern GA, Braekevelt E, Reiner EJ, Painter S (2003) J Great Lakes Res 29:317
48. Smith LM, Schwartz TR, Feltz K, Kubiak TJ (1990) Chemosphere 21:1063
49. McKee P, Burt A, McCurvin D, Hollinger D, Clement R, Sutherland D, Neaves W (1990) Chemosphere 20:1679
50. Sherman RK, Clement RE, Tashiro C (1990) Chemosphere 20:1641
51. Marvin C, Alaee M, Painter S, Charlton M, Kauss P, Kolic T, MacPherson K, Takeuchi D, Reiner E (2002) Chemosphere 49:111
52. Mumma CE, Lawless EW (1975) Survey of industrial processing data: Task I: Hexachlorobenzene and hexachlorobutadiene pollution from chlorocarbon processes. US EPA, Office of Toxic Substances, Washington DC
53. Smith RM, O'Keefe PW, Aldous KM, Hilker DR, O'Brien JE (1983) Environ Sci Technol 17:6
54. O'Keefe P, Meyer C, Hilker L, Aldous K, Jelus-Tyror B, Dillon K, Donnelly R, Horn E, Sloan R (1983) Chemosphere 12:325
55. Harless RL, Lewis RG, Dupuy AE, McDaniel DD (1983) Analysis for 2,3,7,8-tetrachlorodibenzo-*p*-dioxin residues in environmental samples. In: Tucker RE, Young AL, Gray AP (eds) Human environmental risks on chlorinated dioxins and related compounds. Plenum, New York p 161
56. Giesy JP, Jude DJ, Tillit DE, Gale RW, Meadows JC, Zajicek JL, Peterman PH, Verbrugge DA, Sanderson JT, Schwartz TR, Tuchman M (1997) Environ Toxicol Chem 16:713
57. O'keefe PO, Wilson L, Buckingham C, Rafferty L (1990) Chemosphere 20:1277
58. Fehringer NV, Walters SM, Ayers RJ, Kozara RJ, Ogger JD, Schneider LF (1985) Chemosphere 14:909
59. Kuehl DW, Butterworth BC, McBride A, Kroner S, Bahnick D (1989) Chemosphere 18:1997
60. Stalling DL, Smith LM, Petty JD, Hogan JW, Johnson JL, Rappe C, Buser HR (1983) Residues of polychlorinated dibenzo-*p*-dioxins and dibenzofurans in Laurentian Great Lakes fish. In: Tucker RE, Young AL, Gray AP (eds) Human and environmental risks of chlorinated dioxins and related compounds. Plenum, New York p 221
61. Stalling DL, Norstrom RJ, Smith LM, Simon M (1985) Chemosphere 14:627

62. Whittle DM, Sergeant DB, Huestis SY, Hyatt WH (1992) Chemosphere 25:181
63. Braune BM, Norstrom RJ (1989) Environ Toxicol Chem 8:957
64. De Vault D, Dunn W, Bergqvist PA, Wiberg K, Rappe C (1989) Environ Toxicol Chem 8:1013
65. Huestis SY, Servos MR, Whittle DM, van den Heuvel MR, Dixon DG (1997) Environ Toxicol Chem 16:154
66. Giesy JP, Kannan K, Kubitz JA, Williams LL, Zabik MJ (1999) Arch Environ Contam Toxicol 36:432
67. Clark TP, Norstrom RJ, Fox GA, Won HT (1987) Environ Toxicol Chem 6:547
68. Norstrom RJ, Letcher RJ (1997) Role of biotransformation in bioconcentration and bioaccumulation. In: Sijm D, deBrun J, de Voogt P, Wolf P (eds) Biotransformation in environmental risk assessment, Society of Environmental Toxicology and Chemistry – Europe worskhop. SETAC, Brussels, Belgium, p 101
69. Norstrom RJ, Risebrough RW, Cartwright DJ (1976) Toxicol Appl Pharmacol 37:217
70. Hebert CE, Norstrom RJ, Weseloh DV (1999) Environ Rev 7:147
71. Hebert CE, Norstrom RJ, Simon M, Braune BM, Weseloh DV, Macdonald CR (1994) Environ Sci Technol 28:1268
72. CWS Ontario Region Fact Sheet. (2004) The rise of the Double Crested Cormorant on the Great Lakes: winning the war against contaminants.
73. Ludwig JP (2004) Jack-Pine Warbler 62:91
74. Ryckman DP, Weseloh DV, Hamr P, Fox GA, Collins B, Ewins PJ, Norstrom RJ (1998) Environ Monitor Assess 53:169
75. Yamashita N, Tanabe S, Ludwig JP, Kurita H, Ludwig ME, Tatsukawa R (1993) Environ Pollut 79:163
76. Hilscherova K, Blankenship A, Kannan K, Nie M, Williams LL, Coady K, Upham BL, Trosko JE, Bursian S, Giesy JP (2003) Arch Environ Contam Toxicol 45:533
77. Evers DC (1992) A guide to Michigan's endangered wildlife. University of Michigan Press, Ann Arbor
78. Gilbertson M, Reynolds LM (1972) Bull Environ Contam Toxicol 7:371
79. Ewins PJ, Weseloh DVC, Norstrom RJ, Legierse K, Auman HJ, Ludwig JP (1994) Canadian Wildlife Service Occasional Paper 85:1
80. Kubiak TJ, Harris HJ, Smith LM, Schwartz TR, Stalling DL, Trick JA, Sileo L, Docherty DE, Erdman TC (1989) Arch Environ Contam Toxicol 18:706
81. Ryan JJ, Lau BPY, Hardy JA, Stone WB, O'Keefe P, Gierthy JF (1986) Chemosphere 15:537
82. Bishop CA, Ng P, Pettit KE, Kennedy SW, Stegeman JJ, Norstrom RJ, Brooks RJ (1998) Environ Pollut 101:143
83. Hays SM, Aylward LL (2003) Regul Toxicol Pharmacol 37:202
84. EPA (2001) Summary of database of sources of environmental releases of dioxin-like compounds in the United States
85. Smit R (2004) Dioxin emissions from motor vehicles in Australia, 2. National Dioxins Program, Commonwealth of Australia
86. Norstrom RJ, Gilman AP, Hallett DJ (1981) Sci Total Envir 20:217
87. Hebert CE, Shutt LJ, Norstrom RJ (1997) Environ Sci Technol 31:1012
88. Hebert CE, Shutt LJ, Hobson KA, Weseloh DV (1996) Can J Fish Aquat Sci 56:323
89. Owens RW, O'Gorman R, Eckert TH, Lantry BF (2002) The offshore fish community in southern Lake Ontario, 1972–1998. In: Munawar M (ed) The state of Lake Ontario: past, present, and future. Backhuys, Leiden
90. Hebert CE, Norstrom RJ, Zhu J, Macdonald CR (1999) J Great Lakes Res 25:220
91. Frank R, Siron GJ (1980) Sci Total Envir 16:149

92. Deedler B (2003) Summary of 1986 flood including the Tittabawassee River. Tittabawassee River Watch News
93. Lamparski LL, Nestrick TJ, Frawley NN, Hummel RA, Kocher CW, Mahle NH, McCoy JW, Miller DL, Peters TL, Pilepich JL, Smith WE, Tobey SW (1986) Chemosphere 15:1445
94. EPA (2004) Olin Corporation, EPA ID# NYD002123461, Region 2 site fact sheet, March 2004
95. EPA (2004) DuPont Necco Park, EPA ID# NYD980532162, Region 2 site fact sheet, January 2004
96. EPA (2004) Hooker Hyde Park, EPA ID# NYD000831644, Region 2 site fact sheet, January 2004
97. Elder VA, Proctor BL, Hites RA (1981) Environ Sci Technol 15:1237
98. EPA (2004) Hooker 102 St Land, fill, EPA ID# NYD980506810, Region 2 site fact sheet, March 2004
99. EPA (2004) Hooker Chemical – S-Area, EPA ID# NYD980651087, Region 2 site fact sheet, March 2004
100. EPA (2004) Love Canal, EPA ID# NYD000606947, Region 2 site fact sheet, February 2004
101. Kaminsky R, Hites RA (1984) Environ Sci Technol 18:275
102. Howdeshell MJ, Hites RA (1996) Environ Sci Technol 30:969
103. Elonen GE, Spehar RL, Holcombe GW, Johnson RD, Fernandez JD, Erickson RJ, Tietge J, Cook PM (1998) Environ Toxicol Chem 17:472
104. Zacharewski T, Safe L, Safe S, Chittim B, DeVault DS, Wilberg K, Bergqvist P, Rappe C (1989) Environ Sci Technol 23:730
105. Cook PM, Zabel EW, Peterson RE (1997) The TCDD toxicity equivilence approach for characterizing risks for early life-stage mortality in trout. In: Rolland R, Gilbertson M, Peterson RE (eds) Chemically induced alterations in functional development and reproduction of fishes. Proceedings from a session at the Wingspread Conference Center, 21–23 July 1995, Racine, Wisconsin. SETAC, Pensacola, FL, Chap. 2 p 9
106. Smith IR, Marchant B, van den Heuvel MR, Clemons JH, Frimeth J (1994) J Great Lakes Res 20:497
107. Gilbertson M, Fox GA (1977) Environ Pollut 12:211
108. Gilbertson M, Kubiak T, Ludwig J, Fox G (1991) J Toxicol Environ Health 33:455
109. Firestone D (1973) Environ Health Perspect September:59
110. Nosek JA, Sullivan JR, Craven SR, Gendon-Fitzpatrick A, Peterson RE (1993) Environ Toxicol Chem 12:1215
111. Boersma DC, Ellenton JA, Yagminas A (1986) Environ Toxicol Chem 5:309
112. Grasman KA, Scanlon PF, Fox GA (2000) Arch Environ Contam Toxicol 38:244
113. Lorenzen A, Moon TW, Kennedy SW, Fox GA (1999) Environ Health Perspect 107:179
114. Sanderson JT, Norstrom RJ, Elliot JE, Hart LE, Cheng KM, Bellward GD (1994) J Toxicol Environ Health 41:247
115. Ludwig JP, Kurita-Matsuba H, Auman HJ, Ludwig ME, Summer CL, Giesy JP, Tillitt DE, Jones PD (1996) J Great Lakes Res 22:172
116. Hoffman DJ, Rice CP, Kubiak TJ (1996) PCBs and Dioxins in Birds. In: Beyer WN, Heinz GH, Redmond-Norwood AW (eds) Environmental contaminants in wildlife: Interpreting tissue concentrations. CRC, Chap. 7 p 165

Hdb Env Chem Vol. 5, Part N (2006): 151–199
DOI 10.1007/698_5_041
© Springer-Verlag Berlin Heidelberg 2005
Published online: 20 December 2005

# Pesticides in the Great Lakes

Kurunthachalam Kannan[1] (✉) · Jeff Ridal[2] · John Struger[3]

[1] Wadsworth Center, New York State Department of Health
and Department of Environmental Health Sciences,
State University of New York at Albany, Empire State Plaza,
Albany 12201-0509, PO Box 509 New York, USA
*kkannan@wadsworth.org*

[2] St. Lawrence River Institute of Environmental Sciences, 2 Belmont Street,
Cornwall, Ontario K6H 4Z1, Canada
*jridal@riverinstitute.ca*

[3] Environment Conservation Branch, Ecosystem Health Division, Environment Canada,
Burlington, Ontario L7R 4A6, Canada
*john.struger@ec.gc.ca*

**Abstract** Pesticides have been widely and heavily used in agriculture in the Great Lakes
Basin (approximately 93 000 tons were used in 1995 alone). Herbicides account for two-
thirds of the total pesticides used. Herbicide usage in seven of the Great Lakes states
(Illinois, Indiana, Michigan, Minnesota, Ohio, Pennsylvania, and Wisconsin) constituted
approximately 50% of the total usage in the USA. The pesticide use is concentrated in
the corn- and soybean-growing areas of the southern Lake Michigan Basin and western Lake Erie Basin of the Great Lakes. Organochlorines (OC) such as DDT and dieldrin
were the major pesticides used in the Great Lakes Basin prior to 1970s. The past usage of OC insecticides was large enough to cause effects on the Great Lakes ecosystem.
Whereas environmental concentrations of OC pesticides in the Great Lakes Basin have
generally declined during the past 20 years, concerns nevertheless remain, because these
substances persist in the environment and accumulate in the food chain. There continue
to be fish consumption advisories based on unacceptable levels of OC pesticides in sport
and commercial fish from the Great Lakes. Atmospheric transport from agricultural regions in the USA and Canada, where these pesticides were used extensively in the past,
continues to be a source of contamination in the Great Lakes. Hazardous waste sites with
elevated levels of OC pesticides represent another source. In the 1980s and the 1990s, OC
pesticides were replaced with new-generation pesticides, which are more target-specific
and less persistent. Among herbicides, atrazine, metolachlor, cyanazine, acetolachlor, and
alachlor account for about 53% of the total usage, and among insecticides, organophosphates (OP) such as malathion, chlorpyrifos, terbufos, diazinon, and methyl-parathion
account for 72% of the total usage in the late 1990s. Various monitoring programs have
shown that the levels of OC pesticides have declined steadily until the 1980s, and thereafter the rate of this decline has slowed. The relative slow decline or steady state in OC
levels in the 1990s was thought to be due to release/re-suspension and/or recycling of
these compounds through the Great Lakes ecosystem. Atmospheric deposition has become an increasingly significant route of entry of OC pesticides into the Great Lakes
ecosystem, although such depositions have decreased recently, and the lakes are now acting as a source via the degassing of these compounds. Most of the current-use pesticides
are not bioaccumulative; however, because of the high volume of their usage, these compounds are present in Great Lakes waters. The most frequently detected herbicides in
waters include several triazines (atrazine, cyanazine, and simazine), acetanilides (metolachlor and alachlor), and 2,4-D. In addition to streams and rivers, atmospheric transport
is a pathway for current-use pesticides in the Great Lakes. The occurrence of in-use pesticides in surface waters follows broad and complex patterns in land use and associated
pesticide use. In general, concentrations showed an increasing gradient from north to
south, with Superior < Huron < Ontario < Erie. Although most of the current-use pesticides are not bioaccumulative, exposure of aquatic organisms to these compounds can be
deleterious. Current-use pesticides can undergo environmental and biological transformations, although the degradates of the most heavily used herbicides found in surface water
have not been studied widely. In many cases, methods to assess the fate of current-use

pesticides and their degradates in the environment are not available. Future investigations should focus on the fate and effects of current-use pesticides.

**Keywords** Pesticides · Great Lakes · Herbicides · Insecticides · Toxicity


**Abbreviations**

| | |
|---|---|
| 2,4-D | 2,4-Dichlorophenoxyacetic acid |
| 2,4-DB | 2,4-Dichlorophenoxybutyric acid |
| 2,4-DP | 2,4-Dichlorophenoxybutyric acid |
| 2,4,5-T | 2,4,5-Trichlorophenoxyacetic acid |
| 2,3,6-TBA | 2,3,6-Trichlorobenzoic acid |
| 2,4,5-TP | 2,4,5-Trichlorophenoxyacetic acid |
| AOC | Areas of Concern |
| $p,p'$-DDE | 1,1-dichloro-2,2-*bis*(p-dichlorodiphenyl)ethylene |
| DDT | Dichlorodiphenyltrichloroethane |
| FDA | Food and Drug Administration |
| GLNPO | Great Lakes National Program Office |
| IADN | Integrated Atmospheric Deposition Network |
| IJC | International Joint Commission |
| HCB | Hexachlorobenzene |
| $\gamma$-HCH | $\gamma$-Hexachlorocyclohexane |
| MCPA | 4-chloro-2-methylphenoxyacetic acid |
| MCPB | 4(4-chloro-2-methylphenoxy)butyric acid |
| NASS | National Agricultural Statistics Survey |
| NAWQA | National Water Quality Assessment Program |
| OC | Organochlorine |
| OP | Organophosphorus |
| USDA | United States Department of Agriculture |
| USEPA | United States Environmental Protection Agency |
| USGS | United States Geological Survey |


# 1
# Introduction

One-third of the land area in the Great Lakes Basin is used for agriculture (17.3 million ha), with usage concentrated in the southern half of the Basin, with the Lake Erie Basin having the highest concentration. Nearly 75% of the Basin's agricultural land is in the USA; the remainder is in Canada [1]. Agricultural practices are a major source of pesticides in the Great Lakes.

Pesticides have been widely and heavily (approximately 93 000 tons in 1995) used for the control of weeds (as herbicides), insects (as insecticides), and diseases (as fungicides) in the Great Lakes Basin [2, 3], and to a lesser extent in urban settings [4]. In terms of mass applied, herbicides (accounting for two-thirds of the total) are used more extensively than insecticides in Ontario [5] and in the USA [4]. Corn and soybeans are the major crops receiving large amounts of herbicides in the Great Lakes Basin.

Organochlorine (OC) pesticides were widely used in the Great Lakes region in the past [6, 7]. Due to the persistence, toxicity, bioaccumulative properties, and long-range atmospheric transport of these pesticides, their use and production in North America were largely banned in the late 1970s. Organophosphorus (OP) and other insecticides have largely replaced the OC pesticides [8]; however, residues of the OC pesticides continue to be detected in most environmental matrices. The risks to wildlife and human health of pesticide exposure are a matter of public concern, and are the subject of continued scientific research in the Great Lakes Basin. In this chapter, trends in pesticide use, environmental levels, and patterns and chemodynamics of pesticides in the Great Lakes Basin are summarized.

## 1.1
## Overview

According to the National Agricultural Statistics Survey (NASS) of the United States Department of Agriculture (USDA), the Great Lakes Basin is home to diverse agriculture, reporting significant acreage for 36 different crops in 1995. Pesticide use is predominant in the southern Lake Michigan Basin and western Lake Erie Basin of the Great Lakes. While herbicides are used on field crops (e.g., corn, soybeans) grown annually, perennial crops such as tree fruits tend to have more insect and disease problems; accordingly these crops receive higher levels of insecticides and fungicides. Production areas for fruit crops are concentrated in Great Lakes coastal areas and a few inland areas where microclimate is conducive to their growth. For instance, the Niagara Peninsula is one of the major fruit-growing regions in Ontario and has a long history of intensive pesticide use [9]. OC pesticides such as DDT, aldrin, and dieldrin were extensively used on fruit crops from the mid 1940s to 1970s [10].

Compared to agricultural areas, urban areas traditionally have not been recognized as important contributors to pesticide contamination. However, nationwide surveys conducted as a part of the National Water Quality Assessment Program (NAWQA) of the United States Geological Survey (USGS) showed a widespread occurrence of some insecticides commonly used around homes (for lawn care or indoor pest control) and gardens and in commercial and public areas [11]. In fact, several of the insecticides monitored by the NAWQA occurred at higher frequencies and at higher concentrations in urban streams than in agricultural streams. The most commonly detected insecticides in urban streams were diazinon, carbaryl, chlorpyrifos, and malathion, which are among the 10 most highly used insecticides used for homes and gardens in the USA. Some herbicides, including atrazine, simazine, 2,4-D, and prometon, which are used to control weeds in lawns and golf courses and along roads, also occurred frequently in water samples collected from streams and shallow groundwater in urban areas. Thus, urban areas around the Great Lakes also contribute to pesticidal contamination.

Some currently used pesticides that are intermediate in water solubility and persistence can reach fairly high concentrations in water, sediment, and biota, in areas of high use. Pesticides that enter the lakes—whether by direct discharge along the shores, by runoff through tributaries, or from the atmosphere—are retained in the system and become more concentrated over time. For example, atrazine is detected widely in streams in the midwestern USA at concentrations in the range of several hundred ng/L to several tens of µg/L [12]. The half-life of atrazine in deeper lakes was estimated to be greater than 10 years.

The effects of pesticides on wildlife and humans in the Great Lake Basin are well documented. Monitoring of fish, bird, and mammalian species in the Great Lakes during the 1970s provided the first alert of adverse health effects due to exposure to OC pesticides such as DDT. A wide range of effects were observed between the mid-1970s and 1980s, including reproductive failure and population declines, in mink and in certain species of fish and birds in the Great Lakes. Most of these effects were attributed to OC compounds, which include chlorinated industrial chemicals and pesticides. DDT, dieldrin, toxaphene, mirex, and chlordane are no longer used for agricultural purposes in the USA or Canada. Nevertheless, many chlorinated pesticides are transported to the Great Lakes by the process of long-range atmospheric movement. Because of their persistence and historical uses, they continue to occur in the environment, and in wildlife and human tissues. For instance, $p,p'$-DDE concentrations of up to 17 parts-per-million (ppm; wet weight basis) were found in the tissues of bald eagles collected in the upper peninsula of Michigan in 2000 [13]. A $p,p'$-DDE concentration as great as 22 ppm (on a wet weight basis) was found in the eggs of bald eagles collected in 1994 [14].

The unintended effects of pesticides continue to be of concern, and evaluation of the risk to humans and to the environment from current levels of pesticide exposure remains controversial. Exposure is complicated by pesticide mixtures, breakdown products, and strong seasonal concentration pulses. In contrast, most toxicity assessments are based on controlled experiments, evaluating a single contaminant over a limited range of concentrations. Therefore, information on chemodynamics of pesticides is necessary for resource management, conservation, regulation, and policy making at regional and national levels.

## 2
## Trends in Pesticide Usage

Following the regulations on the use of OC pesticides issued in the 1970s, there has been a shift in the type and amount of pesticides used in agriculture [15]. The new-generation pesticides are more target-specific and less persistent. In the mid-1980s, OPs constituted 65% of all insecticides used in the USA. The overall pesticide usage in agriculture is declining, due to the reduction of cropland, regulatory issues, and changes in application rates. For example, atrazine and metolachlor accounted for a large portion of the total herbicide usage in Ontario in the 1980s [16]. Farms in Ontario have decreased their usage of atrazine by two-thirds since 1983. This is significant, since atrazine is one of the most common herbicides in use and also has a higher level of persistence (especially in lake water) than do many other pesticides. However, the use of metolachlor on field corn and soybeans in Ontario increased by 69% between 1983 and 1998 [17].

## 2.1
### Past- and Restricted-Use Pesticides

Organochlorine insecticides were the major class of compounds used in agriculture in the USA and Canada during the 1950s and the 1960s. Among OC insecticides, DDT was the predominantly used compound in agriculture in the USA. Annual production/use of DDT in 1962 in the USA was 80–85 million kg. In 1973, the use of DDT was banned in the USA, except for public-health purposes. In the early 1970s, most uses of DDT were banned in Canada, but the last remaining product registration was not discontinued until 1989. Aldrin and dieldrin ranked second, after DDT, in the USA in the 1960s for the control of soil insects and mosquitoes. Both were manufactured on a commercial scale between 1950 and 1990. Most uses of aldrin and dieldrin were banned in the USA in 1975. In Canada, all food-crop uses of dieldrin were prohibited in 1978, and only limited applications are now permitted. Estimated amount of annual production of aldrin and dieldrin together in 1966 was 9 million kg. The data from the USA indicated that the production was 90% aldrin and 10% dieldrin [18]. Because aldrin is converted into dieldrin in the environment and biota, the use and environmental levels of these two compounds are often combined. Aldrin use was most concentrated in the Midwestern USA for the control of corn worms in Illinois and Iowa. Mirex's use in Canada was banned in 1978. Although it is relatively stable, it slowly breaks down to photomirex (8-monohydromirex), a toxic chemical that contains one chlorine atom fewer than mirex. Annual production of mirex in the USA during 1963–1968 was 300 000–400 000 kg. About 25% of the mirex production was for insecticidal use, and the rest was used as a flame retardant. The flame retardant uses of mirex were curtailed in the 1970s, and it was replaced by other products. Chlordane was widely used in agriculture and homes (primarily for termite control), ever since its development as an insecticide in the mid-1940s. The estimated peak annual use rate of chlordane was 12 million kg in 1971. Although the domestic use of chlordane was curtailed in 1988, chlordane continued to be produced in the USA for export until 1997. Agricultural usage of chlordane has been restricted in Canada since 1970. Heptachlor and heptachlor epoxide have been restricted since 1978 in the USA and since 1970 in Canada. Heptachlor is also an impurity in the technical chlordane mixture. Toxaphene usage peaked in 1972–1975, when this compound was applied at a rate of 30 million kg annually. Toxaphene was prohibited from use in 1983 in the USA, and it was de-registered in 1982 in Canada.

The past usage of OC insecticides was large enough to cause global environmental contamination. Whereas environmental concentrations of OC pesticides in the Great Lakes Basin have generally declined during the past 20 years, and whereas current concentrations in surface water are below the drinking water standards, concerns nevertheless remain, because these substances persist in the environment and accumulate in the food chain. Through the process of biomagnification, concentrations of OC pesticides increase up the food chain. Food-chain magnification of DDT, one of the most bioaccumulative insecticides, in a Great Lakes food web is illustrated in Fig. 1 [19]. OC insecticides can magnify by a factor of several tens to hundreds in predatory organisms relative to their prey. Furthermore, the environmental half-lives of OC pesticides are on the order of several years; this is clear from the fact they are still detectable in the environment, even 30 years after the ban on their use. There continue to be fish consumption advisories based on unacceptable levels of OC pesticides in sport and commercial fish from the Great Lakes. Atmospheric transport from agricultural regions in the USA and Canada, where these pesticides were used extensively in the past, continues to be a source in the Great Lakes [20, 21]. Moreover, a few former production sites of OC pesticides are located in the Great Lakes Basin. Soil, water, biota, and other environmental media in and around these produc-

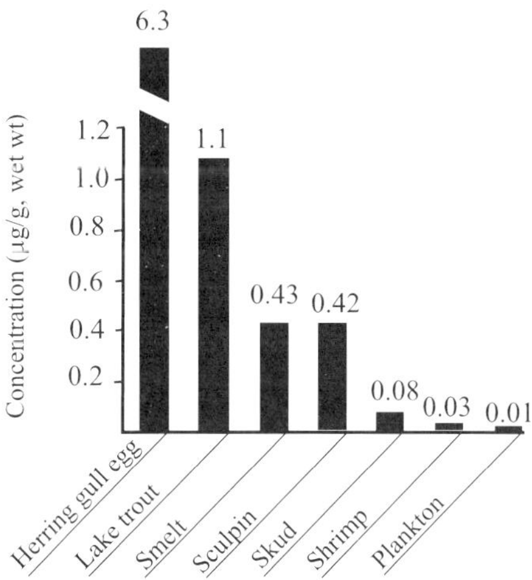


**Fig. 1** Biomagnification of DDT in the Great Lakes food chain. From [19]

tion sites are highly contaminated with the pesticide that was manufactured. For instance, a DDT concentration as great as 32 600 ppm was found in the sediment at the Pine River impoundment, near a former DDT production facility in St. Louis, Michigan [22]. Similarly, mirex was manufactured near Niagara Falls, New York, and State College, Pennsylvania. These production facilities discharged considerable amounts of pesticides directly into the tributaries or into landfill sites located near the Great Lakes. Approximately 30% of the US National Priority sites that list OC pesticides as priority pollutants occur within the Great Lakes region [23]. There are 20 aldrin-, 29 dieldrin-, 23 chlordane-, 44 DDT-, and three mirex-contaminated hazardous waste sites located in the Great Lakes Basin. Hazardous waste sites with elevated levels of these pesticides continue to represent sources to the Great Lakes, either through leaching into groundwater, soil runoff and erosion, or atmospheric transport. A summary of US Areas of Concern (AOC), listing OC pesticides as primary contaminants in the Great Lakes Basin, is shown in Table 1.

The Binational Toxics Strategy between the USA and Canada has identified 12 bioaccumulative substances (referred to as "Level-1" substances) having significant persistency and toxicity to the Great Lakes system, with the goal of reducing the sources of these substances to achieve "naturally occurring" levels [23]. Six of the 12 Level-1 substances are OC pesticides: aldrin/dieldrin, chlordane, DDT, HCB, mirex, and toxaphene. Several other OC pesticides, such as endrin, heptachlor/heptachlor epoxide, hexachlorocyclohexanes, tetra- and penta-chlorobenzenes, and pentachlorophenol, have been identified as "Level-2" substances.

A few OC insecticides that are in use currently—methoxychlor, dicofol, endosulfan, and lindane ($\gamma$-HCH)—are being reassessed by the U.S. Environmental Protection Agency (USEPA). For instance, approximately 45% of the 350 tons of dicofol was used in the USA in 1997 for the control of pests in

**Table 1** Areas of Concern (AOC) that list organochlorine pesticides as chemicals of concern in the US Great Lakes Basin

| State | AOC | OC pesticide |
| --- | --- | --- |
| New York | Buffalo River | Chlordane, DDT |
| | Niagara River | Mirex, Chlordane, DDT, Dieldrin |
| | Oswego Lake | Mirex |
| | Rochester Embayment | Mirex, DDT, Chlordane |
| | St. Lawrence River/Massena | Mirex, DDT |
| Ohio | Black River | DDT |
| | Cuyahoga River | DDT |
| Wisconsin | Menominee River | Several pesticides |
| | Milwaukee Estuary | Several pesticides |

From [23]

cotton. Dicofol was also used for the control of insect pests in fruit-growing areas of the Lake Michigan coast, in western Michigan. Similarly, 35 tons of methoxychlor and 693 tons of endosulfan were used in the USA in 1997. Use of endosulfan in the Great Lakes Basin between 1990 and 1998 accounted for approximately 15% of the total use in the USA [24]. These insecticides are used primarily for the control of pests in apples and cotton. The use of lindane in the USA is restricted to the southern states. However, lindane is one of the 10 most heavily used insecticides in Canada for the protection of livestock, crop seeds, and poultry until 2003. In Canada, lindane use was concentrated in the provinces of Saskatchewan, Alberta, and Manitoba for the control of pests in canola and corn; this use has been recently curtailed [25].

## 2.2
## Current-Use Pesticides

According to the National Agricultural Statistics Service (NASS) of the USDA, approximately 453 million kg of pesticides were used in the United States in the year 2002. The agricultural sector accounts for more than 75% of the total amount of pesticides used. Among the pesticides, herbicides accounted for the largest proportion of total usage ($\sim$ 70%), followed by insecticides (17%), and then fungicides. A survey conducted by the NASS in 2002 (http://www.pestmanagement.info/nass/) suggested that the herbicide usage in seven of the Great Lakes states (Illinois, Indiana, Michigan, Minnesota, Ohio, Pennsylvania, and Wisconsin) constituted approximately 50% of the total usage in the USA. This is primarily because the major corn- and soybean-growing areas are located in the Great Lakes Basin, particularly in Illinois, Indiana, Minnesota, Ohio, and Wisconsin. About 32% of the insecticide usage and 7% of the fungicide usage are accounted for by the seven Great Lakes states in the USA. The 10 pesticide active ingredients that were most commonly used in the USA in 1999 are listed in Table 2.

Organophosphates (OPs) accounted for 72% of the total insecticide usage in the USA in the late 1990s. The 10 top-usage active ingredients in this insecticide class are malathion, chlorpyrifos, terbufos, diazinon, methyl-parathion, phorate, acephate, azinphos-methyl, phosmet, and dimethoate. The use of OP insecticides has declined by 30% since 1980, from an estimated 59 million kg in 1980 to 41 million kg in 1999. However, OP insecticide usage as a percentage of total insecticide used has increased, from 58% in 1980 to 72% in 1999. The use of malathion was increased in 1999 for a USDA-sponsored boll weevil eradication program in cotton-growing areas of the USA.

In the Great Lakes Basin, corn and soybeans together account for approximately 70% of the reported crop acreage. Hence, pesticides used on corn and soybeans represent the largest proportion of pesticides used. Although over 120 pesticide active ingredients are used in the basin, the five top-usage herbicides (atrazine, metolachlor, cyanazine, acetolachlor, and alachlor) account

**Table 2** Most commonly used conventional pesticide active ingredients in the USA in 1999

| Rank | Active ingredient | Range (million lbs) | Type |
| --- | --- | --- | --- |
| 1 | Atrazine | 74–80 | Herbicide |
| 2 | Glyphosate | 67–73 | Herbicide |
| 3 | Metam Sodium | 60–64 | Fumigant |
| 4 | Acetochlor | 30–35 | Herbicide |
| 5 | Methyl bromide | 28–33 | Fumigant |
| 6 | 2,4-D | 28–33 | Herbicide |
| 7 | Malathion | 28–32 | Insecticide |
| 8 | Metolachlor | 26–30 | Herbicide |
| 9 | Trifluralin | 18–23 | Herbicide |
| 10 | Pendimethalin | 17–22 | Herbicide |

Source: USEPA estimates based on USDA/NASS http://www.epa.gov/oppbead1/pestsales

for about 53% of the total volume. These ingredients are primarily applied in the southern Lake Michigan Basin and in the western Lake Erie Basin [23]. Estimates of pesticide usage in the Great Lakes Basin for 1995 are shown in Table 3. The data are estimates derived by applying the 1994/1995 statewide pesticide use data developed by the NASS. The estimates for Ontario have been reported elsewhere [26].

The data for New York State were not available at that time, and therefore were not included in this estimation. The Lake Michigan Basin is the largest of the three basins studied within the Great Lakes Basin (Fig. 2). Since the northern part of the Lake Michigan Basin is forested, the highest pesticide application is in the southern part of the Basin. Atrazine and metolachlor were the most widely applied herbicides, together accounting for 26% of the total pesticide usage. Among insecticides, chlorpyrifos was the most widely applied compound in the Great Lakes states of the USA. This is followed by azinphos-methyl, phosmet, carbaryl, and terbufos. These five insecticides accounted for 5% of the total pesticides applied in the USA and for 35% of the insecticides used in the Great Lakes Basin. The Lake Superior Basin is primarily forested, with little of this land being used for agricultural purposes. Hay is the most abundant crop in the Lake Superior Basin, followed by corn, representing 2.8% of the planted acreage in the basin. In this basin, metam-sodium was the most commonly used fungicide on potatoes and represented 25% of the total pesticide used (Fig. 2).

Similarly to usage patterns in the USA, 75% of the pesticides used in Ontario in 1998 were herbicides [26]. Furthermore, herbicides applied on field corn and soybeans accounted for 65% of the total pesticide used in 1998 in Ontario. Metolachlor, glyphosate, atrazine, dicamba, and bentazon were the major herbicides used in Ontario agriculture in 1998. These five herbicides accounted for 76% of the total herbicide usage in Ontario. Atrazine usage on

**Table 3**  Estimates of total pesticide use in the Great Lakes Basin (in kg) in 1995

| Pesticide | Lake Erie Basin | Lake Michigan Basin | Lake Superior Basin | Ontario |
|---|---|---|---|---|
| Herbicide | 3 669 337 | 3 127 546 | 4201 | 3 919 215 |
| Insecticide | 12 743 | 887 451 | 1088 | 153 049 |
| Fungicide | 51 614 | 1 594 245 | 5759 | 649 805 |
| Other pesticides | 17 624 | 461 829 | 4094 | 492 333 |
| Total | 3 750 953 | 6 071 069 | 15 140 | 5 214 402 |

Source: USDA/NASS 1994–1995 annual reports as summarized in http://www.epa.gov/reg5rcra/ptb/pest/documents/pest_se.pdf. Lake Erie Basin includes Indiana, Michigan, Ohio, and Pennsylvania. Lake Michigan Basin includes Indiana, Illinois, Michigan, and Wisconsin. Lake Superior Basin includes Minnesota, Michigan, and Wisconsin. Data for Ontario are for the whole province, for 1998 [26]

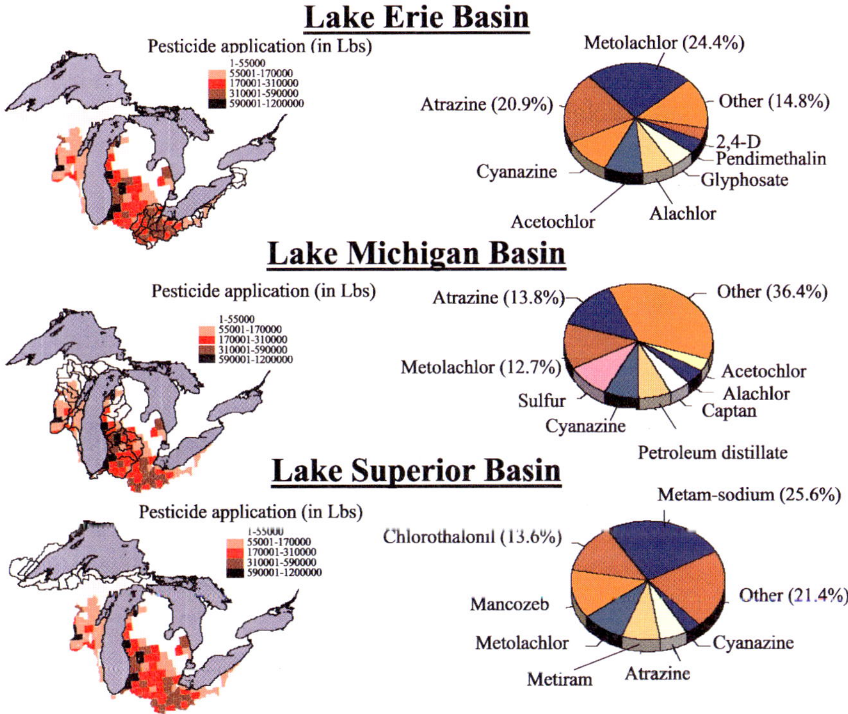


**Fig. 2**  Pesticide usage in the Lake Erie, Lake Michigan, and Lake Superior Basins in 1995: total pesticide usages in Lakes Erie, Michigan, and Superior Basins were 8 280 250, 13 401 918, and 33 422 lbs, respectively. Lake Erie Basin includes Indiana, Michigan, Ohio, Pennsylvania; Lake Michigan Basin includes Indiana, Michigan, Illinois, Wisconsin; Lake Superior Basin includes Minnesota, Michigan, Wisconsin. From [15] (originally from USDA/NASS Chemical survey reports for 1994–1995)

field corn remained stable from 1993 to 1998 ($\sim$ 574 tons). Glyphosate use had increased by four-fold since 1988, with 647 tons used in 1998. Among insecticides, phosmet, acephate, carbaryl, azinphos-methyl, and deltamethrin were the major compounds applied, accounting for 58% of the total insecticide usage. Insecticide use was concentrated in tobacco, fruit, and vegetable cultivation. Mancozeb, metiram, chlorothalonil, and captan were the major fungicides, accounting for 78% of the total use, and were applied primarily on fruit and vegetable crops. Concentrations and loadings of metolachlor, the most intensively used herbicide, at Fort Erie and Niagara-on-the-Lake doubled over the period of 1989–1997 [17], i.e., metolachlor usage in these applications was 60% higher in 1998 than in 1983.

Maps of annual pesticide use in the USA have been compiled for 208 compounds applied in agriculture in the USA (http://ca.water.usgs.gov/pnsp/use92/mapex.html). These maps are based on pesticide usage rates compiled by the National Center for Food and Agricultural Policy (NCFAP) from pesticide use information collected by state and federal agencies over a four-year period (1990–1993 and 1995), and crop acreage data obtained from the 1992 census of agriculture. These maps can be used to identify areas of intensive use of selected pesticides. The usage maps for atrazine (the most commonly used herbicide) and chlorpyrifos (the most commonly used insecticide) are shown in Fig. 3 [27]. Both atrazine and chlorpyrifos are used on corn. Chlorpyrifos use on corn accounts for 44% of this chemical's total use in the USA. Similarly, atrazine use on corn accounts for 84% of this chemical's total use in the USA. Thus, the use of these pesticides is concentrated in the Great Lakes Basin of the USA.

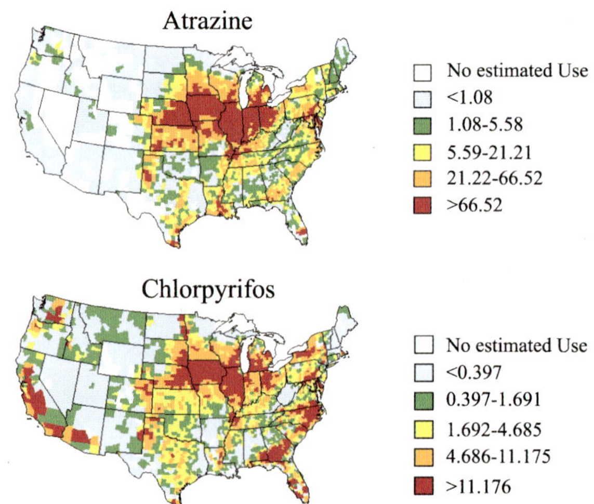


**Fig. 3** Map of estimated atrazine and chlorpyrifos use in the USA in 1995 (use in lbs/mi$^2$/yr). From [27]

**Table 4** Currently used (1990s) pesticides predicted to have the potential to accumulate in sediment and in aquatic biota[1]

| Insecticides | Herbicides | Fungicides or wood preservatives |
|---|---|---|
| Chlorpyrifos[2] | Benfluralin | Dichlone |
| Dicofol[2,3] | Bensulide | PCNB[2] |
| Endosulfan[2,3] | Dacthal[2] | Pentachlorophenol[2] |
| Esfenvalerate | Ethalfluralin | |
| Fenthion | Oxadiazon[2] | |
| Fenvalerate[2] | Pendimethalin | |
| Lindane[2,3] | Triallate | |
| Methoxychlor[2,3] | Trifluralin[2] | |
| Permethrin[2] | | |
| Phorate[4] | | |
| Propargite | | |

Source: http://ca.water.usgs.gov/pnsp/rep/fs09200. [1]Selection criteria: (1) water solubility $< 1.0$ mg/L or $\log K_{ow} > 3$, and (2) soil half-life $> 30$ days. [2]Detected in one or more past studies of stream sediment or biota. [3]Organochlorine insecticide. [4]Analyzed in $> 1000$ total samples by past studies, but not detected.

The key properties of pesticides that control their accumulation in sediment and biota are hydrophobicity and persistence. Generally, pesticides were found to have the potential to accumulate in sediments and in biota if they had a water solubility of $< 1$ mg/L or an octanol-water partition coefficient of ($K_{ow}$) $> 1000$ and a soil half-life of $> 30$ d. Pesticides that are rarely detected in sediment or in biota have high water solubilities and short soil half-lives. A few pesticides that are moderate in hydrophobicity and persistence can also be detected in the environment to some extent. Most of the current-use pesticides have relatively high water solubilities and short soil half-lives, and are not likely to accumulate in biota. However, some currently used pesticides that are intermediate in hydrophobicity and persistence are likely to be detected (Table 4). It is also important to take into account where, and in what amounts, these pesticides are applied.

## 3
## Environmental Levels and Trends

### 3.1
### Past- and Restricted-Use Pesticides

There is a considerable body of literature documenting the accumulation of OC pesticides in the Great Lakes since the late 1960s. Reports documenting the effects of DDT and its metabolites, and the effects of dieldrin, on reproduction and mortality in Lake Michigan herring gulls date back to the mid-1960s [28, 29]. Systematic studies on the environmental occurrence, sources, distribution, and fate of OCs were published in the ensuing years [6, 7, 30]. Currently, several international, national, regional, and local monitoring programs are in place in the Great Lakes region, to monitor sources, trends, distribution, fate, and effects of OC compounds.

The USEPA's Great Lakes National Program Office (GLNPO) and the USGS's Great Lakes Science Center have been conducting joint monitoring programs since 1977 [31] using fish, particularly lake trout and walleye (in Lake Erie only) as biomonitors. Results of fish biomonitoring studies from 1977 to 2003 have been published [22, 32]. Similarly, Environment Canada and the Canadian Wildlife Service have been conducting biomonitoring programs, to detect the trends of persistent organic pollutants in fish and in herring gull eggs, since the 1970s [34, 35]. Trend studies have shown that most of the OC pesticides in salmonid fishes declined rapidly until the 1980s, and they then leveled off in the 1990s [32, 36, 37].

DDT was the major pesticide found in fish and birds in the Great Lakes. On a by-lake basis, lake trout from Lake Michigan contained the highest concentrations of DDT [32, 33] (Fig. 4), followed by those from Lake Ontario. Fish from Lakes Erie and Superior contained the lowest DDT concentrations. No major reduction in DDT concentrations was observed in lake trout col-

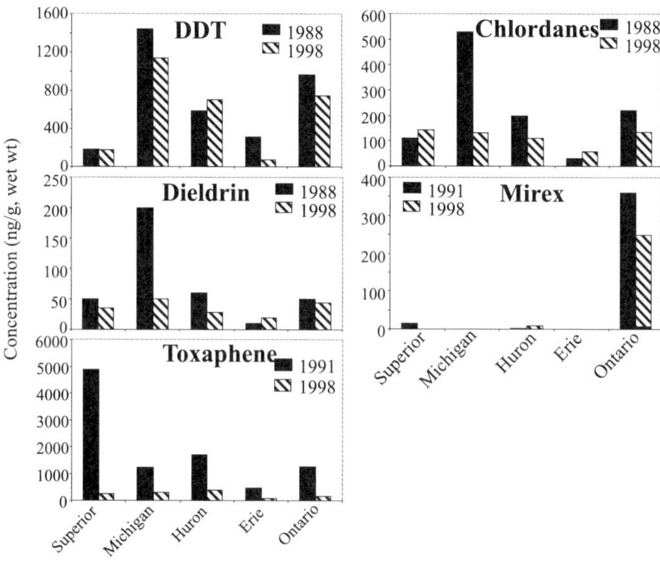


**Fig. 4** Concentrations of organochlorine pesticides in lake trout and walleye (Lake Erie only) collected in 1988 and 1998. For mirex and toxaphene, 1991 concentrations are presented instead of 1988 values. From [32, 33]

lected between 1988 and 1998, although in walleye from Lake Erie, a 4-fold reduction was observed during 1988–1998. The trend in the 1990s was different from that in the 1970s and the early 1980s. Concentrations of DDT declined from 19 200 ng/g in 1970 to 1160 ng/g in 1992 [32]. Total DDT residues in spottail shiners continued to decline at most Great Lakes locations in the 1980s, although yearly declines are less steep than the declines of the 1970s [35]. Similarly, double-crested cormorant eggs collected from Lakes Ontario, Erie, and Superior showed a decline in DDT concentrations from 1970 to 1989, but levels thereafter remained stable [38]. These data reflect a rapid response by Great Lakes biota following a basin-wide ban on the use of DDT imposed by the USA and Canada in 1970–1972. However, the data also indicate that the rate of this decline has now slowed. Thus, the processes responsible for virtual elimination of these residues in the ecosystem will proceed at a slower pace in the future [39].

Similarly to DDT, chlordane and dieldrin concentrations were the highest in Lake Michigan lake trout; the concentrations had declined considerably until 1998. The 1998 concentrations are comparable among the lakes, except for Lake Erie, where the levels are much lower. Dieldrin concentrations in lake trout collected in 1998 were between 19 and 50 ng/g, on a wet weight basis. Lake Ontario is the only Great Lake to have significant levels of mirex in its biota. From 1977 to 1996, the mirex concentrations in coho and chinook salmon weighing > 2 kg exceeded the Food and Drug Administration (FDA) action level of 0.1 mg/kg for this compound. Mirex concentrations in fillets from most salmon under the size of 2 kg are now below the 0.1 mg/kg action level for human consumption [40]. Mirex had declined in Lake Ontario lake trout, but the rate of decline was slow. An increase in mirex concentration in lake trout from 207 ng/g during 1977–1993 to 220 ng/g during 1991–1997 at Oswego, New York, was reported [32, 33]. The FDA action level of mirex in fish is 100 ng/g, on a wet weight basis.

Toxaphene levels are historically the highest in Lake Superior. Toxaphene concentrations declined significantly from 1991 to 1998 [33]. No significant decline in toxaphene concentrations was found in fish from Lake Superior during 1982–1992 [41]. However, concentrations in fish from the other Great Lakes declined after the ban on the use of toxaphene in the USA in 1982.

Concentrations of DDT, chlordanes, and dieldrin in bald eagle eggs collected from Lake Erie in 1989–1994 were lower by 2-, 4-, and 2-fold, respectively, than levels in eggs collected in 1974–1980. However, mirex concentrations did not change during this period [14]. Residue concentrations of DDT, chlordanes, dieldrin, and mirex in the plasma of bald eagles from Lakes Erie and Superior did not show any temporal variation between 1990 and 1996 [14]. Similar to what was observed in fish biomonitoring programs of Environment Canada, DDT concentrations in herring gull eggs have remained relatively stable throughout the 1990s at some Great Lake colonies.

A few significant decreases in levels of dieldrin and heptachlor epoxide were noted in the early 1990s [19].

In general, spatial and temporal monitoring studies of sediment and surface water in the Great Lakes have shown reductions in persistent OC pesticides. However, lindane and dieldrin were ubiquitous in surface waters of the Great Lakes collected during 1997–2000 [17]. Concentrations of OC insecticides in sediment and in fish correspond to land use patterns and past application rates. In agricultural streams, concentrations of OC insecticides were the highest in areas of high past usage. Almost all urban streams had high or medium concentrations of the OC insecticides, as compared with other sites [11].

### 3.1.1
### Trends in Environmental Loadings

Since the cessation of production and use of OC pesticides, direct discharges to the Great Lakes have greatly diminished. The relative steady state in contaminant levels in the 1990s indicates that these chemicals are still being released and/or recycled through the Great Lakes ecosystem. Atmospheric deposition, agricultural land runoff, the slow movement (leaching) of discarded stocks of pesticides and other chemicals from landfill sites and agricultural soils into the Great Lakes via groundwater, and the resuspension of contaminated lake/river sediments continue to be major indirect sources of contamination. These indirect sources are difficult to control, and they contribute slow, but continuous, inputs into the Great Lakes ecosystem. Atmospheric deposition has become an increasingly significant route of entry of contaminants into the Great Lakes ecosystem, especially in the upper Great Lakes.

### 3.1.2
### Atmospheric Transport and Deposition

The three main processes by which atmospheric deposition of pesticides to the lakes occurs are wet deposition (rain and snow), dry deposition (particulates), and air–water gas exchange. For many of the banned pesticides, gas exchange across the air–water interface, in particular, is often the dominant deposition process, when compared with precipitation and dry particle exchange [42–44]. The physicochemical considerations, as well as descriptions of calculation models for atmospheric deposition to lakes, rivers, and the oceans, have been reviewed extensively [45–47].

Atmospheric deposition of persistent toxic chemicals in the Great Lakes was first observed in the late 1970s. Studies by the International Joint Commission (IJC) under the Great Lakes Water Quality Agreement (GLWQA) suggested the role that the atmosphere plays in maintaining high lake concentrations. In 1987, based on the report from the IJC, the GLWQA was revised (Annex 15) to include research and monitoring on the loadings of atmospheric deposition of toxic substances to the Great Lakes.

The USEPA and Environment Canada operate a monitoring network that measures levels of toxic substances in the air of the Great Lakes region. The Integrated Atmospheric Deposition Network (IADN) was created under Annex 15 of the GLWQA. Samples of air and precipitation have been taken since 1990, with a primary goal of measuring several persistent organic pollutants, including 22 chlorinated pesticides [21]. IADN publishes a loadings report every two years, summarizing these measurements and providing estimates for the atmospheric deposition of selected pesticides ($\alpha$- and $\gamma$-HCHs, $\alpha$- and $\gamma$-chlordanes, dieldrin, $p,p'$-DDE, $p,p'$-DDD, and $p,p'$-DDT) in addition to several other compounds to each of the Great Lakes. On the basis of this network and individual studies, there is substantial evidence to support the idea that atmospheric transport and deposition of OC pesticides contribute significantly to, and in some cases are the dominant source of, contaminant inputs in surface waters of the Great Lakes [42]. Application of a potential source contribution model to the IADN data, in order to identify sources, suggested that the regions south and southwest of the Great Lakes are the major sources of pesticides to the Great Lakes [21]. For example, DDT, dieldrin, and chlordanes were shown to be atmospherically transported from areas south of the Great Lakes. However, for certain pesticides such as HCHs, such a southerly trend was not prominent. Similarly, restricted-use OC pesticides, such as endosulfan, were shown to undergo short-range atmospheric transport from Michigan and New York [48].

Chlordanes and the related compounds, *cis*- and *trans*-nonachlors, reflect the compounds' past use in urban centers, where they were used for termite control in houses and commercial buildings. Using a predictive model, the impact of the plume of contaminants from the urban areas such as Chicago, on the atmospheric deposition over Lake Michigan has been examined [49]. The model illustrated that the urban area contributes to the atmospheric deposition of *trans*-nonachlor over a highly variable region of the lake. In an attempt to include the contributions of urban areas to lake-wide loadings, data from IADN's Chicago site were extrapolated onto Lake Michigan loadings [50].

The recent IADN report covers data through 1998 [50], which for the first time have been expressed on a Great Lakes Basin-wide basis, through summation of the total deposition of each pesticide measured over the five Great Lakes (in kg/yr) for the period 1992 to 1998 (Table 5). Total deposition of $\alpha$-HCH showed a trend of decrease, from 950 kg/yr in 1992 to 210 kg/yr in 1998. Dieldrin and DDT had negative total deposition across time, indicating that the lakes are acting as a source of these chemicals to the atmosphere. Among the pesticides studied, lindane showed the highest downward flux, which is explained by current use of this pesticide in agriculture [52]. Atmospheric transport models have shown that lindane usage in the canola fields in the Canadian prairie-provinces serves as a major source of this compound

**Table 5** Total deposition (kg/yr) of selected organochlorine pesticides in the Great Lakes in 1998

| Pesticide | Superior | Michigan | Huron | Erie | Ontario | All lakes |
|---|---|---|---|---|---|---|
| DDT (DDD+DDE+DDT) | 11.8 | 18.9 | 14.7 | − 36 | − 14.8 | − 5.4 |
| Dieldrin | − 250 | — | — | − 70 | − 120 | − 440 |
| $\gamma$-HCH | 7.6 | 200 | 170 | 39 | 36 | 453 |
| $\alpha$-HCH | − 460 | 220 | − 42 | 95 | − 23 | − 210 |

Negative values indicate loss or volatilization and positive values indicate deposition. From [51]

to the Great Lakes ecosystem [25]. All of these estimates are based on data from remote IADN sites, one on each lake, which provide what are considered to be the Great Lakes' background levels.

The only current-use OC pesticide tracked by IADN, other than lindane, is endosulfan. Endosulfan is currently used at approximately 25 000 kg/year in Ontario alone in the mid-1990s [53]. It has a variety of uses in the Great Lakes region, but its heaviest application is in the southern USA, on cotton. In 1997–1998, total atmospheric fluxes of $\alpha$-endosulfan were highest in Lakes Erie and Ontario, in the range of $27–34\,\mathrm{ng/m^2/day}$, whereas values for remote Lake Superior were in the range of $1.7–2.5\,\mathrm{ng/m^2/day}$. Insufficient data were collected to allow estimates for Lakes Michigan and Huron, or for all lakes for endosulfan sulfate.

Through the use of atmospheric half-lives of OC pesticides calculated from the year-to-year trend in atmospheric concentrations measured at IADN sites, virtual elimination dates of these compounds have been estimated; these range from 2010 for $p,p'$-DDT to about 2060 for hexachlorobenzene [54]. There was little, if any, spatial variability in the observed half-lives, suggesting similar loss processes among individual lakes, and also suggesting that the decrease in precipitation concentrations reflects a regional decrease in environmental levels.

A summary of the environmental half-lives of OC pesticides in Lake Ontario air and precipitation is provided in Table 6 [52, 55, 56]. The first-order half-lives for individual OC pesticide concentrations fall within narrow ranges. For example, half-lives of dieldrin in air, precipitation, and water ranged between 2.1 and 4.8 years. Similar decreases have been observed in dissolved concentrations throughout the Great Lakes [17]. Comparable half-lives in gas phase, precipitation, and dissolved phase suggested that the entire Great Lakes system is at or near equilibrium, and that the rate at which the equilibrium is achieved between the various compartments is relatively rapid, likely on the order of months [56].

Trends in atmospheric loadings for OC pesticides during 1992–1998 are shown in Fig. 5. In general, atmospheric depositions to the Great Lakes de-

**Table 6** Comparisons of environmental half-lives (in years) of organochlorine pesticides in gas phase, precipitation, and dissolved phase in Lake Ontario, during 1992–2000

| Pesticide | Gas [55] | Precipitation [56] | Dissolved [52] |
| --- | --- | --- | --- |
| $\alpha$-HCH | $5.0 \pm 0.8$ | NS | $3.2 \pm 0.7$ |
| $\gamma$-HCH | $7.3 \pm 2.5$ | NS | $5.5 \pm 1.2$ |
| HCB | NA | $1.6 \pm 0.3$ | NA |
| Dieldrin | $4.8 \pm 1.2$ | $4.1 \pm 1.6$ | $2.1 \pm 1.0$ |
| $\alpha$-Chlordane | $6.2 \pm 1.8$ | $1.0 \pm 0.1$ | NA |
| $\gamma$-Chlordane | $6.5 \pm 2.1$ | $1.8 \pm 0.4$ | NA |
| $p,p'$-DDT | NS | $1.8 \pm 1.2$ | NA |
| $p,p'$-DDE | $9.2 \pm 4.2$ | $4.9 \pm 1.2$ | NA |
| $p,p'$-DDD | $8 \pm 3.9$ | $2.2 \pm 0.7$ | NA |

NA denotes insufficient data. NS denotes that there is no significant trend. From [52, 55, 56]

creased over this period, although there are exceptions. The flux of $\gamma$-HCH to Lake Ontario, for example, remained within the range 5–10 ng/m$^2$/d during this period, whereas the flux of this compound to Lake Michigan decreased sharply, from 44 in 1992 to 10 ng/m$^2$/d in 1998. In general, dieldrin has consistently volatilized from the Great Lakes at rates ranging from −47 to −5 ng/m$^2$/d. Loadings for some compounds such as $\alpha$-HCH in Lake Superior have reversed, from positive loadings to negative, indicating that the lake

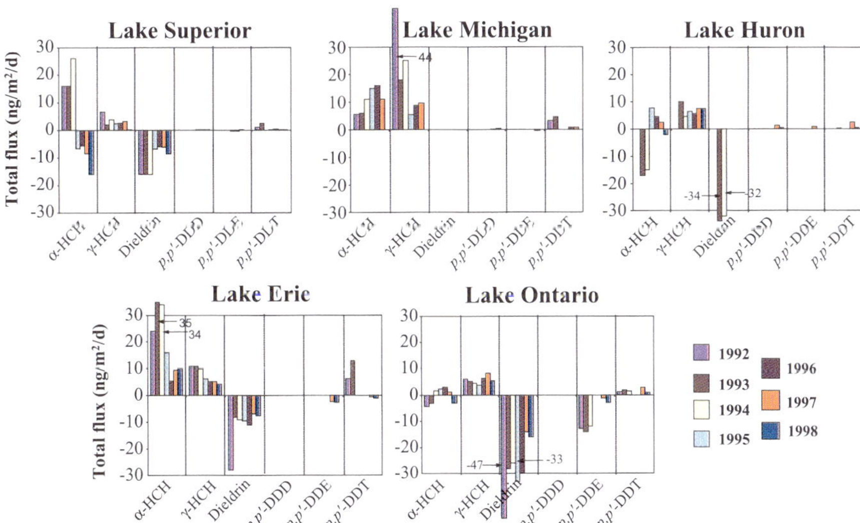


**Fig. 5** Annual average total flux (ng/m$^2$/day) of organochlorine pesticides in the Great Lakes from 1992 to 1998. From [51]

has become a regional source of this compound to the atmosphere in recent years.

Overall, the IADN data indicate that the net loadings of most banned OC pesticides to the Great Lakes are decreasing, and are in some cases approaching zero. In many cases, the dominant factors of gas absorption and volatilization are near steady-state, with the amount going into the lakes equaling the amount coming out [56]. When a loading nears zero, this does not mean that pollutant levels in the lakes are at zero. The pollutants are still present, cycling continuously and evenly between the air and water. In the absence of new sources to the Lakes, concentrations in both air and water, and therefore also in wildlife, are expected to continue to decline, likely at rates determined by the decline of global air concentrations, which in turn is governed by the temperature-mediated volatilization of residues from soil reservoirs [57] and the processes of destruction and long-term burial into the earth.

Historically, less attention has been paid to atmospheric deposition to connecting channels and tributary rivers of the Great Lakes, mainly because of the smaller surface areas presented by rivers for direct atmospheric deposition, compared with deposition and transport to rivers from their watersheds. The atmospheric deposition of chlordanes and HCHs to the St. Lawrence River ranged from $< 1$ kg/yr for $\alpha$-chlordane to $5$ kg/yr for $\alpha$- and $\gamma$-HCHs; these loadings represented $< 1$ to 14% of the atmospheric deposition to the watershed [58].

## 3.2
### Current-Use Pesticides

In comparison with OC pesticides, the current-use OP and carbamate pesticides are relatively more water soluble and less bioaccumulative. However, because of the high volume of their usage, there is a potential for these compounds to be present in the environment. Herbicides such as atrazine, metolachlor, and alachlor have been surveyed in the Great Lakes [59] and along the margins of fields, streams, and tributaries within the Great Lakes Basin [5, 60].

Several surface and groundwater monitoring studies conducted by the USGS's NAWQA program have reported the occurrence of current-use pesticides in surface waters [61]. Ninety-eight pesticides and 20 pesticide transformation products are being routinely monitored in the surface waters of the USA [12]. Of these, 76 pesticides were detected in the surface waters of one or more bodies of water nationwide. Detection frequencies for herbicides and insecticides are compared with patterns of agricultural use (Fig. 6). Absence of a pesticide in the figure does not necessarily mean that the compound does not occur in surface waters, but often means that it was not analyzed. In general, herbicides have been detected more frequently than insecticides, consistent with the greater use of herbicides in the Great Lakes Basin. The most fre-

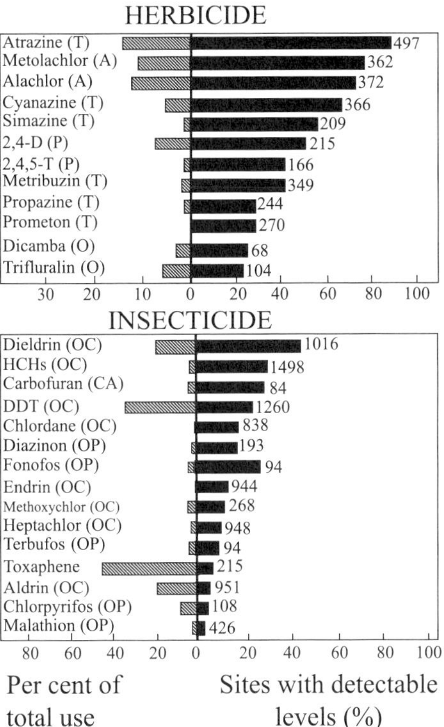


**Fig. 6** Detection frequency for pesticides targeted at 50 or more sites, in 127 studies conducted between 1958 and 1993. The relative agricultural use of each pesticide is shown in the left half of each graph. 2,4,5-T use percentage is based on a 1971 use estimate; values for all other herbicides are based on 1989 use estimates. Organochlorine use percentage is based on 1966 use estimates; values for other pesticides are based on 1990 use estimates. Numbers to the right of the bars represent the number of sites at which the chemical was targeted (T, triazines; A, acetanilides; P, phenoxys; O, others; OC, organochlorines; CA, carbamates; OP, organophosphates). From [12]

quently detected herbicides include several triazines (atrazine, cyanazine, and simazine), acetanilides (metolachlor and alachlor), and 2,4-D. These compounds are among the most used in current agricultural practice. Trifluralin and butylate were detected less frequently, despite relatively high usage. These are volatile compounds, have high $K_{oc}$ (organic carbon partition coefficient) and are usually incorporated into the soil when applied; such factors reduce the likelihood of transport to surface waters [62, 63].

In comparison to herbicides, many of the insecticides studied were rarely detected because of low usage. Some with relatively high usage were rarely detected for other reasons. The most frequently detected insecticides that are in current use are carbofuran and diazinon. Some OC pesticides were among the most frequently detected insecticides. While many of these detections occurred in studies conducted in the 1960s and 1970s, these compounds have also been detected in surface waters in more recent studies. This is due to their storage in the bed sediments of rivers and lakes, and their continued inputs from the atmosphere and soil contaminated from past agricultural applications. Among the OP compounds monitored by the USGS during 1992–1997 in the surface waters of the USA, diazinon, chlorpyrifos, and malathion were detected frequently. OPs were detected most often, and at higher concentrations, in small streams draining urban and agricultural areas, and they were occasionally detected as mixtures of up to nine OPs. Median concentrations of OPs were well below the USEPA's lifetime health advisories, at all of the ground- and surface-water sites evaluated. Of the five OPs that have aquatic-life criteria, the criterion (80 ng/L) for diazinon was exceeded most often [64].

The occurrence of in-use pesticides in surface waters follows broad patterns in land use and associated pesticide use. The patterns are complex, due to the wide range of use practices and processes that govern the movement of pesticides in the hydrologic environment. For instance, the less-frequent occurrence of currently used insecticides compared with herbicides in streams results from the insecticides' relatively low application rates and rapid breakdown in the environment [11]. Urban streams had the highest detection frequencies of historically used insecticides such as DDT, chlordane, and dieldrin, in fish and in sediment [11]. Chlordane and aldrin were widely used for termite control in buildings until the mid-1980s, although their agricultural usage had been restricted in the 1970s.

Environment Canada's Ecosystem Health Division initiated an exploratory survey of current-use pesticides in the Great Lakes in 1994 [65, 66], complementing surveys carried out in tributaries and small streams of Ontario [67, 68]. Whole water samples (20 L) were acidified to pH < 2 immediately upon collection, and were extracted on-board the sampling vessel into dichloromethane. The range of pesticide concentrations measured in the Great Lakes by the Ecosystem Health Division from 1994 to 2000 is shown in Table 7. These surveys were conducted to quantify the concentrations of current-use pesticides, particularly neutral and phenoxy-acid herbicides and OP insecticides, in the Great Lakes Basin. The spatial and seasonal variability, and the fate of some pesticides were discussed in relation to these chemicals' usage, application, and physical and chemical properties [66]. Comparison of concentrations to water-quality criteria for the protection of aquatic life and for drinking water revealed no exceedences.

Thirty-nine pesticides, including analytes and some metabolites, were measured in Great Lakes surface waters between 1994 and 2000. Six pesticides were never detected: barban, diallate-2, triallate, phorate, phosmet, and disulfoton. The highest overall concentrations were measured for atrazine, metolachlor, and simazine. In general, concentrations showed an increasing gradient from north to south, with Superior < Huron < Ontario < Erie. The highest ratios of detections were observed as follows: atrazine (183/188 = 97%);

**Table 7** Range of pesticide concentrations (ng/L) in Great Lake surface waters, 1994–2000

| Pesticide | Lake Ontario (1998, 1999) | Lake Erie (1994, 1998, 2000) | Lake Huron (1999, 2000) | Lake Superior (1996, 1997) |
|---|---|---|---|---|
| **Neutral herbicides** | | | | |
| Atrazine | 12.4–297 | 5.4–1039 | < 3.0[1]–33.6 | < 1.81–4.7 |
| Barban | < 7.6 | < 7.6 | < 7.6 | NA[2] |
| Benzoylprop-ethyl | < 2.1 | < 0.13–0.5 | < 0.13 | NA |
| Butylate | NA | < 0.41–3.5 | < 0.41 | NA |
| D-Atrazine | NA | 3.4–15.7 | 1.1–3.9 | NA |
| Diallate (isomer total) | < 6.5 | < 6.5–13.2 | < 6.5 | NA |
| Diallate-1 | NA | 1.2–2.5 | 0.7–2.9 | NA |
| Diallate-2 | NA | < 0.18 | < 0.18 | NA |
| Diclofop-methyl | < 0.17 | 0.4–2.0 | < 0.17–0.3 | NA |
| D-Simazine | NA | 24.0–281 | INT[3] | NA |
| Metolachlor | 2.5–59.1 | < 5.0–736 | < 0.92–15.3 | < 0.06–0.7 |
| Metribuzin | NA | 4.9–78.7 | < 0.41 | NA |
| Simazine | 5.3–23.9 | < 0.40–43.1 | < 0.40–2.0 | NA |
| Triallate | < 0.7 | < 0.048 | < 0.048 | NA |
| Trifluralin | < 0.045 | < 0.045–0.9 | < 0.045–0.1 | NA |
| **Phenoxy-acid herbicides** | | | | |
| 2,3,6-TBA | < 0.40–9.5 | < 0.40–6.4 | < 0.40–51.1 | < 0.40 |
| 2,4,5-T | < 0.40–1.8 | < 0.37–0.4 | < 0.37–2.4 | < 0.40 |
| 2,4,5-TP | < 0.30–2.4 | < 0.30–4.3 | < 0.34–10.8 | < 0.30 |
| 2,4-D | 2.3–14.5 | < 0.40–84.4 | < 0.29–1.4 | < 0.40–2.5 |
| 2,4-DB | < 0.40–1.3 | < 0.40–24.3 | < 0.40–2.2 | < 0.40–0.9 |
| 2,4-DP | < 0.30 | < 0.28–6.4 | < 0.28–4.9 | < 0.30 |
| Bromoxynil | < 0.30–1.8 | < 0.30–1.2 | < 0.30–2.7 | < 0.30 |
| Dicamba | < 0.30–7.0 | < 0.30–13.9 | < 0.32–37.5 | < 0.30 |
| MCPA | < 0.30–5.4 | < 0.27–4.1 | < 0.27–0.9 | < 0.30 |
| MCPB | < 0.40 | < 0.40 | < 0.40–0.7 | < 0.40 |
| Picloram | < 0.50 | < 0.50–6.1 | < 0.50–2.4 | < 0.50 |
| **Organophosphorus insecticides** | | | | |
| Azinphos-methyl | < 1.2 | < 1.2–45.2 | < 1.2 | NA |
| Chlorpyrifos | < 0.45 | < 0.45–3.8 | < 0.45 | NA |
| Diazinon | < 0.63 | < 0.63–44.4 | < 0.63 | NA |
| Dibrom | < 0.64–7.8 | < 0.64–2.5 | < 0.64 | NA |
| Dimethoate | < 0.65 | < 0.65–34.4 | < 0.65 | NA |
| Disulfoton | < 0.31 | < 0.31 | < 0.31 | NA |
| Ethion | < 0.30 | < 0.30–1.9 | < 0.30 | NA |
| Fonofos | < 0.27 | < 0.27–2.8 | < 0.27 | NA |
| Malathion | < 0.28 | < 0.28–20.8 | < 0.28 | NA |
| Parathion | < 0.51 | < 0.51–3.8 | < 0.51 | NA |
| Phorate | < 0.29 | < 0.29 | < 0.29 | NA |
| Phosmet | < 1.2 | < 1.2 | < 1.2 | NA |
| Terbufos | < 0.17 | < 0.17–14.3 | < 0.17 | NA |

[1]Below the method detection limit. [2]Not analyzed. [3]Interference

simazine (43/46 = 93%); metolachlor (204/223 = 91%); 2,4-D (30/58 = 52%); and 2,4,5-TP (16/58 = 28%). Thirty-two individual pesticides were detected in Lake Erie, 18 in Lake Huron, 12 in Lake Ontario, and only four in Lake Superior. OP insecticides were not frequently detected, and were found mainly in Lake Erie.

### 3.2.1
### Neutral Herbicides

Amounts of triazine and acetanilide herbicides used in Canada and in the USA have greatly increased over the last 30 years. These herbicides are normally applied pre-emergence to bare soil or residual vegetation from the previous year's crop. Atrazine, a triazine herbicide, is primarily used on corn. Metolachlor, one of the most commonly used actetanilides, is primarily used on soybeans. As a group, the commonly used triazine and acetanilide herbicides have moderate to high water solubilities and relatively low soil-sorption coefficients. The water solubility of atrazine is about 33 mg/L, while that of metolachlor is 530 mg/L. Despite these relatively high water solubilities, some triazines and acetanilides persist in soil for up to 90 days, and have moderate to high potentials for loss from fields through runoff, primarily in the dissolved phase [69]. In addition, most are chemically stable in water and are not hydrolyzed. In general, triazines are somewhat more resistant to biodegradation than are acetanilides [70]. Atrazine and metolachlor account for most of the herbicide used in Ontario [71, 72] and in states surrounding the Great Lakes [73–75]. Atrazine and metolachlor have been frequently detected in the Canadian tributaries to Lakes Erie and Ontario [68].

Concentrations of atrazine and metolachlor measured in the surface waters of the Great Lakes from 1994–2000, as well as these two compounds' usage in the basin expressed as kilograms per hectare (kg/ha), are shown in Figs. 7 and 8 [3, 76]. The maximum concentrations of atrazine observed in each lake increased from north to south in the Great Lakes Basin, with Superior (4.7 ng/L) < Huron (33.6 ng/L) < Ontario (297 ng/L) < western Lake Erie (1039 ng/L). The maximum concentrations of metolachlor also increased from north to south across the Great Lakes Basin with Superior (0.7 ng/L) < Huron (15.3 ng/L) < Ontario (59.1 ng/L) < Erie (736 ng/L). The in-lake concentrations of atrazine and metolachlor in Lake Superior are probably the result of atmospheric inputs, as there is very little usage of atrazine and/or metolachlor in the Lake Superior Basin. Both atrazine and metolachlor have been measured in precipitation in the Great Lakes Basin [74, 77]. The concentrations of atrazine and metolachlor measured in the spring, summer, and fall of 1994 exhibited a pattern of seasonality, which corresponded to the timing of application of these pesticides. Spring sampling took place before application, and concentrations were higher in the summer and fall. The median concentration of atrazine in Lake Erie increased from 35.6 to 60.4 ng/L, from April to August 1998. The median

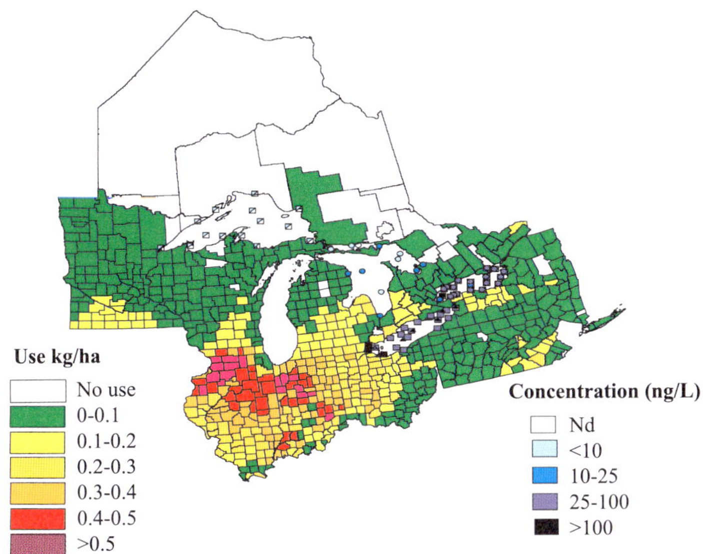


**Fig. 7** Atrazine usage (kg/ha) and concentrations in Great Lakes surface waters (ng/L), 1994–2000

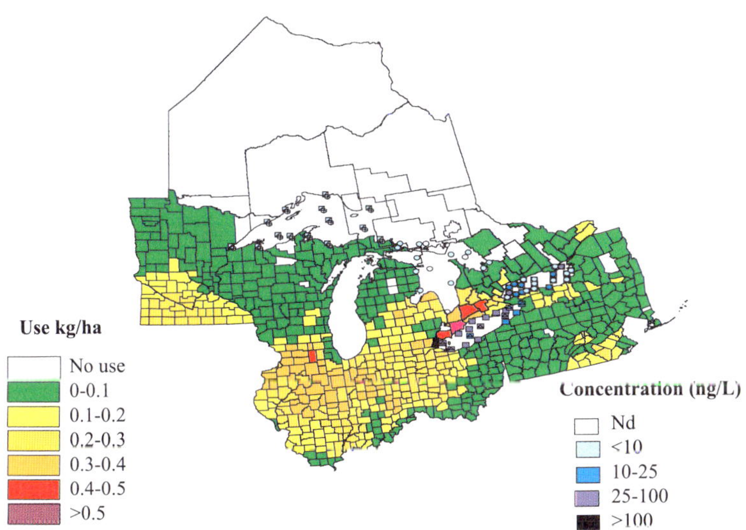


**Fig. 8** Metolachlor usage (kg/ha) and concentrations in Great Lakes surface waters (ng/L), 1994–2000

concentration of metolachlor in Lake Erie increased from 17.0 to 28.7 ng/L, from April to August 1998. Atrazine and metolachlor are usually applied in May, possibly explaining the seasonal increase.

Both atrazine and metolachlor are classified as potential human carcinogens by the USEPA, which has established a maximum contaminant level in drinking water of 3000 ng/L [78]. The maximum concentrations of atrazine and metolachlor measured in the 1994–2000 Great Lakes survey were 1039 ng/L and 736 ng/L, respectively. The effects of long-term, low-level concentrations of atrazine and metolachlor on aquatic ecosystems are largely unknown [79]. The Canadian guidelines for the protection of aquatic life and drinking water (Table 8) were not exceeded for either of these herbicides [80–83].

### 3.2.2
### Organophosphorus Insecticides

OP insecticides came into wide-scale agricultural use in the late 1960s and early 1970s, in replacement of OC pesticides. In general, the total amount of OP insecticides used in agriculture has remained relatively stable over the last 20 years [79]. In Ontario, total agricultural usage of OP compounds has been reported to be 259 000 kg in 1988 [71]. However, OP insecticides, such as diazinon and chlorpyrifos, are also used in urban settings to control insect pests. Some OP insecticides (diazinon, ethion, disulfoton, fonofos, and phorate) have a relatively high potential to move from agricultural fields to runoff water in either the dissolved or sorbed phase, while others (chlorpyrifos and malathion) are unlikely to be transported in runoff [79]. The half-life in surface water for azinphos-methyl and disulfoton is as short as 1–3 days, while it is weeks for diazinon, parathion, and terbufos [84].

Historically, OP insecticides have not been frequently detected in surface waters of the USA [79]. However, occurrence of several OP insecticides in US tributaries to Lake Erie has been reported [60]. Several of the sites monitored included urban storm-water inputs, which might explain the increased presence of diazinon and chlorpyrifos in these tributaries. Diazinon had been measured on numerous occasions in urban tributaries to Lakes Ontario and Erie [68]. Struger [67] also reported frequent detections of azinphos-methyl (guthion), chlorpyrifos, and diazinon in two small streams draining the Niagara fruit belt region of Lake Ontario.

In a 1974 Canadian survey [85], no OP insecticides were detected in the surface waters of the Upper Great Lakes. In a recent study, OP insecticides (dibrom, dimethoate, terbufos, fonofos, diazinon, malathion, chlorpyrifos, parathion, ethion, and azinphos-methyl) were detected only in Lake Erie, except for one detection of dibrom (7.8 ng/L) in Lake Ontario [68].

Usage of diazinon in the Great Lakes Basin expressed as kg/ha [3, 76], and the concentrations measured in surface waters during the current survey (1994–2000), are shown in Fig. 9. Diazinon was detected in two-thirds of the samples (3.5 and 3.9 ng/L) collected in Lake Erie in April 1998, and in all of the samples (range 5.3–44.4 ng/L) collected in the lake in July 1998. Diazinon is usually applied between May and August in agricultural and ur-

**Table 8**  Water-quality guidelines/criteria (ng/L) for the protection of aquatic life and drinking water, for current-use pesticides

|                              | Aquatic life | Drinking water[H] |
|------------------------------|--------------|-------------------|
| **Neutral herbicides**       |              |                   |
| Atrazine                     | 1800[C]      | 5000              |
| Metolachlor                  | 7800[C]      | 50 000            |
| Metribuzin                   | 1000[C]      | 80 000            |
| Simazine                     | 10 000[C]    | —                 |
| Triallate                    | 240[C]       | —                 |
| Trifluralin                  | 200[C]       | 45 000            |
| **Phenoxy-acid herbicides**  |              |                   |
| 2,4-D                        | 4000[C]      | 100 000           |
| MCPA                         | 2600[C]      | —                 |
| **Organophosphorus insecticides** |         |                   |
| Azinphos-methyl              | 10[U]        | 20 000            |
| Chlorpyrifos                 | 41[U]        | 90 000            |
| Diazinon                     | 80[I]        | 20 000            |
| Malathion                    | 100[U]       | 190 000           |
| Parathion                    | 13[U]        | 50 000            |

C: [80], H: [81], I: [82], U: [83]

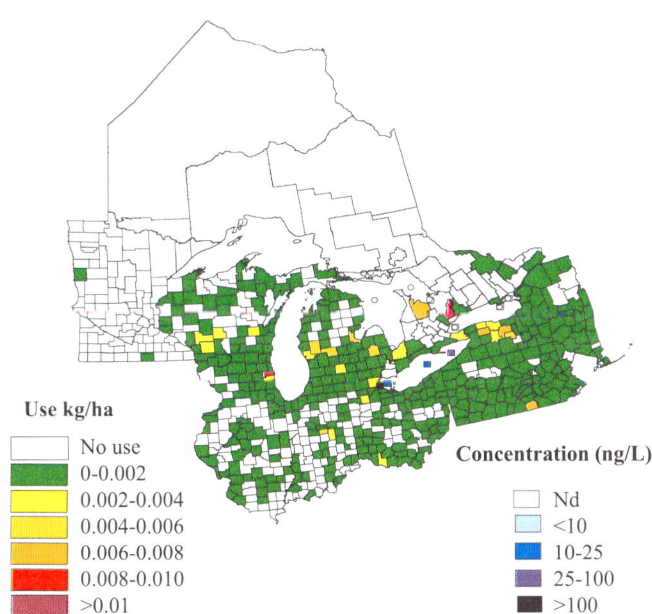


**Fig. 9**  Diazinon usage (kg/ha) and concentrations in Great Lakes surface waters (ng/L), 1994–2000

ban settings, which might account for the higher concentrations and more frequent detections in July. In April 2000, diazinon was detected in three-quarters of samples (range 1.6–9.4 ng/L) collected in Lake Erie. The highest observed concentration of diazinon (44.4 ng/L) was measured in 1998 at a station in the zone of influence of the Maumee River, in the western Lake Erie Basin (Fig. 9).

The USEPA acute and chronic exposure criteria for the safety of fresh-water aquatic biota are both < 100 ng/L for azinphos-methyl, chlorpyrifos, malathion, and parathion (Table 8). Existing guidelines for the protection of aquatic life and for drinking water were not exceeded for any of the OP insecticides analyzed.

### 3.2.3
### Phenoxy-acid Herbicides

The only phenoxy-acid herbicides to have significant current agricultural use in the USA are 2,4-D, esters of 2,4-D, and MCPA [79]. The total amount of phenoxy compounds used in the US agriculture has remained stable over the last two decades [79]. In Ontario, 2,4-D, MCPA, MCPB, mecoprop, and dicamba accounted for about 10% of the herbicide used in agriculture in 1998 [2]. A 1993 pesticide usage survey carried out in Ontario indicated urban applications of phenoxy-acid herbicides similar to agricultural usage [76].

The phenoxy-acid herbicide 2,4-D is used primarily on pasture land, hay, wheat, corn, and barley. However, it is also used heavily for lawn care in urban areas of Ontario, along with dicamba and mecoprop. 2,4-D is applied in several forms, including esters, salts, and the acid form. In water, the esters of 2,4-D may be hydrolyzed rapidly to the acid form, depending on the pH. The acid form of 2,4-D has a water solubility of 890 mg/L, and a reported range of half-lives from 6 to 170 days [70].

In general, phenoxy-acid herbicides do not adsorb strongly to soil, and they have a moderate potential for movement in surface runoff [69]. Their aquatic persistence appears to vary considerably, depending on pH, temperature, season, and the concentrations of suspended sediments and dissolved organic carbon [79]. For example, dicamba is very mobile in soil [86] and may undergo leaching before significant runoff occurs. It is also fairly resistant to biodegradation and hydrolysis, and is unlikely to volatilize [70].

2,4-D and 2,4-DB were detected in all of the lakes, from 1994 to 2000. 2,4-DP, 2,3,6-TBA, dicamba, MCPA, MCPB, bromoxynil, 2,4,5-TP, 2,4,5-T, and picloram were never detected in Lake Superior, but they were detected in at least one sample from each of the other lakes during the study period. Application rates of 2,4-D in the Great Lakes Basin are high in Indiana and in Lake Erie Basin of Canada [3, 76]. Concentrations of 2,4-D in Great Lakes waters exhibited a north-to-south increasing gradient, with Huron < Superior < Ontario < Erie. 2,4-D is applied in urban areas several times between spring and fall, for the control of broadleaf weeds. This may explain its frequent detection during the spring and summer.

Dicamba was not detected in Lake Superior in 1996 and 1997. Its concentration in each of the lakes decreased in the order Lake Huron/Georgian Bay > Erie > Ontario > Superior. Dicamba, like 2,4-D, may also be applied in urban areas several times from spring through fall for broadleaf weed control.

The acute toxicity of phenoxy-acid herbicides to humans and aquatic organisms is relatively low [87]. The USEPA maximum residue level for 2,4-D in drinking water is 70 000 ng/L, and the National Academy of Sciences has recommended a maximum concentration in water for protection of aquatic life of 3000 ng/L [78]. No Canadian guidelines for the protection of aquatic life and drinking water were exceeded (Table 8).

Air samples collected in the Canadian Prairies showed the presence of several currently used herbicides such as triallate, 2,4-D, dicamba, bromoxynil, MCPA, and trifluralin. While triallate was identified to be influenced by local sources, the other herbicides identified in air were thought to be due to regional atmospheric transport [88].

# 4
# Environmental Dynamics of Pesticides in Lake Erie Basin

The Lake Erie–Lake St. Clair drainage basin is a major region of intensive pesticide use in the Great Lakes Basin. The Lake Erie–Lake St. Clair drainage basin has been monitored for pesticide residues by the USGS's NAWQA Program [75, 89]; http://water.usgs.gov/nawqa/summaryreports/. The Lake Erie–Lake St. Clair drainage basin covers 57 757 km$^2$ and includes sections of Michigan, Indiana, Ohio, Pennsylvania, and New York. About 75% of the land area is agricultural with corn, soybeans, and wheat as the major crops. Population density and growth in the Lake Erie Basin are among the highest anywhere in the Great Lakes Basin [75]. About 40% of the population of the Great Lakes Basin lives in the Lake Erie Basin. Thus, pesticide residue trends and patterns in this basin can represent the Great Lakes Basin as a whole. Pesticides have been detected in every stream monitored during 1996–1998 in this basin. Some surface water samples contained a mixture of 18 pesticides; this is among the highest numbers of pesticides detected in any sample nationally.

## 4.1
## Pesticide Usage

Approximately 3.7 million kg of herbicides and more than 12 800 kg of insecticides were applied to agricultural areas in the Lake Erie Basin in 1995 [15]. Herbicides are applied during spring planting to nearly all corn and soybean fields. Similar to the situation in Ontario, metolachlor and atrazine were

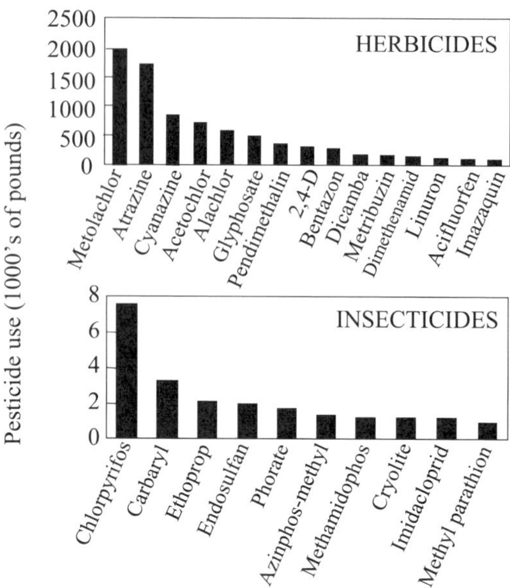

**Fig. 10** Pesticide usage (in 1000s of pounds of active ingredients) in the Lake Erie–Lake St. Clair drainage basins of the Great Lakes Basin during 1994–1995. From [15]

the most heavily used pesticides on soybeans and corn, respectively, in the basin during 1995 [15]. More than 0.9 million kg of metolachlor and 0.77 million kg of atrazine were applied. More than 45 400 kg of active ingredients of cyanazine, acetochlor, alachlor, glyphosate, 2,4-D, pendimethalin, dicamba, metribuzin, and dimethenamid also were applied. Insecticides are occasionally applied to corn in the summer, but they usually are not applied to soybeans. Chlorpyrifos, carbaryl, and ethoprop were the most heavily used insecticides in the basin (Fig. 10).

## 4.2
## Pesticide Detections and Concentrations

Of the 44 pesticides and three degradates analyzed for in surface waters of the Lake Erie Basin, 30 were detected at least once [75]. Atrazine was detected in every water sample, including samples from an area with less than 10% row crops (e.g., corn). The atrazine degradate deethylatrazine and the herbicide metolachlor were detected in 99% of the samples (Fig. 11). Eight of the 47 pesticides and degradates were detected in at least 50% of the water samples. All eight of these are herbicides used on row crops, except for prometon, which is used primarily in urban areas. Diazinon and chlorpyrifos were the most frequently detected insecticides, detected in 43% and 22% of the water samples, respectively. The most heavily applied agricultural pesticide not de-

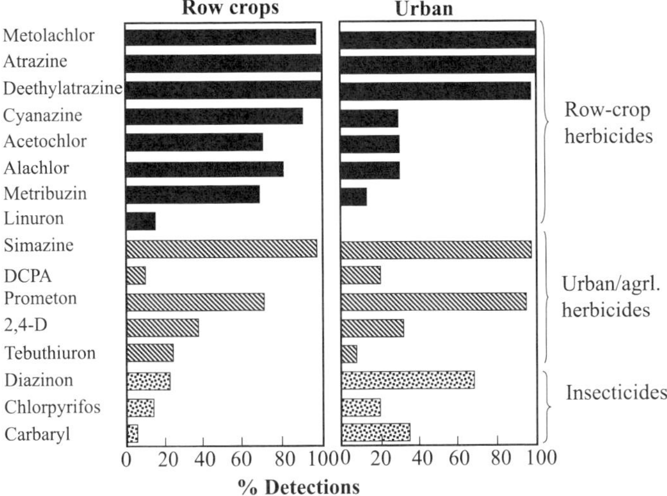


**Fig. 11** Most frequently detected pesticides in drainage basins of Lake Erie Basin stratified by row crop and urban areas. From [89]

tected was the insecticide ethoprop (2100 pounds applied annually). Atrazine and metolachlor had the highest detected concentrations in the basin, at 85 and 78 µg/L, respectively.

## 4.3
## Spatial Distribution

The spatial distribution of pesticides in streams coincides with the types and intensity of land uses in basins. Many herbicides, such as atrazine, are used exclusively or predominantly on row crops, while other pesticides, such as the herbicide prometon and the insecticide diazinon, are usually used in urban areas. Atrazine concentrations were higher in the row-crop basins than in the urban or pasture/forest basins. Diazinon concentrations were higher in the urban basins than in the row-crop or pasture/forest areas.

## 4.4
## Temporal Variations

The highest concentrations of pesticides in streams occur when precipitation causes surface-water runoff and tile drains to transport recently applied pesticides into streams [79, 90]. Most agricultural herbicides are applied either as pre-emergence or early post-emergence in the spring, and the corresponding highest concentrations in streams occur soon after application. In the Lake Erie Basin, the highest herbicide concentrations typically coincided with the first runoff after application and then steadily decreased to pre-

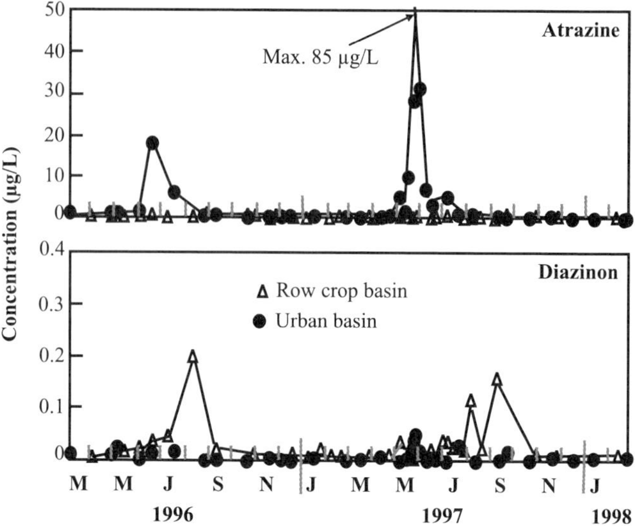


**Fig. 12** Concentrations of atrazine and diazinon in streams draining row-crop land and urban-use land in the Lake Erie Basin: seasonal and spatial trends. From [89]

application values. For some heavily used herbicides such as atrazine, metolachlor, cyanazine, and acetochlor, elevated concentrations persisted for 4–6 weeks after the initial maximum concentration, in the streams in row-crop basins [75]. A typical seasonal pattern for agricultural herbicides, particularly for atrazine concentrations, in the St. Joseph River near Newville, Indiana, applied in the spring is shown in Fig. 12. Maximum measured concentrations of prometon occur in late summer, since it is commonly applied in late summer or early fall in urban areas. Some insecticides primarily used in urban areas, such as diazinon, are applied over a longer period that can extend from late spring to late summer.

## 4.5
## Degradates of Selected Herbicides

Typically, only 0.01–10% of the mass of pesticide compounds applied to fields is detected in streams [91]. The remaining 90–99% of pesticides adsorb to soil, percolate into groundwater, or volatilize [79]. The major degradates of the most heavily used herbicides found in surface water have not been studied widely. Many chemical properties of pesticides affect the amounts transported to streams. In general, acetanilide herbicides are more soluble in water, and thus more mobile than are the triazines [92]. The solubilities of sulfonated degradates of acetanilides (ethane sulfonic acid, or ESA), can be 10 times the solubility of the parent compound [93]. The greater mobilities of the degradates of the acetanilides (amide family) can explain these compounds' greater frequencies of detection and greater concentrations, when compared to the triazines, in some locations. The frequency of detection of degradates ranged from 60% for cyanazine amide to 100% for metolachlor degradates [89]. Acetochlor ESA, acetochlor oxanilic acid, and cyanazine amide had maximum measured concentrations > 4 µg/L, of which acetochlor oxanilic acid had the highest maximum measured concentration, 6.76 µg/L. Concentrations of the triazine and acetanilide degradates had seasonal trends similar to those of their parent compounds. The maximum measured concentrations of the degradates, however, were in samples collected 1–2 weeks after the maximum measured concentration of the parent compound. Concentrations of acetanilide degradates were higher than that of the parent compound for most of the year (except immediately after application). Acetanilide degradates constituted a much larger fraction of the total acetanilide herbicides washing off fields than triazine degradates did for triazine herbicides. These limited data suggest that degradates (especially of acetanilide herbicides) can contribute a significant proportion of the pesticide load to Lake Erie.

In many cases, methods to assess the fate of current-use pesticides and their degradates in the environment are not available, due to lack of information on the transformation products, lack of chemical standards, and lack of methods by which to analyze them. The USGS has been developing water-quality analytical methods for the determination of chloroacetanilide herbicide degradates in water. These methods use solid-phase extraction and high performance liquid chromatography–diode array detection and high performance liquid chromatography–mass spectrometry. Both methods measure the concentration of 10 acetanilide degradates in samples of filtered water. Examples of degradation products of certain chloroacetanilide herbicides are acetochlor ethane sulfonic acid, acetochlor oxanilic acid, alachlor ethane sulfonic acid, alachlor oxanilic acid, dimethenamid ethane sulfonic acid, dimethenamid oxanilic acid, flufenacet ethane sulfonic acid, flufenacet oxanilic acid, metolachlor ethane sulfonic acid, and metolachlor oxanilic acid. These are the degradates of the widely used herbicides acetochlor, alachlor, and metolachlor, which are applied on a variety of crops. Bioaccumulation, degradation, and toxic potentials of current-use herbicides have rarely been investigated. Future investigations should focus on the fate and effects of in-use pesticides in the environment.

## 4.6
### Factors Influencing Pesticide Occurrence

The chemical and physical properties of pesticides affect their behavior in the environment. Solubility in water, soil half-life, and organic carbon partition coefficient ($K_{oc}$) play a major role in determining a chemical's runoff potential. Because families of pesticides have similar chemical structures and physical properties, similar frequencies of detections would be expected when families of pesticides are analyzed. In general, the frequency of detection of pesticides in samples collected from streams in the Lake Erie Basin increased as the amount of pesticides used in those basins increased. Pendimethalin is an exception; although more than 136 200 kg of this herbicide were applied in 1995 in the basin, it was rarely detected. There was also a trend of increasing frequency of detection with increasing runoff potential. Pesticides with medium or low runoff potential could be detected with a higher frequency if greater amounts of pesticide had been applied—as evidenced by detections of cyanazine, alachlor, and acetochlor. In general, the families of pesticides detected more than 10% of the time can be categorized as follows, in decreasing order of detection frequency: triazines, amides, organophosphates, ureas, carbamates, organochlorines, and dinitroanalines. Linuron and tebuthiuron, both members of the urea family, have large runoff potentials yet they have the lowest frequencies of detection among all of those compounds that have a large runoff potential. This may be a reflection of the amount of pesticide applied, the timing of application, or a method of application that hinders runoff. Pendimethalin, the sixth most highly applied pesticide in the basin, has a much lower frequency of detection than the other heavily used pesticides that have medium runoff potentials (alachlor, acetochlor, and cyanazine). This lower frequency is more likely a function of pendimethalin's lower water solubility and higher $K_{oc}$ value relative to the other herbicides with higher frequencies of detection [94]. In addition, varying rates of degradation may affect detection of pesticides.

The environmental factors that influence occurrence and concentrations of pesticides include amount and timing of rainfall after pesticide application, and dilution by water bodies. Another factor that appears to influence pesticide concentrations in streams is soil permeability. Well-drained soils allow water to percolate into the groundwater. As water percolates through soil, some pesticides are filtered by the soil and broken down to degradates by bacteria. In areas with impermeable soils, more water enters streams as surface runoff. Such areas also require tile drains to make the land arable. Because tile drains lessen the underground filtration of the soils, they can transport elevated concentrations of pesticides [95, 96].

## 5
## Atrazine Mass Balance in Lake Michigan

Approximately 29–34 million kg of atrazine are applied per year in the USA, much of which is used in the "corn belt" region that includes the upper Midwest surrounding Lake Michigan. Atrazine is generally applied to soil pre-planting or pre-emergence, but is sometimes also applied to the foliage post-emergence. Atrazine can enter surface waters, including Lake Michigan, through runoff, spray drift, discharge of contaminated groundwater to surface water, wet deposition (dissolution of atrazine vapor in rainfall and washout of particle-bound atrazine), dry deposition (dry settling of particle-bound atrazine), and sorption from the vapor phase. For protection of human health, the USEPA has set a maximum contaminant level of 3 µg/L in drinking water. The USEPA also has set draft ambient aquatic life criteria at 350 µg/L for protection from acute effects and 12 µg/L for protection from chronic effects. The widespread use and environmental fate of this herbicide in the Great Lakes Basin have been the subject of several investigations [59, 97, 98]. The USEPA's GLNPO, as a part of the Lake Michigan Mass Balance (LMMB) study, measured and modeled the concentrations of atrazine in major environmental compartments of the Lake Michigan ecosystem [99]. The LMMB study provides a model for the more reactive (i.e., water soluble), biodegradable compounds that are in current use in the Great Lakes Basin. The model does not include a food-chain component, since atrazine does not bioaccumulate appreciably. A summary of the results is given below. More details can be found elsewhere [99].

In the LMMB study, atrazine and atrazine metabolites (deethyl-atrazine, or DEA, and deisopropyl-atrazine, or DIA) were measured in atmospheric, tributary water-column, and open-lake water column samples from March 1994 through October 1995. Atmospheric vapor, particulate, and precipitation samples were collected from eight stations surrounding Lake Michigan, and from three background stations outside the Lake Michigan Basin. Tributary water-column samples were collected from 11 rivers that flow into Lake Michigan. Open-lake water-column samples were collected from 35 sampling stations in Lake Michigan, two stations in Green Bay, and one station in Lake Huron.

## 5.1
### Atrazine in Atmospheric Components

The predominant atmospheric source of atrazine, DEA, and DIA measured in the LMMB study was precipitation [99]. In atmospheric samples, atrazine was seldom detected in the vapor phase (3.7% frequency). Atrazine was more frequently detected in the particulate phase and in precipitation, with 23% and 50% of sample concentrations reported to be above sample-specific detection limits, respectively. Atrazine was generally not detectable in atmospheric samples during the fall and winter, but atmospheric concentrations peaked during the spring, as a result of agricultural application of the herbicide. Maximum monthly atrazine concentrations in the particulate phase ranged from 160 pg/m$^3$ to 1400 pg/m$^3$, and mean atrazine concentrations during the spring and summer (March through September) ranged from 25 pg/m$^3$ to 370 pg/m$^3$, among the atmospheric sampling stations. In precipitation, maximum monthly atrazine concentrations ranged from 100 ng/L to 2800 ng/L, and monthly volume-weighted mean concentrations during the spring and summer ranged from 19 ng/L to 120 ng/L, among the atmospheric sampling stations. Concentrations of DEA and DIA were correlated with atrazine concentrations and generally followed the same patterns.

In general, concentrations of atrazine in the particulate phase were higher at the atmospheric sampling stations surrounding the southern Lake Michigan Basin than at those stations surrounding the northern basin. This is consistent with the pattern of agricultural land use, which decreases in intensity from south to north in the Lake Michigan region. Atrazine concentrations in precipitation were less reflective of local land-use conditions and suggest long-range transport of the herbicide, in addition to local inputs. Atrazine concentrations in precipitation were not consistently higher surrounding the southern Lake Michigan Basin; in fact, atrazine concentrations in precipitation were often higher at remote sampling stations in the far north than at stations in the southern basin.

## 5.2
## Atrazine in Tributaries

Atrazine was detected above the method detection limit of 1.25 ng/L in 99% of tributary water samples. Maximum atrazine concentrations in Lake Michigan tributaries ranged from 6.4 ng/L in the Manistique River to 2700 ng/L in the St. Joseph River, and mean atrazine concentrations ranged respectively from 3.7 ng/L to 350 ng/L in the two rivers. Concentrations of atrazine in tributaries were strongly influenced by geographical location and corn-crop acreage. Atrazine concentrations were highest in the St. Joseph, Grand, and Kalamazoo Rivers, where agricultural influences were much stronger, and where atrazine use rates were from 0.4 to > 1.1 kg/acre. For these three tributaries with the highest atrazine levels, distinct peaks in atrazine were observed in mid- to late May, corresponding to the agricultural application of the herbicide. Distinct seasonal patterns in atrazine concentration were not observed for the other tributaries.

## 5.3
## Atrazine in the Open-lake Water Column

Within Lake Michigan, atrazine concentrations in open-lake water-column samples were relatively consistent. Average atrazine concentrations at open-lake sampling stations within Lake Michigan ranged from 33 to 48 ng/L. The concentrations in the other Great Lakes are, for comparison: 3–4 ng/L in Lake Superior, 20 ng/L in Lake Huron, 30–110 ng/L in various drainages of Lake Erie, and 70–95 ng/L in Lake Ontario [100]. Atrazine concentrations within Lake Michigan were statistically greater than those measured at the reference station on Lake Huron and were statistically lower than concentrations measured at one Green Bay sampling station. Because the open lake water was well-mixed with respect to atrazine, lake-wide averages could be calculated. Over the course of the study, lake-wide average atrazine concentrations increased from 37 ng/L in April/May 1994 to 39.7 ng/L in March/April 1995. During the same time period, DEA concentrations increased by 14.9%, and DIA concentrations increased by 54%. While atrazine concentrations increased slightly during the study, open-lake average atrazine levels remained more than 50 times below the maximum contaminant level for drinking water, and more than 300 times below the proposed ambient water-quality criterion for protection of aquatic life from chronic effects.

## 5.4
## Mass Balance Analysis

The inventory of atrazine in Lake Michigan is estimated to be 176 tons. Mass balance analysis of atrazine sources to Lake Michigan showed that tributaries accounted for 76% of the total load of 11.8 tons/year [100]. The annual input of atrazine to Lake Michigan is 7% of the inventory. Average annual inputs and losses of atrazine in Lake Michigan are shown in Table 9. Dry deposition, air-water exchange and sedimentation have little effect on atrazine transformation. The half-life of atrazine in the Lake Michigan water column is estimated to be 13.9 years. The long water-column half-life has allowed atrazine to accumulate in the Great Lakes over the last few decades. Model estimates have suggested that the atrazine inventory will show little change in the future, due to the steady state/equilibrium in concentrations. Atrazine does undergo in situ transformation, and the majority of the transformation is conversion of DEA, by biological processes. Approximately 5–10% of the atrazine is lost annually through internal transformation.

**Table 9** Average annual inputs and losses of atrazine (kg/yr) to Lake Michigan

| Source of input or loss | Load (tons/yr) | % of total load |
| --- | --- | --- |
| Tributary | 9.04 | 76 |
| Precipitation | 2.6 | 22 |
| Dry deposition | 0.16 | 1 |
| Air/water exchange | 0.03 | < 1 |
| Outflow | 2.9 | 24 |
| Transformation | 8.89 | 76 |
| Sediment burial | 0.025 | < 1 |

From [100]

Although atrazine has a soil half-life of 30–90 days, transport out of this zone into receiving waters leads to longer half-lives [101]. For pesticides with high water solubilities, such as atrazine, tributary inputs can be a major source, and environmental response to sources is controlled by long hydraulic residence times and slow transformation rates. Despite the fact that only 1% of the applied atrazine is lost by transport to rivers and lakes, and another 1% by aerial transport, the large quantities applied, together with efficient hydraulic transport result in accumulation in aquatic systems.

## 6
## A Case Study of Dynamics of OC Pesticides in Big Creek, Ontario

The Big Creek watershed (68 400 ha), which drains into Lake Erie near Long Point, is the major tobacco-growing area in southern Ontario (Fig. 13). Concentrations of OC pesticides in water, air, and soil in the Big Creek watershed in southwestern Ontario were determined in order to examine the pathways of transport of pesticides from the watershed to Lake Erie [52]. In 1980, the total tobacco growing area was approximately 50% of the area of the watershed. In the 1950s and 1960s, there was heavy use of OC pesticides such as DDT in the tobacco-growing region within the watershed [102]. DDT was used in the watershed until 1971, and dieldrin until 1969. Here we focus on

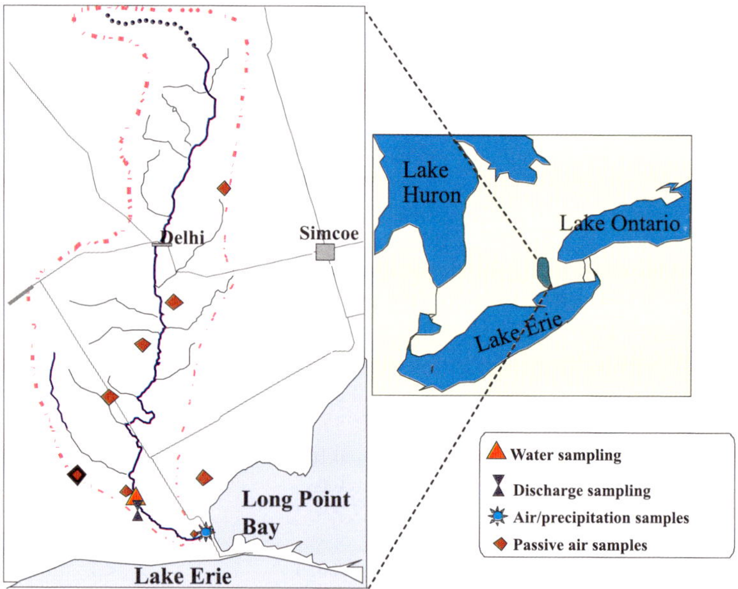


**Fig. 13** Map of Big Creek watershed and location of sampling sites. From [52]

the fate of OC pesticide data to illustrate the continuing watershed sources of these compounds to the Great Lakes tributaries.

## 6.1
## OC Pesticides in Soil

DDT and dieldrin residues were detected in all soil samples from both tobacco-producing and non-tobacco-producing farms sampled within the Big Creek watershed. Soils used for tobacco and orchards had the highest OC pesticide concentrations (0.49–15.5 µg/g dry weight for DDT-related compounds, 0.001–0.044 µg/g dry weight for dieldrin), while concentrations in vegetable-growing and field-crop soils were typically lower. DDT concentrations measured in soils in this region were much greater than those reported for cotton soils in the southeastern USA [62, 103].

## 6.2
## OC Pesticides in Dissolved and Suspended-Solid Phases

DDTs (primarily $p,p'$-DDE, $p,p'$-DDD, and $p,p'$-DDT), dieldrin, and chlordane compounds (primarily *trans*-nonachlor and *cis*-chlordane) were the major OC pesticides in samples of water (dissolved phase) and suspended solids from Big Creek. Total DDT concentrations ranged from 140 pg/L to 1260 pg/L in the dissolved phase, and from 12–71 ng/g dry weight in suspended solids. Total DDT concentrations in the dissolved phase peaked during May and June 2000, coinciding with high flows of the river, heavy rains, and soil erosion (Fig. 14). Particulate-phase DDT concentrations were less variable than those in the dissolved phase; however, concentrations increased during or following very high flows in late May and in October 2000, and again during snow-melt runoff in March 2001.

Elevated DDT concentrations in the particulate phase from Big Creek coincided with elevated suspended solids in the river, reflecting erosion of agricultural soils containing old residues. DDE/ΣDDT ratios (in dissolved phase) were also relatively constant during spring–summer sampling in 2000, but they increased during snow-melt runoff events in February–March 2001. The lack of coincident maxima for DDE/ΣDDT ratios and ΣDDT concentrations suggests that snow-melt runoff accessed different sources than did summer rain events. Rain events could have resulted in erosion of tilled field soils, whereas the snow-melt events occurred prior to tillage and may reflect material deposited in ditches and river banks.

Dieldrin concentrations in dissolved and particulate phases from Big Creek were significantly correlated with ΣDDT (dissolved: $r = 0.90$; $p < 0.001$; particulate: $r = 0.94$; $p < 0.001$), suggesting common sources within the watershed. Dissolved-phase dieldrin concentrations ranged from 76 pg/L to 579 pg/L, and mean and median concentrations were higher than for any

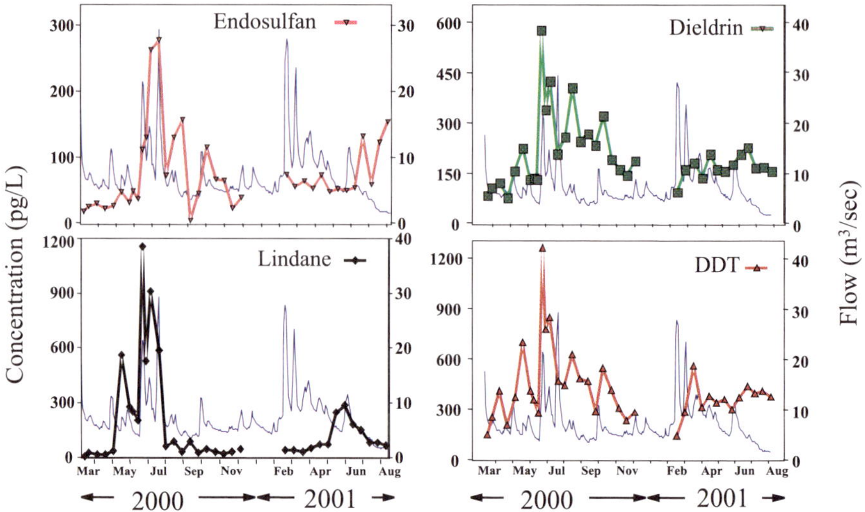

**Fig. 14** Concentrations of endosulfan, lindane, dieldrin, and DDT (pg/L) in the dissolved phase (Big Creek, 2000 and 2001) co-plotted with river flow (m³/s). From [52]

other individual OC pesticides determined in the study. Dieldrin was not a major OC pesticide in suspended solids from Big Creek. This observation is consistent with the greater polarity of dieldrin ($\log K_{\mathrm{ow}} = 5.4$), compared to DDT or chlordanes ($\log K_{\mathrm{ow}}$ of cis-chlordane = 6.16).

## 6.3
## Air and Precipitation Concentrations

The OC pesticide concentrations in air at Big Creek were similar to values measured during the same period at the nearest IADN station, Sturgeon Point (Table 10). These values were typically elevated when compared with the values at Point Petre located on Lake Ontario. DDE and DDT collected in precipitation at Big Creek were also elevated when compared with the IADN sites. This variation was thought to reflect volatilization of DDT residues from soils in the Big Creek watershed.

## 6.4
## Comparison of Fluxes

Fluxes of OC pesticides were estimated by multiplying concentrations (ng/L or µg/m³) with daily flow rates (m³/d). Fluxes of DDT (520 g over 12 months) were greater than for the current-use pesticides such as lindane and endosulfan. Fluxes of DDT and dieldrin were much greater via suspended sediments than via dissolved phase.

**Table 10** Average concentrations (pg/m$^3$) of organochlorine pesticides in air samples and in wet precipitation (ng/L) in Big Creek watershed and at closest IADN sites (Sturgeon Point and Point Petre)

| Location | Year | Months | $\alpha$-Endosulfan | Lindane | $p,p'$-DDE | $p,p'$-DDT |
|---|---|---|---|---|---|---|
| **Vapour phase samples (*pg/m$^3$*)** | | | | | | |
| Big Creek | 2000 | May–Sep | $215 \pm 158$ | $33 \pm 35$ | $37 \pm 31$ | $4.0 \pm 2.2$ |
| Sturgeon Point | 2000 | May–Sep | $184 \pm 117$ | $59 \pm 41$ | $31 \pm 8$ | $11 \pm 4$ |
| | | | $n = 12$ | $n = 12$ | $n = 12$ | $n = 12$ |
| Point Petre | 2000 | May–Sep | $121 \pm 190$ | $20 \pm 15$ | $11 \pm 5$ | $3 \pm 1.6$ |
| | | | $n = 11$ | $n = 11$ | $n = 11$ | $n = 11$ |
| **Precipitation Samples (ng/L)** | | | | | | |
| Big Creek | 2000 | May–Sept | 0.10 | 2.06 | 0.56 | 0.16 |
| Big Creek | 2001 | May–Aug | 1.62 | 2.89 | 0.99 | 0.54 |
| Sturgeon Point | 2000 | May–Sept | 0.93 | 0.90 | 0.026 | 0.077 |
| Sturgeon Point | 2001 | May–Aug | 1.06 | 1.83 | 0.039 | 0.075 |
| Point Petre | 2000 | May–Sept | 1.87 | 1.53 | 0.15 | 0.23 |

The soil and air concentrations of $p,p'$-DDT were used in a soil-air exchange model developed by Harner et al. [103] to estimate $p,p'$-DDT emissions to air from soils under steady-state conditions. The volatilization flux of $p,p'$-DDT from soil to air was larger than fluxes in runoff over the same time period. For example 15.7 kg of DDT was volatilized during the period March–June, if it was assumed that an area of 300 km$^2$ of land was formerly treated with DDT; the steady-state model results were used and calculated monthly, with adjustment for mean monthly temperature. Re-deposition in rainfall was about 15 g of $p,p'$-DDT to the watershed, based on concentrations in precipitation measured at the field station during May–August 2000. The extent of re-deposition of gas-phase DDT compounds on plants and soils was not known.

## 6.5
### Comparison with Results of Previous Studies

In 1970, river water from lower Big Creek was analyzed for DDT and dieldrin [104]. At that time, $\Sigma$DDT concentrations ranged from 4 to 34 ng/L over the period of April–October, while dieldrin concentrations ranged from 0.5 to 18 ng/L. In 2000–2001, total DDT (dissolved + particulate) concentrations ranged from 0.38 to 18.3 ng/L, and dieldrin concentrations ranged from 0.17 to 1.6 ng/L. $p,p'$-DDE/$\Sigma$DDT ratios in 1970 ranged from 0.03 to 0.5 ng/L compared with 0.37 to 0.53 ng/L in 2000. This reflects the much higher proportion of parent $p,p'$-DDT and $o,p'$-DDT in 1970 compared to that in 2000, when $p,p'$-DDE was the predominant form of DDT in the river. Assuming similar flows in 1970 and 2000, input levels of DDT were estimated to be 2 to

4 times lower in recent years, and levels of dieldrin were 2.5 to 20 times lower in 2002 [52].

# 7
# Effects of Pesticides

The analyses of the Great Lakes have resulted in a greater understanding of the potential hazards of releasing persistent OC pesticides, which are both bioaccumulative and toxic. Many of the OC pesticides that have been historical or current problems are no longer manufactured, or their use is heavily restricted. Concentrations of the most problematic compounds such as DDT declined rapidly after the ban, but the current rates of decline are very slow, and it will be a long time before the concentrations in both fish and birds of the Great Lakes environment reach background concentrations [32, 105]. Monitoring results clearly indicate that the trend of decline in concentrations of OC compounds leveled off in the mid-1980s, and resource-management agencies and research institutions are investigating the potential to further reduce sources of contamination in Great Lakes fish and wildlife.

Despite the decline in the levels of OC pesticides, fish-consumption advisories still exist in several sites in the Great Lakes Basin, due to the presence of concentrations that are still high enough to cause human health effects. Current levels of pesticides such as DDT and dieldrin are still high enough to cause effects in predatory species such as bald eagles and mink [106]. For instance, bald eagles nesting along the shorelines of the Great Lakes do not reproduce as well as those nesting on inland bodies of water [107]. The reduced productivities along the shorelines of the Great Lakes and anadromous salmon streams may be due to a number of factors, including exposures to residual concentrations of $p,p'$-DDE.

Concentrations of DDT and dieldrin, both in the open lakes and within the connecting channels, still regularly exceed the aquatic-life criterion guidelines set forth by the New York State Department of Environmental Conservation [108]. For example, the mean concentrations of dieldrin in Lakes Superior and Huron waters are 0.11 and 0.04 ng/L [17], lower than those reported for the other Great Lakes. Despite this, the dieldrin concentrations exceed the NYSDEC criterion of 0.0006 ng/L. Similarly, concentrations of DDT in several locations in Great Lakes are greater than the NYSDEC criterion of 0.007 ng/L. Similarly, DDT and HCH guideline values for the protection of aquatic life are 200 and 100 ng/g on a wet weight basis in fish tissues. Concentrations of DDT and HCH exceed these limits frequently in several sport fish species. Thus, OC pesticides continue to be of concern in the Great Lakes ecosystem.

With regard to current-use pesticides, few pesticides have standards and/or guidelines established by the health organizations. For example, of the 88 pesticides analyzed in the Lake Erie Basin by the NAWQA of the USGS, a maximum permissible concentration has been set for only 14 pesticides, and health advisory lifetime guidelines for the protection of human health are available for just 38 pesticides [12]. Aquatic-life guidelines have been developed for only 18 pesticides. Most of the standards are based on toxicity tests on a single pesticide and do not evaluate the additive or synergistic effects of multiple pesticides. The standards also do not address possible effects of pesticides on endocrine systems, or other subchronic effects [91]. Surveys conducted in the Lake Erie Basin have shown concentrations exceeding the aquatic-life guidelines for the herbicides atrazine, metolachlor, cyanazine, and metribuzin, and for the insecticides chlorpyrifos, carbaryl, and diazinon [12]. Heavily used compounds such as atrazine, metolachlor, cyanazine, and metribuzin were the compounds that most often showed concentrations greater than the guideline values.

Several current-use pesticides, despite their low bioaccumulation potentials, are acutely toxic to aquatic organisms. OPs and carbamates can degrade at a faster rate than do OC pesticides. However, the fate, bioaccumulation, and effects of the metabolites of these classes of compounds are not well studied. Although the concentrations of individual current-use pesticides are generally below the drinking water quality criterion, occurrence of a mixture of several pesticides suggests the need to evaluate the effects of mixtures of pesticides. Furthermore, seasonal pulses in the concentrations of current-use pesticides may affect of organisms that are at sensitive stages of their life cycle. Despite these compounds' lower rates of bioaccumulation, exposure to OPs and carbamates can cause neurobehavioral and immunological effects in wildlife [109, 110]. Reduced semen quality in agrarian population in the Midwest relative to urban population in the USA has been linked to intensive pesticide usage and exposures [111]. A difficult aspect of evaluating potential effects of pesticides is determining varying types and durations of exposure. Exposure assessment is further complicated by pesticide mixtures, breakdown products, and seasonal pulses. Current standards and guidelines do not completely eliminate risks, because: (1) values are not established for many pesticides; (2) mixtures and breakdown products are not considered; (3) the effects of seasonal exposure to high concentrations have not been evaluated; and (4) some types of potential effects, such as endocrine disruption and unique responses of sensitive individuals, have not yet been assessed.

# 8
## Conclusions

Spatial distributions and temporal trends in sediment, water, and fish indicate progress in the reduction of persistent OC pesticides in the Great Lakes Basin. Concentrations of many of the OC pesticides have decreased significantly in Great Lakes wildlife, subsequent to restriction of these compounds' usage in the Great Lakes Basin in the 1970s and the 1980s. Nevertheless, concentrations of several OC pesticides in the Great Lakes are currently either not declining or are declining only slowly. It has been suggested that current concentrations of several OC pesticides are at a steady state, such that the amount lost to the sediments and exported either to the atmosphere or outflowing to the St. Lawrence River is balanced by input from rivers and the atmosphere. Thus, further reductions in the input of OC pesticides and the continued recovery of wildlife populations are dependent, in part, on the control of new inputs.

With regard to current-use pesticides, the waters of the Canadian Great Lakes met all existing guidelines for the protection of aquatic life and drinking water, despite evidence of near-field and far-field influences. While sampling of the open waters of the Great Lakes should be continued on a limited basis to establish ambient conditions, future monitoring should concentrate on the sampling of connecting channels, areas of special concern, and smaller aquatic ecosystems more closely linked to areas of high pesticide use, especially in the lower Great Lakes.

Both seasonal and localized variations in the concentrations of neutral and phenoxy-acid herbicides were observed. These variations can be related to the patterns of use of such herbicides. Multiple pesticides were detected at individual stations in some areas. The frequency of detection of current-use pesticides in the Great Lakes generally increased in the order Superior < Huron and Georgian Bay < Ontario < Erie. The highest concentrations among current-use pesticides were measured for atrazine, metolachlor, 2,4-D, and diazinon in the western basin of Lake Erie, because of the close proximity to areas where these pesticides are applied in both agricultural and urban settings. Existing Canadian and American pesticide guidelines for the protection of aquatic life and drinking water were not exceeded in the Great Lakes.

Monitoring activities should also be focused on sampling of the major pesticides that are in current use. At present, programs are in place to analyze for about half of the pesticides used. Additive and synergistic effects of pesticide mixtures must be examined more closely, since existing guidelines have been developed for individual pesticides only.

## References


1. USEPA (1996) http://www.epa.gov/glnpo/solec/96/landuse/introduction.html Cited 15 June 2005
2. OMAFRA (1999) Pesticide Use in Ontario, 1998. Ontario Ministry of Agriculture, Food and Rural Affairs, Toronto
3. Gianessi LP, Anderson JE (1995) Pesticide use in Illinois, Indiana, Michigan, Minnesota, New York, Ohio, Pennsylvania, and Wisconsin crop production. National Center for Food and Agricultural Policy, Washington

4. Aspelin AL (1994) Pesticides industry sales and usage, 1992 and 1993 market estimates: US Environmental Protection Agency, office of pesticide programs, biological and economic analysis division, economic analysis branch report 733-K-92-001, p 33

5. Frank R, Logan L, Clegg BS (1991) Arch Environ Contam Toxicol 21:585–595

6. Stevens RJJ, Neilson MA (1989) J Great Lakes Res 15:377–393

7. Allan RJ, Ball AJ (1990) Wat Poll Res J Canada 25:387–505

8. Gianessi LP, Puffer C (1992) Insecticide use in US crop production: resources for the future. Quality of the Environment Division, Washington, p 180

9. Statistics Canada (1987) Ontario Agriculture, census Canada 1986. Ministry of Supply and Services, Ottawa, p 281

10. Agriculture Canada (1974) Memorandum regarding proposed revision in acceptable use claims for DDT. Plant Products Division, Ottawa

11. USGS (1999) The quality of our nation's waters—nutrients and pesticides: US Geological Survey circular 1225, p 82. http://water.usgs.gov/pubs/circ/circ1225/html/wq_urban.html. Cited 15 June 2005

12. USGS (2000) Water quality in the Lake Erie–Lake St. Clair drainages. 1996–98 US Geological Survey, circular 1203, Denver

13. Senthilkumar K, Kannan K, Giesy JP, Masunaga S (2002) Environ Sci Technol 36:2789–2796

14. Donaldson GM, Shutt JL, Hunter P (1999) Arch Environ Contam Toxicol 36:70–80

15. Brody TM, Furio BA, Macarus DP (1998) Agricultural pesticide use in the Great Lakes Basin: estimates of major active ingredients applied during 1994–95 for the Lake Erie, Michigan, and Superior Basins: US Environmental Protection Agency, Region V, p 15. http://www.epa.gov/reg5rcra/ptb/pest/documents/pest_use.pdf

16. Kuntz KW, Hanau M (1995) Joint evaluation of upstream/downstream Niagara river monitoring data for the period April 1992 to March 1993. Ecosystem Health Division, Environment Canada, p 84

17. Marvin C, Painter S, Williams D, Richardson V, Rossmann R, van Hoof P (2004) Environ Pollut 129:131–144

18. Jorgensen JL (2001) Environ Health Perspect 109:113–139

19. Environment Canada (1997) Contaminants in herring gull eggs from the Great Lakes: 25 years of monitoring levels and effects. Great Lakes fact sheet, p 12. http://on.ec.gc.ca/wildlife/factsheets/fs_herring_gulls-e.html. Cited on 15 June 2005

20. Harner T, Bidleman TF, Wideman JL, Parkhurst WJ (1999) Environ Pollut 106:323–332

21. Hafner WD, Hites RA (2003) Environ Sci Technol 37:3764–3773

22. USEPA (2003) NPL fact sheets for Michigan. US Environmental Protection Agency, Region V. http://www.epa.gov/region5superfund/npl/michigan/MID000722439.htm. Cited 15 June 2005:

23. USEPA (1998) Great Lakes pesticide report, 1998. US Environmental Protection Agency, Region V. http://www.epa.gov/glnpo/bnsdocs/98summ/pest4.pdf and www.epa.gov/glnpo/aoc. Cited 15 June 2005

24. Buehler SS, Basu I, Hites RA (2004) Environ Sci Technol 38:414–422

25. Ma J, Daggupaty S, Harner T, Li Y (2003) Environ Sci Technol 37:3774–3781

26. Hunter C, McGee B (1999) Survey of pesticide use in Ontario, 1998. Estimates of pesticide used on field crops, fruit and vegetable crops, and other agricultural crops. Policy Analysis Branch, Guelph

27. USGS (1998) Maps of annual pesticide use. National water quality assessment pesticide national synthesis project. US Geological Survey. http://ca.water.usgs.gov/pnsp/use92/mapex.html. Cited 15 June 2005

28. Keith JA (1966) J Appl Ecol 3:57–70
29. Ludwig JP, Tomoff CS (1966) Jack-Pine Warbler 44:77–84
30. Williams DJ, Kuntz KW, L'Italien S, Richardson V (2001) Organic contaminants in the Great Lakes 1992–1998. Intra- and inter-lake spatial distributions and temporal trends. Ecosystem Health Division, Environmental Conservation Branch, Ontario Region, Burlington, Ontario, Report No. EHD/ECB-OR/01–01/I
31. USEPA/USGS (1977) Memoranda of understanding/cooperative agreement for Great Lakes fish monitoring program (GLFMP) between EPA/CRL and USFWS/GLFL, updated July 13, 1989 between USEPA/GLNPO and USFWS/NFRC-GL and June 10, 1996 between USEPA/GLNPO and NBS/GLSC. Ann Arbor, MI and Chicago, IL
32. DeVault DS, Hesselberg R, Rodgers PW, Feist TJ (1996) J Great Lakes Res. 22:884–895
33. Hickey JP, Batterman SA, Chernyak SM (2006) Arch Environ Contam Toxicol (in press)
34. Fox GA, Trudeau S, Won H, Grasman KA (1998) Environ Monit Assess 53:147–168
35. Scheider WA, Cox C, Hayton A, Hitchin G, Vaillancourt A (1998) Environ Monitor Assess 53:57–76
36. Suns KR, Hitchin GG, Toner D (1993) J Great Lakes Res 19:703–714
37. Huestes SY, Servos MR, Whittle DM, Dixon DG (1996) J Great Lakes Res 22:310–330
38. Ryckman DP, Weseloh DV, Hamr P, Fox GA, Collins B, Ewins PJ, Norstrom RJ (1998) Environ Monitor Assess 53:169–195
39. Loganathan BG, Kannan K (1991) Mar Pollut Bull 22:582–584
40. Makarewicz JC, Damaske E, Lewis TW, Merner M (2003) Environ Sci Technol 37:1521–1527
41. Glassmeyer ST, Meyers TR, Devault DS, Hites RA (1996) Environ Sci Technol 31:84–88
42. Hoff RM, Strachan WMJ, Sweet CW, Chan CH, Shackelton M, Bidleman TF, Brice E, Burniston DA, Cussion S, Gatz DF, Harlin K, Schroeder WH (1996) Atmos Environ 30:3505–3527
43. Ridal JJ, Kerman BR, Durham L, Fox ME (1996) Environ Sci Technol 30:845–851
44. McConnell LL, Bidleman TF, Cotham WE, Walla MD (1998) Environ Pollut 101:391–399
45. Baker JE (1997) Atmospheric deposition of contaminants to the Great Lakes and coastal waters. SETAC Press, Pensacola, p 442
46. Bidleman TF (1999) Water Air Soil Pollut 115:115–166
47. Van Pul WAJ, Bidleman TF, Brorstrom-Lunden E, Builtjes PJH, Dutchak S, Duyzer JH, Gryning SE, Jones KC, van Dijk, HFG, Wauchope RD, Buttler TM, Hornsby AG, Augustijn-Beckers PVM, Burt JP (1992) Rev Environ Contam Toxicol 123:1–156
48. Hafner WD, Hites RA (2004) Environ Sci Technol 37:3764–3773
49. Hornbuckle KC, Green ML (2003) Ambio 32:406–411
50. Buehler SS, Basu I, Hites RA (2001) Environ Sci Technol 35:2417–2422
51. USEPA (1998) Atmospheric deposition of toxic substances to the Great Lakes: IADN results through 1998. Environment Canada and the USEPA docket # EPA905-R-01-007
52. Bidleman TF, Harner T, Muir D, Ridal JJ, Ripley B, Strachan WJM, Struger J, Van Vliet L, Waite D (2002) Sources of agrochemicals to the atmosphere and delivery to the Canadian environment. Final report, Toxic Substances Research Consortium, Downsview
53. Harris ML, van den Heuvel MR, Rouse J, Martin PA, Struger J, Bishop CA, Takacs P (2000) Pesticides in ontario: a critical assessment of potential toxicity of agricultural products to wildlife with consideration for endocrine disruption. vol 1. Canadian

Wildlife Service, Environmental Conservation Branch, Ontario Region, Technical report Series No. 340, Ministry of Governmental Services, Cat. No. CW69-5/340E.

54. Cortes DR, Basu I, Sweet CW, Brice KA, Hoff RM, Hites RA (1998) Environ Sci Technol 32:1920–1927

55. Cortes DR, Hoff RM, Brice KA, Hites RA (1999) Environ Sci Technol 33:2145–2150

56. Simcik MF, Hoff RM, Strachan WMJ, Sweet CW, Basu I, Hites RA (2000) Environ Sci Technol 34:361–367

57. Mackay D, Bentzen E (1997) Atmos Environ 31:4045–4047

58. Bergeron M, Poissant L (2000) Contribution relative de l'atmosphere a la masse toxique transportée par le fleuve saint-laurent. Meteorological Service of Canada, Environment Canada, Centre St. Laurent, Montreal

59. Schottler SP, Eisenreich SJ (1994) Environ Sci Technol 28:2228–2232

60. Richards RP, Baker DB (1993) Environ Toxicol Chem 12:13–26

61. Pesticides in surface waters. U.S Geological Survey Fact Sheet FS-039-97. http://ca.water.usgs.gov/pnsp/rep/fs97039/sw4.html, Cited 15 June 2005

62. Kannan K, Battula S, Loganathan BG, Hong C-S, Lam WH, Villeneuve DL, Sajwan K, Giesy JP, Aldous KM (2003) Arch Environ Contam Toxicol 45:30–36

63. Smith AE, Kerr LA, Caldwell B (1997) J Agric Food Chem 45:1473–1478

64. Hopkins EH, Hippe DJ, Frick EA, Buell GR (2000) Organophosphorus pesticide occurrence in surface and ground water of the United States, 1992–97. US Geological Survey Report 00-187

65. Struger J, L'Italien S, Sverko E (2003) In-use pesticide concentrations in surface waters of the Laurentian Great Lakes, 1994–2000. Environment Canada, Ecosystem Health Division, Environmental Conservation Branch-Ontario Region, Burlington, 2003. Report No.: EHD/ECB-OR/03-04/I

66. Struger J, L'Italien S, Sverko E (2004) J Great Lakes Res 30:435–450

67. Struger J (1998) Organophosphorous insecticides and endosulfan in surface waters of the Niagara fruit belt, Ontario, Canada. Presented at the Society of Environmental Toxicology and Chemistry meeting, Charlotte, North Carolina

68. Struger J, Boyd D,Wilson M, Martos P, Ripley B (2002) In-use pesticide concentrations of Canadian tributaries of Lakes Erie and Ontario. International Association of Great Lakes Research Meeting, Winnipeg

69. Goss DW (1992) Weed Technol 6:701–708

70. Muir DCG (1991) Dissipation and transformation in water and sediment. In: Grover R (ed). Environmental Chemistry of Herbicides, Volume 2. CRC Press, Boca Raton, pp 3–88

71. Moxley J (1989) Survey of pesticide use in Ontario, 1988. Economics Information Report 89-08, Ontario Ministry of Agriculture and Food, Legislative Buildings, Queen's Park, Toronto

72. Frank R, Clegg BS, Sherman C, Chapman ND (1990) Arch Environ Contam Toxicol 19:319–324

73. Richards RP, Baker DB, Kramer JW, Ewing DE (1996) J Great Lakes Res 22:414–428

74. Goolsby DA, Thurman EM, Pomes ML, Meyer MT, Battaglin WA (1997) Environ Sci Technol 31:1325–1333

75. Frey JW (2001) Occurrence, distribution, and loads of selected pesticides in streams in the Lake Erie-Lake St. Clair Basin, 1996–98. Water-Resources Investigations Report 00-4169. National Water-Quality Assessment Program (NAWQA), Indianapolis 2001. US Geological Survey

76. OMAFRA (1994) Survey of pesticide use in Ontario, 1993. Ontario Ministry of Agriculture, Food and Rural Affairs, Economics Information Report No. 94-01, Toronto

77. Hall JC, Van Deynze TD, Struger J, Chan CH (1993) J Environ Sci Health B28:577–598
78. Nowell LH, Resek EA (1994) Rev Environ Contam Toxicol 140:1–164
79. Larson SJ, Capel PD, Majewski MS (1997) Pesticides in Surface Waters—Distribution, Trends and Governing Factors. In: Gilliom RJ (ed) Pesticides in the Hydrologic System, Volume 3, U.S. Geological Survey, National Water Quality Assessment Program. Ann Arbor Press, Inc., Chelsea
80. Canadian Council of Ministers of the Environment (1998) Canadian sediment quality guidelines for the protection of aquatic life: introduction and summary tables. CCME EPC-98E. Winnipeg, Manitoba
81. Health Canada (1996) Guidelines for Canadian drinking water quality. Public Works and Government Services. Canada Cat. No. H48–10/1996-1E
82. International Joint Commission (1987) Revised Great Lakes water quality agreement of 1978—agreement with annexes and terms of reference, between the United States and Canada signed at Ottawa October 16, 1983 as amended by protocol signed November 18, 1987. International Joint Commission Canada and United States, Windsor
83. USEPA (1999) National recommended water quality criteria. U.S. Environmental Protection Agency Document # 822-Z-99–001. Washington, DC
84. Howard PH, Boethling RS, Jarvis WF, Meylan WM, Michalenko EM (1991) Handbook of environmental degradation rates. Lewis Publishers, Chelsea, p 725
85. Glooschenko WA, Strachan WMJ, Sampson RCJ (1976) Pest Monit J 10:61–67
86. Norris LA, Montgomery ML (1975) Bull Environ Contam Toxicol 13:1–8
87. Que Hee SS, Sutherland RG (1981) Chemistry, analysis, and environmental pollution, the phenoxyalkanoic herbicides, vol 1. In: CRC Series in Pesticide Chemistry. CRC Press, Boca Raton
88. Waite DT, Bailey P, Sproull JF, Quiring DV, Chau DF, Bailey J, Cessna AJ (2005) Chemosphere 58:693–703
89. Myers DN, Thomas MA, Frey JW, Rheaume SJ, Button DT (2000) Water quality in the Lake Erie–Lake Saint Clair drainages, 1996–1998. National Water Quality Assessment Program, Lake Erie Lake Saint Clair Drainages, USGS Circular 1203
90. Fenelon JM, Moore RC (1998) J Environ Qual 27:884–894
91. Larson SJ, Gilliom RJ, Capel PD (1999) Pesticides in streams of the United States— initial results from the national water-quality assessment program. U.S. Geological Survey Water-Resources Investigations Report 98–4222, p 92
92. Crawford CG (2001) J Am Water Res Assoc 37:1–15
93. Thurman EM, Goolsby DA, Aga DS, Pomes ML, Meyer MT (1996) Environ Sci Technol 30:569–574
94. Panshin SY, Dubrovsky NM, Gronberg JM, Domagalski JL (1998) Occurrence and distribution of dissolved pesticides in the San Joaquin River Basin, California: U.S. Geological Survey Water-Resources Investigations Report 98-4032, p 88
95. Fenelon JM (1998) Water quality in the White River Basin, 1992–96: U.S. Geological Survey Circular 1150, p 34
96. Zucker LA, Brown LC (1998) Agricultural drainage—water quality impacts and subsurface drainage studies in the Midwest. Ohio State University Extension Bulletin 871. Ohio State University, Columbus, OH p 40
97. Thurman EM, Goolsby DA, Meyer MT, Kolpin DW (1991) Environ Sci Technol 25:1794–1796
98. Thurman EM, Cromwell AE (2000) Environ Sci Technol 34:3079–3085
99. USEPA (2005) The Lake Michigan mass balance project. http://www.epa.gov/glnpo/lmmb/substs.html. Cited 15 June 2005

100. Schottler SP, Eisenreich SJ (1997) Environ Sci Technol 31:2616–2625
101. Macdonald RW, Eisenreich SJ, Bidleman TF, Dachs J, Pacyna JM, Jones KC, Bailey RE, Swackhamer DL, Muir DCG (2000) Case studies on persistence and long-range transport of persistent organic pollutants. In: Klecka G, Boethling B, Franklin J et al. (eds) Evaluation of Persistence and Long-range Transport of Organic Chemicals in the Environment. SETAC press, Pensacola, p 245–314
102. Coote DR, Macdonald EM, Dickinson WT, Ostry RL, Frank R (1982) J Environ Qual 11:473–481
103. Harner T, Bidleman TF, Jantunen LM, Mackay D (2001) Environ Toxicol Chem 20:1612–1621
104. Harris CR, Sans WW (1971) Pestic Monit J 5:259–267
105. USEPA (2002) The foundation for global action on persistent organic pollutants: A United States perspective. Office of Research and Development, Washington, EPA/600/P-01/003F
106. Henry KS, Kannan K, Nagy BW, Kevern NR, Zabik MJ, Giesy JP (1998) Arch Environ Contam Toxicol 34:81–86
107. Bowerman WW, Giesy JP, Best DA, Kramer VJ (1995) Environ Health Perspect 103:51–59
108. NYSDEC (1998) Ambient water quality standards and guideline values and ground-water effluent limitations. Division of Water Technical and Operational Guideline Series (1.1.1). New York State Department of Environmental Conservation, Albany
109. Burger J, Kannan K, Giesy JP, Gochfeld M (2002) Effects of environmental pollutants on avian behavior. In: Dell'Omo G (ed) Behavioral Ecotoxicology. Wiley, New York, p 337–376
110. Bishop CA, Boermans HJ, Ng P, Campbell GD, Struger J (1998) J Toxicol Environ Health 55:531–559
111. Swan SH, Kruse RL, Liu F, Barr DB, Drobnis EZ, Redmon JB, Wang C, Brazil C, Overstreet JW (2003) Environ Health Perspect 111:1478–1484

Hdb Env Chem Vol. 5, Part N (2006): 201–265
DOI 10.1007/698_5_042
© Springer-Verlag Berlin Heidelberg 2005
Published online: 22 December 2005

# Toxaphene in the Great Lakes

D. C. G. Muir[1] (✉) · D. L. Swackhamer[2] · T. F. Bidleman[3] · L. M. Jantunen[3]

[1] Aquatic Ecosystem Protection Research Division, Environment Canada,
Burlington, ON L7R 4A6, Canada
*derek.muir@ec.gc.ca*

[2] Environmental Health Sciences and Water Resources Sciences Program,
University of Minnesota, Minneapolis, MN 55455, USA

[3] Centre for Atmospheric Research Experiments, Meteorological Service of Canada,
Environment Canada, Egbert, ON L0L1N0, Canada

**Abstract** Toxaphene is a major persistent organic contaminant in air, water and fish in the Great Lakes. The story of toxaphene in the Great Lakes, like that of most other persistent organochlorine compounds, has only become clear after the ban on the use of this pesticide in the mid-1980s. Although much of the peer-reviewed literature on environmental fate of toxaphene has involved measurements in the Great Lakes, delineating the extent of contamination has proved very challenging because of the difficulties involved with quantifying this multicomponent mixture. The spatial and temporal trends of toxaphene in the Great Lakes are now reasonably well documented. The highest concentrations in fish and lake water are found in Lake Superior. Concentrations of toxaphene declined in lake trout from Lakes Michigan, Huron and Ontario during the 1990s (with half-lives of 5–8 years) but not in Lake Superior. Recent measurements suggest no declines from the mid-1990s to 2000 in all four lakes. Modeling has demonstrated that colder temperatures and low sedimentation rates in Lake Superior, and to some extent in Lake Michigan, conspire to maintain high toxaphene concentrations in the water column. Sediment core profiles from Lake Michigan, Ontario and Superior all show declining inputs in the past 10–20 years, mirroring reduced emissions following toxaphene deregistration in the USA in 1986. Atmospheric transport of toxaphene from agricultural soils in the southern USA continues and modeling results suggest that 70% of the atmospheric inputs to the Great Lakes are due to long-range atmospheric transport and deposition from outside the basin itself. Some degradation is apparent in the Lake Superior food web based on nonracemic enantiomer fractions for selected chlorobornanes, but for most congeners this process is slow and does not result in negative food web biomagnification in Lake Superior. High proportions of hexa- and heptachlorobornanes have been found in some lake sediments and tributary waters, indicating that slow degradation, mainly via dechlorination, is proceeding within the Great Lakes basin.

Toxaphene concentrations probably did not reach levels that would, by themselves, cause effects on salmonid reproduction, survival or growth in the Great Lakes. The levels and effects of toxaphene in fish-eating birds and mammals (such as mink) in the Great Lakes have never been thoroughly investigated; however, it seems likely that exposure levels for birds and mammals would have been lower than for PCBs. In some Great Lakes jurisdictions, concerns remain about human exposure to toxaphene via consumption of Lake Superior lake trout. Given the long half-lives in fish and water, elevated toxaphene is likely to remain a contaminant issue in the Great Lakes until the middle of the twenty-first century.

**Keywords** Toxaphene · Chlorobornanes · Persistence · Deposition · Bioaccumulation

**Abbreviations**

| | |
|---|---|
| Anoxic | lacking oxygen |
| Atropisomer | isomers which are not structurally superimposable |
| BAF | bioaccumulation factor concentration in organism (lipid basis) ÷ concentration in water |
| Benthic | occurring at the bottom of a body of water or in surface sediments |
| Benthic nephloid layer | layer of suspended material in the water column just above the surface sediments |
| BETR model | Berkeley-Trent North America emissions model |
| BSAF | biota-sediment accumulation factor concentration in organism (lipid basis) ÷ concentration in sediment (organic carbon basis) |
| Chiral | having optically active centers such that molecule that is not superimposable on its mirror image |
| Chlorinated camphenes | minor component of technical toxaphene |
| Chlorinated bornane | major bicyclic component of technical toxaphene |
| $\Sigma$CHB | total chlorobornanes (sum of individual congeners) |
| Congener | member of the same chemical class (chlorobornane for instance) |
| $\delta^{13}$C | $[(R_{sample}/R_{standard}) - 1] \times 1000$ where $R$ = ratio of carbon 13 to carbon 12 isotopes |
| $\delta^{15}$N | $[(R_{sample}/R_{standard}) - 1] \times 1000$ where $R$ = ratio of nitrogen 15 to nitrogen 14 isotopes |
| $\Delta H_{ex}$ | enthalpy of surface-to-air exchange |
| ECD | electron capture detector |
| Enantiomer | either of a pair of chemical compounds whose molecular structures have a mirror-image relationship to each other |
| Epilimnion | upper surface layer overlying the thermocline in a stratified lake |
| ER | enantiomer ratio = A/B wehre A and B are the (+) and (–) enantiomer concentrations, respectively |
| EF | enantiomer fraction. EF = A/(A + B) |
| Flux | amount deposited or emitted over area and time (generally units of $g\,m^{-2}\,time^{-1}$) |
| Gas absorption | entry of gaseous pollutant into the water column due to a fugacity ratio ($f_W/f_A$) of < 1 between water and air |
| Fugacity | escaping tendency of a substance |
| Fugacity ratio | $f_W$ and $f_A$ are the fugacities in water and air, respectively. $f_W/f_A = 1$ at equilibrium |
| GC | gas chromatography |

| | |
|---|---|
| GC-MS | gas chromatographic mass spectrometry |
| GC-ECNI-MS | gas chromatographic electron capture negative ion MS |
| geminal chlorines | two chlorines located on the same ring carbon atom of the chlorobornane |
| $H$ | Henry's law constant |
| HCH | hexachlorocyclohexane |
| Homolog | Group of chemicals within the same class having the same mass but different structure e.g. octachlorobornanes |
| Hypolimnion | lower deep water layer of essentially uniform temperature in a stratified lake |
| HYSPLIT model | hybrid single-particle Lagrangian integrated trajectory model available on the National Oceanic and Atmospheric Administration (NOAA) Air Resource Laboratory website (www.arl.noaa.gov/ready/hysplit4.html) |
| IADN | Integrated Atmospheric Deposition Network |
| Inventory | in the context of sediment total quantity per unit area $(\mathrm{g\,m^{-2}})$ |
| $K_{OA}$ | octanol–air partition coefficient |
| $K_{OW}$ | octanol–water partition coefficient |
| $K_{AW}$ | air–water partition coefficient |
| kt | kilotonnes |
| Oligotrophic | lake having very low levels of dissolved nutrients (the Great Lakes for instance) |
| Ombrotropic bog | deriving nutrients solely from the atmosphere and isolated from ground water flow |
| Parlar numbers | chlorobornane congener numbers assigned by the research group led by Dr. H. Parlar (Technische Universität München, Freising-Weihenstephan, Germany) |
| PCBs | polychlorinated biphenyls |
| Pelagic | occurring in the open lake water column |
| PSCF model | potential source contribution function model |
| Racemic | having equal amounts of enantiomers |
| Salmonids | member of family Salmonidae (lake trout for example) |
| Technical toxaphene | the original chlorinated camphene product used as an insecticide |
| TMF | trophic magnification factor antilog of the slope of regression of the (log) contaminant versus $\delta^{15}N$ in organisms |
| Volatilization | to enter the atmosphere in the vapor state |
| $VP_L$ | sub-cooled liquid vapor pressure |
| WS | water solubility |
| XAD | a styrene-divinylbenzene copolymer resin used in water and air sampling |

# 1

# Introduction

Toxaphene was introduced as an insecticide in 1947 and became one of the most widely used chlorinated pesticides in the United States and in the world during the 1970s. The analysis and environmental fate of toxaphene has been reviewed in the Handbook of Environmental Chemistry [1]. A comprehensive review of nomenclature, environmental measurements and toxicology of toxaphene based on the literature to the late 1990s was published by de Geus et al. [2]. A detailed review article by Saleh [3] summarized the environmental toxicology and chemical measurements on toxaphene up to the late 1980s. In this chapter the focus is on toxaphene contamination in the Laurentian Great Lakes of North America. Rice and Evans [4] reviewed the work on toxaphene in the Great Lakes to the early 1980s, and a summary of more recent data was done by Swackhamer et al. [5]. Much of the literature on environmental measurements, sources and the fate of toxaphene published in the past 20 years is based on work in the Great Lakes basin. While our goal is to summarize the key findings and conclusions on toxaphene in the context of the Great Lakes, many of these conclusions may be broadly applicable to the global fate of this chemical.

## 1.1
## Composition of Technical Product

Toxaphene (CAS No. 8001-35-2) is a complex substance due to its bicyclo [2.2.1]heptane or norbornane skeleton. It is synthesized by chlorination of $\alpha$-pinene under ultraviolet (UV) light. Polychlorobornanes ($C_{10}H_{18-n}Cl_n$, $n = 5 - 12$) are formed as the main components in a Wagner-Meerwin-type rearrangement reaction [6]. Chlorinated camphenes are minor components, formed where the Wagner-Meerwin rearrangement is hindered. The average chlorine content is 67–69% [7]. Structures of two of the most environmentally important components of toxaphene are shown in Fig. 1. A total of 16 640 chlorinated bornane congeners are theoretically possible from chlorination [1]; however, due to steric considerations and manufacturing conditions, the technical product consists almost entirely of hexa- to decachlorobornanes which theoretically comprise about 10 770 compounds [1]. Two-dimensional

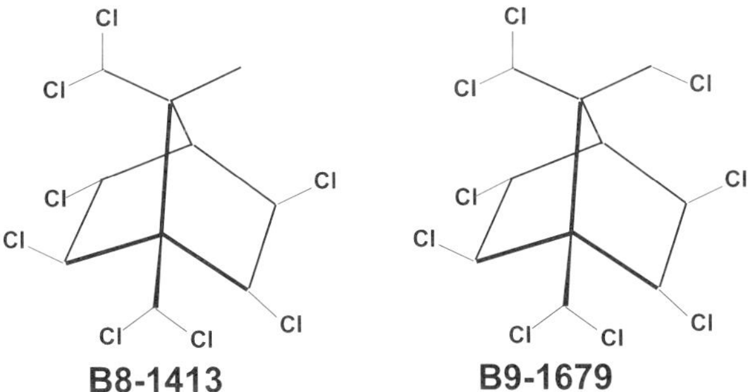



**Fig. 1** Structures of major recalcitrant chlorobornanes in toxaphene, B8-1413 (Parlar 26) and B9-1679 (Parlar 50). Nomenclature of these chlorobornanes is described by Andrews and Vetter [9]

high-resolution gas chromatography (GC) has been shown to separate over 1000 components [8]. The peak area percentage of all components identified, measured using an electron capture detector (ECD), amounts to about 50% of the total toxaphene area [2, 3]. Most analyses of toxaphene in the Great Lakes region have been of the technical product and thus a single value of "total" toxaphene has typically been reported along with the total hexa- to nonachlorobornanes. When "toxaphene" concentrations are reported herein, they are based on quantification using the technical product. The sum of chlorobornanes quantified using individual standards are reported as total chlorobornanes or $\Sigma$CHB. In this paper we use the Andrews and Vetter nomenclature system for congeners, which has become widely used in recent literature [9].

## 1.2
## Properties

Key physical-chemical properties for environmental modeling of toxaphene and selected major congeners are presented in Table 1. Vapor pressures (VP$_L$, subcooled liquid, 20 °C) for hepta- and nonachlorobornanes range from $3.41 \times 10^{-3}$ Pa to $2.53 \times 10^{-5}$ Pa [10] and octanol–water partition coefficients ($K_{OW}$) of hexa- to nonachlorobornanes from 5.1 to 6.7 [11]. These fall within the VP range of $3 \times 10^{-3}$ to $1 \times 10^{-5}$ Pa for most penta- to octachlorobiphenyls [12] and log $K_{OW}$ values of 4.9 to 6.7 are found for dichloro to pentachloro-PCBs [13]. Water solubilities and air–water partition coefficients ($K_{AW}$) have been measured for technical toxaphene only [14, 15]. While WS of individual chlorobornanes can be estimated from measured log $K_{OW}$ values [16], the estimated values are much lower ($0.01–0.09$ mg L$^{-1}$) than reported for the technical product (WS $= 0.63$ mg L$^{-1}$; Table 1). Calculated log $K_{AW}$s for individual hepta- to decachlorobornane congeners (using measured VP [10] and estimated WS) range from $-1.4$ to $-2.3$, so they are much higher than the measured value of $-3.91$ for technical toxaphene [15]. Thus we have chosen not to report estimated WS and log $K_{AW}$ values in Table 1 and caution readers about applying physical property estimation models for individual congeners until additional direct measurements are available.

## 1.3
## Sources of Toxaphene in North America

### 1.3.1
### Toxaphene Production and Usage

The history of toxaphene production and use in the United States has been chronicled by Li [17], who used a combination of production/use statistics

**Table 1** Physical-chemical properties of technical toxaphene and selected congeners

| | Technical toxaphene | B6-923 | B7-10C1 | B7- 499 | B7-515 | B8-1413 | B8-1414 | B8-1945 | B8-806/ 809 | B8-2229 | B9-1679 | B9-1025 | B10-1110 | Refs. |
|---|---|---|---|---|---|---|---|---|---|---|---|---|---|---|
| Parlar # | | | | 21 | 32 | 26 | 40 | 41 | 42 | 44 | 50 | 62 | 69 | |
| Empirical formula | $C_{10}H_{10}Cl_8$ | $C_{10}H_{12}Cl_6$ | $C_{10}H_{11}Cl_7$ | $C_{10}H_{11}Cl_7$ | $C_{10}H_{11}Cl_7$ | $C_{10}H_{10}Cl_8$ | $C_{10}H_{10}Cl_8$ | $C_{10}H_{10}Cl_8$ | $C_{10}H_{10}Cl_8$ | $C_{10}H_{10}Cl_8$ | $C_{10}H_9Cl_9$ | $C_{10}H_9Cl_9$ | $C_{10}H_8Cl_{10}$ | |
| Mol. Weight (MW) $(g\,mol^{-1})$ | $414^1$ | 345 | 380 | 390 | 380 | 414 | 414 | 414 | 414 | 414 | 448 | 448 | 483 | |
| VP $(Pa)^2$ 20 °C | $8.9\times10^{-4}$ | $3.82\times10^{-3}$ | $2.48\times10^{-3}$ | $3.05\times10^{-3}$ | $1.58\times10^{-3}$ | $1.57\times10^{-3}$ | $8.18\times10^{-4}$ | $8.18\times10^{-4}$ | $8.32\times10^{-4}$ | $8.04\times10^{-4}$ | $4.73\times10^{-4}$ | $3.28\times10^{-4}$ | $1.18\times10^{-4}$ | [10] |
| WS $(mg\,L^{-1})^3$ 20 °C | 0.63 | — | — | — | — | — | — | — | — | — | — | — | — | [14] |
| Log $K_{AW}^4$ 20 °C | – 3.91 | — | — | — | — | — | — | — | — | — | — | — | — | |
| Log $K_{OW}$ | 5.50 | — | — | 5.49 | 5.23 | 5.52 | 5.57 | 5.19 | 5.78 | 5.75 | 5.84 | 5.96 | 6.64 | [11] |
| Log $K_{OA}$, 20 °C (Calculated)$^5$ | 9.41 | — | — | — | — | — | — | — | — | — | — | — | — | |

[1] Assuming average composition $C_{10}H_{10}Cl_8$
[2] Sub-cooled liquid VPs of congeners from Bidleman et al. [10]; VP of solid technical toxaphene from Murphy et al. [14]
[3] WS of solid technical toxaphene from Murphy et al. [14]
[4] Log $K_{AW}$ from Jantunen and Bidleman [15]
[5] $K_{OA}$ calculated from $K_{OW}/K_{AW}$

and surrogate data from cropland distribution to apportion toxaphene usage across the country by $1/4°$ latitude $\times 1/6°$ longitude grids. Production began in 1946 by Hercules, Inc. in Georgia (now Nor-Am Chemical Co.), which was the principal manufacturer until ceasing production in 1981. After the end of the 1960s, toxaphene was also produced by three other companies, Tenneco Chemicals in New Jersey, Riverside Chemicals in Texas and Vicksburg Chemical in Mississippi. Total toxaphene production was 720 kilotonnes (kt), of which 490 kt was used in the United States and the rest was exported. Peak production of 55 kt occurred in 1974. Usage can be divided into four periods [17]: a) 1947–1971. Toxaphene application was mainly to cotton with an annual consumption of 11.5 kt and cumulative use during this period was 290 kt. b) 1972–1975. The U.S. DDT ban in 1972 stimulated an increase in toxaphene use to $30 \, \text{kt} \, \text{y}^{-1}$ on cotton, soybeans, corn and other field crops, with a total use of 150 kt. c) 1976–1982. Toxaphene applications declined to $7.6 \, \text{kt} \, \text{y}^{-1}$ and a total use of 45 kt. d) 1983–1986. Toxaphene registrations were canceled in October, 1982 and existing stocks were allowed to be applied for specific purposes through 1986. Use as a herbicide to control sicklepod weed was permitted during this period. Total use was 1.75 kt.

The bulk of toxaphene usage was in the southern part of the United States. Twelve states, covering the Southeast, Delta States, Southern plains and Appalachian agricultural regions, accounted for 81.7% (38.2 kt) of the total toxaphene consumption (46.75 kt) in the period 1977–1986. A substantial quantity of toxaphene was also applied in the northern USA; 16.2% (7.57 kt) of total national use in the Northeast, Corn Belt, Lakes States and Northern Plains agricultural regions. Corn Belt and Lakes States that border on the Great Lakes consumed 7.2% of the total use, or 3.37 kt. Toxaphene was used on livestock at rates of $1 \, \text{kt} \, \text{y}^{-1}$ from 1972–1976 and $0.2 \, \text{kt} \, \text{y}^{-1}$ from 1977–1986, and in the United States and Canada at unspecified rates as a fish poison to clear rough fish from lakes before introducing game fish.

Toxaphene was also used in the Mexico–Central America region. The plant in Nicaragua that supplied toxaphene for use on cotton in Central America closed in 1991. Total production at this plant was 60 kt from 1974–1979, 21 kt from 1980–1985, 4.2 kt in 1986–1987 and 1.2 kt in 1988–1990 [17, 18]. Toxaphene use in Mexico took place from the early 1950s until its ban in 1993. Total consumption in Mexico was estimated as 68 kt, with peak application of 3 kt in 1974. Use in 1981 was 1.8 kt, and it declined to zero in 1995 [17].

Toxaphene usage estimates were made by MacLeod et al. [19] as input to their Berkeley-Trent North America (BETR North America) emissions model. Total usage for all of North America from 1945–2000 was 534 kt, with the southern United States and northern Mexico accounting for 70% of the total. These estimates are quite close to those of Li [17] for cumulative consumption in the United States and Mexico (558 kt) and in the Great Lakes states between 1977–1986 (3.37 kt).

## 1.3.2
## Use and Sources within the Great Lakes Watershed

MacLeod et al. [19] estimated the total use of toxaphene in the Great Lakes basin at 4100 t, representing about 1–4% of the total use. Historical investigation of records for the Lake Michigan watershed revealed that 224 t of toxaphene was used in the Green Bay watershed between 1950 and 1980, with most used as a pesticide on cropland but small amounts on livestock and in lakes as a piscicide. It has been suggested that even if there were only a 1% runoff into Lake Michigan, this would represent a large fraction of the estimated inventory of toxaphene in the lake, 11 103 kg [20] (see Sect. 3.3). Studies in the 1980s identified volatilization as the major loss pathway of toxaphene while runoff losses were only important immediately after application [21, 22]. Nevertheless, volatilization, transport and deposition, combined with direct runoff, of toxaphene applied within the Basin could have been important especially for Lake Michigan and the lower Great Lakes (Erie and Ontario) which receive agricultural inputs. Raff and Hites [23] measured the concentration of toxaphene on suspended sediments of the Mississippi River from Minnesota to Louisiana. Although the highest concentrations occurred in the lower Mississippi, residues of $0.4$–$1.8\,\mathrm{ng\,g^{-1}}$ were found at river stations in Minnesota and Wisconsin, which receive considerable amounts of agricultural runoff. Toxaphene at $1.0\,\mathrm{ng\,g^{-1}}$ was also found in suspended sediment of Lake Itasca, which feeds the headwaters of the Mississippi River. Since there is very little agricultural activity in the Lake Itasca watershed, the source of toxaphene for the lake is probably atmospheric deposition. MacLeod et al. [19] attributed 30% of the atmospheric inputs to the Great Lakes to within-basin sources using the BETR model. This was based on the assumption that primary emissions of toxaphene were distributed 70% into soil, 25% into air, and 5% into freshwater as the chemical was applied. The remaining 70% was attributed to long-range atmospheric transport and deposition from the southern US and northern Mexico.

Struger [24] investigated the use of toxaphene in Canada from historical pesticide registration records and estimated that 11 600 kg were used annually in 1977 and 1978, with 10–15% used in the province of Ontario and thus mainly within the Great Lakes watershed. At this time registered uses included orchard soil application and use on livestock and in farm buildings. Assuming an annual use of 1100–1700 kg in Ontario during the 1960s and 1970s, 20–50 t may have been used in total within Canadian Great Lakes agricultural watersheds. Small amounts (525 gals of a formulated product) were used as a lampricide in tributaries of Lake Superior during the mid-1950s [25]. All uses in Canada were discontinued as of 1982.

## 1.4
## Soil Residues

Li et al. [17] apportioned usage of toxaphene throughout the United States on a $1/6°$ latitude $\times 1/4°$ longitude grid and used these gridded data to estimate soil residues [26]. At the beginning of 2000, 29 kt of toxaphene was estimated to reside in agricultural soils. The highest residues were predicted to occur in the southern states, with average concentrations of 100 to $1250\,\text{ng}\,\text{g}^{-1}$, while those in most Great Lakes states were $< 10\,\text{ng}\,\text{g}^{-1}$ but ranged up to $100\,\text{ng}\,\text{g}^{-1}$ in New York and Pennsylvania. To our knowledge, no confirmation of these predictions for the Great Lakes states has been made by direct measurements.

Only a few surveys have been done to measure toxaphene residues in agricultural soils. Carey et al. [27, 28] surveyed cropland in the United States in 1971, collecting 1473 soil samples in 37 states that covered the eastern, southern and mid-western portion of the country and four western states: California, Oregon, Washington and Idaho. Results have been summarized by Li et al. [26]. Only 33 of the sites reported toxaphene use and 92 soil samples were positive. The maximum and arithmetic mean of positive samples were 36330 and $281\,\text{ng}\,\text{g}^{-1}$ dry weight, respectively. Soil surveys have recently been done at farms in Alabama in 1996 and 1999–2000 [29–31], Louisiana and Texas in 1999 [29, 30] and Georgia and South Carolina in 1999 [32]. Total toxaphene concentrations in the latter studies ranged from $< 3$–$6500\,\text{ng}\,\text{g}^{-1}$ (Table 2), and residues appeared to be log-normally distributed [29].

**Table 2** Toxaphene residues in agricultural soils of the United States, $\text{ng}\,\text{g}^{-1}$ dry weight

| State | Year | N sites | Range | Arithmetic mean | Geometric mean | Refs. |
|---|---|---|---|---|---|---|
| Alabama, | 1996–2000 | 62 | $< 3$–6500 | 688[a] | 92[b] | [29] |
| Louisiana | | | | | | |
| Texas | | | | | | [31] |
| Georgia | 1999 | 16 | 9.5–269 | 83 | 55 | [32] |
| South Carolina | 1999 | 16 | 3.3–2500 | 277 | 72 | [32] |
| USA | 1971 | 1473 | 180–36330[a] | 281[a] | NR[c] | [27, 28] |

[a] Positive samples only
[b] All samples, $1.5\,\text{ng}\,\text{g}^{-1}$ assumed for samples $< 3\,\text{ng}\,\text{g}^{-1}$
[c] NR: not reported

## 1.5
## Emissions into the Atmosphere

### 1.5.1
### Measurements of Volatilization from Soil

Release of toxaphene into the air from contaminated soils was investigated by Bidleman and Leone [29, 30], who collected air samples at 40 cm height above agricultural soils at 16 farms in the southern United States in 1999–2000. Total toxaphene concentrations in the air samples ranged from < 300 to 42 000 pg m$^{-3}$, with arithmetic and geometric means of 8400 and 3600 pg m$^{-3}$. A significant positive correlation ($p < 0.001$, $r^2 = 0.73$) was found between the air and soil concentrations and fugacity calculations indicated a potential for net soil-to-air exchange. These emissions give rise to elevated ambient air concentrations in the southern United States, which in turn are advected northward to the Great Lakes (Sect. 2.2).

### 1.5.2
### Modeled Emissions

Harner et al. [33] formulated a fugacity-based model of soil–air exchange and applied the model to estimate emissions of toxaphene from Alabama soils. Key drivers of the model were concentrations of toxaphene in the soils and the octanol–air partition coefficient ($K_{OA}$) as a descriptor of volatility. The latter was estimated from the water solubility and Henry's law constant of technical toxaphene (Table 1). Predicted annual mean concentrations in ambient air were 89 pg m$^{-3}$ when arithmetic mean soil concentrations were used in the model, and 26 pg m$^{-3}$ using geometric mean soil concentrations. The model underestimated the toxaphene concentrations measured in ambient air of northern Alabama during 1996–1997 (arithmetic mean = 176 pg m$^{-3}$, geometric mean = 151 pg m$^{-3}$, concentration adjusted to 288 K = 103 pg m$^{-3}$) [34]. The predicted release of toxaphene to the atmosphere from soils in Alabama was 3 to 11 kt y$^{-1}$.

Toxaphene emissions for the entire United States were estimated by Li et al. [26], using application rates in the current year and residues carried over in the soil from past years, the latter estimated by assuming a 10-yr dissipation half life in soils. After the final year of toxaphene use (1986), only soil residues contributed to the emissions. Emission factors were calculated on a 1/4° latitude ×1/6° longitude grid using the soil–air exchange and canopy models of Sholtz et al. [35, 36]. About 80% of the emissions were from the southeastern, delta, and Appalachian states. The total quantity of toxaphene emitted to the atmosphere in 2000 was estimated as 364 tonnes.

## 1.5.3
## Composition of Toxaphene Residues in Southern U.S. Soil and Air

GC-mass spectrometry total ion chromatographic profiles of the $Cl_7$ to $Cl_9$ toxaphene residues in soil and overlying air samples at a Texas farm and in a technical toxaphene standard are compared in Fig. 2 [30]. Relative to the soil, air samples contained lower proportions of the later-eluting, less volatile components. Alterations caused by selective degradation are also evident. Peaks 5–8 are octachlorobornanes with similar vapor pressures, 0.00080 to 0.00092 Pa at 20 °C [10], so differences in the proportions of the peaks are probably not due to volatility. Peak 5 matches the retention time of B8-531. Peak 6 contains mainly B8-1414 + B8-1945 [37], which were more or less resolved depending on the sample. They are considered together here. Peak 7 consists largely of coeluting B8-806 and B8-809 with minor proportions of other congeners [37]. Peak 8 contains about 50% B8-2229 and several octachlorinated compounds [37]. The retention time of authentic B8-2229 matches the right shoulder of peak 8 more closely than the main peak [30], and James and Hites [38] show an unidentified peak eluting after B8-806/B8-809 and B8-2229. Here, the area of both the shoulder and main peak were included in peak 8.

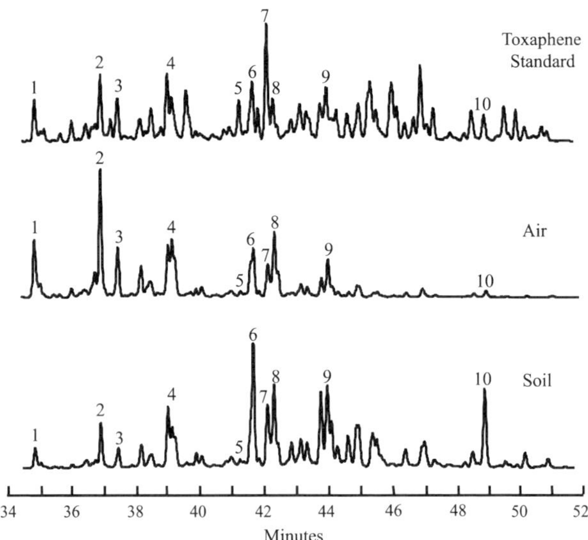

**Fig. 2** Total ion chromatographic profiles of the $Cl_7$ to $Cl_9$ toxaphene residues in soil and overlying air samples at a Texas farm and a technical toxaphene standard [30]. Peak identities are tentatively assigned: 1 – B7-1001, 2 – B8-1413/unknown hepta, 3 – unknown octa, 4 – B7-515, 5 – B8-531, 6 – B8-1414 + B8-1945, 7 – B8-806/809; 8 – mainly B8-2229, 9 – B9-1679, 10 – B9-2206

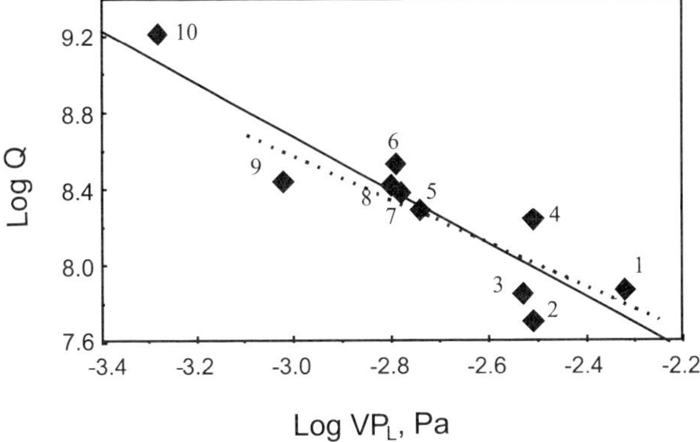


**Fig. 3** Correlation of soil/air concentration ratios ($Q$) with sub-cooled liquid vapor pressure ($VP_L$) for 10 chlorobornane congeners or congener groups which span the volatility range in technical toxaphene [30] or a sampling event in Alabama [34]

Peaks 5 and 7 were depleted in both the soil and overlying air. Similar degradation profiles were seen in southern ambient air samples from Alabama [34], Texas and Arkansas [38], and in Great Lakes air (Sect. 3.1.3). Their main components, B8-531, B8-806 and B8-809, are predicted to be relatively labile on the basis of molecular structures [39, 40].

The toxaphene residues in soil emission experiments at 15 farms covering Alabama, Louisiana and Texas were examined in detail by calculating the soil/air concentration ratios ($Q$) for ten congeners or congener groups (Fig. 2) which span the volatility range in technical toxaphene [30]. $\log Q$ was negatively correlated with $\log VP_L$ in 26 sampling events ($p < 0.0001$ to $0.015$, $r^2 = 0.54$ to $0.89$). An example plot for a sampling event in Alabama is shown in Fig. 3.

# 2
# Toxaphene in the Air of North America

## 2.1
## Levels before Deregistration

As indicated in Sect. 1.1, toxaphene was heavily used in the southern United States and historical atmospheric concentrations were also high. Reported concentrations in North American air before the 1986 final deregistration of toxaphene are summarized in Table 3. Averages at a particular location were typically several thousand $pg\,m^{-3}$ in the southern states and $100\,pg\,m^{-3}$ or higher in mid-western and northern states and in Canada. Rice et al. [41] were the first to establish evidence for a strong south-to-north concentration

**Table 3** Toxaphene levels in air of North America before deregistration, $\mathrm{pg\,m^{-3}}$

| Location | Year | Range | Mean | Refs. |
|---|---|---|---|---|
| *Southern USA* | | | | |
| Stoneville, MS | 1967–1968 | $6.2 \times 10^4$–$1.34 \times 10^6$ | — | [56] |
| | 1972–1974 | $< 1000$–$1.54 \times 10^6$ | $1.67 \times 10^5$ | [57] |
| Sapelo Island, GA | 1975 | 1700–5200 | 2800 | [58] |
| Organ Pipe Natl. Park, AZ | 1974 | 2700–7000 | 4600 | [59] |
| Hayes, KS | 1974 | 83–2600 | 1100 | [59] |
| North Inlet estuary, SC | 1977–1979 | 320–1900 | 750 | [60] |
| Columbia, SC | 1977–1985 | $400$–$6.6 \times 10^5$ | 8500, 1237[a] | [44] |
| Greenville, MS | 1981 | $1000$–$2.2 \times 10^5$ | 6600 | [41] |
| *Midwestern/Northern USA and Canada* | | | | |
| Northwest Territories | 1974 | 40–130 | 92 | [59] |
| Newfoundland | 1977 | 57–160 | 110 | [61] |
| St. Louis, MO | 1981 | 370–5200 | 1300 | [41] |
| Bridgeman, MI | 1981 | 10–1200 | 360 | [41] |
| Beaver Is., Lake Michigan | 1981 | 20–280 | 94 | [41] |
| *Other Measurements* | | | | |
| Bermuda | 1973–1974 | $< 20$–3300 | 700 | [58] |
| Between Bermuda and USA | 1973–1974 | $< 40$–1600 | 530 | [58] |

[a] Normalized to 288 K

gradient in air in 1981, a year before toxaphene deregistration. This study, carried out by sampling concurrently at four sites, found arithmetic mean concentrations of $6600\,\mathrm{pg\,m^{-3}}$ at Greenville, Mississippi, $1300\,\mathrm{pg\,m^{-3}}$ at St. Louis, Missouri, $360\,\mathrm{pg\,m^{-3}}$ at Bridgeman, southern Michigan and $94\,\mathrm{pg\,m^{-3}}$ at Beaver Island, Lake Michigan. Determination of toxaphene in the period of these studies was done using packed-column gas chromatography (GC) and electron capture detection (GC-ECD) and therefore could have overestimated concentrations compared to measurements by capillary GC-ECD and GC-electron capture negative ion mass spectrometry (GC-ECNI-MS).

## 2.2
### Levels after Deregistration

Measurements since the 1986 deregistration are summarized in Table 4 as ranges and arithmetic mean concentrations. Air samples in these studies were collected with traps containing glass fiber filters backed up with polyurethane foam or XAD-2 resin. With the exception of results from Hoff et al. [42, 43], who used capillary GC-ECD, all determinations were done using GC-ECNI-MS. Where possible, average values adjusted to a common temperature of 288 K are also given. These were obtained by plotting the natural logarithm of

**Table 4** Toxaphene levels in air of North America after deregistration, $pg\,m^{-3}$

| Location | Year | Range | Mean | Refs. |
|---|---|---|---|---|
| *Southern USA* | | | | |
| Columbia, SC | 1994–1995 | 39–408 | 190, 130[a] | [44] |
| Muscle Shoals, AL | 1996–1997 | 39–611 | 176, 102[a] | [34] |
| Rohwer, AR | 2000–2001 | 170–5600 | 1600, 950[a] | [38, 45] |
| | 2002–2003 | 69–4100 | 1400, 710[a] | |
| Lubbock, TX | 2000–2001 | 27–1600 | 280, 160[a] | [38] |
| Cocodrie, LA | 2002–2003 | 12–240 | 61, 36[a] | [45] |
| *Mexico – Central America* | | | | |
| Belize, Central America | 1995–1996 | — | 32 | [54] |
| Chiapas, Mexico | 2000–2001 | 319–771 | 505 | [53] |
| *Great Lakes States and Eastern Canada* | | | | |
| Egbert, ON | 1988–1989 | < 0.1–580 | 26, 16[a] | [42, 43] |
| Sable Island, NS | 1988–1989 | 20–71 | 35 | [66] |
| Green Bay, WI | 1988–1989 | 15–109 | 49 | [52] |
| Bloomington, IN | 2003–2004 | 3.0–360 | 60 | [45] |
| | | | 39[a] | |
| L. Huron, Michigan, Erie and Ontario | 1990 | 14–65 | 33 | [52] |
| L. Ontario (Point Petre) | 1992 | 1.5–10 | 4.9, 7.1[a] | [51] |
| | 1995–1997 | 0.9–10 | 3.8, 3.7[a] | [51] |
| Open L. Ontario | 1998–2000 | 12–53 | 20 | [67] |
| L. Superior (Eagle Harbor) | 1996–1997 | < 1–63 | 7.6, 6.4[a] | [49] |
| Open L. Superior | 1996–1997 | 4.9–41 | 20 | [46] |
| | | | 30[a] | |
| Open L. Superior | 1997–1998 | 3–54 | 17, 15[a] | [50] |
| Open L. Huron and Erie | 1996–1997 | 7.8–39 | 23 | [46] |
| L. Michigan (Sleep. Bear Dunes) | 1998 | 19–70 | 36 | [50] |
| L. Michigan (Sleep. Bear Dunes) | 2000–2001 | 1–36 | 10, 11[a] | [38, 45] |
| | 2002–2003 | 0.9–110 | 23, 27[a] | |
| Open L. Michigan | 1997–1998 | 2–57 | 14, 15[a] | [50] |

[a] Normalized to 288 K

concentration [44] or partial pressure ($p$) [34, 38, 45, 46] versus the reciprocal temperature; for example:

$$\ln p = \Delta H_{ex}/RT + C \tag{1}$$

In Eq. 1, $\Delta H_{ex}$ is the apparent enthalpy of surface-to-air exchange, $T$ is the temperature (K), $R = 8.31\ \mathrm{Pa\,m^3\,mol^{-1}\,K^{-1}}$ and $C$ is a constant [47, 48]. Reported values of $\Delta H_{ex}$ ($\mathrm{kJ\,mol^{-1}}$) range from 46–85 at sites in Michigan, Indiana, Texas and Arkansas [38, 45], 40–64 in northern Alabama [34], 91 at Egbert, Ontario [42], 47 at Eagle Harbor, Lake Superior [49], 63–76 over open lakes Michigan and Superior [46, 50], and 44 at Point Petre, Lake Ontario [51]. For comparison, the enthalpy of water-to-air exchange for technical toxaphene is $61\,kJ\,mol^{-1}$ [46], and enthalpies of vaporization for toxaphene congeners range from $78\text{--}92\,kJ\,mol^{-1}$ [10].

Average concentrations at 288 K in the southern states were $103\text{--}950\,pg\,m^{-3}$ in most locations, but only $36\,pg\,m^{-3}$ was measured at Cocodrie, LA where cleaner air from the Gulf of Mexico may have diluted local emissions [45]. Levels in Great Lakes air were typically an order of magnitude lower than those in the southern states. Hoff et al. [42] noted that air samples collected in southern Ontario had elevated toxaphene concentrations when air trajectories came from the southern states and the Caribbean. A similar association with air transport from the south was subsequently found by others [45, 49, 50, 52]. This is consistent with a continuing source of toxaphene emanating from contaminated soils in the southern states and possibly farther south. Toxaphene concentrations averaged $505\,pg\,m^{-3}$ in Chiapas, southern Mexico during 2000–2001 [53], while the mean in Belize, Central America $(32\,pg\,m^{-3})$ [54] was similar to Great Lakes levels. James and Hites [38] and Hoh and Hites [45] repeated the Rice et al. [41] transect experiment in 2000–2001 and 2003–2004. Toxaphene concentrations at 288 K were $160\,pg\,m^{-3}$ at Lubbock, Texas, $710\text{--}950\,pg\,m^{-3}$ at Rohwer, Arkansas, $25\text{--}39\,pg\,m^{-3}$ at Bloomington, Indiana and $11\text{--}27\,pg\,m^{-3}$ at Sleeping Bear Dunes on Lake Michigan, confirming the southern states as an ongoing source region.

## 2.3
### Transport and Dispersal Models

Voldner and Schroeder [55] formulated the first model of atmospheric transport and deposition of toxaphene to the Great Lakes. Toxaphene usage in 1976 and 1980 was apportioned among states in the southern and northern United States, and it was assumed that 50% of the toxaphene applied was emitted into the atmosphere. Transport was estimated using a Lagrangian model and deposition was considered to take place through precipitation, dry particle removal and air–lake gas exchange. The model predicted annual average air concentrations over the Great Lakes in the range of $400\text{--}700\,pg\,m^{-3}$, which is within the range of air concentrations in the midwestern states, Great Lakes region and eastern Canada measured before toxaphene deregulation (Table 3; [56–61]). Of the estimated 7.7 kt of toxaphene emitted in 1976 and 3.6 kt in 1980, 60% was deposited within North America and the remainder was advected out of the continent from the east coast. Total deposition to the five Great Lakes was estimated to be 0.078 kt in 1976 and 0.034 kt in 1980, and deposition to the entire Great Lakes basin was about twice these quantities. This represented about 1–2% of emissions.

MacLeod et al. [19] applied the Berkeley-Trent North American mass balance contaminant fate model (BETR North America) [62] to derive a dynamic mass budget for toxaphene between 1945 to 2000. BETR North America divides the continent into 24 discrete ecological regions. Within each region, contaminant transport and fate is described using a seven-compartment fugacity model that includes a vertically stratified atmosphere, vegetation, soil, freshwater, freshwater sediments and near-shore coastal water. Modeled toxaphene concentrations were consistent with measured values for air, freshwater sediments and soil.

Of the $5.34 \times 10^5$ t of toxaphene that were used in North America (includes the United States and Mexico), 1.54 t (3%) was estimated to remain in active circulation as of the year 2000, $6.5 \times 10^4$ t (12%) had been removed from the continental environment by advection of air and water and burial in soils and sediments, and $4.54 \times 10^5$ t (85%) had been removed by degradation reactions. Most of the toxaphene in active circulation resides in the soils of the southern U.S. and Mexico (83%). The efficiency of toxaphene transfer to the Great Lakes from its usage within the basin was estimated to be 78 to 220 times greater than transfer from the southern United States. However, load-

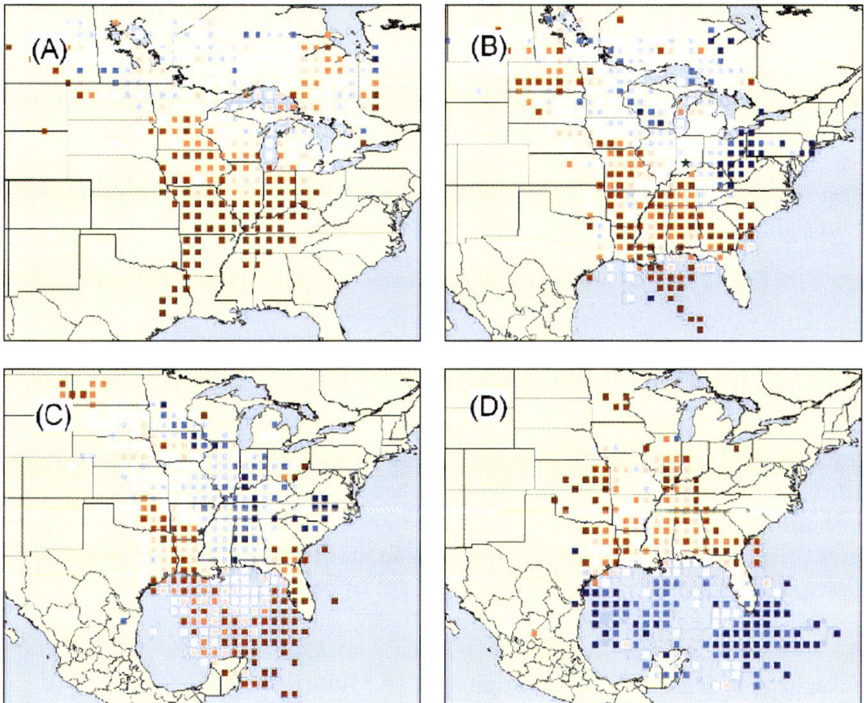

**Fig. 4** Individual Potential Source Contribution Function maps for toxaphene: (**A**) Michigan; (**B**) Indiana; (**C**) Arkansas; (**D**) Louisiana from Hoh and Hites [45]. *Dark red* represents PSCF values from 0.91 to 1.0, *with shades* of *pink ranging* down to 0.61 by tenths, and *white* represents from 0.41 to 0.60. *Dark blue* represents 0–0.10, *with shades* of *blue ranging up* to 0.40 by tenths. The *green star* in each map represents the sampling site. [Reproduced with permission from ACS]

ings to local usage were less than 30% of the total due to the relatively small amount of toxaphene applied. Approximately 70% of the loadings to remote lakes in the basin came from continental-scale transport from adjacent regions and the southern states.

McDonald and Hites [63] determined toxaphene residues in tree bark samples collected from the southern and northern United States and Canada. Highest residues were found in the south and they decreased northward. The distribution of residues in tree bark and ambient air followed a radial dilution inverse square relationship with a source region centered near 30–35°N, 90–95°W. Calculations based on the octanol–air partition coefficient ($K_{OA}$) of toxaphene and assuming 1% lipophilic material in bark suggested that bark and air were approximately in equilibrium. Hoh and Hites [45] implemented a potential source contribution function (PSCF) model to examine sources of toxaphene along back trajectories predicted by the HYSPLIT model. As illustrated in Fig. 4, the PSCF results show that for Michigan and Indiana (see Fig. 4a,b), toxaphene was emanating from sampling sites from the south, presumably from the cotton-growing regions centered in Arkansas. Atmospheric toxaphene was high at the Arkansas site, indicating that it was a source region. The PSCF map for this site (Fig. 4c), however, indicated that there may be toxaphene source regions to the southwest of this location in Louisiana, Texas and Oklahoma, and that the Gulf of Mexico itself may be a toxaphene source for the Arkansas site. The PSCF map for the Louisiana site (Fig. 4d) indicates that the Gulf of Mexico is not a toxaphene source. Toxaphene sources to the Louisiana site are north of the site in states ranging from Texas to Georgia.

## 3
## Historical Trends of Toxaphene Deposition in the Great Lakes Basin

Toxaphene is atmospherically transported to the Great Lakes region where it undergoes air–water gas exchange, and to a lesser extent wet and dry deposition [64, 65]. In the past, the Great Lakes have been a sink for toxaphene because they have large surface areas and are cold [5, 65]. This section examines the present-day situation of toxaphene inputs through gas exchange and precipitation as well as the historical record of deposition in sediments.

### 3.1
### Toxaphene in Great Lakes Air, Precipitation and Surface Water

### 3.1.1
### Air

Recent (post-deregistration) measurements of toxaphene in air over the Great Lakes and surrounding regions are summarized in Table 4. Some of these studies were done on board ship and concurrently with water sampling [46, 50, 67], while others were done from land-based stations near the lakes [38, 42, 43, 45, 49–52] or from ships with no water collection [52].

Toxaphene concentrations in Great Lakes air are dependent on air transport pathways and temperature. As noted in Sect. 2.2, higher levels have been associated with air transport from the south. Short-term studies have found higher toxaphene levels in air during warmer months. McConnell et al. [52] reported average air concentrations at Green Bay, Wisconsin of 21 and 59 pg m$^{-3}$ in February 1988 and June 1989. Higher concentrations were found over Lake Superior in August 1996 ($28 \pm 10$ pg m$^{-3}$) than in May 1997 ($12 \pm 4.6$ pg m$^{-3}$) [46]. Similar air concentrations were found over Lake Ontario in July 1998 ($19 \pm 4$ pg m$^{-3}$) and June 2000 ($25 \pm 20$ pg m$^{-3}$), although there was only a 5–8 degree difference in air or water temperatures during those months [59]. The temperature dependence of toxaphene partial pressure in Great Lakes air has been determined in several studies (Sect. 2.2).

Whereas measurements of toxaphene in water by different laboratories agreed within a factor of two or better (Sect. 3.1.3), those in air at a particular site varied by factors of 2–5. This may be in part because the different studies in Table 4 were not run concurrently and, as noted above, toxaphene levels in air are more variable due to changing air pathways and temperatures. A study was done as part of the Integrated Atmospheric Deposition Network (IADN) program to evaluate interlaboratory performance for determining toxaphene. Eleven groups participated; all were experienced in toxaphene analysis and all used GC-MS methodology. The laboratories were supplied with a technical toxaphene solution as an unknown and an air sample extract from the southern United States [68]. The average recovery for "total toxaphene" in the technical toxaphene unknown was 113%, but only six laboratories came within ±30% of the target value and the interlaboratory relative standard deviation was 40%. The RSD for the air sample extract was 65%. The interlaboratory performance in the determination of single toxaphene congeners was about the same as for total toxaphene.

### 3.1.2
### Toxaphene in Great Lakes Precipitation

Precipitation has not been extensively investigated as an input pathway for toxaphene to the Great lakes. An early report (W. Swain cited in Rice and Evans [4]) stated that samples collected near Lake Huron in 1980–1981 had toxaphene concentrations ranging from 7–108 ng L$^{-1}$. Rice and Evans [4] reported a concentration of 9 ng L$^{-1}$ in a composite sample from the southeastern shore of Lake Michigan for the same period. A Great Lakes atmospheric deposition workshop proposed a basin-wide consensus concentration of 0.2 ng L$^{-1}$ for the period 1989–1991 based on measurements in northwestern Ontario [69]. Elsewhere, Harder et al. [70] detected toxaphene in South Car-

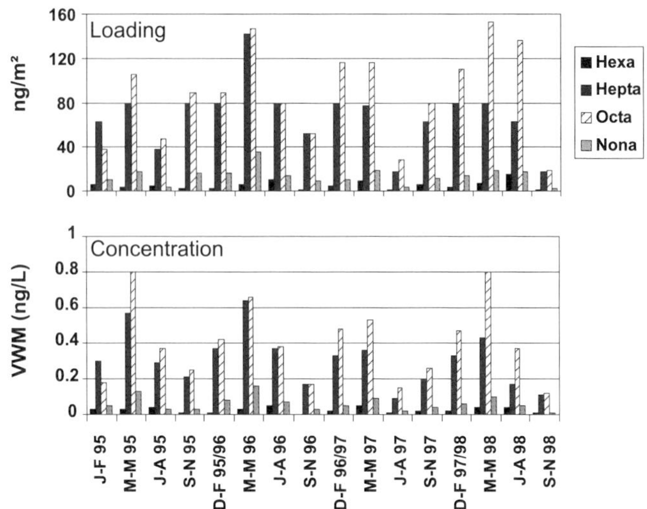



**Fig. 5** Seasonal toxaphene volume weighted concentrations (ng L$^{-1}$) and loadings (ng m$^{-2}$) at Point Petre near Lake Ontario from 1995 to 1998 [71]

olina rain, and noted that concentrations were highest in late summer/early fall, which they attributed to use patterns in the area at the time.

Burniston et al. [71] determined toxaphene in precipitation at the IADN Point Petre sampling site on Lake Ontario continuously from 1995 to 1998. Maximum concentrations and fluxes for all homologs generally occurred in the spring (March–April–May) each year (Fig. 5). Hepta- and octachlorobornanes were the predominant homolog groups. B8-789 was the predominant congener in the precipitation samples, accounting for 48% and 46% of $\Sigma$CHB in 1997 and 1998 respectively. This octachlorobornane had a maximum volume weighted mean of 0.21 ng L$^{-1}$ in the spring of 1997 and a low of 0.054 ng L$^{-1}$ in the autumn of 1998. Other congeners of significance were B8-2229 (P44) > B8-810 > B8-1471 > B8-1414/B8-1945 (P-40/P-41). $\Sigma$CHB concentrations averaged 50% of toxaphene quantified with a single response factor for technical toxaphene.

Mean toxaphene concentrations in precipitation at Point Petre in 1995–98 did not vary significantly over the four-year period. However, they appear to be about 1/10 of those reported by Rice and Evans [4], reflecting the cessation of agricultural use of toxaphene in 1984. Toxaphene was nevertheless one of the most prominent organochlorines in Great Lakes precipitation in the 1990s, exceeding all except total HCH and PCBs at Point Petre [72, 73]. While air concentrations of toxaphene achieve maxima during the summer [38, 42], levels in precipitation at Point Petre peaked earlier, possibly reflecting transport on airborne particles and revolatilization from newly tilled soils in southern agricultural areas prior to traveling to the Great Lakes from source regions and greater partitioning into the gaseous phase at the higher summer temperatures [71].

### 3.1.3
### Surface Water

Measurements of toxaphene in surface water of the Great Lakes between 1996–2000 have all been made by passing 80–300 L of water through a glass fiber filter and an XAD-2 resin column, followed by determination using GC-ECNI-MS [5, 46, 50, 65, 74]. Comparisons of dissolved toxaphene, quantified by all investigators using technical toxaphene, are given in Table 5. Toxaphene concentrations are highest in Lake Superior (910–1120 pg L$^{-1}$), intermediate in Lakes Michigan and Huron (380–470 pg L$^{-1}$), and lowest in Lakes Erie and Ontario (81–230 pg L$^{-1}$). The agreement among different research groups is remarkably good (20%) for Lake Superior, but results vary by about a factor of two for Lakes Erie and Ontario. Lower concentrations of toxaphene in these lakes may have led to a more difficult analytical situation. Also, there was a 3–7 y difference between the two sets of measurements.

**Table 5** Toxaphene levels in Great Lakes surface water, pg L$^{-1}$

| Lake | Year | Samples | Range | Mean | S.D. | Refs. |
|---|---|---|---|---|---|---|
| Superior | 1996 | 5 | — | 1120 | 180 | [5, 65] |
| | 1996–1997 | 26 | 520–1440 | 918 | 220 | [46] |
| | 1997–1998 | 23 | 650–1100 | 910 | 150 | [50, 76] |
| | 1997 | 7 | 680–1020 | 801 | 116 | [78] |
| | 1998 | 15 | 490–1000 | 696 | 142 | [74] |
| | 2002 | 9 | 590–860 | 718 | 82 | [77] |
| Huron | 1993 | 3 | — | 470 | 250 | [5, 65] |
| Michigan | 1994–1995 | 64 | | 380 | 120 | [5, 65] |
| | 1997–1998 | 27 | 210–670 | 410 | 150 | [50, 76] |
| Erie | 1993 | 2 | — | 230 | 7 | [5, 65] |
| | 1996 | 1 | — | 96 | — | [67] |
| Ontario | 1993 | 8 | | 170 | 70 | [5, 65] |
| | 1998–2000 | 14 | 48–120 | 81 | 22 | [67] |
| Nipigon | 1998 | 5 | 78–145 | 95 | 28 | [77] |
| Siskiwit (Isle Royale) | 1998 | 1 | — | 210 | — | [74] |

Glassmeyer et al. [75] estimated a 6-yr half life of toxaphene in Lake Ontario water based on the trend in fish residues, so there might have been a real decrease in water concentrations between sampling times. Swackhamer and Symonik [76] found a decline in Lake Superior water concentrations from 1996–1998, corresponding to a half-life of approximately three years. However, Muir et al. [77] did not observe a decline in toxaphene in samples collected at the same open lake stations from May 1997 to May 2002 (Table 5). These discrepancies may be due to the short time frames of the assessments, and to year-to-year variations in temperature [76]. Muir et al. [74] found no significant differences in mean toxaphene concentrations between the nearshore and offshore sites. Concentrations at 100 m depth were similar to levels at 4 m at the two sites where the deep samples were collected. In the two deepest sites in Lake Michigan, concentrations taken at mid-hypolimnion in early June were not significantly different to surface water concentrations [80]. However, dissolved concentrations within the benthic nephloid layer at both sites were two-fold higher than the hypolimnion samples, likely due to differences in water–particle partitioning caused by greater particulate mass concentrations. Taken together, these results suggest rather uniform concentrations throughout the water column in late spring–early summer.

Significant seasonal dynamics in the dissolved phase concentrations were observed in a study of Lakes Superior and Michigan [76]. Sampling took place over two years and included spring, summer, and fall conditions. There was significant variability in dissolved concentrations with season that corresponded to whether the lakes were stratified or not. During stratified conditions, there was significant volatilization loss from the epilimnion, and then concentrations increased when the lakes vertically mixed in the fall. Concentrations during unstratified periods were a third to a half greater in concentration than under stratified conditions.

Lower toxaphene concentrations have been found in smaller lakes within the Great Lakes watershed. Muir et al. [77] found toxaphene levels averaging $95\,\mathrm{pg\,L^{-1}}$ in Lake Nipigon (Table 5). Total toxaphene in Lake Siskiwit (Isle Royale) surface water sampled in May 1998 was $210\,\mathrm{pg\,L^{-1}}$ [74]. Chlorobornane congeners were not determined in this sample; however, the heptachlorobornanes were the predominant homolog group ($110\,\mathrm{pg\,L^{-1}}$) followed by octachlorobornanes ($76\,\mathrm{pg\,L^{-1}}$).

### 3.1.4
**Toxaphene Congeners in Great Lakes Water and Air**

In studies of toxaphene in Lakes Superior [46] and Ontario [51, 67], two persistent congeners, B8-1413 and B9-1979, were determined in air and six other congeners in water. B8-1413 and B9-1979 were of particular interest because they were among the most prominent chlorobornanes in Great Lakes fish [74, 81, 82]. It should be noted that although peaks matching the retention times and ion ratios of these congeners have been quantified, they may not be single compounds in air or water samples. Shoeib et al. [37] demonstrated this by using multidimensional gas chromatography with electron capture detection to examine the composition of toxaphene peaks in technical toxaphene and air samples collected on the north shore of Lake Ontario. Measurements of eight major congeners in water and air, including B8-1413 and B9-1979, are summarized in Table 6. B7-1001 and B8-789 were the major congeners in Lake Superior surface water based on three cruises (1997, 1998 and 2002). Results reported by Jantunen and Bidleman [46] for B8-1413 and B9-1679 from expeditions in August 1996 and May 1997 were somewhat higher (especially for B8-1413 in 1997) than those reported by Muir et al. [74]. B8-1413 and B9-1679 were $0.60 \text{ pg L}^{-1}$ and $1.2 \text{ pg L}^{-1}$ in Lake Ontario during July 1998 and June 2000, accounting for 0.74% and 1.5% of the total toxaphene ($801 \text{ pg L}^{-1}$). For comparison, B8-1413 is 0.47% and B9-1697 is 1.2% of technical toxaphene (studies summarized by Bidleman et al. [68] Jantunen and Bidleman [46]).

Single congeners in air samples taken over water averaged $2.5 \text{ pg m}^{-3}$ for B8-1413 and $1.5 \text{ pg m}^{-3}$ for B9-1679 in August 1996 (Lakes Superior, Huron and Erie), and $0.44 \text{ pg m}^{-3}$ for B8-1413 and $0.24 \text{ pg m}^{-3}$ for B9-1679 in May 1997 (Lakes Superior and Huron) [46]. Total toxaphene levels were $30 \text{ pg m}^{-3}$ in August and $11 \text{ pg m}^{-3}$ in May. As percentages of total toxaphene, B8-1413 was 8.4% in August and 4.0% in May, and B9-1679 was 5.2% in August and 2.2% in May.

For Lake Ontario in July 1998 and June 2000, concentrations of B8-1413, B9-1679 and total toxaphene averaged 2.8, 2.2 and $20 \text{ pg m}^{-3}$. The two congeners as percentages of total toxaphene were 14% for B8-1413 and 11% for B9-1679. Shoeib et al. [51] found somewhat lower percentages of B8-1413 (7.1%) and B9-1679 (6.1%) in air at Point Petre, Lake Ontario during 1995–1997. Shoeib et al. [51] also quantified eight other congeners in air (Table 6). The concentrations of total toxaphene reported by Shoeib et al. [51] were lower by a factor of two or more than other studies which they attributed to the use of response factors for individual GC peaks rather than the single response factor used by other investigators. Average concentrations of B8-1413 and B9-1679 reported by Shoeib et al. [51] at Point Petre for 28 samples from 1995 to 1997 were ten times lower than found by Jantunen and Bidleman [46] during a July 1998 cruise on Lake Ontario. This difference is probably due to the average temperatures of the sampling periods. Jantunen and Bidleman [46] also found approximately six-fold higher concentrations of B8-1413 and B9-1679 over Lakes Superior and Huron in August compared to early May.

B8-1413 and B9-1679 were significantly ($p < 0.001$ to $< 0.05$) enriched in all air samples collected over the lakes compared to technical toxaphene. Alabama air sampled in 1996–1997 also showed significant ($p < 0.05$) enrichment of these two congeners (as percentages of total toxaphene, B8-1413 $= 1.9 \pm 2.0\%$ and B9-1679 $= 3.3 \pm 3.4\%$) [15], although the enrichment was less than for Great Lakes air. Vetter and Scherer [40] showed that these two

**Table 6** Major toxaphene congeners in Great Lakes air and surface water

| Lake | Date | N | B7-1001 | B8-1413 | B8-789 | B8-1414/1945 | B8-806/809 | B8-2229 | B9-1679 | B9-1025 | Refs. |
|---|---|---|---|---|---|---|---|---|---|---|---|
| *Water*, pg L$^{-1}$ | | | Mean ± SD | Mean ± SD | Mean ± SD | Mean ± SD | Mean ± SD | Mean ± SD | Mean ± SD | Mean ± SD | |
| Superior | Aug-96 | 14 | — | 2.2 ± 0.9 | — | — | — | — | 8.6 ± 4.0 | — | [46] |
| | May-97 | 12 | — | 4.8 ± 2.2 | — | — | — | — | 18 ± 3.8 | — | [46] |
| | May-97 | 7 | 19 ± 5.0 | 1.4 ± 0.4 | 32 ± 8.7 | 16 ± 4.3 | 9.6 ± 2.8 | 17 ± 4.8 | 6.0 ± 1.5 | < 0.1 | [78] |
| | May-98 | 15 | 17 ± 3 | 1.2 ± 0.3 | 25 ± 5.2 | 14 ± 2.2 | 8.5 ± 1.2 | 14 ± 2.0 | 5.0 ± 1.1 | < 0.1 | [74] |
| | May-02 | 9 | 6.8 ± 0.8 | 0.7 ± 0.9 | 94 ± 14 | 9.6 ± 1.1 | 5.0 ± 0.7 | 22 ± 3.0 | 5.5 ± 0.9 | 0.4 ± 0.1 | [77] |
| Ontario | 1998&2000 | 14 | — | 0.60 ± 0.21 | — | — | — | — | 1.2 ± 0.5 | | [67] |
| Nipigon | July-98 | 5 | 1.4 ± 0.5 | < 0.1 | 7.0 ± 4.6 | 1.5 ± 0.7 | 0.5 ± 0.2 | 3.8 ± 1.6 | 0.5 ± 0.3 | < 0.1 | [77] |
| *Air*, pg m$^{-3}$ | | | | | | | | | | | |
| Superior, Huron, Erie | Aug-96 | 7 | — | 2.5 ± 1.0 | — | — | — | — | 1.5 ± 0.60 | — | [46] |
| Superior, Huron | May 97 | 7 | — | 0.44 ± 0.34 | — | — | — | — | 0.24 ± 0.05 | — | [46] |
| Ontario | 1995–1997 | 28 | 0.59 ± 0.34 | 0.27 ± 0.16 | 0.06 ± 0.04 | 0.20 ± 0.15 | 0.17 ± 0.12 | 0.35 ± 0.26 | 0.23 ± 0.16 | — | [51] |
| Ontario | July 98 & June 00 | | — | 2.8 ± 1.5 | — | — | — | — | 2.2 ± 0.76 | — | [46] |

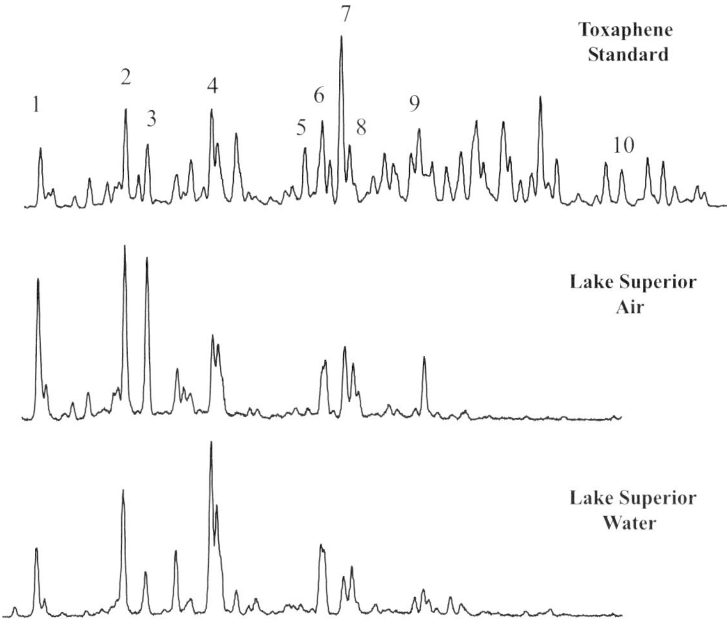



**Fig. 6** GC-MS chromatograms of a technical toxaphene and air and water extracts from Lake Superior collected in 1997 [46]. Tentative identifies of Peaks #1–10 are given in Figure 2

congeners have stable structures compared to other polychlorinated bornanes due to the placement and orientation of the chlorine atoms on the carbon backbone.

Total ion chromatograms of the $Cl_7$ to $Cl_9$ toxaphene residues in water and air of Lake Superior are compared to a toxaphene standard in Fig. 6, and single-ion chromatograms of the individual homologs are shown by Jantunen and Bidleman [46]. Peaks in Fig. 6 are numbered in the same way as those in Fig. 2. The toxaphene residues in water and air have similar patterns, both dominated by the lighter earlier eluting congeners and showing transformations from the standard mixture, particularly for peaks 5 and 7. Shoeib et al. [37, 51] found similar depletions in air samples from Point Petre, Lake Ontario.

Areas of peaks tentatively identified as B8-531, B8-1414 + B8-1945, B8-806/809 and B8-2229 (peaks 5–8 Fig. 2) were normalized to peak B8-1414/1945 = 1 (Peak 6), and the resulting profiles are shown in Fig. 7 for Lake Superior air and water, air sampled above farm soil in the southern United States [30], and a toxaphene standard. The profiles in Lake Superior surface water and air appear more similar to the toxaphene that has outgassed from farm soil than to the technical toxaphene standard, but there are differences in the profiles that cannot be explained. Chlorobornanes B8-531 and B8-806/809 were signifi-

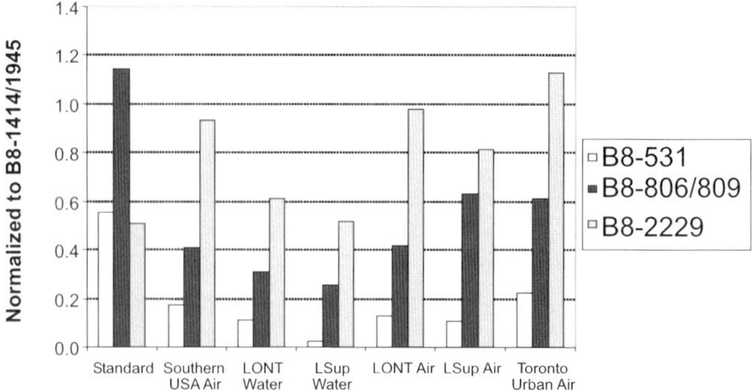



**Fig. 7** Ratios of peaks tentatively identified as B8-531, B8-1414 + B8-1945, B8-806/809 and B8-2229 (peaks 5–8 Figure 2) relative to peak B8-1414/1945 = 1 (Peak 6), for technical toxaphene, southern USA, and water and air samples from the Great Lakes [29, 30]

cantly ($p < 0.01$) depleted in air and water compared to technical toxaphene, and depleted in the water relative to the air ($p < 0.05$). B8-2229 is significantly enriched in the air of Lake Superior and the southern United States ($p < 0.001$), but depleted in Lake Superior water ($p < 0.05$). It may be that the relative volatilities of these toxaphene components from water differ more than their vapor pressures, which are very similar (Sect. 1.3.2). Toxaphene is undergoing net volatilization from Lake Superior (Sect. 3.2). This process is

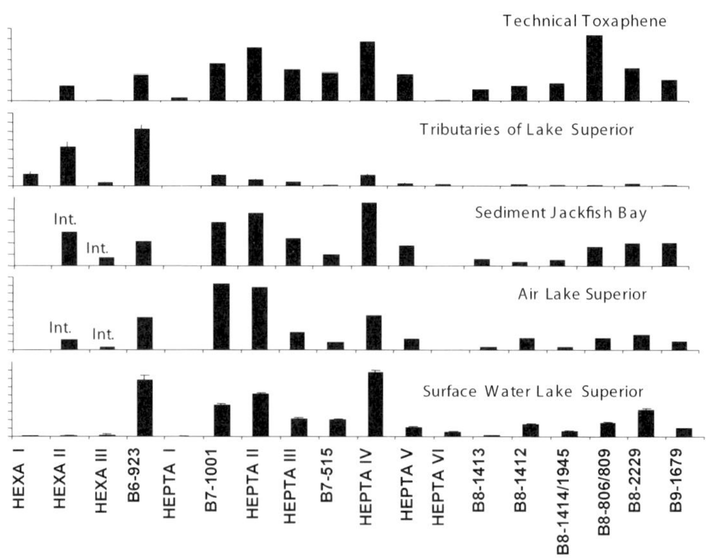



**Fig. 8** Chlorobornane congener profiles in Lake Superior tributary water, air, sediment and surface waters [91]. The "int" label for some hexachloro-peaks indicates that they are partially interfered with by other co-eluting compounds

controlled by the Henry's law constant, which has been measured only for technical toxaphene [15] but not for single congeners.

Muir et al. [83] examined chlorobornane profiles in river water from six tributaries of Lake Superior. Tributary waters had high proportions of B6-923 and other unidentified hexa- and heptachlorobornanes (Fig. 8). So far as is known, toxaphene has not been used in these tributaries, which suggests that atmospherically deposited toxaphene is being slowly transformed and released as mainly hexa/hepta-CHBs in watershed soils, wetlands, and river sediments. The Nipigon River, the major tributary of Lake Superior, which drains Lake Nipigon, had a congener pattern similar to Lake Superior water, while water from smaller forested watersheds showed the enrichment in hexa-CHBs. Ruppe et al. [84] have identified several of the hexa- and heptachloro-compounds in Lake Ontario sediment which appear to be among those reported in Fig. 8.

### 3.1.5
### Chlorobornane Enantiomers in Great Lakes Water and Air

Enantiomer ratios (ERs) or fractions (EFs) of chiral pollutants such as chlordane, $\alpha$-HCH and atropisomer PCBs have been used to determine sources, pathways and fates of these compounds in the abiotic environment [85–87]. Biological processing of chiral compounds often results in significant deviation from the EF value present in the technical mixture. Most of the compounds in toxaphene, and all of the major congeners identified in environmental samples to date, are chiral. Chiral separations of chlorobornanes have been carried out, but for abiotic samples (as illustrated in Fig. 2) they are very challenging because of the complexity of the chlorobornane mixture [1, 2, 88]. The ER of a heptachlorobornane in sediments from a lake treated with toxaphene and from contaminated estuarine sediments has been shown to be nonracemic, indicating that dechlorination of technical toxaphene components is occurring by biodegradation [89, 90].

Karlsson et al. [91] found that ERs of B7-515, B8-1412, B8-1414, and B8-806/809 were close to racemic (ER = 1) in air and water of Lake Superior (Table 7). B8-1945 was nonracemic with ERs of 1.13–1.20 in air above Lake Superior and 1.18–1.21 in the surface water. ERs for B8-1945 was also nonracemic in tributaries. The results suggest that B8-1945 is undergoing biodegradation within the Lake Superior water column and watershed. The similar ERs in air and water supports the hypothesis that volatilization of toxaphene from lake water into air is the main source of the airborne toxaphene over the lake in August [46] (Sect. 3.2). ERs of B8-1945 in air from an Alabama farm field were racemic [92] and thus the southern source signal is weak over Lake Superior. The sediment sample from Stn 403 (Jackfish Bay) was close to racemic for all compounds except for B8-806/9, indicating that limited biodegradation was occurring.

**Table 7** Enantiomer Ratios of selected hepta and octachlorobornane congeners in air, open lake and tributary waters, and sediments in Lake Superior [91]. Parlar numbers for each congener are shown in parentheses

|  | N | B7-515 (P32) | B8-1412 — | B8-1414 (P40) | B8-1945 (P41) | B8-806/809 (P42a/42b) |
|---|---|---|---|---|---|---|
| Congener Stnd[a] |  | $0.96 \pm 0.04$ | n.i.s.[b] | $0.99 \pm 0.01$ | $0.92 \pm 0.01$ | $1.0 \pm 0.02$ |
| Tech. Tox. |  | 1.03 | 0.98 | 1.06 | 0.98 | 0.98 |
| Air | 2 | 0.95–1.02 | 0.99 | 0.89–0.98 | 1.13–1.20 | 0.94–1.01 |
| Lake water | 5 | 0.97–1.03 | 0.98–1.01 | 0.97–0.99 | 1.18–1.21 | 0.93–1.03 |
| Sediments | 1 | n.d. | 1.04 | 1.05 | 1.01 | 0.92 |
| Tributaries | 9 | 0.98–1.04 | 0.98–1.05 | 0.93–1.04 | 0.97–1.26 | 0.87–0.98 |

[a] ERs of congeners detected in a toxaphene congener standard
[b] n.i.s. = not in standard

## 3.2
## Air–Water Gas Exchange of Toxaphene

Loadings of persistent organic pollutants (POPs) to the Great Lakes take place by air–water gas exchange, precipitation and dry deposition of particles. Gas exchange dominates in many situations and differs from other loading pathways in being a reversible process where deposition and volatilization occur simultaneously. The dynamic mass balance model described by Swackhamer et al. [65] indicates that the net exchange direction of toxaphene in Lake Superior was depositional during the 1960s and 1970s due to high air concentrations during this time of heavy usage. Usage and air concentrations declined in the 1980s and 1990s, and the exchange shifted in the direction of net volatilization [65]. This section describes studies of toxaphene gas exchange that have been carried out since the mid-1990s.

## 3.2.1
## Gas Exchange Calculations

The net direction of gas exchange is estimated from measured concentrations of dissolved toxaphene in water and gaseous toxaphene in air ($C_W$ and $C_A$, mol m$^{-3}$). Expressed in fugacity terminology [46]:

$$f_W = C_W H \tag{2}$$

$$f_A = C_A R T_A \tag{3}$$

$$f_W/f_A = C_W H / C_A R T_A \tag{4}$$

where $f_W$ and $f_A$ are the fugacities in water and air (Pa), $H$ is the Henry's law constant of technical toxaphene at the temperature of the water

(Pa m$^3$ mol$^{-1}$) [15], $R$ is the gas constant (Pa m$^3$ mol$^{-1}$ K$^{-1}$), and $T_A$ is the temperature of the air (K). The water/air fugacity ratio $f_W/f_A = 1$ at equilibrium; values $< 1$ and $> 1$ imply net deposition and volatilization, respectively. Fluxes are calculated from gradients in water and air fugacity [46] or concentration [50, 64, 65] across the air–water interface. Most flux calculations are based on the two-film model, which requires mass transfer coefficients for the air and water films as functions of wind speed and the Henry's law constant [46, 48, 93]. The sign convention used here for flux is positive for net deposition ($f_W/f_A < 1$) and negative for net volatilization ($f_W/f_A > 1$). This follows the convention used by Hoff et al. [72], James et al. [50] and Swackhamer et al. [65] and is opposite to the designation used by Jantunen and Bidleman [46].

### 3.2.2
### Fugacity Gradients and Fluxes of Toxaphene

Gas exchange studies of total toxaphene in Lakes Superior, Michigan and Ontario were conducted between 1996–2000 [46, 50, 67] using concurrently measured water and air concentrations. The air and water data from these studies are discussed in Sects. 2.2, 3.1.1 and 3.1.2, and are summarized in Tables 3, 5, and 6. An earlier mass budget for toxaphene [65] estimated gas exchange from measured water concentrations and historical air data from Hoff et al. [64]. All studies based their fugacity and flux calculations on the temperature-dependent Henry's law constant for technical toxaphene [15].

A summary of toxaphene atmospheric flux estimates to the Great Lakes is given in Table 8. The reported fluxes for a particular lake in Table 8 differ considerably in magnitude and sometimes even in direction. Seasonal changes in flux direction are due to variations in water temperature (influencing the Henry's law constant) and air concentration. Changes in flux magnitude are driven by air concentration and by wind speed, which affects the air–water mass transfer coefficients.

Mean water/air fugacity ratios ($f_W/f_A$, Eq. 4) in Lake Superior were $2.0 \pm 0.6$ in August 1996 and $2.3 \pm 0.7$ in May 1997 [46]. Fugacity ratios in Lake Ontario averaged $0.56 \pm 0.17$ in July 1998 and $0.22 \pm 0.066$ in June 2000 [67]. The errors associated with the fugacity ratio calculations for discrete air sampling events in these studies were $\pm 30\%$, propagated from the relative standard deviations of the water concentration ($\pm 22\%$) and the Henry's Law constant ($\pm 20\%$). Mean values of $f_W/f_A$ were significantly different from 1 (air–water equilibrium) at $p < 0.01$. Results indicate that toxaphene was undergoing net volatilization in Lake Superior and net absorption in Lake Ontario during these months.

In studies of Lakes Superior and Michigan during spring through fall of 1997 and 1998 [50], fugacity ratios were not specifically calculated but net fluxes of toxaphene were estimated to be from water to air (Table 8), imply-

**Table 8** Net gas exchange fluxes[a] of toxaphene in the Great Lakes, $ng\,m^{-2}\,d^{-1}$

| Time period | Superior | Huron | Michigan | Erie | Ontario | Refs. |
|---|---|---|---|---|---|---|
| Spring 1993 | — | – 1.2 | — | 3.3 | 4.2 | [65] |
| Summer 1993 | — | 13 | — | 25 | 27 | [65] |
| Fall 1993 | — | – 14 | — | – 2.5 | 0.4 | [65] |
| Winter 1993 | — | – 0.5 | — | 4.2 | 5.4 | [65] |
| Annual 1993 | — | – 0.6 | — | 7.2 | 9.2 | [65] |
| Spring 1994–1995 | — | — | 0.5 | — | — | [65] |
| Summer 1994–1995 | — | — | 18 | — | — | [65] |
| Fall 1994–1995 | — | — | 9.8 | — | — | [65] |
| Winter 1994–1995 | — | — | 1.3 | — | — | [65] |
| Annual 1994–1995 | — | — | 2.3 | — | — | [65] |
| Spring 1996 | – 11 | — | — | — | — | [65] |
| Summer 1996 | 4.4 | — | — | — | — | [65] |
| Fall 1996 | – 32 | — | — | — | — | [65] |
| Winter 1996 | – 13 | — | — | — | — | [65] |
| Annual 1996 | – 13 | — | — | — | — | [65] |
| August 1996 | – 11 | — | — | — | — | [46] |
| May 1997 | – 4.2 | — | — | — | — | [46] |
| Spring 1997 | – 19 | — | – 9.4 | — | — | [50] |
| Summer 1997 | – 26 | — | – 14 | — | — | [50] |
| Spring 1998 | – 9 | — | – 2.8 | — | — | [50] |
| Summer 1998 | – 33 | — | – 28 | — | — | [50] |
| July 1998 | — | — | — | — | 3.7 | [67] |
| June 2000 | — | — | — | — | 6.2 | [67] |

[a] Positive = net deposition, negative = net volatilization

ing $f_W/f_A > 1$. Both sets of Lake Superior measurements are consistent with the mass budget of Swackhamer et al. [65] which showed that toxaphene was volatilizing from Lake Superior during the 1990s. Swackhamer et al. [65] also predicted net toxaphene absorption in Lake Ontario, as was found by Jantunen and Bidleman [67]. However, James et al. [50] calculated a net volatilization flux of toxaphene in Lake Michigan for 1997–1998, whereas Swackhamer et al. [65] concluded that there was a slight net gas absorption in 1994–1995 (Table 8). The opposite conclusions probably result from different assumptions about the atmospheric concentration of toxaphene and the use of event vs. seasonally averaged temperature data [50]. Considering all studies (Table 8), Lakes Erie and Ontario are net recipients of toxaphene, Lake Superior is losing more toxaphene than it is gaining and Lake Huron is near air–water equilibrium.

The yearly cycle of toxaphene flux and annual loadings to Lake Superior by gas exchange have been estimated in two studies. Swackhamer et al. [65] estimated toxaphene fluxes in the four seasons, assuming air concentrations of 60 $pg\,m^{-3}$ in summer and 10 $pg\,m^{-3}$ in the other three seasons, after Hoff et al. [64]. Highest net volatilization fluxes occurred in fall and winter, dropped off in spring and were lowest (with reversal from volatilization to deposition, Table 8) in summer. The annual net output due to volatilization was 390 kg.

Jantunen and Bidleman [46] estimated monthly and annual fugacity ratios and fluxes of toxaphene in Lake Superior. Their calculations used monthly average air concentrations of toxaphene over the lake, estimated from their parameters of Eq. 1 (Sect. 2.2). The concentration of toxaphene in surface water was assumed constant over the year, since measurements in August 1996 and May 1997 were not statistically different. However, we now know

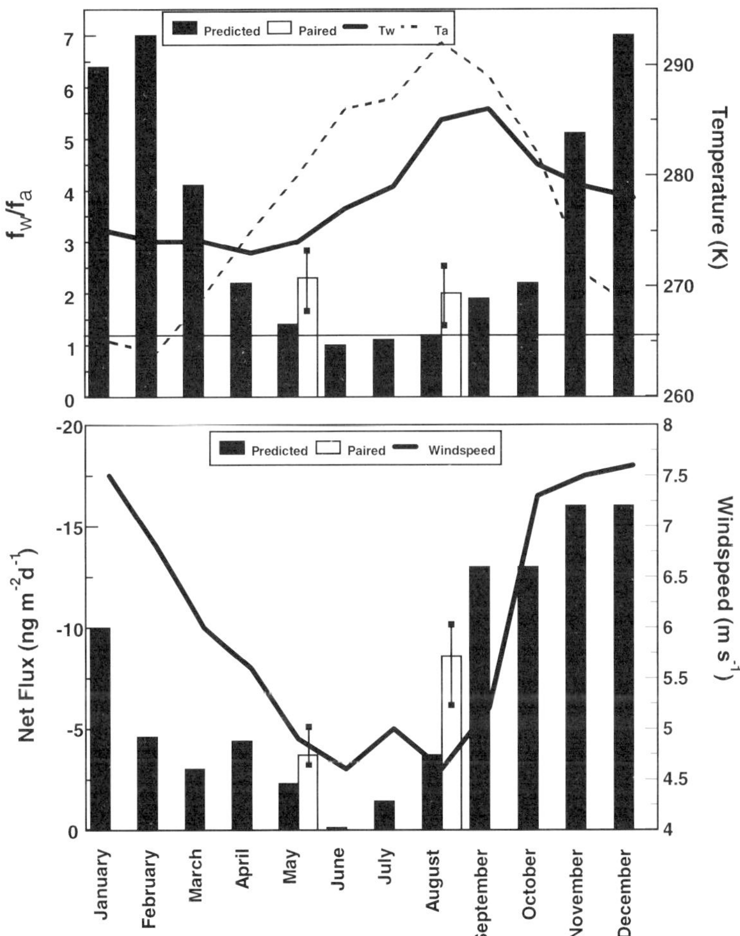


**Fig. 9** Estimated monthly and annual fugacity ratios ($f_w/f_a$) and fluxes of toxaphene in Lake Superior from Jantunen and Bidleman [46] with signs changed to indicate negative flux as volatilization. Monthly averaged $f_w/f_a$ and net fluxes are compared with those estimated from paired air and water measurements in August 1996 and May 1997

that the surface concentrations in Lake Superior vary considerably depending on the season [76], which would affect both the Swackhamer et al. [65] and Jantunen and Bidleman [46] model outputs. Daily water temperatures and hourly wind speeds were averaged by month. Results are shown in Fig. 9, which also compares the monthly averaged $f_W/f_A$ and net fluxes with those estimated from paired air and water measurements in August 1996 and May 1997. Differences are due to monthly average air temperatures, which were higher than those measured during the shipboard experiments. This resulted in higher predicted air concentrations for the two months and correspondingly lower $f_W/f_A$. The predicted monthly $f_W/f_A$ ranged from 1.0–7.0 and this indicates that toxaphene was undergoing net volatilization from the lake during most months and achieved air–water equilibrium in summer. Values of $f_W/f_A$ were highest in the winter (despite colder water temperatures) and lowest in the summer, driven by atmospheric concentrations that were approximately four times lower in the winter than summer. Monthly averaged net volatilization fluxes ranged from $-0.1$ to $-16\,\mathrm{ng\,m^{-2}\,d^{-1}}$. Higher fluxes occurred from September to January ($-10$ to $-16\,\mathrm{ng\,m^{-2},d^{-1}}$) than February to August ($-0.1$ to $-4.6\,\mathrm{ng\,m^{-2}\,d^{-1}}$). The net loss of toxaphene due to volatilization over the year was 220 kg, or approximately 2% of the toxaphene burden in the lake. This is slightly lower than the volatilization output of 390 kg estimated by Swackhamer et al. [65].

Toxaphene fluxes to Lake Ontario due to precipitation inputs were estimated by Burniston et al. [71] based on continuous monthly measurements from 1995 through 1998. Average annual fluxes ranged from $1.7$–$2.2\,\mathrm{ng\,m^{-2}\,d^{-1}}$ or $11$–$14\,\mathrm{kg\,yr^{-1}}$ for Lake Ontario which is significantly higher than the $3.5\,\mathrm{kg\,yr^{-1}}$ estimated by Swackhamer et al. [65] using data from Hoff et al. [64]. Net gas absorption of Lake Ontario was estimated by Swackhamer et al. [65] at $65.5\,\mathrm{kg\,y^{-1}}$ for Lake Ontario. Thus precipitation inputs were about 17–21% of gas exchange volatilization fluxes, at least for Lake Ontario. Comparable data are not available for Lakes Michigan and Superior. Precipitation is probably more important than previously assumed in the overall mass balance of toxaphene in the Great Lakes.

## 3.3
## Levels, Transformation and Historical Profiles in Sediment and Peat Cores

### 3.3.1
### Dated Sediment Cores

Profiles of toxaphene in dated sediment cores have been used to infer historical deposition and degradation of toxaphene in Lake Superior and the nearby inland lakes [50, 94, 95], Michigan [94, 96] and Ontario [94, 97–99]. There is sufficient detail from these studies to infer the history of inputs of toxaphene

**Table 9** Concentrations and fluxes and year of maximum deposition of toxaphene in dated sediment cores from the Great Lakes

| Lake | Core location | Site name | Date collected | Surface conc'n ($ng\,g^{-1}\,dw$) | Surface flux ($ug\,m^{-2}\,yr^{-1}$) | Maximum flux ($\mu gm^{-2}\,yr^{-1}$) | Year of max deposition | % per year decline from maxima | Average open lake decline[1] % per year | Refs. |
|---|---|---|---|---|---|---|---|---|---|---|
| Superior | 47°05'N, 90°40'W | Basswood | 1991 | 2.8 | 0.97 | 0.97 | 1990 | 0.0 | 4.6 ± 2.8 | [94] |
| | 47°20'N, 89°15'W | NOAA-3 | 1994 | 12 | 1.1 | 1.6 | 1974 | − 1.6 | | [94] |
| | 47°20'N, 89°15'W | NOAA-3 | 1991 | 19 | 1.4 | 2.5 | 1982 | − 4.9 | | [94] |
| | 47°00'N, 87°00'W | LS5 | 1997 | 3.1 | 0.2 | 2.6 | 1991 | − 15 | | [50] |
| | 47°50'N, 87°50'W | LS12 | 1997 | 24 | 2.3 | 4.7 | 1989 | − 6.4 | | [50] |
| | 47°10'N, 89°45'W | LS17 | 1997 | 8.7 | 0.5 | 1.6 | 1985 | − 5.7 | | [50] |
| | 48°48'N, 86°59'W | Jackfish B. | 1998 | 3.42 | 1.0 | 6.9 | 1986 | − 7.1 | | [95] |
| | 47°35'N, 86°56'W | Site 8c | 1998 | 2.58 | 0.26 | 0.87 | 1989 | − 7.8 | | [95] |
| | 48°34'N, 87°34'W | Site 404 | 1998 | 0.46 | 0.05 | 0.65 | 1969 | − 3.2 | | [95] |
| Michigan | 42°45'N, 87°00'W | LM-18 | 1991 | 15 | 2.4 | 3.2 | 1978 | − 1.9 | 3.1 ± 1.2 | [94] |
| | 44°42'N, 86°41'W | LM-47s | 1992 | 22 | 5.2 | 6.6 | 1980 | − 1.8 | | [94] |
| | 45°06'N, 86°23'W | LM-68k | 1992 | 39 | 8.5 | 9.8 | 1989 | − 4.4 | | [94] |
| | 45°01'N, 85°34'W | HMS1 | 1998 | 5 | 0.70 | 11 | 1972 | − 3.6 | | [96] |
| | 44°50'N, 85°37'W | HMS2 | 1998 | 28 | 0.41 | 28 | 1972 | − 3.8 | | [96] |
| Ontario | 43°26'N, 79°24'W | Site 1007 | 1998 | 11 | 4.5 | 13 | 1977 | − 3.1 | 3.2 ± 0.6 | [98] |
| | 43°20'N, 79°12'W | Site 20 | 1993 | 6 | 7.4 | 32 | 1968 | − 3.1 | | [97] |
| | 43°38'N, 78°43'W | Site 600 | 1993 | 10 | 3 | 10 | 1968 | − 2.8 | | [97] |
| | 43°41'N, 78°10'W | Site 601 | 1993 | 16 | 4 | 14 | 1969 | − 3.0 | | [97] |
| | 43°45'N, 77°55'W | Site 602 | 1993 | 11 | 15 | 28 | 1982 | − 4.2 | | [97] |
| | 43°39'N, 77°19'W | Site 603 | 1993 | 12 | 9 | 22 | 1972 | − 2.8 | | [97] |
| | 44°09'N, 76°37'W | Site 607 | 1993 | 10 | 6 | 22 | 1970 | − 3.2 | | [97] |
| | 44°02'N, 76°52'W | Site 608 | 1993 | 10 | 8 | 17 | 1981 | − 4.4 | | [97] |
| | 43°44'N, 76°45'W | Site 609 | 1993 | 5 | 10 | 28 | 1971 | − 2.9 | | [97] |
| | 43°22'N, 79°21'W | Site 13 | 1991 | 14 | 3.9 | 6.9 | 1979 | − 3.6 | | [94] |
| | 43°35'N, 78°00'W | Site 43 | 1991 | 16 | 6.0 | 10 | 1971 | − 2.0 | | [94] |
| | 43°32'N, 76°54'W | Site E30 | 1990 | 15 | 5.0 | 14 | 1971 | − 3.4 | | [94] |
| Siskiwit | 46°50'N, 91°00'W | | 1991 | 8.9 | 2.5 | 5.1 | 1971 | − 2.5 | 1.5 | [94] |
| Outer Is. | 47°05'N, 90°30'W | | 1991 | 4.0 | 0.65 | 0.72 | 1972 | − 0.5 | | [94] |

[1] Excluding site LS5 which had a very high % decline from a maxima in 1991.

to three of the five Great Lakes reasonably accurately. Surface concentrations and fluxes as well as maximum fluxes and year of maximum deposition of toxaphene are presented in Table 9.

### 3.3.2
### Historical Onset and Maximum Toxaphene Deposition

The onset of toxaphene inputs into the Great Lakes occurs in sediment horizons dated to the mid-1940s and is quite consistent in all three lakes in which cores have been analyzed by at least three research groups. However, the dates of maximum concentrations and fluxes vary over a 15–20 year period (Table 9). In Lake Ontario maximum inputs occurred in the 1970s (mean of 12 cores = 1973). This corresponds well to the production and use of toxaphene in the US which reached a maximum in 1974. Maximum fluxes in Lake Ontario are also reasonably uniform, averaging $18\ \mu g\ m^{-2}\ yr^{-1}$. Maxima in cores from the central basins of Lakes Michigan and Superior generally occur in the mid-to-late 1980s and fluxes were significantly lower than in Lake Ontario [94]. In Lake Michigan, Pearson et al. [94] found maxima in 1978 and 1980 for cores in the southern (42.75°N) and middle basins (44.7°N), respectively, while a core in the northern sector (45.1°) had a maximum in 1989 and two-fold higher surface concentrations than the more southerly sites. Schneider et al. [96] found earlier maxima (1972) for toxaphene in two cores from Grand Traverse Bay (GTB), a steep-walled embayment on the northeast coast of Lake Michigan with high sediment particle focusing. A near-shore core from GTB had higher toxaphene fluxes than outer GTB and open lake sites indicating possible local inputs from agricultural watersheds and from nearby Traverse City MI. Averaging the open lake sites and outer GTB gives a maximum flux of $7.6\ \mu g\ m^{-2}\ yr^{-1}$. Maximum toxaphene fluxes in Lake Superior occurred in the 1980s (mean of nine cores = 1984). The average maximum flux was $1.9\ \mu g\ m^{-2}\ yr^{-1}$ excluding a core from Jackfish Bay, an inlet on the north shore of Lake Superior, which, similar to inner GTB, receives industrial and municipal effluents as well as drainage from streams draining the northwestern Ontario boreal forest. Two small lakes, Siskiwit (WI) and Outer Island (WI), near the western end of Lake Superior, had maxima in the early 1970s [94], as did Clay Lake in northwestern Ontario [95].

### 3.3.3
### Dated Peat Cores

In a unique study, Rapaport and Eisenreich [100] determined the time trends in accumulation of toxaphene in dated peat cores collected in 1982–1984 from ombrotrophic bogs (deriving nutrients solely from the atmosphere and isolated from ground water flow) in eastern North America. Three of their peat core sites were located to the west (at Marcell, MN), north (Diamond, ON) and east (Alfred, ON) of the Great Lakes basin, while other sites were further east in Québec, Maine and Nova Scotia. Maximum toxaphene fluxes in the cores near the Great Lakes basin ranged from $4-7\,\mu g\,m^{-2}\,yr^{-1}$. These maxima occurred in surface or near-surface slices dated to the late 1970s, except at Diamond (ON) where an earlier maxima (mid-1960s) was observed. Maximum fluxes in these peat cores were greater than for Lake Superior but generally lower than observed for Lakes Ontario and Michigan (Table 9). Although some post-depositional mobility of toxaphene occurred, most of the peat cores reflected the known historical use pattern of toxaphene in the USA. The results also illustrate the significant deposition of toxaphene to terrestrial watersheds within the Great Lakes basin, which exceeded PCB and DDT deposition by two- to four-fold during the 1970s [101].

### 3.3.4
### Investigations of Pulp Mills as Possible Toxaphene Sources

The persistently high concentrations of toxaphene observed in Lake Superior lake trout in the 1990s and apparent lack of decreases in sediment cores in northern Lake Michigan led to speculation that there might be significant within-basin sources of toxaphene. A scientific workshop held in 1996 concluded, based on the evidence available, that there was potential for regional specific sources, particularly in Lake Superior [102], and recommended further work on sediments to get a clearer picture of toxaphene inputs. Dichlorobornanes had been observed in kraft pulp and paper mill effluent in New Zealand [103] and pentachlorinated toxaphene-like compounds were reported in fish and effluent near pulp and paper mills in Japan [104]. Toxaphene-like products were produced from chlorination of camphene in sunlight [105]. All this led to the hypothesis that inadvertent toxaphene production could be occurring from chlorination of terpene precursors present in pulp mill effluents. In the late 1990s several groups investigated this hypothesis and concluded that pulp mills were not sources of toxaphene. Shanks et al. [106] measured toxaphene concentrations in river sediments from Great Lakes tributaries in Wisconsin and Michigan. They found higher toxaphene concentrations in rivers within agricultural areas where toxaphene had been used as a pesticide compared to upstream and non-agricultural areas. Sediments downstream of pulp and paper mills did not have elevated levels. Rappe et al. [107, 108] analyzed sample splits from the same study and reached similar conclusions, although they obtained quite different concentrations (about 0.01 of those reported by Shanks et al. [106]). [The discrepancy could be explained if Rappe et al. used individual congeners rather than technical toxaphene to quantify the small number of peaks observed]. Rappe et al. [107] also detected the persistent dechlorination products of toxaphene "hex-sed" (B6-923) and "hepsed" (B7-1001) upstream and downstream of one of the mills. Rappe et al. [109] also found that chlorobornanes in Baltic Sea sediments did not show a trend with distance from former chlorine bleaching mills. Rappe et al. [108] did not detect hexa- to decachlorobornanes in influent and effluent of eight forest industry related facilities in the US midwest and in Canada. In another study, Swackhamer and Engstrom [110] measured toxaphene concentrations in dated sediment cores from an historic holding pond for Kraft bleach effluent, and compared them to a control site. Concentrations of toxaphene were not different between the sites for the time period that the pond was used to hold effluent. Muir et al. [95] did not detect elevated toxaphene in dated sediment cores from depositional areas downstream of two bleach Kraft pulp and paper mills in northern Ontario. Historical profiles and fluxes in the cores were similar to those in Great Lakes profundal sediments. Taken together these recent studies clearly show that recent sediments near bleached pulp and paper mills in the Great Lakes basin have no toxaphene signal that can be attributed to the mill production processes.

### 3.3.5
### Recent Trends in Deposition

Sediment core profiles in Lake Ontario show a mean decline of $2.8 \pm 0.8$-fold in toxaphene fluxes over a 20–25 year period which corresponds to $3.2 \pm 0.6\%$ per year (Table 9; [97, 98]). The average decline was similar in Lake Michigan, averaging $3.1 \pm 1.2\%$ per year. Rapid declines were observed for toxaphene deposition in GTB (15–68 times over 26 years), but the percent decline per year was actually similar to the open lake sites due to the long post-maximum time interval [96]. Annual percent declines in toxaphene in Lake Superior were generally more rapid than in Michigan, Ontario or inland lakes (Table 9), averaging $4.6 \pm 2.8\%$ per year. A mid-lake core LS5 showed a more rapid decline (15% per year from 1991 to 1997) but this may be unrepresentative because of the short time period post-maximum toxaphene input (Table 9; [94]). The sediment core flux results suggest similar rates of response to reduced toxaphene inputs in all lakes. The later maxima for toxaphene in profundal sediments, particularly in Lake Superior, may reflect low sedimentation rates and efficient recycling of carbon which has maintained a low but continuous input of toxaphene. Smaller inland lakes near western Lake Superior had earlier maxima more closely tracking the known use of toxaphene. In the case of northern Lake Michigan, terrestrial inputs from past agricultural use on both sides of the lake may have continued to input toxaphene after atmospheric inputs declined. Only a low proportion of toxaphene in the water column (1–5%); [94] is associated with particles in the Great Lakes. Thus redistribution and final burial of toxaphene in profundal sediments via sedimenting particles after entry into the water column can take many years. For example, in small lakes receiving a single toxaphene addition in the early 1960s, maximum concentrations in sediment horizons dated to the early 1970s, ten years after treatment [89, 111].

### 3.3.6
### Toxaphene Inventory in Sediments

The approximate total burdens of toxaphene in the three Great Lakes can be calculated from the average inventories in the lakes [94, 97]. Inventories are $500\,\mu g/m^2$ in Lake Ontario [97], about $6\,\mu g/m^2$ in Lake Superior and range from 75 to $430\,\mu g/m^2$ in Lake Michigan. Pearson et al. [94] estimated toxaphene burdens of 6500, 10 200, and 6000 kg in the sediments of Lakes Superior, Michigan and Ontario, respectively. Howdeshell and Hites [97] estimated a higher burden in Lake Ontario (10 000 kg) based on a large number of cores. These burdens are much more similar among the three Great Lakes than are PCBs (Lake Superior, 4900 kg; Lake Michigan, 75 000 kg; Lake Ontario, 110 000 kg [111]) which further supports the hypothesis that toxaphene inputs have been much more influenced by atmospheric deposition than local sources. While inventories showed less than two-fold spatial variations in Superior and Ontario, there were six-fold higher inventories in northern than southern Lake Michigan, which points to the importance of local sources in the northern sector of the lake.

### 3.3.7
### Degradation in Sediment Inferred from Homolog Profiles

Howdeshell and Hites [97] noted an increase in the ratio of the hexachloro to heptachloro homolog groups in Lake Ontario sediments, that they attributed to in situ degradation. Pearson et al. [94] also found that homolog composition in Lake Ontario and Michigan cores were generally enhanced in the hexa- and heptachlorinated homologs (hexa + hepta homologs comprised 45–70% of the total toxaphene) as compared to technical toxaphene sources (Strobane and Hercules). However, the homolog compositions in GTB sediment did not appear to change significantly throughout the core [96]. The lack of significant changes in toxaphene homolog composition suggests that limited degradation or remobilization has occurred within these cores. The overall degradation of toxaphene is clearly very low in Great Lakes sediments, with half-lives estimated to range from 40 to > 100 years [94]. Howdeshell and Hites [97] estimated a doubling time of 40 yrs for the ratio of hexa- to heptachloro homologs, which gives an indication of the extent of dechlorination because hexachloro homologs are low in technical toxaphene. Rose et al. [112] found doubling times of 26 and 84 yr respectively for hexa- and heptachloro-homolog groups in Lochnagar, a small Scottish Lake.

### 3.3.8
### Degradation in Sediment Inferred from Congener Profiles

While toxaphene degradation can be difficult to discern at the homolog level, at the congener level distinctive increases in the hexachlorobornane B6-923 and heptachlorobornane B7-1001, as well as their dechlorination products, are discernable in Great Lakes sediments. Both these congeners lack dichloro-(geminal) substituents on any one carbon atom [113, 114]. Structures with gem-dichloros are more easily degraded. Major precursors include the C2-geminal substituted congeners, B7-515 and B8-806/B8-809, which are among the major components of the technical toxaphene mixture. Based on studies of sediments of lakes treated with toxaphene [89, 111, 114] as well as on laboratory studies [113, 115, 116], there is strong evidence for slow in situ dechlorination of chlorobornanes in anoxic sediments. Dechlorination occurs primarily at geminal chlorine atoms [6, 115, 116]. Dechlorination at the C2 carbon proceeds more rapidly than at C8 or C10 positions for the major

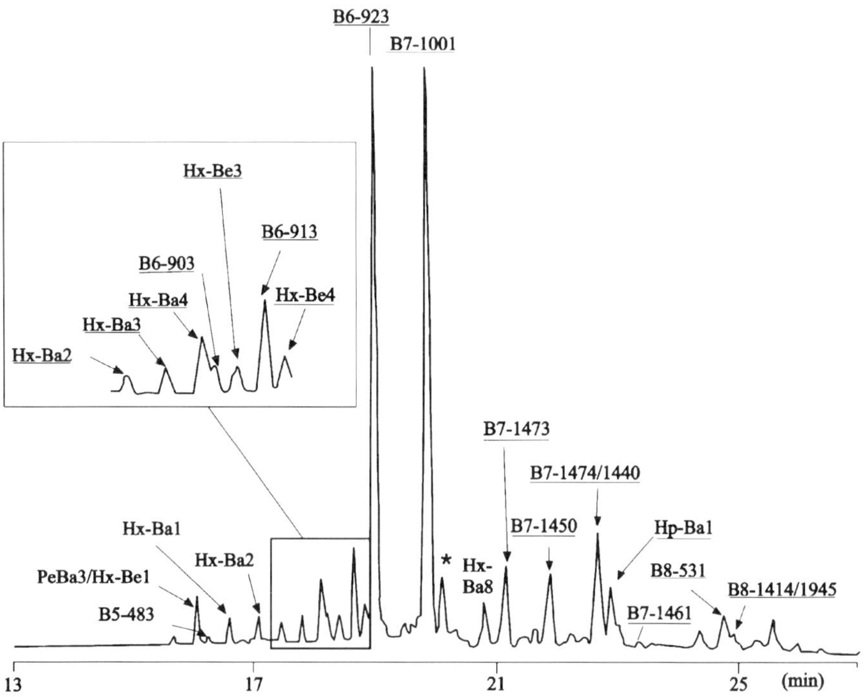


**Fig. 10** Gas chromatography-electron-capture negative-ion high-resolution mass spectrometry multiple-ion chromatogram (sum of penta-octachlorobornanes) of a sediment sample from the Mississauga Basin of Lake Ontario (2–3 cm depth) showing predominance of B6-923 and B7-1001 [84]. Other congener identities are defined in ref [84]. [Reproduced with permission from SETAC]

chlorobornanes. Ruppe et al. [115] found that the bridge carbons (C8, C9) of the B9-1679 (P50), a major chlorobornane in environmental samples, were dechlorinated by an anaerobic bacterium faster than the geminal chlorines at the C10 position.

Ruppe et al. [84] demonstrated that the toxaphene profile in the sediment sample from western Lake Ontario was dominated by the known major metabolites B6-923 and B7-1001 (Fig. 10). These authors were also able to identify transformation products of B7-1001. They concluded that only small amounts of B7-1001 and B6-923 had been metabolized based on the ratios to their transformation products identified in vitro using an anaerobic bacterial culture.

The presence of high proportions of B7-1001 and B6-923 relative to total chlorobornane congeners ($\Sigma$CHB) in sediment has been shown to be indicative of high concentrations of past direct emissions from agricultural use [83], use as a piscicide [111] or from manufacturing facilities [90, 117]. However, lakes receiving only atmospheric inputs also show increasing proportions of these residues with sediment depth [112]. The ratios of B6-923 and B7-1001 to $\Sigma$CHB in Lakes Superior, Nipigon and Ontario surface sediments ranged from 0.002–0.23 and 0.002–0.38, respectively, and were similar to their ratios of approximately 0.003 and 0.30, respectively, in technical toxaphene (Table 10). Higher ratios were generally seen subsurface due to gradual conversion of prominent geminal C2-substituted congeners to B6-923 and B7-1001. However, this process varies widely among locations. In Lake Superior, B6-923 and B7-1001 constituted 100% of $\Sigma$CHB in sediment horizons of a nearshore site (Stn 403) dating to 50 years post-deposition, but at

**Table 10** Proportion of persistent chlorobornane degradation products B6-923 and B7-1001 in total chlorobornanes in Great Lakes sediments

| Lake | Year | Station | Lat/long | Depth (cm) | B6-923 | B7-1001 | Refs. |
|---|---|---|---|---|---|---|---|
| Superior | 1998 | 403 | 48°48′N, 86°59′W | 0-0.5 | 0.009 | 0.06 | [95, 98] |
|  |  |  |  | Maximum (1986) | 0.002 | 0.11 | [95, 98] |
|  | 1998 | 80 | 47°35′N, 86°56′W | 0 0.5 | 0.002 | 0.002 | [95, 98] |
|  |  |  |  | Maximum (1989) | 0.001 | 0.001 | [95, 98] |
|  | 1996 | 169 |  | 0–1 cm | 0.007 | 0.069 | [78] |
|  | 1996 | 157 |  | 0–1 cm | 0.003 | 0.079 | [78] |
|  | 1997 | 152 |  | 0–1 cm | 0.008 | 0.034 | [78] |
|  | 1997 | 157 |  | 0–1 cm | 0.002 | 0.068 | [78] |
|  | 1997 | 169 |  | 0–1 cm | 0.002 | 0.032 | [78] |
|  | 1997 | 201 |  | 0–1 cm | 0.009 | 0.061 | [78] |
| Ontario | 1998 | 1007 | 43°26′N, 79°24′W | 0–1 cm | 0.09 | 0.3 | [84] |
|  | 1998 | 1034 | 43°34.6′, 79°12′ | 0–1 cm | 0.23 | 0.38 | [99] |
|  |  |  |  | maxima | 0.18 | 0.35 | [99] |

St 80, B6-923 and B7-1001 were near detection limits and constituted < 1% of $\Sigma$CHB throughout the core [95]. The nearshore site, Jackfish Bay, receives pulp and paper mill effluent and municipal effluents as well as drainage from streams draining the northwestern Ontario boreal forest, and this carbon input may enhance anaerobiosis compared to low sedimentation at open lake sites. Doubling times for relative abundance of B6-923 and B7-1001 in a core from Jackfish Bay (Lake Superior) in Canada were $8 \pm 2$ and $27 \pm 3$ yr, respectively [95]. Marvin [99] found that B6-923 and B7-1001 constituted 100% of $\Sigma$CHB in sediments dated to the 1940s and 50s from the central Niagara basin of Lake Ontario, suggesting that the earliest deposited residues had been significantly transformed. Rose et al. [112] found doubling times of 17 and 12 yrs respectively in an isolated Scottish Lake. The results for toxaphene-treated lakes, those with mixed sources such as Ontario and cores from lakes such as Superior, which receive toxaphene by atmospheric deposition, all point to the gradual conversion of chlorobornanes to B6-923 and B7-1001 and then slowly to even less chlorinated residues such as pentachlorobornanes [84], although with a wide variation in rates. The ultimate fate of the toxaphene inventory in Great Lakes sediments is thus slow dechlorination over several hundred years to "virtual elimination" or levels at detection limits.

# 4
# Levels and Trends in Biota

Toxaphene was first reported in Great Lakes fish in 1974 and this was confirmed in the late 1970s [4]. The early work on temporal and spatial trends of toxaphene in Great Lakes fish has been summarized by Rice and Evans [4]. Until the early 1980s this work was conducted using packed column GC-ECD and therefore may have been subject to interferences from PCBs and chlorinated pesticides. In the early 1980s Ribick et al. [118] developed an extraction and clean-up method for toxaphene in fish tissues which has been the basis of all toxaphene determinations in Great Lakes biota since then. Swackhamer et al. [119] developed an GC-ECNIMS method for quantifying toxaphene which has been used with modifications for almost all subsequent quantification of total toxaphene. Here we focus mainly on more recent studies of toxaphene in Great Lakes biota.

## 4.1
### Toxaphene in Great Lakes Vertebrates other than Fish

Surveys of waterfowl eggs collected in 1977–78 showed that toxaphene was detectable in several species, especially in mergansers (*Mergus serrator*), although levels were lower than PCBs by a factor of ten or more [120]. Ohlendorf et al. [121] included herons (*Ardeo herodia*) from Lake St. Clair in a survey of organochlorine residues in dead herons in the eastern and southern USA. Toxaphene was detected in only a limited number of samples. Surprisingly, no further work on toxaphene in Great Lakes birds appears to have been conducted and there are no reports of toxaphene measurements in Great Lake mammals. The difficulty of analyzing toxaphene by packed column GC combined with substantial metabolism of toxaphene by birds and mammals undoubtedly contributed to the low detection frequency of toxaphene in these early studies.

Information on the concentrations of toxaphene in avian species, generally, and in fish-eating birds in particular, is limited. The metabolism of toxaphene has been studied in chickens [122] but its metabolism by other avian species has not been investigated [2, 3]. Chlorobornane congener patterns have been characterized in penguins (*Phygoscelis adeliae*) [6, 123], skuas (*Catharacta* sp) [124] and albatross (*Diomedea* sp) [125], but not in species frequenting the Great Lakes. Earlier work reported toxaphene in guillemots and osprey from the Baltic Sea and the Barents Sea [126, 127]. The major chlorobornanes in birds are the recalcitrant B8-1413, B8-1412, B9-1679 and B9-1025 [123–125].

Struger et al. [128] confirmed "trace" amounts of toxaphene in snapping turtles from Hamilton Harbour in the Great Lakes using GC-mass spectrometry but did not present concentrations.

## 4.2
## Spatial Trends in Great Lakes Fishes

Fish collection surveys in the late 1970s and early 1980s showed that toxaphene was present in all Great Lakes fish, with highest concentrations in lake trout (*Salvelinus namaycush*) and coho salmon (*Oncorhynchus kisutch*) [4, 129–131]. Results for toxaphene and major congeners in fish from the Great Lakes measured during the 1990s are presented in Table 11. Lake trout along with sculpins (*Myoxocephalus thompsoni* or *Cottus cognatus*) and rainbow smelt (*Osmerus mordax*), have been the most commonly analyzed species and were therefore selected for presentation in Table 11.

## 4.2.1
## Toxaphene in Lake Trout

Glassmeyer et al. [49, 132] found that toxaphene concentrations in lake trout and rainbow smelt were ranked Lake Superior > Michigan > Huron > Ontario in samples collected in 1992. Whittle et al. [81] and Swackhamer [133] also found highest toxaphene concentrations in Lake Superior lake trout and lowest concentrations in lake trout from Lake Erie collected in 1998 and 1999–2000, respectively. Whittle et al. [81] used GC-ECD for quantifying toxaphene and selected chlorobornanes while Glassmeyer et al. [49, 132] and Swackhamer [133] used GC-ECNIMS. Although there is an eight-year difference in collection time, agreement between the GC-ECD and GC-ECNIMS results from the three studies was reasonably good (coefficients of variation range from 33–77%, which is similar to those in the Great Lakes Fish Monitoring Program [133], and there is no particular trend with time. There appears to be a distinct difference in toxaphene concentrations between the Apostle Islands and Keweenaw Point in Lake Superior, and between the eastern and western basins in Lake Erie. The latter may be due to differences in the historic loadings to the two basins in Lake Erie, while the former has been observed for many other analytes and is attributed to differences in foodwebs of localized lake trout population.

Differences among and within labs are common for toxaphene analyses. Elevated PCBs in Lake Ontario fish could conceivably interfere with ECD analysis of some chlorobornane congeners. Quantification by GC-ECNIMS, which has been used by all other studies of toxaphene in fish during the 1990s, is not without its uncertainties as well. Glassmeyer et al. [49] concluded that their original data [132] were three-fold too high due to a difference in internal standards used in the quantification procedure. Only their most recent results, which are in agreement with several other studies, are reported here.

Toxaphene concentrations are also higher in Lake Superior and Lake Huron lake trout than in this species from inland lakes in northwestern and central Ontario [81, 134]. Lake trout from Lake Nipigon had the highest mean (wet weight) concentrations, followed by Lake Simcoe (north of Toronto) and Sandy Beach Lake (west of Thunder Bay) (Table 11).

Muir et al. [74] found that toxaphene concentrations in lake trout from Siskiwit Lake on Isle Royale were about 15-fold lower than in Lake Superior lake trout when compared on a wet weight or lipid weight basis (Table 11). The Lake Superior trout were larger (mean wt 2300 g) than those from Siskiwit (mean = 2000 g) and the weight/age ratio was about 12% higher (grams/yr), indicating a higher growth rate, although these differences do not explain the much higher levels in the Lake Superior fish.

Toxaphene homolog composition in lake trout varied among lakes [49, 133]. In the 1992 data, there was a similar proportion of octachlorobornanes (45–50%) in Superior, Michigan, and Huron, a proportion that is similar to the technical toxaphene standard. Lake trout from Lake Ontario had a much lower proportion of octa- ($\sim$ 30%), and higher hexa- (18%) and hepta- (35%), suggesting a more "weathered" source of chlorobornanes. In the 2000 data, the octachlorobornanes in the upper three lakes was 52–60%, and in the lower two lakes was 42–47%. The contribution of the hexa- and heptachlorobornanes was much greater in the lower lakes, with 47–48% in the lower lakes to 15–25% in the upper lakes. The reason for these differences is not clear although, it is consistent with the high proportions of hexa- and heptachlorobornanes in Lake Ontario sediments (Table 10).

**Table 11** Comparison of reported concentrations of total toxaphene and selected congeners in fish from the Great Lakes and nearby lakes during the 1990s (ng g$^{-1}$ wet wt)

| Lake | Location | Species | Year | Tissue[1] | % Lipid | Total toxaphene | B8-1413 | B9-1679 | B9-1025 | GC method | Refs. |
|---|---|---|---|---|---|---|---|---|---|---|---|
| Superior | Apostle Is | Lake trout | 1992 | W | 23 | 1300 | — | — | — | ECNIMS | [49] |
| Superior | Apostle Is | Lake trout | 1993 | W | 22 | 2373 | — | — | — | ECNIMS | [5] |
| Superior | Apostle Is | Lake trout | 1998 | W | 12 | 889 | 6.3 | 50 | 56.4 | ECNIMS | [74] |
| Superior | Eastern basin | Lake trout | 1998 | W | 17 | 1926 | 17 | 110 | 65 | ECD | [81] |
| Superior | Apostle Is | Lake trout | 1992 | W | 19 | 1300 | — | — | — | ECNIMS | [49] |
| Superior | KP | Lake trout | 1994 | M | 9.1 | 392 | — | — | — | ECNIMS | [139] |
| Superior | KP | Lake trout | 1999 | W | 11 | 673 | — | — | — | ECNIMS | [133] |
| Superior | Apostle Is | Lake trout | 2000 | W | 12 | 2493 | — | — | — | ECNIMS | [133] |
| Superior | Apostle Is | Sculpin | 1993 | W | 5 | 225 | — | — | — | ECNIMS | [5] |
| Superior | Apostle Is | Sculpin | 1998 | W | 4.7 | 116 | 4.5 | 5.1 | 3.2 | ECNIMS | [74] |
| Superior | Eastern basin | Sculpin | 1998 | W | — | 546 | — | — | — | ECD | [81] |
| Superior | KP[3] | Sculpin | 1994 | W | 2.8 | 325 | — | — | — | ECNIMS | [139] |
| Superior | Apostle Is | Smelt | 1998 | W | 4.9 | 179 | 5.0 | 7.1 | 2.8 | ECNIMS | [74] |
| Superior | Eastern basin | Smelt | 1998 | W | — | 291 | — | — | — | ECD | [81] |
| Superior | Apostle Is | Smelt | 1992 | W | 5.6 | 90 | — | — | — | ECNIMS | [49] |
| Superior | KP[3] | Smelt | 1994 | W | 4.3 | 205 | — | — | — | ECNIMS | [139] |
| Superior | | Lamprey | 1997 | M | 5.2 | 1220 | — | — | — | ECD | [136] |
| Ontario | Oswego | Lake trout | 1992 | W | 20 | 140 | — | — | — | ECNIMS | [49] |
| Ontario | West Basin | Lake trout | 1998 | W | — | 639 | — | — | — | ECD | [81] |
| Ontario | N. Hamlin | Lake trout | 1999 | W | 17 | 169 | — | — | — | ECNIMS | [133] |
| Ontario | Oswego | Lake trout | 2000 | W | 20 | 521 | — | — | — | ECNIMS | [133] |
| Ontario | West Basin | Sculpin | 1998 | W | — | 245 | — | — | — | ECD | [81] |
| Ontario | West Basin | Smelt | 1994 | W | 5.8 | 30 | — | — | — | ECNIMS | [49] |

**Table 11** continued

| Lake | Location | Species | Year | Tissue[1] | % Lipid | Total toxaphene | B8-1413 | B9-1679 | B9-1025 | GC method | Refs. |
|---|---|---|---|---|---|---|---|---|---|---|---|
| Ontario | West Basin | Smelt | 1998 | W | — | 66 | — | — | — | ECD | [81] |
| Ontario | West Basin | Lamprey | 1997 | M | 5.0 | 150 | — | — | — | ECD | [136] |
| Michigan | GTB[4] | Burbot | 1997–98 | W | 7.8 | 1200 | — | — | — | ECNIMS | [138] |
| Michigan | GTB[4] | Lake trout | 1997–98 | W | 14 | 443 | — | — | — | ECNIMS | [138] |
| Michigan | Charlevoix | Lake trout | 1993 | W | 20 | 810 | — | — | — | ECNIMS | [49] |
| Michigan | Sturgeon Bay | Lake trout | 1992 | W | 18 | 350 | — | — | — | ECNIMS | [49] |
| Michigan | Saugatuck | Lake trout | 1992 | W | 19 | 490 | — | — | — | ECNIMS | [49] |
| Michigan | Sturgeon Bay | Lake trout | 1999 | W | 14 | 901 | — | — | — | ECNIMS | [133] |
| Michigan | Saugatuck | Lake trout | 2000 | W | 15 | 1123 | — | — | — | ECNIMS | [133] |
| Michigan | 6 tributaries | Coho salmon | 2000 | M | 4 | 192 | — | — | — | ECNIMS | [133] |
| Michigan | GTB[4] | Salmon | 1997–98 | W | 4.5 | 264 | — | — | — | ECNIMS | [138] |
| Michigan | GTB[2] | Sculpin | 1997–98 | W | 3.5 | 228 | — | — | — | ECNIMS | [138] |
| Michigan | Saugatuck | Smelt | 1994 | W | 5.7 | 50 | — | — | — | ECNIMS | [49] |
| Michigan | GTB[4] | Whitefish | 1997–98 | W | 6.5 | 164 | — | — | — | ECNIMS | [138] |
| Huron | Rockport | Lake trout | 1992 | W | 17 | 340 | — | — | — | ECNIMS | [49] |
| Huron | | Lake trout | 1998 | W | — | 365 | — | — | — | ECD | [81] |
| Huron | Port Austin | Lake trout | 1999 | W | 16 | 467 | — | — | — | ECNIMS | [133] |
| Huron | Rockport | Lake trout | 2000 | W | 15 | 676 | — | — | — | ECNIMS | [133] |
| Huron | Swan River | Chinook salmon | 2000 | M | 4 | 395 | — | — | — | ECNIMS | [133] |
| Huron | | Sculpin | 1998 | W | — | 312 | — | — | — | ECD | [81] |
| Huron | Rockport | Smelt | 1994 | W | 6.0 | 200 | — | — | — | ECNIMS | [49] |
| Huron | | Smelt | 1998 | W | — | 119 | — | — | — | ECD | [49] |
| Huron | | Lamprey | 1997 | M | 4.6 | 300 | — | — | — | ECD | [136] |

**Table 11** continued

| Lake | Location | Species | Year | Tissue[1] | % Lipid | Total toxaphene | B8-1413 | B9-1679 | B9-1025 | GC method | Refs. |
|---|---|---|---|---|---|---|---|---|---|---|---|
| Erie | | Lake trout | 1998 | W | — | 81 | — | — | — | ECD | [81] |
| Erie | | Smelt | 1998 | W | — | 16 | — | — | — | ECD | [81] |
| Erie | | Walleye | 1992–94 | W | 9.0 | 130 | — | — | — | ECNIMS | [132] |
| Erie | Dunkirk | Walleye | 1999 | W | 8 | 31 | — | — | — | ECNIMS | [133] |
| Erie | Middle Bass Isl | Walleye | 2000 | W | 7 | 232 | — | — | — | ECNIMS | [133] |
| Erie | Trout Run | Coho salmon | 2000 | M | 6 | 107 | — | — | — | ECNIMS | [133] |
| Erie | | Lamprey | 1997 | M | 6.2 | 220 | — | — | — | ECD | [136] |
| Nipigon | | Lake trout | 1998 | W | — | 1009 | — | — | — | ECD | [81] |
| Siskiwit | | Lake trout | 1998 | W | 4.6 | 60 | 0.5 | 3.3 | 2.7 | ECNIMS | [74] |
| Siskiwit | | Lake whitefish | 1998 | W | 6.0 | 232 | 0.8 | 3.9 | 0.5 | ECNIMS | [74] |
| Siskiwit | | Yellow perch | 1998 | W | 9.7 | 770 | 4.4 | 48 | 33 | ECNIMS | [74] |
| Sandy Beach | | Lake trout | 2000 | W | 15 | 149 | 14.6 | 5.5 | 6.9 | ECNIMS | [134] |
| Eva | | Lake trout | 2000 | W | 16 | 73 | 6.4 | 2.0 | 2.3 | ECNIMS | [134] |
| Simcoe | | Lake trout | 1998 | W | — | 134 | — | — | — | ECD | [81] |
| Opeongo | | Lake trout | 1998 | W | — | 89 | — | — | — | ECD | [81] |

[1] W = whole fish, M = muscle
[2] Glassmeyer et al. [132] also reported results for lake trout from 1992–1994 but in a subsequent paper [49] the authors indicated that their earlier results, which were 3-fold higher, were based on a more stable set of response factors and therefore judged the most accurate
[3] KP = Keweenaw Peninsula
[4] GTB = Grand Traverse Bay, Lake Michigan

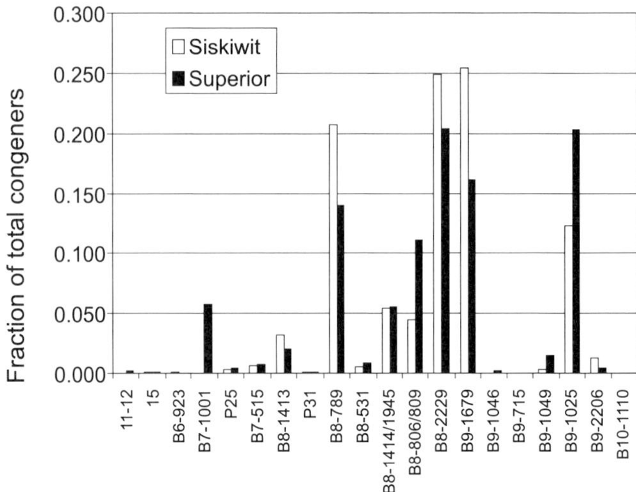

**Fig. 11** Proportions of congeners of the sum of 20 chlorobornane/camphene congeners in lake trout from Lake Superior and Siskiwit Lake (Isle Royale) in samples collected during June 1998 from each lake. B7-1001 was not determined in the lake trout from Siskiwit Lake [74]

Congener specific data are available only for Lake Superior, Siskiwit Lake and inland lakes [74, 81, 134, 135]. Congeners B8-1413, B8-2229, B9-1679 and B9-1025 generally increased in prominence and in concentration (ng/g lipid wt) up the food chain. B8-789, which differs from B9-1025 by one chlorine at position 9c, was also prominent in lake trout and may be a dechlorination product of the latter congener. The prominent hexachlorobornane, B6-923, was not detected in lake trout and levels of B7-1001 were near detection limits (Fig. 11). Whittle et al. [81] and Kiriluk et al. [135] reported values for B8-1413, B9-1679 and B9-1025 in lake trout from Lake Superior that were in general agreement with other measurements (Table 11). The same toxaphene congeners predominated in Lake Siskiwit although lake trout samples had 25% higher proportions of congeners B8-789 and B9-1679 and a 30% lower proportion of B9-1025. This may be due to dietary differences, as discussed in Sect. 5.

### 4.2.2
### Other Fish Species

Spatial trends of toxaphene have also been studied using smelt [49, 131] and sea lamprey (*Petromyzon marinus*) [136]. Smelt from western Lake Superior had about two-fold higher toxaphene than those in southern Lake Michigan and three-fold higher than in Lake Ontario (Table 11), paralleling trends in lake trout. Sea lamprey are parasitic organisms that feed on blood and tissues of large fish such as lake trout. Toxaphene concentrations were four-fold higher in lamprey from Lake Superior compared to those collected in Lake Huron. Overall, the lamprey showed the same spatial trend of toxaphene, as seen in smelt and lake trout (Table 11). Fisk and Johnston [137] found 350-fold higher toxaphene levels in eggs of walleye (*Stizostedion vitreum*) from Lake Superior compared to eggs from walleye in Lake Manitoba, a large shallow lake in central Manitoba (Canada).

Stapleton et al. [138] found that burbot (*Lota lota*) from Grand Traverse Bay had higher levels of toxaphene than lake trout, followed by salmon > sculpin > lake whitefish (*Prosopium cylindraceum*). Mean concentrations of toxaphene in lake trout from GTB were within the range (means of 350–810 ng/g wet wt) observed by Glassmeyer et al. [49] for three sampling locations in Lake Michigan.

Kucklick et al. [139] noted that bloater (*Coregonus hoyi*) in Lake Superior were among the most highly contaminated forage fish species with respect to toxaphene ($1100 \pm 260$ ng/g ww), higher than sculpin, smelt or herring. Stapleton et al. [138] found that bloaters in GTB displayed very low toxaphene concentrations ($92 \pm 32$ ng/g ww) compared to those in Lake Superior.

In Lake Siskiwit both piscivores (northern pike) and planktivores (lake whitefish (*Coregonus clupeaformis*), pigmy whitefish (*Prosopium coulter*)) had very similar total toxaphene concentrations to lake trout, while yellow perch (*Percopsis omiscomaycus*) had much higher concentrations of toxaphene than lake trout (Table 11) [74]. Indeed one sample of yellow perch had 1050 ng/g ww total toxaphene.

Toxaphene in skin-on fillets of coho salmon from Lake Michigan were approximately twice those of Lake Erie [133]. Concentrations in Chinook salmon from Lake Huron were greater than those in coho from Lake Michigan.

## 4.3
### Temporal Trends of Toxaphene in Fish

Glassmeyer et al. [49, 132] found that toxaphene concentrations in lake trout and rainbow smelt declined significantly ($p < 0.05$) in all lakes except Lake Superior. Results from their study, which compared samples from western Lake Superior between 1977 and 1992, are combined with data from Swackhamer [133] and Muir et al. [74] in Fig. 12 to give a 24-year perspective on temporal trends of total toxaphene. No decline in toxaphene concentrations is evident over this time period. This agrees with water concentrations, which were estimated to be 1200 ng/L in 1977, based on concentrations in lake trout [49], and measured at 1120 ng/L in 1996 and 720 ng/L in 2002 (Table 9), representing a decline of 1–2% per year.

The results show that toxaphene concentrations are similar in lake trout from Lakes Michigan and Superior and have not declined significantly since the 1980s. For the period 1982 to 1992, Glassmeyer et al. [49] found declines of 6 and 8% per year in Lakes Huron and Ontario, respectively, corresponding to

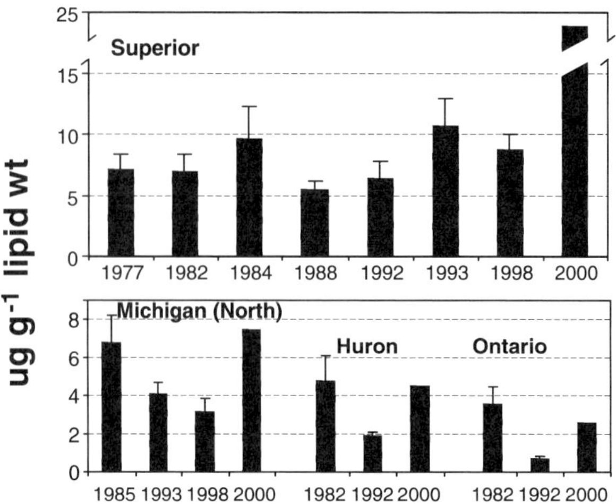


**Fig. 12** Trends in concentrations of toxaphene (means ±90% confidence intervals) in lake trout from Lakes Superior, Michigan (Charlevoix/Grand Traverse Bay area), Huron and Ontario. Results from Glassmeyer et al. [49] (1977–1993) are combined with more recent results from Stapleton et al. [138], Swackhamer et al. [133] and Muir et al. [74]

half-lives of 7.7 and 4.3 years. Recent data from Swackhamer [133] are generally higher than values reported by Glassmeyer et al. [49]and Muir et al. [74], which suggests that there is uncertainty about the rate of decline caused by analytical differences and changes in food web structure over time.

Kiriluk et al. [135] confirmed the lack of decline of toxaphene in Lake Superior lake trout over the period 1980–1990 using measurements of specific chlorobornane congeners from archived samples. The homolog and congener composition of toxaphene in Lake Superior lake trout has changed over time [81, 82, 135]. Recalcitrant congeners B9-1679 and B9-1025 have increased in proportion while B8-1414/B8-1945 and B8-2229 declined relative to $\Sigma$CHB. Whittle et al. [81] have noted that, based on changes in carbon isotope ratio ($\delta^{13}$C) signatures in the Lake Superior food web, lake trout have shifted from a diet dominated by sculpin and smelt to one dominated by lake herring (*Coregonus artedii*). This may be a factor in prolonging the elevated toxaphene in Lake Superior relative to the other lakes, because herring have higher toxaphene than sculpin. Interestingly, percent lipid in the adult lake trout from Lake Superior that are selected for temporal trend studies has declined as well since the early 1990s.

# 5
# Bioaccumulation and Food Web Transfer

## 5.1
## Effects of Lipid, Size and Season on Toxaphene Levels in Biota

Toxaphene concentrations are significantly influenced by fish size/age, % lipid and trophic position [74, 138–140]. Muir et al. [74] found toxaphene and major congeners in lake trout were significantly correlated ($p < 0.05$) with % lipid, age, length and weight, but not with nitrogen isotope ratios ($\delta^{15}N$). This study utilized results from 75 lake trout averaging $7.0 \pm 1.6$ years of age. Lipid was significantly correlated with concentrations of total toxaphene and all congeners except B9-1679. All octa- and nonachlorobornane concentrations were significantly correlated with each other, reflecting similar bioaccumulation patterns in lake trout [74]. Stapleton et al. [140] found that toxaphene concentrations increased with size class in sculpin and were highest in the largest class ($> 180$ mm). Toxaphene in sculpin was also significantly correlated with lipid content ($r^2 = 0.44$), but the significance of the relationship of PCBs with lipid in the same sculpin was much greater ($r^2 = 0.60$).

Swackhamer et al. [5, 141] measured toxaphene in phytoplankton (10–100 µm size fraction), *Mysis*, *Bythotrephes*, and *Diporeia* samples collected as part of the US EPA Lake Michigan Mass Balance Study. Samples were collected from 11 master sites throughout the lake on seven cruises covering all four seasons, from June 1994 to October 1995. Concentrations varied by season and by year, with year having the strongest effect on a dry-weight basis. Lipid normalization reduced the between-year variability, but increased the seasonal variability. Seasonal changes in lipid occurred more quickly than toxaphene could reach equilibrium within the organisms, thus increasing variability in concentration across time. Mean values by year, expressed both on a dry-weight and lipid-normalized basis, are shown in Table 12. Concen-

**Table 12** Average concentrations $\pm 95\%$ confidence limits (dry weight and lipid normalized) of toxaphene in the lower Lake Michigan food web, 1994 and 1995 [5, 141]. Data are from 11 stations from each of 3 cruises in 1994 and 4 cruises in 1995.

| Organism/ trophic level | 1994 ng/g, ww | ng/g, OC/lipid normalized | 1995 ng/g, ww | ng/g, OC/lipid normalized |
|---|---|---|---|---|
| Phytoplankton | $66 \pm 13$ | $307 \pm 71$ | $40 \pm 6$ | $284 \pm 66$ |
| Mysis | $110 \pm 12$ | $928 \pm 188$ | $75 \pm 6$ | $762 \pm 200$ |
| Bythotrephes | $173 \pm 24$ | $1460 \pm 189$ | $155 \pm 20$ | $1787 \pm 372$ |
| Diporeia | $432 \pm 39$ | $4410 \pm 1205$ | $397 \pm 35$ | $3249 \pm 911$ |

trations, regardless of lipid normalization, were in the order phytoplankton <
*Mysis* < *Bythotrephes* < *Diporeia*.

Stapleton et al. [140] found that toxaphene varied seasonally in zooplankton (> 153 µm) in Grand Traverse Bay in northeastern Lake Michigan. Toxaphene concentrations in bulk zooplankton and mysid shrimp (*Mysis relicta*) were higher in April than in August/September. This variation corresponded to variations in lipid content of the zooplankton which ranged from 0.75% in June to 2.5% in August. Although lipid content of benthic amphipods (*Diporeia hoyi*) also increased from May to August, toxaphene did

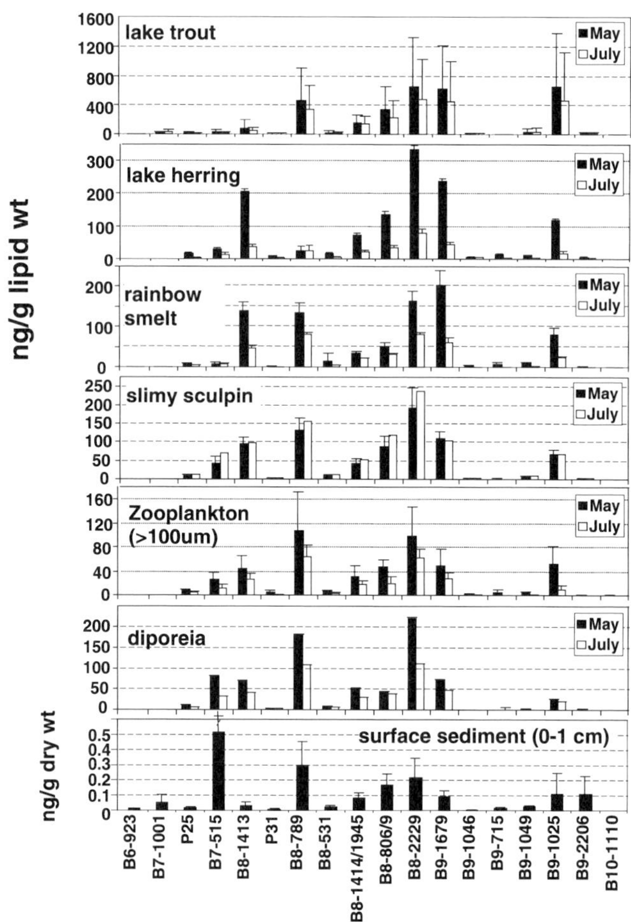

**Fig. 13** Chlorobornane congeners in the benthic and pelagic biota from Lake Superior. Bars represent arithmetic means ± SD, ng/g lipid wt, except for sediment (ng/g dry wt). Sediment concentrations in the lower panel are based on the average of top two slices from cores taken in Jackfish Bay (near shore) and at Site 80 (mid lake) in May 1998. Adapted from Muir et al. [74]

not vary systematically. Toxaphene levels in alewife and bloater also varied by up to 50% from April to September but over a two-year period did not show systematic variation with lipid content or sampling date.

Muir et al. [74] found that toxaphene and major congener concentrations were significantly higher in May than October in lake herring, rainbow smelt, *Mysis*, and zooplankton (Fig. 13) on both a wet weight or lipid weight basis. The difference was particularly striking for B8-1413, B8-2229 and B9-1679 in lake herring, which were all about 5–6 times higher in the spring collection on a lipid adjusted basis. Seasonal differences were also found in the benthic food web with higher toxaphene and congener concentrations in benthic dwelling *Diporeia* and slimy sculpin collected in spring sampling. The congener pattern in *Diporeia* and sculpin resembled that in sediments (Fig. 13), with high proportions of B8-789 and other octachlorobornanes than observed in water or zooplankton. However, the seasonal differences for *Diporeia* and slimy sculpin were not as great as they were for pelagic food web species (Fig. 13), and in the case of sculpin, they were not significantly different between May and October. This may reflect greater seasonal changes in the composition of the zooplankton community ($> 102\,\mu$m) in the pelagic zone compared with the more stable benthic community.

## 5.2
## Bioaccumulation Factors

Bioaccumulation factors for toxaphene in lake trout are remarkably constant, ranging (log BAF) from 6.6 to 7.1 in the Great Lakes and Lake Siskiwit (Table 13). These values are calculated mainly from results for water from Swackhamer et al. [5, 65]. Glassmeyer et al. [49] concluded that their log BAFs for toxaphene in lake trout from four of the Great Lakes were within experimental error, and assumed a single value of 6.77 in order to back-calculate water concentrations of toxaphene. Their estimated water concentrations, using this BAF method, were identical to predicted values in Lake Michigan, and within 20% of the value in Lake Superior estimated by Swackhamer et al. [65] using a dynamic mass balance model. These BAF values for toxaphene are a factor of 2–4 lower than for PCBs in lake trout, but similar to results for OC pesticides with $K_{OW}$ values in the same range, such as the DDT group and chlordane isomers [141]. This reflects the greater recalcitrance of PCBs, which becomes more apparent for food webs that include mammals and birds, where toxaphene is generally not as prominent a contaminant due to metabolism by top predators [2].

BAFs for the chlorobornane congeners B8-1413, B9-1679 and B9-1025 were higher than total toxaphene, and among the highest of all chlorobornanes detected, reflecting the recalcitrance of these compounds. Muir et al. [74] found a significant correlation of BAFs with log $K_{OW}$ for 19 chlorobornanes

**Table 13** BAFs and BMFs for toxaphene and other major organochlorines in Great Lakes food webs[1]

| Lake | Toxaphene | B8-1413 | B9-1679 | B9-1025 | $\Sigma$PCB | Refs. |
|---|---|---|---|---|---|---|
| Log bioaccumulation Factor[1] | | | | | | |
| Huron | 6.6 | — | — | — | — | [49] |
| Michigan | 6.9 | — | — | — | — | [49] |
| Michigan | 6.9 | — | — | — | 7.3 | [138] |
| Ontario | 6.6 | — | — | — | — | [49] |
| Superior | 6.7 | — | — | — | — | [49] |
| Superior | 7.1 | 8.7 | 8.0 | 9.7 | 7.8 | [74] |
| Superior | 7.0 | — | — | — | — | [5] |
| Siskiwit | 6.8 | — | — | — | — | [74] |
| Biomagnification factor (BMF)[2] | | | | | | |
| Huron | 1.5 | — | — | — | — | [49] |
| Michigan | 4.7 | — | — | — | — | [49] |
| Ontario | 6.9 | — | — | — | — | [49] |
| Siskiwit | 0.6 | 0.15 | 0.4 | 0.68 | 0.5 | [74] |
| Superior | 2.7 | 0.49 | 3.7 | 8.1 | 5.5 | [74] |
| Superior | 2.4 | — | — | — | — | [5] |
| Superior | 4.0 | — | — | — | — | [49] |
| Superior | 0.94 | — | — | — | 1.2 | [139] |
| Michigan | 5.2 | — | — | — | — | [138] |
| Sediment-biota accumulation factor[3] | | | | | | |
| Michigan | 5.2 | — | — | — | — | [5] |
| Michigan | 9.6 | — | — | — | 2.1 | [138] |
| Superior | 15 | 35 | 13 | 4.3 | 3.1 | [74] |

[1] Log bioaccumulation factor = concentration in lake trout (lipid weight) $\div$ concentration in water. Toxaphene water concentrations from Swackhamer et al. [65] except for Muir et al. [74]. PCB water concentrations from Pearson et al. [146] for Lake Michigan and Swackhamer [141] for Lake Superior

[2] Biomagnification factor (BMF) = concentration in lake trout (lipid weight) $\div$ by average concentration in smelt [49, 74, 139] or sculpin [5]. For Siskiwit Lake the BMF was calculated for lake trout – lake whitefish [74]

[3] Sediment-biota accumulation factor (BSAF) = concentration in *Diporeia* $\div$ in surface sediment (organic carbon normalized). BMF results for Siskiwit Lake calculated using lake trout. Sediment concentrations for Michigan from Grand Traverse Bay [96] and for northern Lake Michigan from Pearson et al. [94]

(Fig. 14). The BAFs for toxaphene congeners in lake trout were generally greater than log $K_{OW}$ values (BAF/$K_{OW}$ > 1). A simple predictive model of bioconcentration versus $K_{OW}$ would underestimate toxaphene concentrations in Lake Superior by 1–2 orders of magnitude. This reflects the importance of the dietary biomagnification pathway, especially for the more highly chlorinated congeners (octa- and nonachlorobornanes).

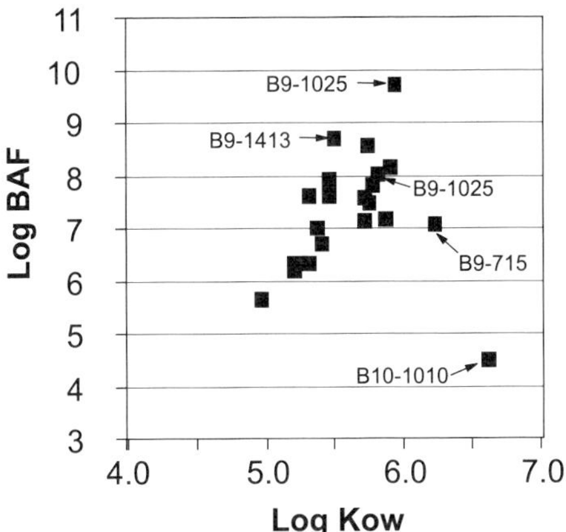


**Fig. 14** Bioaccumulation factor (BAF) in Lake Superior lake trout versus octanol-water partition coefficient ($K_{OW}$) for 19 toxaphene congeners plus B9-715 and B10-1010 [74]. Results for 6 congeners (P31, B8-1413, B9-1679, B9-1046, B9-715 and B10-1010) were estimated assuming water concentrations at the detection limit. The coefficient of variation $r^2 = 0.35$; $P < 0.01$

The BSAFs for toxaphene for *Diporeia*-sediment organic carbon range from 5.2 to 15 compared to 2.1–3.1 for $\Sigma$PCBs (Table 13). The toxaphene BSAFs were well above the theoretical range of equilibrium BSAF values of 1–4 (depending on biota lipid and sediment organic carbon) for nonmetabolizable chemicals in biota in intimate contact with sediment [142, 143]. This indicates that toxaphene congeners are highly bioavailable relative to PCBs. Water concentrations of toxaphene congeners were relatively high compared to the more recalcitrant PCBs in Lake Superior. Total toxaphene exceeds $\Sigma$PCB by about seven-fold in Lake Superior and two-fold in Lake Michigan [5, 74, 141]. Several individual congeners, such as B8-789, B8-1414 and B8-2229, had concentrations in the 10–25 pg/L range, about ten times higher than concentrations of CB153 or other hexachlorobiphenyls [74]. Thus uptake from water is proportionally more important for toxaphene congeners than for PCBs. However, the nonequilibrium BAFs and BSAFs also illustrate the importance of biomagnification pathways rather than equilibrium partitioning for toxaphene congeners, even in amphipods.

## 5.3
## Food Web Magnification

Several authors have shown that concentrations of total toxaphene (lipid-adjusted) increases significantly with trophic and specifically with $\delta^{15}$N in Great Lakes food webs [5, 74, 138, 139]. The $\delta^{15}$N values are known to increase by about 3.5‰ from prey to predator. Thus trophic levels can be assigned

**Table 14** Comparison of trophic magnification factors (TMFs) for toxaphene congeners and recalcitrant organochlorines in food webs of large oligotrophic lakes[1]

| Lake/Site | Location | Toxaphene | B8-1413 | B9-1679 | $p,p'$-DDE | CB153 | $\Sigma$PCB | Refs. |
|---|---|---|---|---|---|---|---|---|
| Great Lakes | | | | | | | | |
| Superior | Apostle Is. | 1.30 ± 0.23 | 0.88 ± 0.22 | 1.29 ± 0.17 | 1.55 ± 0.48 | 1.54 ± 0.47 | 2.34 ± 1.16 | [74] |
| Superior | Keweenaw Penn. | 1.07 ± 0.12 | — | — | 1.40 ± 0.19 | — | 1.27 ± 0.16 | [139] |
| Michigan | Grand Traverse Bay | 1.23 ± 0.11 | — | — | — | — | 1.40 | [138][2] |
| Other studies | | | | | | | | |
| Laberge | Yukon (Canada) | 1.33 ± 0.26 | 1.42 ± 0.35 | 1.36 ± 0.30 | 1.41 ± 0.33 | 1.36 ± 0.29 | 1.28 ± 0.21 | [147, 148][3] |
| Kusawa | Yukon (Canada) | 1.34 ± 0.23 | 1.43 ± 0.31 | 1.41 ± 0.32 | 1.65 ± 0.51 | 1.59 ± 0.46 | 1.42 ± 0.29 | [147, 148][3] |
| Fox | Yukon (Canada) | 1.08 ± 1.10 | 1.00 ± 1.02 | 0.95 ± 1.17 | 1.32 ± 0.21 | 1.35 ± 0.25 | 1.08 ± 0.04 | [147, 148][3] |
| Peter | Nunavut (Canada) | 1.37 ± 0.27 | 1.27 ± 0.18 | 1.26 ± 0.18 | 1.45 ± 0.37 | 1.42 ± 0.32 | — | [149][3] |
| Baikal | Russia | — | — | — | — | — | 1.30 | [150][4] |

[1] All results calculated from the relationship log [OC lipid wt conc'n] = constant + A×[$\delta^{15}$N]. TMF = $10^A$ ± 95% confidence intervals.

[2] Toxaphene TMF calculated from mean values given by Stapleton et al. [138] in supplementary data for their paper and for PCBs from Fig. 6 in their paper

[3] Results from Kidd et al. [148, 149] include unpublished data on concentrations of B8-1413 and B9-1679. In these studies toxaphene, B8-1413 and B9-1679 were determined by GC-ECD and confirmed by GC-negative ion MS

[4] The Lake Baikal food web included pelagic whitefish (*Coregonus autumnalis migratorius*) as the top predator

to individual organisms by determining their $\delta^{15}$N. Because % lipid also increases, it is necessary to use lipid-adjusted concentrations. The slope of regression of the (log) contaminant versus $\delta^{15}$N has been termed a bioaccumulation rate [144], and its antilog the trophic magnification factor (TMF) [145].

TMFs for toxaphene and selected congeners in Lakes Superior and for total toxaphene in Grand Traverse Bay of Lake Michigan are compared in Table 14. TMF values for toxaphene range from 1.07 to 1.40 in three separate studies and are within the range observed for lake trout food webs in northern lakes in Canada [147, 148] and for Lake Baikal in Russia [150].

Most octa- and nonachlorobornanes were found to have TMF values of $> 1$ in the Lake Superior pelagic food web, indicating significant food chain biomagnification [74]. TMF values for B9-1679 were similar to total toxaphene and about 80% of those for recalcitrant chemicals, $p'$-DDE and CB153 (Table 14). TMFs for B-1413 were $< 1$ in Lake Superior. The reasons for the lack of biomagnification of B8-1413 are less clear. B8-1413 is considered recalcitrant based on its 2,3,5,6-endo-exo-substitution and it biomagnifies in marine food webs [40, 123]. Lipid weight concentrations of B8-1413 were lower in lake trout than in forage fish, suggesting it is being biodegraded.

Where stable isotope data are not available, food web magnification can also be assessed using BMFs from prey to predator (Table 13). BMFs for smelt (as a representative forage fish) to lake trout in Lakes Superior, Huron, Michigan and Ontario ranged from 0.94 to 6.9. Values greater than 1 are indicative of very slow elimination of toxaphene. The low BMF value was for lake trout analyzed as fillets, while all other results (BMF range of 1.5–5.2) are based on whole fish. A BMFs $< 1$ was also found in Lake Siskiwit, where lipid-based concentrations in whitefish were higher than in lake trout. This was also the case for white sucker, another possible forage species. Muir et al. [74] found that lake trout from Lakes Superior and Siskiwit had almost identical $\delta^{13}$C and $\delta^{15}$N signatures, but lipid levels in lake trout in Lake Siskiwit were about three-times lower than in the Lake Superior lake trout. Lake trout in Siskiwit were likely feeding on zooplankton based on their $\delta^{13}$C signature. The highest toxaphene in Lake Siskiwit fishes was in yellow perch which were feeding on prey species with a benthic or littoral $\delta^{13}$C signature. Thus elevated toxaphene was associated with more benthic feeding in this lake. Campbell et al. [151] found that toxaphene levels in lake trout from Bow Lake, a small alpine lake, were correlated with $\delta^{13}$C and lipid, and not with $\delta^{15}$N, due to a short pelagic-influenced food chain. Similar to Lake Siskiwit, the toxaphene BMF for zooplankton prey species (*Hesperodiaptomus arcticus*) to lake trout in Bow Lake was $< 1$.

## 5.4
### Transformations in the Food Web – Enantiomer Ratios

Enantiomer measurements reported by Karlsson et al. [91, 152] showed that biotransformation of chlorobornanes was occurring both at low and high trophic levels. Most congeners were found to be racemic in lake water and sediment, except for B8-1414 and B8-806/809 (see Sect. 3.1.5). The ERs of all congeners in technical toxaphene were racemic (1.0).

Nine of 11 congeners that could be routinely determined by chiral GC had nonracemic EFs in lake trout (EF values significantly greater or less than 0.5) (Fig. 15). Smelt, lake herring and *Diporeia* also had nine or ten congeners with significantly nonracemic EFs, while zooplankton had the fewest nonracemic congeners and lower EF values than the other species. The mean EFs of chlorobornane in lake trout were weakly correlated with the EFs in lake herring ($P = 0.07$) and rainbow smelt ($P = 0.03$). This result suggests that dietary transfer of the nonracemic congeners is occurring and that the pattern seen in lake trout is not entirely due to biotransformation in the trout. In the case of B9-715, herring and (especially) rainbow smelt appear to be able to effectively degrade this congener, based on its high EF value (Fig. 15), low levels in lake trout and a negative TMF value [74].

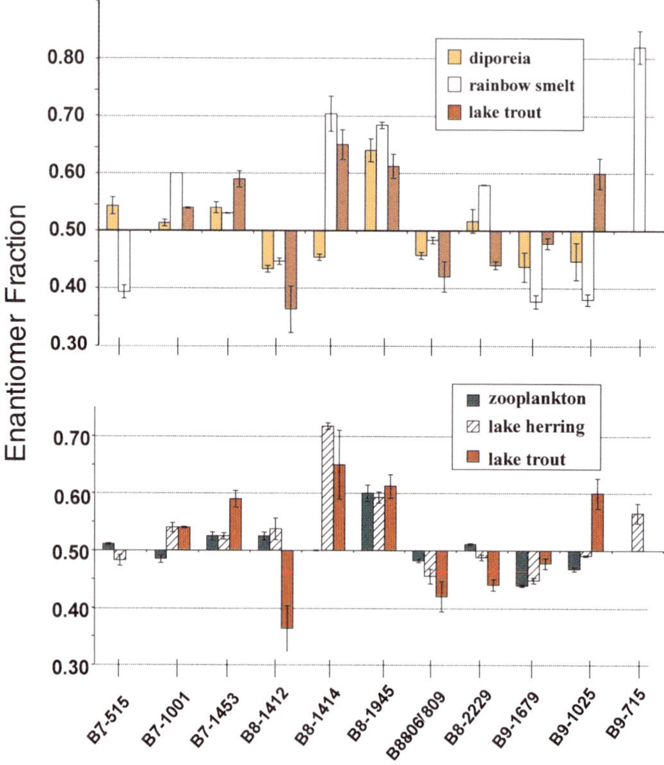


**Fig. 15** Enantiomer fractions (EFs) of 11 chlorobornane congeners in Lake Superior food web samples (adapted from Karlsson et al. [152]). EFs less than 0.5 are shown with *negative bars* to indicated an excess of the (−) enantiomer. *Vertical lines* represent standard deviations. If *no bar* is present the EF value is approximately 0.5 i.e. racemic

The observation of nonracemic EFs even in invertebrates was unexpected given the reported recalcitrance of many of these chlorobornanes. Even chlorobornanes substituted at the 2-endo, 3-exo, 5-endo, and 6-exo positions (including B8-1412 and B9-1679) had nonracemic EFs in zooplankton and in *Diporeia*. The chlorobornane EFs in zooplankton and *Diporeia* were also strongly correlated ($P = 0.008$). This correlation implies that both groups of organisms may have a common source of nonracemic congeners. This could be bacteria or phytoplankton in settling particles in the water column.

Overall, the enantiomer results are consistent with observations that species at higher trophic levels have greater capabilities of enantioselective biodegradation [6, 87, 88]. However, the EFs in lake trout, rainbow smelt or lake herring were not correlated with the enthalpy of formation of the congeners, $\Delta H_f$. Congeners with 2,3,5,6-endo, exo-substitution have higher $\Delta H_f$, and these congeners are prominent in marine mammals and birds [123]. Congeners with both endo-exo substitution and with geminal chlorines at C2 and C5 had significant EFs and high BAFs, BSAFs and BMFs in the Lake Superior food web. Only a few of the geminal-substituted congeners, such as B9-715, behaved as expected from structural considerations [91, 152].


# 6
# Implications for Wildlife and Human Health

## 6.1
## Effects on Fish

A substantial body of literature exists on the aquatic and mammalian toxicology of toxaphene [2, 3] and this will not be reviewed further here. There has been very little study of the possible biological consequences of prolonged elevated toxaphene concentrations on fish and fish-eating wildlife in the Great Lakes. However, Delorme et al. [153] reported a study which implies that continued high levels of toxaphene in lake trout may have biological consequences in terms of fish recruitment. They found that survival of wild adult lake trout and white sucker treated with toxaphene at 7 μg/g (wet wt) was decreased compared with controls over a four-year period following their release into a small, oligotrophic lake in northwestern Ontario. The fish were captured, treated with a single intraperitoneal injection of toxaphene and then released. Survival of lake trout was not decreased after a 3.5 μg/g ip injection. Fertilization rates and survival to swim-up of white sucker eggs and fry from females treated with toxaphene decreased relative to controls in both years, while lake trout were unaffected. The study estimated a mean residence time of about 350 days for toxaphene in adult lake trout and over 600 days for B8-1413 and B9-1679, which is consistent with the high BAFs and TMFs. Long half-lives for toxaphene were also found in white suckers.

The increased mortality of lake trout observed by Delorme et al. [153] may have occurred at spawning. Mayer et al. [154] observed increased mortality of brook trout (*Salvelinus fontinalis*) during spawning at similar concentrations to those used by Delorme et al. This mortality may be specific for salmonids because no increased mortality was observed for channel catfish or fathead minnows under similar exposure conditions [155].

Bone hydroxyproline has been shown to be depressed in fathead minnows during exposure to toxaphene [155]. This effect has not been studied in Great Lakes fish; however, lake trout collected from Lake Laberge (Yukon), which had similar toxaphene levels to those in Lake Superior [147, 148], had significantly lower levels of hydroxyproline than lake trout from nearby Lake Kusawa or from a small northwestern Ontario lake [156]. This effect did not seem to be related to toxaphene exposure, however, because fish treated intraperitoneally with toxaphene showed no differences in mean levels of hydroxyproline or calcium [157]. A broader survey including fish from lakes in Yukon and Northwest Territories did not show a correlation between toxaphene levels and collagen or hydroxyproline [157]. Thus, bone collagen did not prove to be a strong indicator of toxaphene exposure in lake trout in contrast to results for laboratory studies.

Delorme et al. [153] concluded that high body burdens of toxaphene in lake trout could result in subtle population-level effects by decreasing growth and thereby decreasing recruitment in the reproducing population. The range of mean concentrations of toxaphene in lake trout (500–600 mm) in Lake Superior during the 1980s was 1.0–1.5 μg/g wet wt according to Glassmeyer et al. [49], so most individual lake trout were below the no-effect concentrations. Any effects on survival due to toxaphene would need to consider the combined effects of other contaminants, notably PCBs, which were prominent contaminants in lake trout in Lake Superior in the 1980s [131], and would therefore be difficult to discern.

## 6.2
### Fish Consumption Advisories

The elevated toxaphene levels in Lake Superior have social and economic impacts as well. The 2003 "Guide to Eating Sport Fish" published by the Ontario Ministry of Environment and Energy [158] indicates fish consumption advisories for Lake Superior lake trout due to toxaphene levels. As a result of the high toxaphene levels, the Ontario Ministry of Natural Resources has restricted sale with Ontario (but not the export) of large lake trout harvested from Lake Superior as of 1995. An assessment of toxaphene levels in Lake Superior fish fillets by Health Canada concluded that consumption of lake trout, salmon, longnose sucker and whitefish muscle from northwestern areas of the lake should be limited to one or two meals per month (55–135 g/week) based on a provisional tolerable daily intake of 0.2 μg/g body wt/day [159].

These results have significant economic implications for commercial fishers and for aboriginal peoples in the region who depend on the fishery as a major food source. In the United States no consumption of fish with greater than 4.8 µg/g wet weight is recommended based on noncancer health endpoints or 0.18 µg/g (ww) based on cancer endpoints [160]. As of December 1998 there were six fish consumption advisories for toxaphene in the USA, but they were all for locations in the southern states [160]. No toxaphene advisories for Great Lakes fish consumption have been issued by the Great Lakes states.


# 7
## Lessons Learned and Future Trends

The story of toxaphene in the Great Lakes, like that of most other persistent organochlorines, only became clear after the use had peaked and bans were implemented. Delineating the extent of toxaphene contamination proved very challenging due to the difficulties involved with quantifying this multicomponent mixture. Current methodology gives reasonably good agreement for total toxaphene but problems still exist in separating and quantifying individual congeners.

The spatial distribution in water, sediments and top predator fishes in the Great Lakes are now reasonably well studied. Temporal trends of toxaphene are well documented in lake trout but are less certain in other media. Concentrations in the water column appear to be more important to understanding the fate and bioaccumulation of toxaphene than for other contaminants such as PCBs. Sediment core profiles from Lake Michigan, Ontario and Superior all show declining inputs in the past 10–20 years, mirroring reduced emissions following the toxaphene ban in the USA in 1984. The sediment profiles do not reflect the water concentrations, especially in Lake Superior, due to the relatively low proportion of toxaphene in the water column on particles combined with low sedimentation rates. Concentrations appear to have declined in lake trout and smelt in all lakes except Superior [49, 81] during the 1990s, although recent measurements (in 2000) show higher levels [133]. High water concentrations in Lake Superior continue and, therefore, seem to be responsible for the continued elevation of toxaphene in fish by maintaining high exposure to benthic and pelagic organisms. Modeling by Swackhamer et al. [65] demonstrated that colder temperatures and low sedimentation rates in Lake Superior, and to some extent in Lake Michigan, conspire to maintain high water concentrations.

Atmospheric transport of toxaphene from the southern USA continues [38, 45]. MacLeod et al. [19] attributed 70% of the atmospheric inputs to the Great Lakes to long-range atmospheric transport and deposition from the southern US and northern Mexico. The dynamic mass balance model [65] indicates that the net exchange direction of toxaphene in Lake Superior was depositional during the 1960s and 1970s due to high air concentrations during this time of heavy usage. Usage and air concentrations declined in the 1980s and 1990s, and the exchange shifted in the direction of net volatilization [65]. Lakes Michigan and Ontario, however, remain depositional because lower water concentrations favor net absorption of gaseous toxaphene. Glassmeyer et al. [49] predict half-lives in water of 18–31 years for Lake Superior, 5–8 years for Lake Michigan and about 8.5 and 6 years for Lakes Huron and Ontario, respectively, based on data from the 1980s and early 1990s. Recent water column measurements confirm the extended persistence of toxaphene in the Lake Superior water column (Table 5). Some degradation is apparent in the Lake Superior food web based on nonracemic ERs for selected chlorobornane, but for most congeners this process is slow and does not result in negative food web biomagnification in Lake Superior. The observation of high proportions of hexa- and heptachlorobornanes in lake sediments and tributary waters is another indication of the slow degradation, mainly via dechlorination, that is proceeding within the Great Lakes basin.

Toxaphene concentrations probably did not reach levels that would, by themselves, cause effects on salmonid reproduction, survival or growth in the Great Lakes. Whether their prolonged high concentrations attenuated effects of other prominent contaminants such as PCBs is not known. The levels and effects of toxaphene in fish-eating birds and mammals (such as mink) in the Great Lakes have never been thoroughly investigated. Given the capability of most homeotherms to degrade chlorobornanes it seems likely that exposure levels for birds and mammals would have been lower than for PCBs. In some Great Lakes jurisdictions concerns remain about human exposure to toxaphene via consumption of Lake Superior lake trout. Given the long half-lives in fish and the water, elevated toxaphene is likely to remain a contaminant issue until the middle of the twenty-first century.

# References


1. Vetter W, Oehme M (1999) In: Paasivirta J (ed) New types of persistent halogenated compounds (Handbook of Environmental Chemistry Series). Springer, Berlin Heidelberg New York, pp 237–287
2. de Geus H-J, Besselink H, Brouwer A, Klungsrøyr J, McHugh B, Nixon E, Rimkus G, Wester PG, de Boer J (1999) Environ Health Persp 107(S1):115–144
3. Saleh MA (1991) Rev Environ Contam Toxicol 118:1–85
4. Rice CP, Evans MS (1984) In: Niriagu JO, Simmons MS (eds) Toxic contaminants in the Great Lakes. Wiley, New York, pp 163–194
5. Swackhamer DL, Pearson RF, Schottler SP (1998) Chemosphere 37:2525–2561
6. Buser HR, Müller MD (1994) Environ Sci Technol 28:119–128
7. Pollack GA, Kilgore WW (1978) Residue Rev 50:87–140
8. Korytár P, van Stee LIP, Leonards PEG, de Boer J, Brinkman UATh (2003) J Chromatogr A 994:179–189
9. Andrews P, Vetter W (1995) Chemosphere 31:3879–3886

10. Bidleman TF, Leone AD, Falconer RL (2003) J Chem Eng Data 48:1122–1127

11. Fisk AT, Cymbalisty CD, Tomy GT, Stern GA, Muir DCG (1999) Chemosphere 39:2549–2562

12. Falconer RL, Bidleman TF (1994) Atmos Environ 28:547–554

13. Mackay D, Shiu WY, Ma K-C (2005) Handbook of physical-chemical properties and environmental fate for organic chemicals, 2nd edn. CRC, Boca Raton, FL.

14. Murphy TJ, Mullin MD, Meyer JA (1987) Environ Sci Technol 21:155–162

15. Jantunen LM, Bidleman TF (2000) Chemosphere Global Change Sci 2:225–231

16. Meylan WM, Howard PH, Boethling RS (1996) Environ Toxicol Chem 15:100–106

17. Li YF (2001) J Geophys Res 106:17919–17927

18. Castillo LE, De La Cruz E, Ruepert C (1997) Environ Toxicol Chem 16:41–41

19. MacLeod M, Woodfine D, Brimacombe J, Toose L, Mackay D (2002) Environ Toxicol Chem 21:1628–1637

20. DePinto J, Fiest T, Smith D (1997) Presented at the EPA-Paper Industry Meeting on Toxaphene, May 1997, Chicago, IL, USA. Limno-Tech Inc., Ann Arbor, MI

21. Glotfelty DE, Leech MM, Jersey J, Taylor AW (1989) J Agric Food Chem 37:546–551

22. Willis GH, McDowell LL, Murphree CE, Southwick LM, Smith S (1983) J Agric Food Chem 31:1171–1177

23. Raff JD, Hites RA (2004) Environ Sci Technol 38:2785–2791

24. Struger J (1998) Historical registration status of toxaphene in Canada (Ontario). Internal report. Environment Canada, Burlington, ON

25. Speirs JM (1954) Experimental ammocoete poisoning programme, 1954. Appendix No. 5, Annual Report to the Great Lakes Research, Committee

26. Li YF, Bidleman TF, Barrie LA (2001) J Geophys Res 106:17929–17938

27. Carey AE, Gowen J, Tai H, Mitchell W, Wiersma G (1978) Pestic Monit J 12:117–136

28. Carey AE, Douglas P, Tai H, Mitchell W, Wiersma G (1979) Pestic Monit J 13:17–22

29. Bidleman TF, Leone AD (2004) Environ Pollut 128:49–57

30. Bidleman TF, Leone AD (2004) Environ Toxicol Chem 23:2337–2342

31. Harner T, Wideman JL, Jantunen LMM, Bidleman TF, Parkhurst WJ (1999) Environ Pollut 106:323–332

32. Loganathan, 2003

33. Harner T, Bidleman TF, Mackay D (2001) Environ Toxicol Chem 20:1612–1621

34. Jantunen L, Harner T, Bidleman TF (2000) Environ Sci Technol 34:5097–5105

35. Scholtz MT, McMillan AC, Slama C, Li YF, Ting N, Davidson K (1997) Pesticides emission modelling. Development of a North American pesticide inventory. Report CGEIC-1997-1, 1–242, Canadian Global Emissions Interpretation Centre, Toronto, ON, Canada

36. Scholtz MT, Voldner E, McMillan AC, Van Heyst BJ (2002) Atmos Environ 36:5005–5013

37. Shoeib M, Brice KA, Hoff RM (2000) Chemosphere 40:201–211

38. James RR, Hites RA (2002) Environ Sci Technol 36:3474–3481

39. Heimstad ES, Smalas AO, Kallenborn R (2001) Chemosphere 43:665–674

40. Vetter W, Scherer G (1999) Environ Sci Technol 33:3458–3461

41. Rice CP, Samson PJ, Noguchi GE (1986) Environ Sci Technol 20:1109–1116

42. Hoff RM, Muir DCG, Grift NP, Brice KA (1993) Chemosphere 27:2057–2062

43. Hoff RM, Muir DCG, Grift NP (1992) Environ Sci Technol 26:266–275

44. Bidleman TF, Alegria H, Ngabe B, Green C (1998) Atmos Environ 32:1849–1856

45. Hoh E, Hites RA (2004) Environ Sci Technol 38:4187–4194

46. Jantunen LM, Bidleman TF (2003) Environ Toxicol Chem 22:1229–1237

47. Hoff R, Brice KA, Halsall CJ (1998) Environ Sci Technol 32:1793–1798

48. Wania F, Haugen JE, Lei YD, Mackay D (1998) Environ Sci Technol 32:1013–1021

49. Glassmeyer ST, Brice KA, Hites RA (1999) J Great Lakes Res 25:492–499

50. James RR, McDonald JG, Symonik DM, Swackhamer DL, Hites RA (2001) Environ Sci Technol 35:3653–3660

51. Shoeib M, Brice KA, Hoff R (1999) Chemosphere 5:849–871

52. McConnell LL, Bidleman TF, Cotham WE, Walla MD (1998) Environ Pollut 101:391–399

53. Alegria H, Bidleman TF, Salvador-Figueroa M (2005) Environ Pollut, in press.

54. Alegria HA, Bidleman TF, Shaw TJ (2000) Environ Sci Technol 34:1953–1958

55. Voldner EC, Schroeder WH (1989) Atmos Environ 23:1949–1961

56. Stanley CW, Barney JE (1971) Environ Sci Technol 5:430–435

57. Arthur RD, Cain JD, Barrentine BF (1976) Bull Environ Contam Toxicol 15:129–134

58. Bidleman TF, Olney CE (1975) Nature 257:475–476

59. Bidleman TF, Rice CP, Olney CE (1976) In: Windom HL, Duce RA (eds) Marine pollutant transfer, Chapter 13. D.C. Heath & Co, Lexington, MA, pp 323–351

60. Bidleman TF, Christensen EJ (1979) J Geophys Res 84:7857–7862

61. Bidleman TF, Christensen EJ, Billings WN, Leonard R (1981) J Marine Res 39:443–464

62. MacLeod M, Woodfine DG, Mackay D, McKone T, Bennett D, Maddalena R (2001) Environ Sci Pollut Res 8:156–163

63. McDonald JG, Hites RA (2003) Environ Sci Technol 37:475–481

64. Hoff RM, Bidleman TF, Eisenreich SJ (1993) Chemosphere 27:2047–2055

65. Swackhamer DL, Schottler S, Pearson RF (1999) Environ Sci Technol 33:3864–3872

66. Bidleman TF, Cotham WE, Addison RF, Zinck ME (1992) Chemosphere 24:1389–1412

67. Jantunen L, Bidleman TF (2004) Unpublished data. Meteorological Service of Canada, Environment Canada, Toronto, ON

68. Bidleman TF, Cussion S, Jantunen LM (2004) Atmos Environ 38:3713–3722

69. Eisenreich SJ, Strachan WMJ (1992) Estimating atmospheric deposition of toxic substances to the Great Lakes. Workshop Report, Jan 31–Feb 2, 1992. Great Lakes Protection Fund and Environment Canada, Burlington, ON

70. Harder HW, Christensen EC, Mathews JR, Bidleman TF (1980) Estuaries 3:142–147

71. Burniston DA, Strachan WMJ, Wilkinson RJ (2005) Environ Sci Technol 39:7005–7011

72. Hoff RM, Strachan WMJ, Sweet CW, Chan CH, Shackleton M, Bidleman TF, Burniston DA, Cussion S, Gatz DF, Harlin K, Schreoder WH (1996) Atmos Environ 30:3505–3527

73. Simcik MF, Hoff RM, Strachan WJ, Sweet CW, Basu I, Hites RA (2000) Environ Sci Technol 34:361–367

74. Muir DCG, Whittle DM, De Vault DS, Bronte CR, Karlsson H, Backus S, Teixeira C (2004) J Great Lakes Res 30:316–340

75. Glassmeyer ST, De Vault DS, Meyer TR, Hites RS (2000) Environ Sci Technol 34:1851–1855

76. Swackhamer DL, Symonik D (2004) Environ Sci Technol, in review

77. Muir DCG, Backus S, Teixeira C, Whittle M (2003) Toxaphene, OC pesticides and PCBs in Lake Superior and Lake Nipigon. Unpublished data. National Water Research Institute, Burlington, ON

78. Burniston DA, Strachan W (2001) Toxaphene in Lake Superior water and sediments, 1996–1997. Unpublished data. National Water Research Institute, Burlington, ON

79. Golden KA, Wong CS, Jeremiason JD, Eisenreich SJ, Sanders G, Hallgren J, Swackhamer DL, Engstrom DR, Long DT (1994) Water Sci Technol 28:19–31

80. Pearson R, Swackhamer DL (2002) Toxaphene in Lake Michigan waters. Unpublished data. School of Public Health, University of Minnesota, Minneapolis, MN

81. Whittle DM, Kiriluk RM, Carswell AA, Keir MJ, MacEachen DC (2000) Chemosphere 40:1221–1226

82. Whittle DM, Karlsson H, MacEachen C, Keir M, Muir D, De Vault D (2000) Organohal Compd 47:125–128

83. Muir DCG, Stern G, Karlsson H (1999) Organohal Compd 41:565–568

84. Ruppe S, Neumann A, Braekevelt E, Tomy GT, Stern GA, Maruya KA, Vetter W (2004) Environ Toxicol Chem 23:591–598

85. Falconer RL Bidleman TF, Gregor DJ, Semkin R, Teixeira C (1995) Environ Sci Technol 29:1297–1302

86. Jantunen L, Bidleman TF (1998) Arch Environ Toxicol Contam 35:218–228

87. Wong CS, Mabury SA, Whittle DM, Backus SM, Teixeira C, DeVault DS, Muir DCG (2004) Environ Sci Technol 38:84–92

88. de Geus H-J, Baycan-Keller R, Oehme M, de Boer J, Brinkman UATh (1998) J High Res Chromatogr 21:39–46

89. Vetter W, Bartha R, Stern G, Tomy G (1999) Environ Toxicol Chem 18:2775–2781

90. Maruya KA, Wakeham SG, Vetter W, Lee RF, Francendese L (2000) Environ Toxicol Chem 19:2198–2203

91. Karlsson H, Muir D, Strachan W, Backus S, DeVault D, Whittle M (1999) Organohal Compd 41:597–600.

92. Karlsson H, Muir DCG, Bidleman TF (2000) Toxaphene congener ER values for Alabama air. Unpublished data, Meteorological Service of Canada, Environment Canada, Toronto, ON

93. McConnell LL, Bidleman TF (1995) Sci Total Environ 159:101–117

94. Pearson R, Swackhamer D, Eisenreich S, Long D (1997) Environ Sci Technol 31:3523–3529

95. Muir D, Karlsson H, Kohli M, Wang X, Backus S, Lockhart L, Wilkinson P (2000) Organohal Compd 47:256–259

96. Schneider AR, Stapleton H, Cornwell J, Baker JE (2001) Environ Sci Technol 35:3809–3815

97. Howdeshell MJ, Hites RA (1996) Environ Sci Technol 30:220–224

98. Muir DCG, Backus S, Comba M (2000) Toxaphene congeners in Lake Ontario and Lake Superior sediment cores. Unpublished data. National Water Research Institute, Burlington, ON

99. Marvin C (2004) Toxaphene congeners in Lake Ontario sediment cores. Unpublished data. National Water Research Institute, Burlington, ON

100. Rapaport RA, Eisenreich SJ (1986) Atmos Environ 20:2367–2379

101. Rapaport RA, Eisenreich SJ (1988) Environ Sci Technol 22:931–941

102. Eisenreich SJ (1996) Toxaphene in the Great Lakes: Concentrations, trends and sources. Workshop Proceedings. US Environmental Protection Agency, Chicago, IL, p 18

103. Stuthridge TR, Wilkins AL, Langdon AG, Mackie KL, McGarlane PN (1990) Environ Sci Technol 24:903–908

104. Jarnuzi G, Matsuda M, Wakimoto T (1992) Kanyo Kagaku 2:364–365

105. Larson RA, Marley KA (1986) Formation of toxaphene-like contaminants during simulated paper pulp bleaching. Water Resources Center, Univ Illinois, Chicago, IL, WRC 205

106. Shanks KE, McDonald JG, Hites RA (1999) J Great Lakes Res 25:383–394

107. Rappe C, Haglund P, Andersson R, Buser H (1998) Organohal Compd 35:291–294

108. Rappe C, Haglund P, Andersson R, Buser H (1998) Organohal Compd 35:287–291

109. Rappe C, Andersson R, Bonner M, Cooper K, Fiedler H, Lau C, Howell F (1997) Chemosphere 34:1297–1304
110. Swackhamer D, Engstrom D (2000) Toxaphene concentrations in dated sediment cores from a holding pond for kraft bleach effluent. Unpublished data. School of Public Health, University of Minnesota, MN
111. Miskimmin BM, Muir DCG, Schindler DW, Stern GA, Grift NP (1995) Environ Sci Technol 29:2490–2495
112. Rose NL, Backus S, Karlsson H, Muir DCG (2001) Environ Sci Technol 35:1312–1319
113. Fingerling G, Hertkorn N, Parlar H (1996) Environ Sci Technol 30:2984–2992
114. Stern GA, Loewen MD, Miskimmin BM, Muir DCG, Westmore JB (1996) Environ Sci Technol 30:2251–2258
115. Ruppe S, Neumann A, Vetter W (2003) Environ Toxicol Chem 22:2614–2621
116. Buser HR, Haglund P, Müller MD, Poiger C, Rappe C (2000) Chemosphere 40:1213–1220
117. Maruya KA, Vetter W, Wakeham SG, Lee RF, Francendese L (2000) In: Lipnick RL, Hermens JLM, Jones KC, Muir DCG (eds) Persistent, bioaccumulative and toxic chemicals: fate and exposure (ACS Symp Ser 772). American Chemical Society, Washington, DC
118. Ribick MA, Dubay GR, Petty JD, Stalling DL, Schmitt CJ (1982) Environ Sci Technol 16:310–318
119. Swackhamer DL, Charles J, Hites RA (1987) Anal Chem 59:913–917
120. Haseltine SD, Heinz Gh, Reichel WL, Moore JF (1981) Pesticide Monit J 15:90–97
121. Ohlendorf HM, Swineford DM, Locke LN (1981) Pesticide Monit J 14:125–135
122. Saleh MA, Skinner RF, Casida JE (1979) J Agric Food Chem 27:731–737
123. Vetter W, Klobes U, Luckas B (2001) Chemosphere 43:611–621
124. Weichbrodt M, Vetter W, Scholtz E, Luckas B, Reinhardt K (1999) Int J Environ Anal Chem 73:309–328
125. Muir DCG, Jones PD, Karlsson H, Koczansky K, Stern GA, Kannan K, Ludwig JP, Reid H, Robertson CJR, Giesy JP (2002) Environ Toxicol Chem 21:413–423
126. Jansson B, Vazz T, Blomkvist G, Jensen S, Olsson M (1979) Chemosphere 8:181–190
127. Jansson B, Andersson R, Asplund L, Litzen K, Nylund K, Sellström U, Uvemo U, Wahlberg C, Wideqvist U, Odsjo T, Olsson M (1993) Environ Toxicol Chem 12:1163–1174
128. Struger J, Elliott JE, Bishop CA, Obbard ME, Norstrom RJ, Weseloh DV, Simon M, Ng P (1993) J Great Lakes Res 19:681–694
129. Clark JR, DeVault D, Bowden RJ, Weishaar JA (1984) J Great Lakes Res 10:38–47
130. Schmitt CJ, Ribick MA, Ludke JL, May TW (1983) National Pesticide Monitoring Program: Organochlorine residues in freshwater fish, 1976–1979. Fish and Wildlife Service Resource Publ 152. U.S. Department of the Interior, Washington, DC
131. De Vault DS, Hesselberg R, Rodgers PW, Feist TJ (1996) J Great Lakes Res 22:884–895
132. Glassmeyer ST, De Vault DS, Meyer TR, Hites RS (1997) Environ Sci Technol 31:84–88
133. Swackhamer DL (2004) Great Lakes Fish Monitoring Program: 1999–2000 Data. Final Report to US EPA Great Lakes National Program Office (April, 2004). US EPA, Chicago, IL
134. Muir D, Anderson R, Droulliard K, Evans M, Fairchild W, Guildford S, Haffner D, Kidd K, Payne J, Whittle M (2002) Toxic substances research initiative project, Final Report. Health Canada, Ottawa, p 65
135. Kiriluk RM, Whittle DM, Keir MJ, Carswell AA, Huestis SY (1997) Chemosphere 34:1921–1932
136. MacEachen DC, Russell RW, Whittle DM (2000) J Great Lakes Res 26:112–119
137. Fisk AT, Johnston TA (1998) J Great Lakes Res 24:917–928

138. Stapleton HM, Materson C, Skubina J, Ostrom P, Ostrom NE, Baker JE (2001) Environ Sci Technol 35:3287–3293
139. Kucklick JR, Baker JE (1998) Environ Sci Technol 32:1192–1198
140. Stapleton HM, Skubinna J, Baker JE (2002) J Great Lakes Res 28:52–64
141. Swackhamer DL (1995) Toxaphene concentrations in Lake Michigan food web organisms, unpublished data. School of Public Health, University of Minnesota, MN
142. Di Toro DM, Zarba CS, Hansen DJ, Berry WJ, Swartz RC, Cowan CE, Pavlou SP, Allen HE, Thomas NA, Paquin P (1991) Environ Toxicol Chem 10:1541–1583
143. Ankley GT, Cook PM, Carlson AR, Call DJ, Swenson JA, Corcoran HF, Hoke RA (1992) Can J Fish Aquat Sci 49:2080–2085
144. Broman D, Naf C, Rolff C, Zebuhr Y, Fry B, Hobbie J (1992) Environ Toxicol Chem 11:331–345
145. Fisk AT, Hobson KA, Norstrom RJ (2001) Environ Sci Technol 35:732–738
146. Pearson RF, Hornbuckle K, Eisenreich SJ, Swackhamer DL (1995) Environ Sci Technol 30:1429–1436
147. Kidd KA, Schindler DW, Muir DCG, Lockhart WL, Hesslein RH (1995) Science 269:240–242
148. Kidd KA, Schindler DW, Hesslein RH, Muir DCG (1997) Can J Fish Aquat Sci 55:869–881
149. Kidd KA, Hesslein RH, Koczanski K, Stephens GR, Muir DCG (1998) Environ Pollut 102:91–103
150. Kucklick JR, Bidleman TF, McConnell LL, Walla MD, Ivanov GP (1994) Environ Sci Technol 28:31–37
151. Campbell LM, Schindler DW, Muir DCG, Donald DB, Kidd KA (2000) Can J Fish Aquat Sci 57:1258–1269
152. Karlsson H, Muir D, Teixeira C, Strachan W, Backus S, DeVault D, Whittle M, Bronte C (2000) Organohal Compd 45:121–124
153. Delorme PD, Lockhart WL, Mills KH, Muir DCG (1999) Environ Toxicol Chem 18:1992–2000
154. Mayer FL, Mehrle PM, Dwyer WP (1975) Toxaphene effects on reproduction, growth and mortality of brook trout. EPA-600/3-75-013. US Environmental Protection Agency, Duluth, MN
155. Mayer FL, Mehrle PM, Dwyer WP (1977) Toxaphene: Chronic toxicity to fathead minnows and channel catfish. EPA-600/3-77-069. US Environmental Protection Agency, Duluth, MN
156. Lockhart WL, Muir DCG (1996) In: Synopsis of research conducted under the 1994/95 Northern Contaminants Program, Environmental Studies No. 73. Indian and Northern Affairs Canada, Ottawa, p 199–207
157. Delorme PD (1995) PhD Thesis. University of Manitoba, Winnipeg, MB Canada
158. Ontario Ministry of the Environment (2003) Guide to eating Ontario sport fish: 2003-2004, 22th edn. Queens Printer for Ontario, Toronto, Canada
159. Health Canada (1995) Report on human health significance of toxaphene in Lake Superior fish species. Bureau of Chemical Safety, Health Canada, Ottawa, ON
160. US Environmental Protection Agency (1999) Toxaphene update: Impact on fish advisories. EPA823-F-99-018. US Environmental Protection Agency, Washington DC, p 6

Hdb Env Chem Vol. 5, Part N (2006): 267–306
DOI 10.1007/698_5_043
© Springer-Verlag Berlin Heidelberg 2006
Published online: 5 January 2006

# Polychlorinated Naphthalenes in the Great Lakes

Paul A. Helm[1] (✉) · Kurunthachalam Kannan[2] · Terry F. Bidleman[3]

[1] Environmental Monitoring and Reporting Branch,
Ontario Ministry of the Environment, 125 Resources Road, West Wing,
Toronto, Ontario, M9P 3V6, Canada
*paul.helm@ene.gov.on.ca*

[2] Wadsworth Center, New York State Department of Health
and Department of Environmental Health and Toxicology,
State University of New York at Albany, Empire State Plaza, P.O. Box 509,
Albany, New York, 12202-0509, USA
*kkannan@wadsworth.org*

[3] Centre for Atmospheric Research Experiments,
Science and Technology Branch, Environment Canada, 6248 Eighth Line,
Egbert, Ontario, L0L 1N0, Canada
*terry.bidleman@ec.gc.ca*

**Abstract** This review examines the sources and occurrence of polychlorinated naphthalenes (PCNs) in the Great Lakes environment and summarizes current knowledge of the toxicity, analysis, and environmental chemistry and fate of these compounds. PCNs have entered the Great Lakes through the production and use of Halowax technical mixtures, as trace contaminants in Aroclor PCB mixtures, and through industrial processes such as chlor-alkali production and waste incineration. Air concentrations of PCNs were highest in urban areas, and congener profiles indicate that evaporative emissions relating to past uses are the dominant sources, but combustion processes also contribute. Sediment measurements indicate that the highest concentrations are in the Detroit River, and congener profiles indicate Halowax contamination from past inputs. Fish from this area had the highest reported concentrations in the Great Lakes region, followed by lake trout from Lake Ontario. Estimates of dioxin toxic equivalents (TEQ) of PCNs indicate their contributions are as important as the dioxin-like PCBs in some aquatic species, and more important in air and sediments. No time trend information for PCNs in the Great Lakes exists, and further spatial assessment, and TEQ comparisons of PCNs, PCDD/Fs, and PCBs in additional fish species, should be undertaken.

**Keywords**  Analysis · Chloronaphthalenes · PCNs · TEQ · Toxicity


## Abbreviations

| | |
|---|---|
| BCF | Bioconcentration factor |
| BMF | Biomagnification factor |
| ECD | Electron capture detector |
| FID | Flame ionization detector |
| GC | Gas chromatograph |
| HPLC | High pressure liquid chromatograph |
| MS | Mass spectrometer |
| $K_{OA}$ | Octanol–air partition coefficient |
| $K_{OW}$ | Octanol–water partition coefficient |
| $K_{AW}$ | Air–water partition coefficient |
| $K_{P}$ | Particle–gas partition coefficient |
| $p_{L}^{\circ}$ | Subcooled liquid vapor pressure |
| PCNs | Polychlorinated naphthalenes |
| PCDDs | Polychlorinated dibenzo-$p$-dioxins |
| PCDFs | Polychlorinated dibenzofurans |
| PCBs | Polychlorinated biphenyls |
| POPs | Persistent organic pollutants |
| monoCNs | Monochloronaphthalenes |
| diCNs | Dichloronaphthalenes |
| triCNs | Trichloronaphthalenes |
| tetraCNs | Tetrachloronaphthalenes |

pentaCNs  Pentachloronaphthalenes
hexaCNs   Hexachloronaphthalenes
heptaCNs  Heptachloronaphthalenes
octaCN    Octachloronaphthalene
REP       Relative potency
TEF       Toxic equivalency factor
TEQ       Toxic equivalents


# 1
# Introduction

Polychlorinated naphthalenes (PCNs) are a class of persistent organic pollutants (POPs) which were detected in the environment in the 1970s and in the Great Lakes in the early 1980s. Since 1995 there have been more focused efforts in determining the extent of PCN contamination in air, sediments, and biota in the Great Lakes region. There are 75 possible congeners of PCNs with from one to eight chlorine atoms substituted onto the aromatic naphthalene molecule (Table 1 and Fig. 1). PCNs are structurally similar to polychlorinated dibenzo-*p*-dioxins (PCDDs), dibenzofurans (PCDFs), and biphenyls (PCBs), particularly those PCBs having planar character with no chlorines in the *ortho* (2,6) positions (Fig. 1). PCNs were first synthesized in 1833 by Laurent using molten naphthalene and chlorine gas [1].

Production-scale processes and uses of PCNs as flame retardants and dielectrics in capacitors were patented in the early 1900s [2–4], but PCNs also found uses in cable coverings, dye-making, and as fungicides in wood [5–7]. Production is thought to have ceased worldwide as substitutes such as PCBs and other plastics replaced PCNs in many applications. However, regulations controlling PCN use and manufacture are non-existent in many countries, and there have been recent accounts of in-use products containing PCNs [8–10]. In addition to commercial formulations, PCNs are produced in combustion processes and are present in PCB mixtures.

Concern related to PCNs began in workplace settings with occurrences of chloracne, jaundice, liver disease, and death noted from prior to scale-up of production until PCN substitutes became popular (reviewed by Hayward (1998) [12]). Occupational exposures were reported as recently as 1989 [8]. Environmental contamination by PCNs was first observed in the 1970s in sediments near industrial sites [13], in air, water, and soil near sites of PCN manufacture and use [14, 15], and in fish [16] and fish-eating birds [17, 18].

Interest in PCNs was renewed in the 1990s as congener-specific analytical methods became more routine and precise. PCNs are now known to be as widespread in the environment as most other halogenated compounds, as indicated by their occurrence in the ambient air [19–21] and biota [22–24] of

**Table 1** IUPAC numbering and substitution of tri- to octaCN congeners[a]

| Number | Chlorine substitution | Number | Chlorine substitution | Number | Chlorine substitution |
|---|---|---|---|---|---|
| Trichloronaphthalenes | | 35 | 1,2,4,8 | 59 | 1,2,4,5,8 |
| | | 36 | 1,2,5,6 | 60 | 1,2,4,6,7 |
| 13 | 1,2,3 | 37 | 1,2,5,7 | 61 | 1,2,4,6,8 |
| 14 | 1,2,4 | 38 | 1,2,5,8 | 62 | 1,2,4,7,8 |
| 15 | 1,2,5 | 39 | 1,2,6,7 | | |
| 16 | 1,2,6 | 40 | 1,2,6,8 | Hexachloronaphthalenes | | |
| 17 | 1,2,7 | 41 | 1,2,7,8 | | |
| 18 | 1,2,8 | 42 | 1,3,5,7 | 63 | 1,2,3,4,5,6 |
| 19 | 1,3,5 | 43 | 1,3,5,8 | 64 | 1,2,3,4,5,7 |
| 20 | 1,3,6 | 44 | 1,3,6,7 | 65 | 1,2,3,4,5,8 |
| 21 | 1,3,7 | 45 | 1,3,6,8 | 66 | 1,2,3,4,6,7 |
| 22 | 1,3,8 | 46 | 1,4,5,8 | 67 | 1,2,3,5,6,7 |
| 23 | 1,4,5 | 47 | 1,4,6,7 | 68 | 1,2,3,5,6,8 |
| 24 | 1,4,6 | 48 | 2,3,6,7 | 69 | 1,2,3,5,7,8 |
| 25 | 1,6,7 | | | 70 | 1,2,3,6,7,8 |
| 26 | 2,3,6 | Pentachloronaphthalenes | | 71 | 1,2,4,5,6,8 |
| | | | | 72 | 1,2,4,5,7,8 |
| Tetrachloronaphthalenes | | 49 | 1,2,3,4,5 | | |
| | | 50 | 1,2,3,4,6 | Heptachloronaphthalenes | | |
| 27 | 1,2,3,4 | 51 | 1,2,3,5,6 | | |
| 28 | 1,2,3,5 | 52 | 1,2,3,5,7 | 73 | 1,2,3,4,5,6,7 |
| 29 | 1,2,3,6 | 53 | 1,2,3,5,8 | 74 | 1,2,3,4,5,6,8 |
| 30 | 1,2,3,7 | 54 | 1,2,3,6,7 | | |
| 31 | 1,2,3,8 | 55 | 1,2,3,6,8 | Octachloronaphthalene | | |
| 32 | 1,2,4,5 | 56 | 1,2,3,7,8 | | |
| 33 | 1,2,4,6 | 57 | 1,2,4,5,6 | 75 | 1,2,3,4,5,6,7,8 |
| 34 | 1,2,4,7 | 58 | 1,2,4,5,7 | | |

[a] Substitution at numbered positions on the naphthalene molecule shown in Fig. 1. Taken from [6, 11]

polar regions. They exhibit toxicity in a manner similar to other planar chlorinated aromatic compounds like PCDD/Fs and the relative potencies of several congeners have been determined in relation to 2,3,7,8-tetrachlorodibenzo-*p*-dioxin (TCDD) [25–28]. Furthermore, PCNs have been shown to bioaccumulate in food webs in the Baltic Sea and Great Lakes [29–32].

This review focuses on the occurrence, sources, and environmental chemistry (physical-chemical properties, analysis, and environmental pathways) of PCNs and places observations from the Laurentian Great Lakes basin of North America in context with studies from other regions around the world. We summarize the significant body of research conducted since reviews published in the late 1990s [6, 7, 12] and finish by highlighting continuing areas of PCN occurrence, behavior, and effects which are in need of understand-

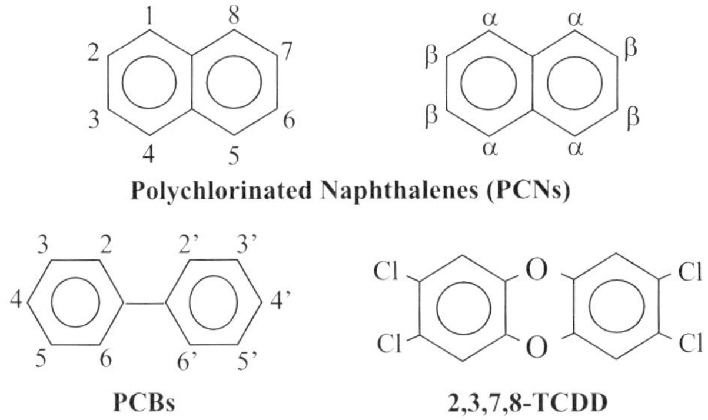

**Polychlorinated Naphthalenes (PCNs)**

**PCBs**                    **2,3,7,8-TCDD**

**Fig. 1** Chemical structures of polychlorinated naphthalene (PCN), polychlorinated biphenyl (PCB), and 2,3,7,8-tetrachlorodibenzo-*p*-dioxin (TCDD)

ing, particularly with the likelihood that these compounds will be included in international agreements restricting their production and use.

## 2
## Production and Use

The production and uses of PCNs have been reviewed previously [5–7, 12, 33, 34] and thus are only briefly described here. PCNs were produced commercially as complex technical mixtures with trade names which included Halowaxes (USA), Seekay Waxes (UK), Nibren Waxes (Germany), and Clonacire Waxes (France). The synthesis involved the chlorination of molten naphthalene using chlorine gas and metal chlorides (iron(III) or antimony(V)) as catalysts. Reaction temperatures ranged from 80–200 °C depending on the degree of chlorination required, and proceeded as nucleophilic and electrophilic reactions favoring substitution in the $\alpha$-(1,4,5,8) positions on the naphthalene molecule (Fig. 1) [6].

Production volumes are not well known, but are unlikely to have exceeded 10% of total PCB production [35], and are estimated at $50–150 \times 10^3$ t [6, 12]. In the US, annual production was approximately 3200 t in 1956 declining to near 2300 t by 1973 [34]. Koppers Company (Pittsburgh, USA), the main US manufacturer, ended US production in 1977, and the remaining US production had ended by 1980 [33]. It is thought that there is no current production worldwide, although a technical PCN mixture and PCN-containing products were recently imported into Japan [10].

In North America, Halowax mixtures were produced and sold with varying chlorine content. Halowaxes 1031 and 1000 were predominantly mono- and dichloronaphthalenes (mono- and diCNs), with 22–26% chlorine content, and used as engine oil additives (15–18% of the Halowax market) and in fabric production (10% of the market). Halowaxes 1001 and 1099, mainly tri- and tetraCNs with 50–52% chlorine, were the highest production mixtures accounting for 65% of Koppers production in 1972. These mixtures, which contain 10% di-, 40% tri-, 40% tetra-, and 10% pentaCNs, were impregnants for automobile capacitors. Approximately 8% of the Halowax market was for the 1013 (10% tri-, 50% tetra-, 40% pentaCNs; 56% chlorine) and 1014 (20% tetra-, 40% penta-, 40% hexaCNs; 62% chlorine) mixtures, which were used in electroplating and as binders in graphite electrodes for chlorine production. Halowax 1051 (10% hepta-, 90% octaCN) accounted for less than 1% of the market and its use was unknown [33, 34]. Other uses for PCNs included cable coverings, separators in batteries, and binders in ceramics. The flame retardant, insecticidal/fungicidal, and waterproofing properties of PCNs were also useful in the paper, wood, and textile industries. PCNs were additives to gear and cutting oils, in dye production, and the casting of alloys [5–7, 12, 33, 34].

## 3
## Toxicity of PCNs

The toxic effects of PCNs have been reviewed by Kover (1975), Crookes and Howe (1993), and Hayward (1998) [12, 33, 34]. We briefly summarize acute and chronic toxicity, and toxic potentials of PCNs relative to 2,3,7,8-TCDD.

### 3.1
### Acute Toxicity Associated with Occupational Exposure in Humans and Accidental Exposure of Domestic Animals

PCNs are highly toxic to humans and animals. Occupational and accidental exposures of humans to PCN mixtures have been reported by several investigators since the early 1900s (e.g. [36–40]). The toxic effects observed in humans following exposure to PCNs include typical skin lesions, referred to as "chloracne", and liver degeneration referred to as "yellow atrophy". In fact, PCNs were the first agents to cause widespread outbreaks of chloracne in the USA and European countries. The use of PCNs as substitutes for natural waxes and rubber in Germany during World War I led to the first large outbreaks of chloracne [5]. Chloracne and yellow atrophy were documented among individuals employed at a cable manufacturing facility using Halowax in West Chester County, New York, in the 1940s [40]. There were a number of mortalities during the 1930s and 1940s from acute yellow atrophy of the liver among workers exposed to PCNs [41]. Other health effects associated with subchronic exposures to PCNs include jaundice, abdominal pain, edema, ascites, and decrease in liver size.

Several cases of accidental exposure of cattle to high levels of PCNs have been documented. During the 1940s and 1950s, "bovine hyperkeratosis" or

"X-disease" in cattle was a major concern in the USA. The disease was caused by accidental ingestion of PCNs used as lubricants in machines used to make pelletized feed [38]. X-disease was also associated with the use of PCNs in wood preservatives and wax for binding twine. The primary effect of PCN poisoning in cattle is interference with the biotransformation of carotene to vitamin A, which results in vitamin A deficiency, followed by inflammation of the oral mucosa, lacrimation, excessive salivation, and irregular food consumption. In advanced stages, thickening of skin and loss of hair, referred to as hyperkeratosis is observed. A combination of penta- and hexaCN at a dose of 5.5 mg kg$^{-1}$ body weight (bw) given orally for 5 days, reduced plasma vitamin A significantly. A single oral dose of hexaCN at 11 mg kg$^{-1}$ bw caused mortality in cattle within 2 weeks [42]. Other domestic animals are less susceptible to PCN poisoning than cattle as cats, dogs, swine, sheep, and chicken fed with higher chlorinated PCNs elicited typical symptoms of X-disease, but required a tenfold higher dose than for cattle.

In rats, guinea pigs and other test animals, acute yellow atrophy and other liver changes have been observed in feeding and inhalation experiments. Exposure of rabbits to hepta- and octaCNs at 500 mg kg$^{-1}$ bw resulted in death of all individuals and one of three individuals exposed to pentaCNs. No mortality occurred in rabbits given mono-, di- or tetraCNs for 7 days [43]. Most of the toxic effects discussed above were associated with penta- and hexaCNs.

Chickens fed with penta- and hexaCNs at 100 mg kg$^{-1}$ in the diet produced "chick edema" disease [44]. Penta- and hexaCNs fed to turkeys at 20 mg kg$^{-1}$ in feed for 40 days resulted in 50% mortality. At a concentration of 5 mg kg$^{-1}$ feed, PCNs caused 6.5% mortality and 33% reduction in body weight [44]. Histological examination revealed liver damage in exposed birds. Reproductive problems were also observed in chickens fed with a Halowax 1014 mixture at doses greater than 100 mg kg$^{-1}$ bw.

The toxic effects associated with PCN exposures in humans and wildlife are, in general, characteristic of effects due to chlorinated hydrocarbons such as 2,3,7,8-TCDD. For instance, chloracne, vitamin A depletion, edema and liver damage have been observed in animals exposed to TCDD. The human toxicity and mechanistic relationship of PCNs to TCDD may be useful in understanding these classes of compounds. Particularly, acute and subacute exposures of humans and cattle to PCNs may provide important clues to the toxic effects at high levels for other dioxin-like compounds.

## 3.2
## Chronic Exposure Studies

PCNs, particularly penta- and hexaCNs, induce hepatic ethoxyresorufin O-deethylase (EROD) and aryl hydrocarbon hydroxylase (AHH) activities [45, 46] and oxidative stress resulting in increased lipid peroxidation, decreased hepatic vitamins A and E, and decreased catalase and superoxide dismutase activities in exposed laboratory animals [46]. Exposure of pregnant rats to 1,2,3,4,6,7-hexaCN (CN66) at $1 \mu g \, kg^{-1} \, bw \, day^{-1}$ on days 14–16 of gestation, resulted in rapid onset of lutinizing and follicle-stimulating hormones and spermatogenesis in male offspring [47]. These results suggested the potential of PCN congeners to disrupt the endocrine system.

Chronic exposure of birds to PCNs has been linked to edema in chickens [44] and both lethality and EROD induction in chicken and eider duck embryos [48]. These effects are similar to that observed for TCDD.

Little is known regarding the effects of PCNs on aquatic and benthic organisms. A few available studies suggest that effects similar to those observed in birds also occurred in fish. Microinjection of yolk-sac embryos with Halowax 1014 induced hepatic EROD activity in rainbow trout [49] while dietary exposure of three-spined sticklebacks (*Gasterostus aculeatus*) and rainbow trout fry [49, 50] to PCNs induced EROD activity and hepatic lipid accumulation [51]. Dietary exposure of Baltic salmon (*Salmo salar*) to Halowaxes at $0.1-10 \mu g \, g^{-1}$ caused EROD induction in a dose-dependent manner. Furthermore, hepatotoxicity and gonadal abnormalities were observed in fish [52]. Halowax mixtures 1014 and 1013 were shown to be toxic to medaka (*Oryzias latipes*) at various early life stages [53]. Water-borne exposure of juvenile mud crab (*Rhithropanopeus harrisii*) to Halowax 1099 at $20-100 \mu g \, L^{-1}$ for 5 days resulted in increased respiratory rates [54]. Crookes and Howe (1993) cite a report in which sheepshead minnow eggs exposed to 1-monoCN from fertilization to 28 days post-hatch showed no effects on survival and growth up to $0.39 \, mg \, L^{-1}$ but all fish died within 28 days at concentrations $\geq 0.79 \, mg \, L^{-1}$.

## 3.3
## Comparative Toxicity to TCDD

HexaCNs have shown a pattern of toxicity in humans and animals that closely resembles that of 2,3,7,8-TCDD. They are both potent chloracnegens in humans (Hayward [12]), produce similar symptoms such as edema, in chickens, and both PCNs and TCDD deplete vitamin A in cows and rats. The profile of biological responses elicited by PCNs suggests that at least a portion of the toxic responses are mediated through an aryl hydrocarbon receptor (AhR)-dependent mechanism of action [26, 27], similar to TCDD. *In vitro* bioassays, which measure AhR-dependent reporter gene activation or enzyme induction, have been used to characterize the potencies of PCN congeners relative to TCDD. Hanberg et al. [26] examined microsomal enzyme induction potency of certain hexa- and heptaCNs using rat hepatoma H4IIE cells, and showed their potencies relative to TCDD were as great as 0.003 [26]. Engwall et al. [48] found that the toxicity of certain hexaCN congeners to chicken and eider duck embryos was similar to that of mono-*ortho* PCB congeners, but less than that of the non-*ortho* PCB congeners 77 and 126 [48]. The availability of more purified PCN congeners encouraged further assessment of the relative potencies (REPs) of PCNs. REPs of the most potent PCNs varied from 0.001 to 0.004 in the *in vitro* rat hepatoma H4IIE cell assay [27]. Villeneuve et al. (2000) used three *in vitro* assays to characterize the dioxin-like potency of 18 PCN congeners. The PLHC-1 fish hepatoma assay was relatively insensitive compared to EROD and luciferase assays using rat hepatoma cells. Hexa- and pentaCNs were the most potent congeners with REPs between $10^{-3}$ and $10^{-7}$ [28]. These REPs were similar to those of some PCB congeners. A summary of the relative potencies of PCN congeners is given in Table 2. To date, approximately 30 congeners have been tested for their potencies relative to TCDD. PCN congeners which have been shown to be more potent include CN52/60, 54, 57, 56, 66/67, 68, 69, 63, 70, and 73. Four Halowax mixtures have also been tested for their ability to induce AhR-mediated activity in H4IIE-luc cells [55]. Halowax 1051, which has a high chlorine content and is composed primarily of hepta- and octaCNs, was the most potent of the Halowax mixtures tested.

PCNs have not been tested for their potential to cause cancer and reproductive effects in test animals.

**Table 2** Comparison of potencies of PCN congeners derived using rat hepatoma cells at $EC_{50}$ relative to 2,3,7,8,-TCDD

| PCN # | H4IIE EROD [a] | H4IIE-EROD [b] | H4IIE-luc [a] | H4IIE-luc [c] | H4IIE-DR-CALUX [d] |
|---|---|---|---|---|---|
| 1 | | | | | $1.7 \times 10^{-5}$ |
| 2 | | | | | $1.8 \times 10^{-5}$ |
| 5 | $3.1 \times 10^{-9}*$ | | $3.5 \times 10^{-8}$ | $2.0 \times 10^{-7}*$ | $3.5 \times 10^{-5}$ |
| 10 | | | | | $2.7 \times 10^{-4}$ |
| 48 | | | | | $4.1 \times 10^{-5}$ |
| 50 | | | | | $6.8 \times 10^{-5}$ |
| 54 | $7.6 \times 10^{-5}$ | | $< 6.4 \times 10^{-4}$ | $1.7 \times 10^{-4}$ | $5.8 \times 10^{-5}$ |
| 56 | $2.2 \times 10^{-5}$ | | $4.6 \times 10^{-5}$ | | – |
| 57 | $1.6 \times 10^{-6}$ | | $3.5 \times 10^{-6}$ | | |
| 66 | $6.3 \times 10^{-4}$ | | $2.6 \times 10^{-3}$ | $3.9 \times 10^{-3}$ | $1.2 \times 10^{-3}$ |
| 67 | $2.9 \times 10^{-4}$ | $2.0 \times 10^{-3}$ | | $1.0 \times 10^{-3}$ | $4.8 \times 10^{-4}$ |
| 68 | | $2.0 \times 10^{-3}$ | | $1.5 \times 10^{-4}$ | $4.9 \times 10^{-4}$ |
| 70 | $2.1 \times 10^{-3}$ | | $9.9 \times 10^{-3}$ | $5.9 \times 10^{-4}$ | $2.8 \times 10^{-3}$ |
| 71 | | $7.0 \times 10^{-6}*$ | | | |
| 72 | | | | | $6.0 \times 10^{-5}$ |
| 63 | | $2.0 \times 10^{-3}$ | | | |
| 64 | | $2.0 \times 10^{-5}$ | | | |
| 69 | | $2.0 \times 10^{-3}$ | | | $1.1 \times 10^{-4}$ |
| 73 | $4.6 \times 10^{-4}$ | $3.0 \times 10^{-3}$ | $6.9 \times 10^{-4}$ | $1.0 \times 10^{-3}$ | $5.2 \times 10^{-4}$ |

[a] [28]; [b] [25, 26]; [c] [27]; [d] [56]; * High uncertainty

# 4
# Environmental Chemistry of PCNs

## 4.1
## Physicochemical Properties

Physicochemical properties are necessary for describing chemical fate, transport, and bioaccumulation. The most useful of these are the subcooled liquid vapor pressure ($p_L^\circ$, Pa), water solubility ($\mu g\,L^{-1}$), and the octanol–water ($K_{OW}$), octanol–air ($K_{OA}$) and air–water ($K_{AW}$) partition coefficients. The available measured data on these properties for PCN congeners are summarized in Table 3.

Measurements of $p_L^\circ$ as a function of temperature were made by capillary gas chromatography for 17 PCN congeners [57]. From these data and reported relative retention indices [58], $p_L^\circ$ values of all 75 PCNs were estimated [57]. $K_{OA}$ values were determined as a function of temperature for 24 di- through hexaCNs using a gas saturation technique [59]. $K_{OA}$ for other di- to hexaCNs were estimated from: $\log K_{OA} = -1.275 \log p_L^\circ + 6.09$ [57]. Other estimation methods for these properties have also been presented for PCNs [60–62].

PCNs are planar molecules and their volatility properties more closely resemble those of the approximately planar mono- and non-*ortho* PCBs than the multi-*ortho* PCBs [57]. In a plot of the heats of vaporization ($\Delta_{vap}H$, kJ mol$^{-1}$) of PCNs versus the Le Bas molar volume, the regression line fell close to the same line for mono-*ortho* PCBs and intermediate of multi-*ortho* and non-*ortho* PCBs. A plot of log $K_{OA}$ versus log $p_L^\circ$ was virtually identical to one for mono- and non-*ortho* PCBs, and distinct from the regression line for multi-*ortho* PCBs. The combination of $K_{OA}$ and $p_L^\circ$ allowed the activity coefficients in octanol to be estimated, which ranged from 1–7 and were negatively correlated with the Le Bas molar volume of the PCN molecule. No such correlation was found for multi-*ortho* PCBs.

No direct measurements of $K_{AW}$ are available for PCNs. Estimates were made from $K_{AW} = K_{OW}/K_{OA}$ using homolog average values for $K_{OW}$ and $K_{OA}$ [63]. Resulting log $K_{AW}$ values at 12 °C were $-2.57$ for triCNs and $-3.00$ for tetraCNs. This approach was applied on a single congener basis to all PCNs for which both $K_{OA}$ and the few measured $K_{OW}$ values were available (Table 3). Log $K_{AW}$ at 25 °C for di- to hexaCNs range from $-2.69$ to $-1.57$ and show no apparent trend with homolog.

## 4.2
## Analytical Methods

Reports of PCNs in the environment surfaced in the early 1970s [13, 68]. Analytical methods used at the time were reviewed by Brinkman and Reymer

**Table 3** Physicochemical properties of PCNs

| PCN | Water solubility[a,b] ($\mu g\,L^{-1}$) | $\log K_{OW}$[a,b,c,d] | $\log K_{OA}$[e,f] (at 25 °C) | $\log p_L^{\circ}$[g] (at 25 °C) | $\log K_{AW}$[h] (at 25 °C) |
|---|---|---|---|---|---|
| 1 | 2870 | 3.95 | | 0.747 | |
| 2 | 924 | 4.04 | | 0.402 | |
| 3 | 137 | 4.47 | 6.75 | −0.521 | −2.28 |
| 4 | | | 6.68 | −0.460 | |
| 5 | 312 | 4.78 | 6.67 | −0.453 | −1.89 |
| 6 | 396 | 4.67 | 6.67 | −0.453 | −2.00 |
| 7 | | | 6.67 | −0.453 | |
| 8 | | | 6.67 | −0.453 | |
| 9 | 450 | 4.30 | 6.99 | −0.703 | −2.69 |
| 10 | 474 | 4.61 | 6.70 | −0.478 | −2.09 |
| 11 | | | 6.68 | −0.463 | |
| 12 | 240 | 4.81 | 6.68 | −0.463 | −1.87 |
| 13 | | | 7.49 | −1.102 | |
| 14 | | | 7.43 | −1.054 | |
| 15 | | | 7.43 | −1.054 | |
| 16 | | | 7.42 | −1.045 | |
| 17 | | | 7.42 | −1.045 | |
| 18 | | | 7.58 | −1.169 | |
| 19 | | | 7.31 | −0.955 | |
| 20 | | | 7.29 | −0.943 | |
| 21 | 65 | 5.47 | 7.29 | −0.943 | −1.82 |
| 22 | | | 7.45 | −1.069 | |
| 23 | | | 7.46 | −1.077 | |
| 24 | | | 7.31 | −0.955 | |
| 25 | | | 7.37 | −1.002 | |
| 26 | 17 | 5.12 | 7.48 | −1.093 | −2.36 |
| 27 | 4.2 | 5.87 | 8.37 | −1.790 | −2.50 |
| 28 | 3.7 | 5.77 | 8.24 | −1.688 | −2.47 |
| 29 | | | 8.24 | −1.688 | |
| 30 | | | 8.24 | −1.688 | |
| 31 | | | 8.60 | −1.967 | |
| 32 | | | 8.38 | −1.796 | |
| 33 | | | 8.01 | −1.504 | |
| 34 | | | 8.01 | −1.504 | |
| 35 | | | 8.38 | −1.796 | |
| 36 | | | 8.16 | −1.622 | |
| 37 | | | 8.01 | −1.504 | |
| 38 | | | 8.38 | −1.796 | |
| 39 | | | 8.24 | −1.688 | |
| 40 | | | 8.39 | −1.807 | |
| 41 | | | 8.53 | −1.917 | |
| 42 | 4.1 | 6.29 | 7.85 | −1.382 | −1.57 |
| 43 | 8.2 | 5.86 | 8.23 | −1.682 | −2.37 |

**Table 3** (continued)

| PCN | Water solubility[a,b] ($\mu g\,L^{-1}$) | log $K_{OW}$[a,b,c,d] | log $K_{OA}$[e,f] (at 25 °C) | log $p_L^\circ$[g] (at 25 °C) | log $K_{AW}$[h] (at 25 °C) |
|---|---|---|---|---|---|
| 44 | | | 8.09 | −1.572 | |
| 45 | | | 8.25 | −1.695 | |
| 46 | | | 8.59 | −1.959 | |
| 47 | 8.1 | 5.81 | 8.08 | −1.558 | −2.27 |
| 48 | | | 8.32 | −1.752 | |
| 49 | | | 9.30 | −2.520 | |
| 50 | | 7.00 | 8.97 | −2.260 | −1.97 |
| 51 | | | 8.97 | −2.260 | |
| 52 | | | 8.76 | −2.098 | |
| 53 | | 6.80 | 9.11 | −2.369 | −2.31 |
| 54 | | | 9.05 | −2.323 | |
| 55 | | | 9.16 | −2.411 | |
| 56 | | | 9.35 | −2.561 | |
| 57 | | | 9.09 | −2.350 | |
| 58 | | | 8.88 | −2.192 | |
| 59 | | | 9.22 | −2.456 | |
| 60 | | | 8.76 | −2.098 | |
| 61 | | | 8.88 | −2.192 | |
| 62 | | | 9.09 | −2.350 | |
| 63 | | | 9.98 | −3.052 | |
| 64 | | | 9.75 | −2.873 | |
| 65 | | | 10.05 | −3.107 | |
| 66 | | 7.70 | 9.67 | −2.804 | −1.97 |
| 67 | | | 9.67 | −2.804 | |
| 68 | | | 9.75 | −2.873 | |
| 69 | | 7.50 | 9.75 | −2.873 | −2.25 |
| 70 | | | 10.09 | −3.134 | |
| 71 | | | 9.83 | −2.936 | |
| 72 | | | 9.83 | −2.936 | |
| 73 | | 8.20 | | −3.556 | |
| 74 | | | | −3.609 | |
| 75 | 0.08 | 7.77 | | −4.165 | |

[a] [64]; [b] [65]; [c] [66]; [d] [67]; [e] [59]; [f] Calculated from log $K_{OA} = -1.275$ log $p_L^\circ + 6.09$ (see g); [g] [57]; [h] Calculated from log $K_{OW}$ − log $K_{OA}$

(1976), noting analytical challenges primarily due to interfering organochlorine (OC) pesticides and PCBs when using gas chromatography (GC) with electron capture detection (ECD), and from the lack of individual standards [5]. Methods of isolating PCNs from other chlorinated compounds included column chromatography using silicic acid to separate PCNs and PCBs from the OC pesticides [69]; however, separation of the PCNs from the PCBs remained problematic. Carbon-based column chromatography was found to be a possible method for separating planar aromatic compounds from other OC substances [70]. Chemical separation of PCNs and PCBs via perchlorination to octaCN and decaCB [71, 72], and dechlorination to the naphthalene and biphenyl skeletons [73, 74] was explored. However, quantification was limited to total PCNs and PCBs. The coupling of mass spectrometry to GC allowed for homolog-specific quantification of PCNs [14, 15]. No interference from PCBs was found in tests with an Aroclor and individual PCBs.

The foundations of current PCN analytical methods were established in the early 1980s with the analysis of PCNs by capillary (high resolution) GC and MS detection following carbon clean up and fractionation [75]. Typical methods are summarized in Jakobsson and Asplund (1999) [7] and are essentially those used for other halogenated compounds. Following solvent extraction of environmental matrices or sampling media, extracts are cleaned and fractioned by column chromatography or high performance liquid chromatography (HPLC), often with carbon [31, 75–77], 2-(1-pyrenyl)ethyldimethylsilylated silica (PYE) [29, 78, 79], or both [80] to isolate PCNs from other OC compounds. Instrumental analysis is conducted using capillary column GC and MS detection, either low- or high-resolution using electron impact (EI) or negative ionization modes [7, 81].

The availability of analytical standards of individual PCN congeners has hampered quantification. Although nearly all congeners have been synthesized (see citations in [6, 7]), only about 25 of the tri- to octaCNs are commercially available. Erickson et al. (1978) used the relative molar response for available PCN standards and assumed response values for the penta- to heptaCNs for MS quantification by homologs [14, 15]. This approach was further developed later as Heidmann (1988) used correction factors generated from relative responses by GC-flame ionization detection (FID) and GC-MS (in EI mode; EI-MS) to quantify PCNs [82]. Wiedmann & Ballschmiter (1993) [11] found the molar response in EI-MS could be normalized to the ionization cross section of the molecule to give a constant value of injected mass to signal intensity, and then quantified individual congeners/peaks in fly ash and Halowax 1014 [11]. Characterization of technical mixtures by GC FID and subsequent use of the mixture as a quantitative standard [76, 83] or the use of molar responses based on pure standards by EI-MS have been shown to give consistent results for total PCNs in an interlaboratory comparison [81]. However, results for individual congeners were more varied suggesting further refinements are needed. Only recently have isotopically labeled congeners become available allowing for method development based on isotope dilution similar to routine PCDD/F analysis.

More recent efforts in analytical improvements have focused on maximizing the separation of PCN congeners by capillary gas chromatography. Järnberg et al. investigated capillary GC columns of varying polarity and structural selectivity, finding that the 5% phenyl-dimethylpolysiloxane stationary phase was the most practical [58]. Retention data have been published for this phase [58, 77, 83] and for the 1701 ([14%-cyanopropyl-phenyl]-methylpolysiloxane) phase [77]. However, a number of coelutions remain unresolved on most stationary phases, particularly within the more toxic and bioaccumulative penta- and hexaCN homologs. Investigators have examined a variety of capillary GC columns attempting to resolve these coelutions [84–86]. Imagawa and Yamashita (1997) found that hexaCNs 64/68 and hexaCNs 71/72 were separated using $\alpha$-cyclodextrin and $\beta$-cyclodextrin columns, respectively, but hexaCNs 66/67 were not resolved [84]. Helm et al. (1999) found that another $\beta$-cyclodextrin-based column was able to resolve all coeluting hexaCN congeners (Fig. 2) and pentaCN congeners [86]. Resolution of CN66/67 has also been achieved by HPLC using 2-(1-pyrenyl)-ethyldimethylsilylated (PYE) silica gel stationary phase [85, 87]. The PYE phase has also been incorporated into a multidimensional HPLC fractionation scheme with final analysis by capillary GC-high resolution mass spectrometry resulting in resolution of CN64/68 and 71/72 [80].

The results of the second interlaboratory study of PCN analytical methods using environmental matrices, undertaken by the US National Institute of Standards and Technology (NIST), should provide an indication of the comparability of published environmental PCN data and show where additional method enhancements are needed. Further method development efforts in the analysis of PCNs, and other complex mixtures are likely to focus on improving efficiencies by optimizing run times and separation. Examples may include time-of-flight mass spectrometry and multidimensional GC. New methods, such as isotope ratio mass spectrometry, may contribute to further source apportionment of complex mixtures [88].

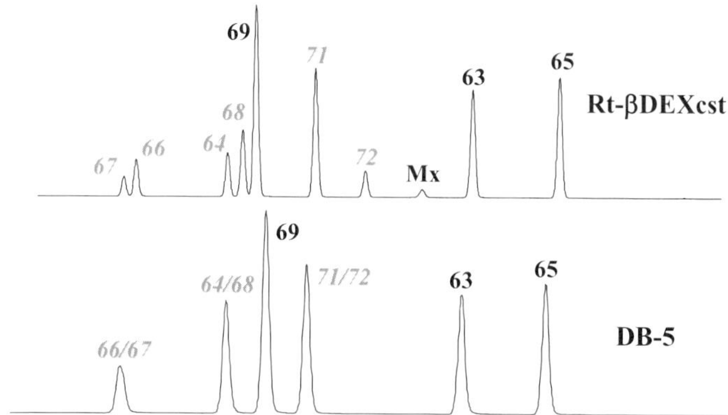


**Fig. 2** Chromatograms illustrating the separation of all of the hexaCN congeners on the Rt-$\beta$DEXcst (Restek, Bellefonte PA, USA) compared to the DB-5 (Agilent Technologies (J & W Scientific), Palo Alto CA, USA) capillary GC columns [86]

## 4.3
## Fate and Transport Processes

The fate of POPs such as PCNs in the environment is governed by their physical chemical properties. The extent to which POPs transform or degrade in the environment, the rate at which they exchange between environmental compartments, and the degree to which they travel once released are important considerations in assessment processes under national laws and international treaties. The few studies in this area regarding PCNs are summarized below.

### 4.3.1
### Degradation

Biological degradation of PCNs in the environment has not been studied in detail. Järnberg et al. (1999) added Halowax 1014 to a natural lake sediment fortified with supernatant from municipal sewage sludge in a 28-day aerobic degradation experiment [89]. No change in the congener composition of the tetra- to hexaCNs was found. Microbial degradation of the monoCNs has received more attention and is summarized in Crookes and Howe (1993) [33].

Photochemical degradation in solution and the atmosphere has been examined. Smog chamber experiments indicated an atmospheric half-life of 2.7 days for 1,4-diCN under hydroxyl radical attack [90]. Half-lives for hydroxyl radical reactions have been estimated for PCNs using quantitative-structure activity relationships (calculated by the Atmospheric Oxidation Program for Windows (AOPWIN), US EPA; described in Meylan & Howard (1993) [91]). Half-lives range from 12–57 days for removal of the tri- to hexaCNs, similar to experimental findings of 9–34 days for removal of tri- to pentaCBs [92].

In solution, PCNs absorb ultraviolet (UV) light at > 300 nm resulting in photochemical degradation to predominantly dechlorination and dimerization products [93–95]. Reaction rates in methanol were found to decrease with the degree of chlorination for the mono- to tetraCNs [94], while Gulan et al. (1974) found that reaction products were observed more rapidly for higher chlorinated PCNs [96]. Ruzo et al. (1975) suggested that this difference may result from the greater extinction coefficients for the higher chlorinated PCNs leading to absorption of more incident radiation than the lower chlorinated congeners, or that lower chlorinated congeners may act as sensitizers when mixtures are irradiated [94]. Congeners with chlorines substituted in the 1,8 or "peri" positions of the naphthalene molecule (e.g., 1,8-diCN and 1,3,5,8-tetraCN) gave a higher proportion of dechlorination products and reacted considerably faster than other congeners within the same homolog [94]. Depletions of 1,8-substituted congeners were observed in Halowax 1014 profiles irradiated by sunlight in methanol and in sediments from lakes in Sweden receiving PCNs from atmospheric deposition [89]. The authors also noted a shift in the Halowax profile to lower chlorinated congeners. UV irradiation of octaCN in hexane resulted in sequential dechlorination producing hepta-, hexa-, penta-, and tetraCNs over the short duration of experiments.

Based on limited studies, it seems that photochemical reactions may be the predominant degradation processes of the higher chlorinated PCNs in the environment.

### 4.3.2
### Intermedia Exchange

The distribution of POPs between the gaseous and particle phases is an important consideration in understanding the atmospheric transport, deposition, and degradation of such contaminants in the environment. Harner and Bidleman (1997, 1998) found that PCNs were distributed between these two phases in the urban atmosphere of Chicago, USA, and that the more chlorinated congeners (hexa- to octaCNs) were predominantly in the particle phase (average 70–98%) [76, 97]. Harner and Bidleman (1998) described this distribution of PCNs on particles using the octanol–air partition absorption model [97] first described by Finizio et al. (1997) [98]. The $K_{OA}$ model was shown to resolve apparent differences within compound classes and between compound classes when regressing vapor pressure with the observed particle–gas partition coefficient ($K_P$) [99]. Log–log regressions of the observed $K_P$ with $K_{OA}$ were statistically equivalent for the PCNs and the PCBs, indicating they exhibit similar partitioning behavior [99]. In Arctic aerosols, PCN partitioning was also suitably described by the $K_{OA}$ model, while including a soot portion within the model resulted in over-prediction of partitioning [100].

Diffusive exchange across the air–water interface is an important loading and loss pathway to and from Great Lakes water bodies (e.g., [101, 102]). Air–water exchange is estimated using a fugacity-based two-film model and has been examined for PCNs in Lake Ontario. Using Henry's Law constants calculated from $K_{OA}$ and $K_{OW}$ for the triCN and tetraCN homologs, Helm et al. (2003) estimated fugacity ratios ($f_W/f_A$; fugacity in water [$f_W$]/fugacity in air [$f_A$]) for tri- and tetraCNs based on concentrations in air and water in June 2000 [63]. The $f_W/f_A$ was > 1 for triCNs indicating net volatilization. For the tetraCNs, the $f_W/f_A$ were mostly > 1 but were closer to equilibrium values ($f_W/f_A = 1$) than the triCNs. An $f_W/f_A < 1$ was found for tetraCN in one air sample collected in the western end of Lake Ontario in closer proximity to urban and industrial areas. This indicates that elevated levels in urban air (discussed in 6.1) may be influencing the net direction of air–water exchange in areas of the lake [63], as has been shown for PCBs where net absorption occurs in the southern part of Lake Michigan due to the influence of the airshed of Chicago, USA [103, 104].

The exchange of PCNs between air and vegetation [105] and air and ambient surfaces [106] has also been explored. Similar uptake and depuration rates were found for PCNs and PCBs in experiments using pasture vegetation in indoor air, and plant-side resistance governed air–pasture and pasture–air exchange. However, PCB uptake was more rapid than PCNs of similar $K_{OA}$ during diffusion into the leaf interior. The authors suggested that these differences may result from variations in degree of planarity, polarity, or chlorination at a given $K_{OA}$. Although depuration coefficients were similar, less PCN (10%) remained at the end of the experiment than PCBs (20%) [105]. Levels of PCNs in ambient air at a semirural location in the UK were found to be in part controlled by temperature-dependent air–surface cycling and advection of air influenced by ongoing diffuse emissions [106]. This finding was supported by a lower ratio of PCNs in air measured in mid-afternoon (15:00 h) to early-morning (03:00 h) than for OC pesticides, and by slopes from Clausius–Clapeyron relations giving observed enthalpies of phase change $(\Delta H)$ that were considerably lower than $\Delta H$ values determined from $p_L^\circ$ and $K_{OA}$ measurements [106].

### 4.3.3
### Long-Range Transport

PCNs are sufficiently persistent to undergo long-range atmospheric transport as indicated by measured and estimated half-lives of $> 2$ days for atmospheric hydroxyl radical attack, and the ability to undergo condensation–volatilization cycling between surfaces combined with advective transport, as discussed above. In the Great Lakes region, higher air concentrations are found near the more populated/industrialized centers (discussed in Sect. 6.1) while concentrations decrease and homolog and congener profiles shift to a greater abundance of the more volatile, less-chlorinated PCNs. Surveys in Europe show the greatest PCN concentrations in UK cities [106–108], in urban and industrialized areas of Poland, and in Moscow [108]. Low to non-detectable levels were found at remote sites in Ireland, Iceland, and Norway, and generally in the northern and southern parts of Europe [108]. Long range transport from central European source regions resulted in higher concentrations of PCNs collected in air [19, 21] and snow [21] in the Eastern Arctic. Elevated concentrations were shown to be associated with air masses traveling from source regions by calculated 5-day back trajectories. [19]. Higher levels of PCNs were found at an island in the southern Baltic Sea compared to a station in northern Sweden close to the Arctic Circle. The greatest levels at the Baltic station occurred during periods of warm temperatures and when air transport was from the southwest and west, while lowest levels at the northern station occurred during cold weather and winds from the east [109].

Low PCN concentrations were found at remote air monitoring stations at Dunai, Russia and Alert, Canada [19, 20] and at Tagish, a subarctic site in western Canada [20]. LRT is also implied by the occurrence of PCNs in Arctic [22–24] and Antarctic biota [23].

## 4.4
## Bioaccumulation and Biomagnification

Due to the hydrophobic nature of PCNs (log $K_{OW}$ 3.9–8.2), they tend to bioconcentrate and biomagnify in the aquatic food web. Although very few laboratory exposure studies have examined bioconcentration factors (BCF) of PCNs in aquatic organisms, they are thought to be similar to those of PCBs with corresponding $K_{OW}$ values. Based on an equilibrium partitioning model, BCFs for PCNs have been estimated to vary from 200 for monoCNs to $> 10^6$ for octaCN [33]. In general, concentrations of PCNs in aquatic organisms have been several hundred to thousands of times greater than that in surrounding water. PCN levels in algae were 24- to 140-fold greater than in the surrounding water [110] while shrimp concentrated various PCN mixtures from water by a factor of 63 to 257 [111]. BCFs between 100 and 600 have been reported for PCNs in fish from Tokyo Bay [33].

The BCF of PCNs in fish and other aquatic organisms generally increases with the degree of chlorination. However, hepta- and octaCNs seem to behave differently, probably due to their extreme lipophilicity and/or steric hinderance due to their larger size ($> 9.5$ Å) [67]. The extreme lipophilicity may prevent these congeners from desorbing from particles or lipids. Further, due to their planarity, PCNs tend to bind strongly to sediments. Biota-sediment accumulation factors (BSAF; lipid-normalized contaminant concentration in tissues divided by organic carbon-normalized contaminant concentration in sediment) for PCNs in aquatic organisms from a site in Georgia, USA was 0.0012–0.00007 [77]. Corresponding BSAFs for PCBs varied from 0.48 to 1.7 [112]. These results suggest that PCNs tend to sorb more strongly than PCBs to sediments and thereby sediments could be a major sink for PCNs in the environment.

Uptake efficiencies of PCN congeners 66/67, 71, 73, and 75 administered to pike through diet were between 63 and 78%, except for CN75 which had an uptake efficiency of 35% [113]. As mentioned earlier, the uptake of CN75 may be inhibited due to its larger molecular surface area and extreme hydrophobicity. Bioaccumulation studies of PCNs indicate that several congeners may be metabolized by higher trophic animals in the food web [6]. PCN congeners which do not have unsubstituted adjacent carbons (42, 52, 58, 60, 61, 64, 66, 67, 68, 69, 71, 72, 73, 74, and 75) have biomagnification factors generally greater than one [6]. Mammals seem to have the ability to metabolize mono- to tetraCNs. Ruzo et al. (1976) studied the metabolism of 1,2,3,4-tetraCN and 1,2,3,4,5,6-hexaCN in pigs, finding that the tetraCN was hydroxylated on the adjacent ring but did not detect metabolites of the hexaCN [114]. Cornish and Block (1958) showed that 79% of monoCN, 93% diCN, and 45% tetraCN administered to rabbits were excreted as metabolites in the urine while the penta-, hepta-, and octaCNs did not form metabolites [43]. Clearance half-lives of 1,2,3,4-tetraCN (CN27) and octaCN in rainbow trout ranged from 8–27 and 15–62 days, respectively [115].

The biomagnification potential of each PCN congener varies considerably, probably as a result of metabolism. Biomagnification factors (BMFs; lipid-normalized [lw] concentration in predator/concentration in prey) of various isomers in benthic algae and zebra mussels from the St. Clair River, Michigan, generally ranged from 3 to 10 for the tetra- to hexaCNs [32]. The overall BMF between $\Sigma$PCNs in algae and zebra mussels was $\sim 4.9$. In fish from Lake Ontario, the BMF for $\Sigma$PCN for lake trout and their prey weighted to diet was 4.2 (Helm et al. unpublished results, 2005). In general, the highest BMFs tend to be congeners 42, 52/60, and 66/67, and increase with increasing degree of chlorination, however, this varies with location and diet [6, 32].

# 5
# Sources of PCNs to the Great Lakes Environment

PCNs have entered and are still entering the Great Lakes environment from point source and diffuse releases to the atmosphere and surface waters due to the historical use and disposal of the Halowax technical PCN mixtures, the past and ongoing use of Aroclor PCB formulations, the chlor-alkali industry, and combustion-based processes. Erickson et al. (1978) found elevated concentrations of PCNs in air near a PCN production facility (range 25 to 2900 ng m$^{-3}$) and near capacitor manufacturers (range non-detect to 33 ng m$^{-3}$) [14, 15]. A chloronaphthalene isomer was also measured at concentrations of 80 and 3400 ng m$^{-3}$ in home and school basements near the Love Canal dumpsite in Niagara Falls, New York [116]. Effluent discharges from chemical, manufacturing, and textile industries also represent point sources. PCN concentrations were higher downstream of capacitor manufacturers [14] and were detected in the effluent of a manufacturing plant in Ontario [117]. Very high sediment concentrations in the Detroit River area result from the impact of chemical and manufacturing industries along the river [118–120]. Effluents and sludges from chlor-alkali facilities that used graphite electrodes are implicated as sources of PCN contamination [77, 79, 121]. The use of Halowaxes as binders in the electrodes, highly chlorinated Aroclor 1268 as electrode lubricants, or reaction of chlorine with the electrode substrates are possible sources of the PCNs [77, 79]. Similarly, effluent from magnesium production, which also used graphite electrodes, was a source of PCNs [16]. Sewage treatment plant effluents have not been measured for PCNs but are a likely source since PCNs are present at levels ranging from 50–190 ng g$^{-1}$ dry weight (dw) in sewage sludges in the UK [122].

Thermal industrial processes such as metal refining and waste incineration also release PCNs into the environment. PCNs are present in the flue gas [123–125] and fly ash [11, 79, 126–129] of municipal solid waste incinerators (MSWI). PCNs were present in the stack emissions from aluminum smelting at a metal reclamation plant [130] and in the slag from a copper ore smelting process [131]. Fly ash taken from industries or waste processes located within the Great Lakes region, including an iron sintering plant, a cement kiln, a hospital waste incinerator, and a municipal solid waste incinerator, each contained PCNs [126]. The burning of wood for domestic purposes will also be a diffuse atmospheric source of PCNs in North America, as is the case in the UK [132].

Diffuse sources contributing PCNs to the atmosphere through evaporation or to waters from surface runoff include spills of engine and/or cutting oils, disposal of oil, capacitor inlays, and cable insulation, and leaching from in-use capacitors, insulated cables, treated paper and wood products, and sealants/lubricants. Weistrand et al. (1992) found PCNs were leaching from old electronic equipment in a laboratory [9]. Kover (1975) and Crookes & Howe (1993) suggested that landfills are likely PCN sources since many of the materials containing them (capacitors, cables) are disposed of in landfills [33, 34]. PCNs were present at ng L$^{-1}$ levels in water percolating from a city dump in Sweden [79] and up to μg L$^{-1}$ levels in ground water near an illegal landfill in Spain [133]. Amending agricultural fields with municipal biosolids (sewage sludge) containing PCNs has been shown to elevate soil concentrations [134] and could be a diffuse source through volatilization or surface run-off, although this has not been investigated.

The Aroclor PCB formulations produced in North America have been shown to contain PCNs [126, 135–137]. Haglund et al. (1993) found ΣPCN concentrations (sum of di- to octaCNs) to range from 4–170 μg g$^{-1}$ in Aroclors [136], while Aroclor mixtures analyzed by Yamashita et al. (2000) were found to contain lower levels (5–67 μg g$^{-1}$) [137] for tri- to octaCNs. In these mixtures, PCN concentrations increased as the chlorine content increased. The levels and patterns of PCNs have also been reported in Clophen PCB mixtures [136, 137] and mixtures from France (Phenochlors), Russia (Sovol), and Japan (Kanechlors) [137]. Falandysz (1998) estimated that PCB usage results in approximately 100 t of PCNs in PCB applications [6], while Yamashita et al. (1999) suggested this value to be 169 t [137], both less than 1% of estimated global commercial PCN production. Feedstock biphenyl contamination with naphthalene in the PCB production process is the most likely source of PCNs [138]. PCNs from technical PCB mixtures enter the environment along with PCBs from spills, leaks, disposal, and fugitive emissions.

Homolog distributions and congener-specific profiles of source-related samples are used to indicate the possible source of PCNs found in environmental samples (discussed below in Sect. 5.1). The homolog profiles determined in various sources that are likely to contribute to Great Lakes contamination, including the Halowaxes, Aroclor mixtures, and fly ashes, are summarized in Fig. 3. Variation of homolog and congener profiles between batches of Halowax and Aroclor mixtures [126, 137], as well as changes in profiles during transport and partitioning in the environment, present challenges to source apportionment.

PCNs were emitted to the Great Lakes environment over an extended period of time, and although inputs from production and ongoing usage are probably minimal, the relative importance of current unintentional (i.e., combustion) emissions, recycling of past inputs, and other possible current sources (municipal sewage effluents and biosolids applications) remains unknown. Further assessment of potential source areas within the Great Lakes is needed to determine if in-place PCN "hot-spots" exist which impact the local or regional ecosystem. The potential of on-going industrial usage and related sources should be investigated, especially since recent imports of manufactured materials into Japan containing PCNs apparently originated from Canada [10].

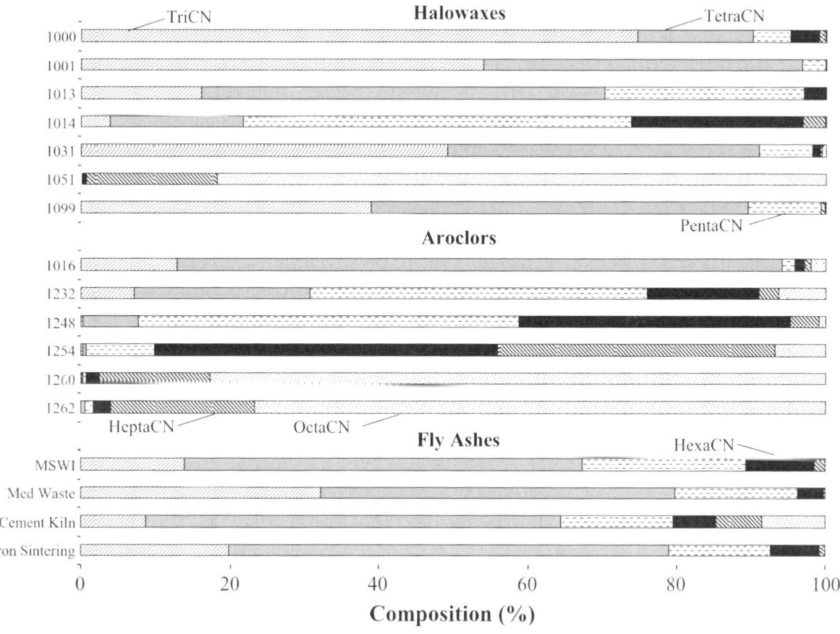



**Fig. 3** Percent homolog composition of tri- to octaCNs in source-related samples from the Great Lakes region for Halowaxes, Aroclors, and industrial fly ashes from a municipal solid waste incinerator (MSWI), a medical waste incinerator (Med Waste), a cement kiln, and an iron sintering plant [126, 137, 139]

## 5.1
### Indicators of Combustion-Related PCN Sources

Several investigations have been undertaken to characterize and distinguish between types of PCN sources. Congener-specific analytical methods (discussed in Sect. 4.2) have generated congener profiles for PCNs in Halowax technical mixtures [83, 127, 139], PCB mixtures [79, 137], and in incinerator and other combustion emissions [89, 123, 126–129]. Some congeners are not detected or are trace components of technical PCN and PCB mixtures, but are formed in combustion processes, while others are present at significant proportions in technical mixtures as well as in combustion sources. These are summarized in Table 4. The presence and enhancement of these congeners in sample profiles are indicative of combustion source contributions.

A number of PCN studies in the Great Lakes region have used various indicators to note contributions of combustion sources of PCNs. Congeners CN44, 29, and 54 were present in sediments in the upper Detroit River area, indicating some contributions of combustion sources although the profiles were dominated by a Halowax signature [119]. Similar findings were observed in air over the lakes using ratios of CN44/(CN44+CN42) and CN54/(CN54+CN61) [63]. A range of combustion-related congeners, including CN44, 29, 35, 52/60, 50, 51, 54, and 66/67, contributed more to air profiles at a site in north Toronto, Canada, relative to a downtown location indicating strong contributions from combustion sources (Fig. 4) [126]. Principle components analysis was used as a pattern-recognition technique, confirming that samples strongly influenced by combustion sources had patterns more reflective of waste incineration. Air samples most indicative of evaporative sources resembled Halowaxes 1000, 1099, and 1013 [126]. Harner et al. (2005) used enrichment factors calculated for selected combustion-related PCN congeners to indicate that air at background locations within the basin are influenced to a greater extent by combustion sources than urban sites [140].

**Table 4** List of combustion-related "marker" PCN congeners[a]

| Congeners present in combustion sources and absent/traces in technical mixtures | | Major combustion congeners also present in technical mixtures | |
|---|---|---|---|
| 13 | 29 | 17/25 | 50 |
| 18 | 44 | 27/30 | 51 |
| 20 | 54 | 35 | 52/60 |
| 26 | 70 | 36/45 | 66/67 |
| | | 39 | 73 |

[a] [79, 123, 126–129]

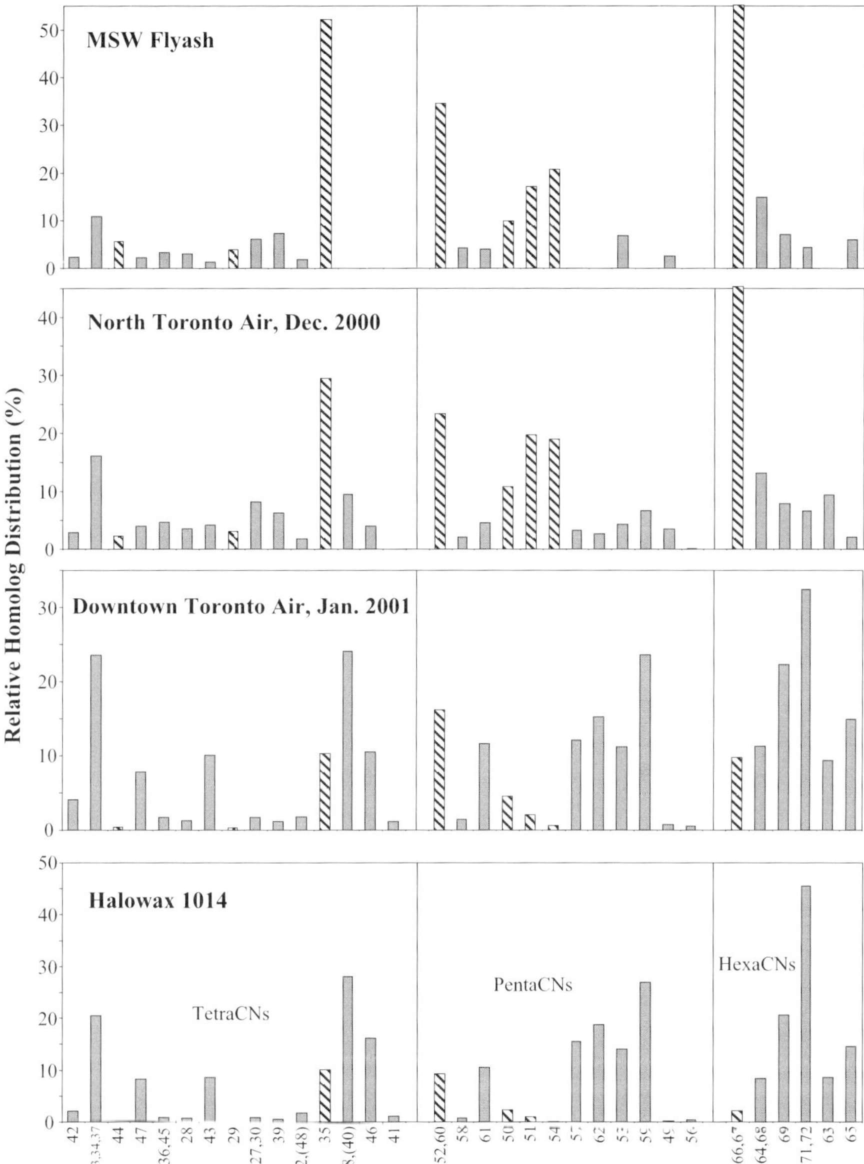


**Fig. 4** Congener profiles (tetra- to hexaCNs, expressed as relative contribution to their respective homolog group) in selected air samples collected from north Toronto and downtown Toronto, Canada compared to MSWI fly ash and Halowax 1014 [126]. The *hatched bars* represent congeners that are produced in greater abundance during combustion processes (Table 4)

Elsewhere, Jaward et al. (2004) used a ratio of CN28/43/29 to the sum of 13 PCN congeners as a combustion indicator in passive air samples collected across Europe, finding that contributions of CN28/43/29 were < 5% in most locations, but were elevated at locations in Spain (28%) and Hungary (20%) [108]. Meijer et al. (2001) determined that the contribution of combustion-related congeners (CN29, 52/60, 51, 54, and 66/67) to their homolog sums increased significantly with year to more recent times in archived UK soils [134].

# 6
# Levels and Profiles of PCNs in the Great Lakes Environment

Monitoring studies of PCNs have been conducted in several regions such as Sweden, the Baltic Sea, Germany, the UK, and the Arctic, particularly since the early 1990s. This section summarizes findings from such studies for the Great Lakes region where PCNs have been measured in air, water, sediment, and biota. Comparisons of PCN concentrations in air, fish, and sediments, although not comprehensive, provide some insight into the spatial distribution of contamination by PCNs across the basin (Fig. 5).

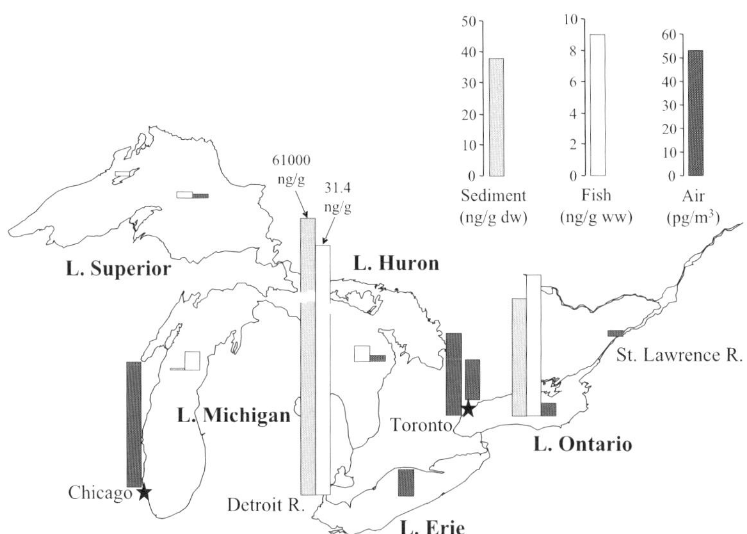


**Fig. 5** Spatial distribution of PCNs in sediment [118–120], predator fish [149], and air [63, 76, 126] in the Great Lakes region. *Bars* represent maximum sediment concentrations (tri-octaCNs, ng g$^{-1}$ dw), maximum fish concentrations (tri-octaCNs, ng g$^{-1}$ ww), and mean air concentrations (tetra-octaCNs, pg m$^{-3}$)

## 6.1
## Abiotic Environment

### 6.1.1
### Urban Air

Ambient air measurements for PCNs have been conducted in urban, suburban, semi-rural, rural, and remote locations around the world, as well as in urban areas (Chicago, USA; Toronto, Canada) and over or near the waters of the Great Lakes region (Table 5). The highest concentrations were generally found in urban areas. Harner and Bidleman (1997) reported a mean $\Sigma$PCN concentration (sum of tri- to octaCNs) of $68\,pg\,m^{-3}$ (excludes one outlier) in air collected in Chicago in February 1995, while two samples collected in north Toronto, Ontario in March 1995 were lower with concentrations of 12 and $22\,pg\,m^{-3}$ [76]. Mean $\Sigma$PCN levels were higher at a downtown site ($51\,pg\,m^{-3}$) than the north Toronto location ($28\,pg\,m^{-3}$) for samples collected periodically from 1998 to 2001 [126]. These are similar to mean levels of $60\,pg\,m^{-3}$ reported in an urban area in Germany [141], but lower than the mean concentrations of 138 and $160\,pg\,m^{-3}$ found in urban Manchester, UK, air [107], and of 152 and $110\,pg\,m^{-3}$ in a semirural area near Lancaster, UK (in 1995 and 2001, respectively) [106, 142], and $85\,pg\,m^{-3}$ in Chilton, UK [142].

Homolog distributions and congener patterns may provide indications of contributing sources of PCNs (Sect. 5.1). The triCNs were most abundant in Chicago air followed by tetraCNs, and a similar distribution was found in north Toronto [76, 97]. The authors suggested the homolog profiles resembled those of the widely used Halowax 1001 and 1099 mixtures. In the subsequent Toronto study, the PCN homolog distributions in air varied between the two sampling sites. The tetraCNs were most abundant in air at the downtown location and the pentaCNs contributed more to the homolog distribution than at the north Toronto site; the triCNs were prevalent in air at the north Toronto location [126]. These observations indicated that the downtown site was probably closer to and more directly influenced by evaporative emission sources.

Helm and Bidleman (2003) found that PCN congener patterns varied between the downtown and north Toronto sites in a detailed examination [126]. The patterns were similar at the downtown site for all sampling periods with only minor contributions of combustion-related congeners (Sect. 5.1). However, the patterns varied at the north Toronto site, particularly during sampling periods with lower temperatures, with an apparent enrichment of combustion-related congeners [126]. In the most extreme case it was estimated that combustion sources accounted for approximately 50% of PCNs in air. It was suggested that PCNs were predominantly from evaporative emissions to the air in downtown Toronto, much of which was built and industrialized during the peak usage period for PCNs, while current sources

**Table 5** Summary of concentrations (pg m$^{-3}$) of ΣPCNs in air in the Great Lakes region and comparison to other regions

| Region/location | N | Tri-OctaCN Mean | Range | Tetra-OctaCN Mean | Range | Refs. |
|---|---|---|---|---|---|---|
| **Great Lakes** | | | | | | |
| Chicago, 1995 | 15 | 68[a] | 23–378 | 53 | 16–80 | [76] |
| North Toronto, 1995 | 2 | 17 | 12–22 | | | [76] |
| Downtown Toronto, 2000–01 | 6 | 51 | 31–78 | 35 | 18–59 | [126] |
| North Toronto, 1998–2001 | 12 | 28 | 7–84 | 17 | 5–49 | [126] |
| Lake Ontario, 1998 | 5 | | | 5.5 | 3.1–7.9 | [63] |
| Lake Ontario, 2000 | 5 | 12.3 | 8.5–17.8 | 5.6 | 3.2–9.2 | [63] |
| St. Lawrence River, 1998 | 4 | | 2.7 | 1.3–5.2 | | [63] |
| Near Cornwall ON, 1999 | 1 | 6.5 | – | 3.5 | – | [63] |
| Lake Erie | 1 | | 11.5 | – | | [63] |
| Lake Huron | 4 | | 2.4 | 1.0–3.0 | | [63] |
| Lake Superior | 9 | | 1.6 | 0.5–2.6 | | [63] |
| **Europe** | | | | | | |
| Augsburg, Germany, 1992–93 | 47 | 60 | – | | | [141] |
| Manchester UK, 1998–99 | 2 | | 138–160[b] | | | [107] |
| Hazelrigg UK, 1995 | 27 | 152 | 73–223 | | | [106] |
| Hazelrigg UK, 2001 | 42 | 110 | 31–310 | | | [142] |
| Chilton UK, 2001 | 41 | 85 | 31–180 | | | [142] |
| Mace Head, Ireland, 2000 | 25 | 15 | 1.7–55 | | | [142] |
| Urban, rural, remote sites, 2002 | 71 | – | 0.2–220[c] | | | [108] |
| **Remote and other locations** | | | | | | |
| Alert, Canada, 1994–95 | 14 | 0.69 | < 0.01–1.3 | | | [20] |
| Dunai, Russia, 1994–95 | 11 | 0.82 | < 0.01–2.9 | | | [20] |
| Tagish, Canada, 1994–95 | 12 | 0.38 | < 0.01–1.7 | | | [20] |
| Alert, Canada, 1993–94 | 5 | 3.5 | 0.91–8.0 | | | [19] |
| Dunai, Russia, 1993 | 3 | 0.84 | 0.31–1.2 | | | [19] |
| Barents Sea, 1996 | 2 | 40.4 | 32.0–48.7 | | | [19] |
| Eastern Arctic Ocean, 1996 | 10 | 11.6 | 1.6–16.0 | | | [19] |
| Norwegian Sea, 1996 | 2 | 7.1 | 5.8–8.3 | | | [19] |
| Ny–Ålesund, Norway, 2001 | 6 | 35.0 | 27.1–48.3 | | | [21] |
| Tromsø, Norway, 2003 | 10 | 25.0 | 8.8–47.0 | | | [21] |
| Hoburgen, Sweden, 1990–91 | 7 | | | 5.1 | 3.0–10.0 | [109] |
| Ammarnäs, Sweden, 1990–91 | 7 | | | 1.6 | 0.5–3.8 | [109] |

[a] Excludes high value of 378 pg m$^{-3}$
[b] Range of values for two pooled extracts
[c] Sum of 13 tri-pentaCN congeners

such as combustion from industries now located in the northern parts of Toronto, and fewer evaporative sources in the less dense, newer residential areas, account for the more varied patterns at this site [126].

Urban air concentrations were multiplied by relative potencies (REPs) (listed in Table 2) to calculate the contribution of PCNs to dioxin toxic equivalents (TEQ) and to compare the contributions of the dioxin-like PCBs using REPs from Giesy et al. (1997) [143]. On average, PCNs contributed 64% of total PCN and dioxin-like PCB (PCN+PCB) TEQ in downtown air and 48% in north Toronto. The PCN contribution to PCN+PCB TEQ in Chicago air was similar (68%) when recalculated using the same REPs [97, 126]. Although PCNs are as important as PCBs in air on a TEQ basis, polychlorinated dioxins and furans remain the dominant contributors of TEQ in downtown Toronto air [126].

## 6.1.2
### Air over the Great Lakes

The spatial distribution of PCNs in the air over the Great Lakes has been examined by collecting samples during several cruises on the lakes between 1996 and 2000, including Lake Ontario and the St. Lawrence River, Lake Superior, and during transit through Lakes Erie and Huron [63]. Comparisons were made for the tetra- to octaCNs (summarized in Fig. 5 and Table 5). PCN concentrations were greater over the more populated and industrialized lower Great Lakes (Lakes Ontario and Erie) than the upper Great Lakes (Lakes Huron and Superior) and the upper St. Lawrence River. Over Lake Ontario, higher concentrations were associated with wind directions from the west and proximity to the western portion of the lake, which is more intensely developed [63].

Spatial differences in the homolog profiles were noted on a regional scale over the lakes, and locally over Lake Ontario. The penta- and hexaCNs contributed more and the tetraCNs less, on a mass basis, to the distribution of PCNs over Lake Ontario than over Lake Superior. This indicates that more sources are within the Lake Ontario region while transport influences the distribution of the lighter tetraCNs further from these sources [63]. Similarly, in the July 1998 cruise of Lake Ontario and the upper St. Lawrence River, air over the St. Lawrence River had higher proportions of tetraCNs (88–91% of $\Sigma$PCN) than over-lake air (82–83%) [63]. The air over the lake also showed greater proportions of penta- and hexaCNs. The over-lake distributions probably result from the influence of nearby urban sources, while enrichment of the tetraCNs, which are favored during transport due to their lower vapor pressures, occurs over the St. Lawrence River further from source areas.

Evaporative sources of PCNs to air over the lakes were dominant, although minor contributions of combustion sources were indicated by the presence of combustion-related congeners CN44, 29, and 54 [63].

### 6.1.3
### Water

Very few measurements of PCNs in ambient waters have been attempted or reported. In a stream near a PCN production facility in the USA, PCN concentrations ranged from 600 to 1400 ng L$^{-1}$ with mono- to pentaCNs quantified [14]. Near capacitor production facilities, a concentration of 5500 ng L$^{-1}$ (mono-, di-, and tetraCNs) was found in the outfall of one facility but was below detection in stream water. A concentration of 600 ng L$^{-1}$ (tri- and tetraCNs) was found downstream of a second facility [14]. In two Spanish rivers, mean mono- and diCN concentrations of 650–760 and 150–260 ng L$^{-1}$, respectively, were found [144].

Only limited studies have been undertaken of PCNs in water in the Great Lakes region. PCNs were measured in water from Lake Ontario in June, 2000, with triCN concentrations (12.1–19.0 pg L$^{-1}$) and tetraCN concentrations (4.3–7.7 pg L$^{-1}$) reported [63]. PCNs were not detected in the waters of Owen Sound Harbour (MDL of 10 ng L$^{-1}$), although an industrial effluent in the area was found to contain PCNs [117].

### 6.1.4
### Soil and Sediment

PCNs have been measured in very few soils in North America. Erickson et al. (1978) found total PCNs at concentrations of 130–2300 ng g$^{-1}$ near a PCN manufacturing facility in the USA, and values of non-detect to 7.3 ng g$^{-1}$ and 2.7–470 ng g$^{-1}$ near two capacitor manufacturing plants [14]. Elevated concentrations of PCNs (several parts-per-million), primarily Halowax 1014, have been found in soils of a former industrial facility located along the Detroit River [145]. While the occurrence of PCNs in background soils in the Great Lakes region has not been determined, their presence is likely as they tend to be sorbed to organic matter as with other OC compounds. As well, soils in the region amended with municipal sewage sludges (or biosolids) may have elevated PCN concentrations compared to unamended soils. Archived sludge-amended soils in the UK had PCN levels which were 1.5–6 times higher than in control soils [134] over the time period for which archived soils were analyzed. PCNs in the amended soils ranged from 9.4 ng g$^{-1}$ in 1976 then decreased to 2.5 ng g$^{-1}$ in 1990 while control soils decreased from 6 ng g$^{-1}$ to 0.4 ng g$^{-1}$ over the same time period. The relative contributions of combustion congeners were found to increase over this period [134].

Sediments have been analyzed for PCNs in a limited number of areas within the Great Lakes. PCNs were not found in sediments from six Michigan harbors (approximate detection limit 30 ng g$^{-1}$) [146] and were detected but not quantified in Lake Ontario surface sediments [147]. Table 6 summarizes more recent measurements of PCNs in sediments in the Great Lakes

**Table 6** Summary of ΣPCN concentrations (ng g$^{-1}$ dw) in sediments from selected areas of the Great Lakes region

| Location | Sediment | ΣPCN[a] | Refs. |
|---|---|---|---|
| Lake Michigan | Surface | 0.31–0.79 | [119] |
| Lake St. Clair | Suspended sediment | 3.5 | [120] |
| Detroit River Area | | | |
|   Upper Detroit River[b] | Surface, or 0–30 cm | 0.1–187 | [119] |
|   Upper Detroit River | Surface | ND–8.5 | [118] |
|   Upper Detroit River | Suspended sediment | 1.2–9.2 | [120] |
|   Rouge River | Surface | 18.2 | [119] |
|   Trenton Channal | Surface | 68–61 000 | [118] |
|   Trenton Channal | Suspended sediment | 7350–8200 | [120] |
|   Lower Detroit River (West)[c] | Surface | 0.4–1600 | [118] |
|   Lower Detroit River (West) | Suspended sediment | 299 | [120] |
|   Lower Detroit River (East) | Suspended sediment | 6.0–8.0 | [120] |
| Lake Ontario | Surface | 20.7–38.0 | d |

[a] Tri-octaCN
[b] Upper Detroit River defined as upstream of Grosse Ile
[c] Lower Detroit River defined as downstream of Grosse Ile, West is the US side, and East is the Canadian side
[d] Helm et al. unpublished results, 2005

basin. The highest concentrations (up to 61 μg g$^{-1}$ dw) have been detected in surface and suspended sediments within the Detroit River in the Trenton Channel [118, 120]. These levels are similar to those found in soil and sediment near a chlor-alkali facility at a Superfund site in Georgia contaminated with PCNs (18–23 μg g$^{-1}$ dw) [77]. Sediment levels were lower in the upper Detroit River (above Grosse Ile) and through the eastern sections of the river. Only two of the lakes have sediment values reported for PCNs. Surface sediment samples collected from Lake Michigan (near Muskegon, MI) had concentrations of < 1 ng g$^{-1}$ dw [119], while surface sediment samples from across Lake Ontario ranged from 21 to 38 ng g$^{-1}$ dw (Helm et al. unpublished results, 2005). In comparison, PCN concentrations in sediments from background locations, subject only to atmospheric deposition, were 0.23 ng g$^{-1}$ dw in Lake Storvindeln (Sweden) [79] and ranged from 2.7 pg g$^{-1}$ dw in the surface layer to a subsurface maximum of 12.2 pg g$^{-1}$ dw in a core from Esthwaite Water (UK) [148].

PCNs were determined in a homolog-specific [118] and congener-specific [119, 120] manner in the Detroit River studies. The penta- and hexaCNs homologs were found to be prevalent (Fig. 6), and particularly congeners 59, 69 and 71/72, leading Kannan et al. (2001) to suggest that the Halowax mixtures were the dominant source [119]. However, they noted that the presence of congeners 54, 44, and 29 also indicated the contributions of combustion sources. The observed profiles differ from those in sediments associated with contamination from a chlor-alkali production facility [77]. In Lake Ontario sediments, octaCN was most abundant, followed by the heptaCNs and declining by degree of chlorination with the triCNs contributing the least to ΣPCN (Fig. 6) (Helm et al. unpublished results, 2005). These profiles do resemble those associated with chlor-alkali production [77].

Many of the sediment samples from the Great Lakes were analyzed using congener-specific methods, facilitating comparisons between PCNs, dioxin-like PCBs, and PCDD/Fs on a TEQ basis using REPs for PCNs from Table 2. In sediments from three locations in the upper Detroit River, contributions to TEQ were greatest for PCNs (42–84%) relative to PCBs and PCDD/Fs [119]. In suspended sediments from the Trenton Channel, PCN contributions to TEQ were tenfold higher than that of PCDD/Fs and PCBs [120]. As a result of sediment contamination, fish in the same area have been found to have relatively higher TEQ contributions from PCNs [149]. In Lake Ontario sediments, PCNs contributed approximately 85% of the TEQ relative to the dioxin-like PCBs (Helm et al. unpublished results, 2005).

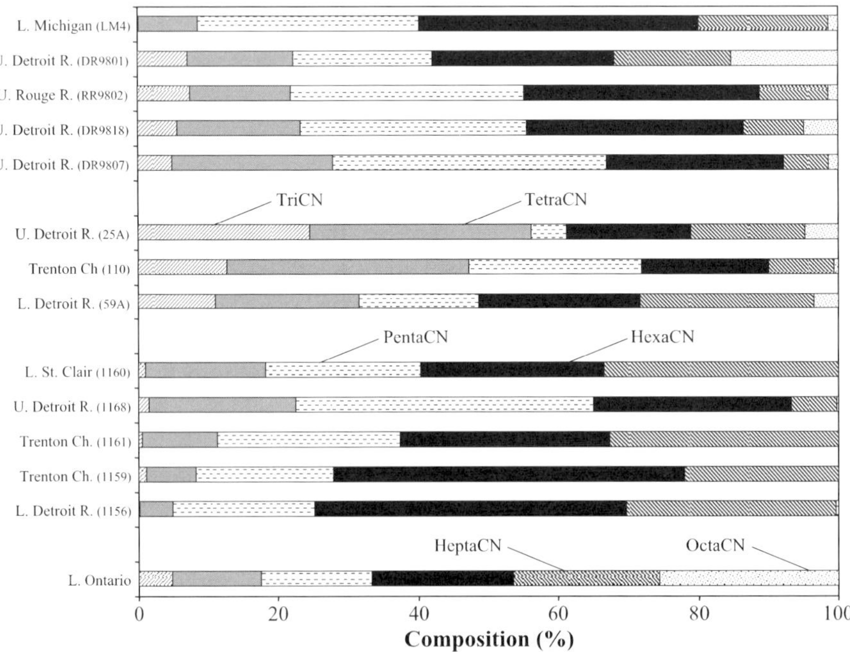


**Fig. 6** Percent homolog composition of tri- to octaCNs in selected sediments and suspended sediments from the Great Lakes [118–120]

## 6.2
## Biotic Environment

### 6.2.1
### Wildlife

Due to their persistence and lipophilic properties, PCNs tend to bioaccumulate in humans and wildlife. Occurrence of PCNs in wildlife tissues, particularly in birds, was reported as early as the 1970s [18, 150]. The development of congener-specific methods of PCN analysis has led to several reports of their occurrence in biota in the Baltic Sea region [6, 29, 31, 151, 152], the Great Lakes ( [119, 149], Helm et al. unpublished results, 2005), the Mediterranean Sea [153] and polar regions [23, ?, 24]. The occurrence of PCNs in biota from polar regions indicates their ability to undergo long-range transport (Table 7).

Concentrations of PCNs in biota from the Great Lakes watershed are compiled in Table 8. The greatest concentrations occurred in carp and walleye from the Detroit River (up to $31.4\,\mathrm{ng\,g^{-1}}$ wet weight [ww]) [149], followed by lake trout from Lake Ontario (up to $9.0\,\mathrm{ng\,g^{-1}}$ ww) (Helm et al. unpublished results, 2005), and eggs of double-crested cormorant eggs (up to $2.4\,\mathrm{ng\,g^{-1}}$ ww) [154]. In comparing TEQ contributions, PCNs contributed 2–57% of the TEQ (relative to PCB) in Detroit River fish [149], 1–22% in several species from the St. Clair, Saginaw, and Raisin Rivers [32], and 12–22% in Lake Ontario lake trout (Helm et al. unpublished results, 2005). The contributions to TEQ by PCDD/Fs in these samples were not determined in these studies. In herring gull and double-crested cormorant eggs, PCNs contributed on average approximately 1% to total (PCN, PCB, and PCDD/F) TEQs [154].

The profiles of PCN homologs (Fig. 7) and congeners in biological matrices vary depending on the location and species. However, the percent

**Table 7** $\Sigma$PCN concentrations ($\mathrm{pg\,g^{-1}}$ ww) in biota from polar regions

| Location | Species | N | $\Sigma$PCN[a] | Refs. |
|---|---|---|---|---|
| Pangnirtung, Canada | Beluga whale blubber | 6 | 33.4–356 | [22] |
| Pangnirtung, Canada | Ringed seal blubber | 6 | 29.4–60.9 | [22] |
| Prince Leopold & Devon Islands, Canada | Northern fulmar eggs | 10 | 114–236 | [24] |
| Ross Sea, Antarctica | Krill | Pool | 1.5 | [23] |
| Ross Sea | Icefish | 2 | 2.1–4.7 | [23] |
| Ross Sea | Silverfish | 1 | 86 | [23] |
| Ross Sea | Weddell seal blubber | 1 | 77 | [23] |
| Ross Sea | South polar skua liver | 1 | 2550 | [23] |
| Alaska, United States | Polar bear | 5 | < 0.1–945 | [23] |

[a] Tri-octaCNs

**Table 8** Reported concentrations of ΣPCNs (pg g$^{-1}$ ww) in biota from the Great lakes and its watershed

| Location/species | Tissue | ΣPCN [a] | Refs. |
|---|---|---|---|
| Lake Superior | | | |
| Double-crested cormorant | Egg | 380–2000 | [154] |
| Herring gull | Egg | 83–1300 | [154] |
| Lake trout | Whole | 340–370 | [149] |
| Lake whitefish | Fillet | 240 | [149] |
| Coho salmon | Fillet | 240 | [149] |
| Lake Michigan | | | |
| Lake trout | Fillet | 1200 | [149] |
| Siskiwit Lake, Isle Royale, Lake Superior | | | |
| Lake trout | Fillet | 41–140 | [149] |
| Lake trout | Whole | 250 | [149] |
| Lake Huron | | | |
| Double-crested cormorant | Egg | 520–2400 | [154] |
| Herring gull | Egg | 390–720 | [154] |
| Lake trout | Fillet | 980 | [149] |
| Lake whitefish | Fillet | 1100 | [149] |
| Detroit River | | | |
| Carp | Whole | 1310–26 400 | [149] |
| Walleye | Whole | 31 400 | [149] |
| Michigan inland lakes and rivers | | | |
| Various | | 19–430 | [149] |
| Raisin River | | | |
| Largemouth bass | Fillet | 220 | [32] |
| Smallmouth bass | Fillet | 11–28 | [32] |
| Round goby | Whole | 260–1140 | [32] |
| Zebra mussel | Whole | 1120–1750 | [32] |
| Phytoplankton | Whole | 12 200 | [32] |
| Amphipod | Whole | 1100 | [32] |
| Benthic algae | Whole | 780 | [32] |
| Saginaw River | | | |
| Round goby | Whole | 190–320 | [32] |
| Zebra mussel | Whole | 360 | [32] |
| St. Clair River | | | |
| Smallmouth bass | Fillet | 32 | [32] |
| Round goby | Whole | 29–130 | [32] |
| Zebra mussel | Whole | 22–53 | [32] |
| Phytoplankton | Whole | 1370 | [32] |
| Benthic algae | Whole | 2 | [32] |

composition of tri- and tetraCNs generally decrease from lower to higher trophic organisms, while penta- and hexaCN proportions increase at higher trophic levels [32]. Furthermore, the congener profiles tend to shift to fewer dominant congeners which, having no adjacent unsubstituted carbons, fa-

**Table 8**  (continued)

| Location/species | Tissue | $\Sigma$PCN[a] | Refs. |
|---|---|---|---|
| Lake Ontario | | | |
|   Lake trout | Whole | 1210–9000 | [b] |
|   Walleye | Whole | 690–2480 | [b] |
|   Lake Whitefish | Whole | 410–1020 | [b] |
|   Rainbow smelt | Whole | 120–620 | [b] |
|   Slimy sculpin | Whole | 960–1270 | [b] |
|   Alewife | Whole | 90–210 | [b] |
|   *Mysis relicta* | Whole | 240–460 | [b] |
|   *Diporeia hoyi* | Whole | 1370–1540 | [b] |
|   Net plankton | Whole | ND-20 | [b] |

[a] Tri-octaCN
[b] Helm et al. unpublished results, 2005

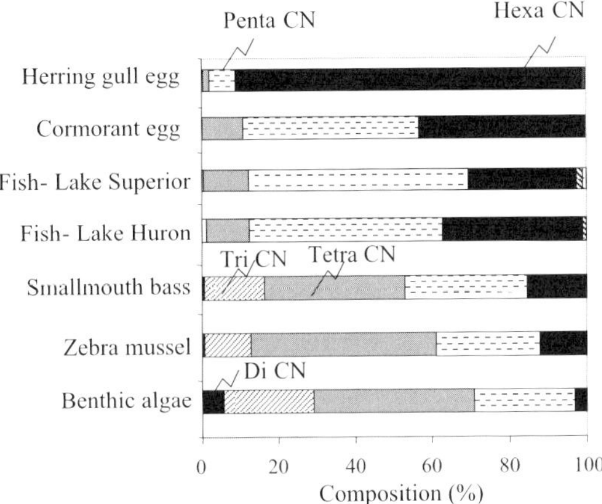


**Fig. 7**  Percent homolog composition of tri- to octaCNs in selected biota from the Great Lakes region [32, 149, 154]. Smallmouth bass, zebra mussels, and benthic algae were collected from the St. Clair River

vor bioaccumulation (discussed in Sect. 4.4). However, exposures to different PCN sources, location, and other environmental variables influence the observed congener profiles [31, 32, 149].

## 6.2.2
## Humans

The only study that reported the accumulation of PCNs in humans in the Great Lakes region is by Williams et al. (1993) [85]. Adipose fat in people from sev-

**Table 9** Residue concentrations of $\Sigma$PCNs (ng g$^{-1}$ lipid weight) in human adipose tissue from several countries

| Location | Range | Refs. |
| --- | --- | --- |
| Ontario, Canada | 0.1–6.2 | [85] |
| Sweden | 1.0–3.9 | [159] |
| Tokyo, Japan | 2.8–17.0 | [158] |
| Matsuyama, Japan | 3.2–10.0 | [160] |
| Osaka, Japan | 2.0–250 | [160] |
| Saratov, Russia | 4.1–14.0 | [161] |
| Alma-Ata, Kazakhstan | 2.0–18.0 | [161] |
| Manheim, Germany | 18.0–35.0 | [161] |
| Rheda-Wiedenbrück, Germany | 1.1–8.0 | [161] |
| Stralsund, Germany | 0.9–5.1 | [161] |

eral municipalities in Ontario contained PCNs at concentrations ranging from 0.1 to 6.2 ng g$^{-1}$ lipid weight (lw). Congeners 66/67 were predominant in human fat. Based on the analysis of foodstuffs, oil and fat were suggested to be the major sources of human exposure to PCNs in Spain [155]. However, studies have indicated that the main route of exposure, in general, is probably the ingestion of fish [156]. Asplund (1994) suggested a relation between high fish intake to high levels of PCN concentrations in blood plasma [157]. PCNs have been measured in human adipose tissues (Table 9) and breast milk from several other regions [158–161]. The levels in the milk are strongly correlated to milk fat content and reflect the accumulated levels in the adipose tissues. Witt and Niessen (2000) measured PCNs in adipose tissue of children in Germany and Russia and found that the concentrations vary from 0.9 to 34.6 ng g$^{-1}$ lw and were exclusively tetra-, penta-, and hexaCNs [161]. The distribution of the congeners varied between samples and locations. Weistrand and Norén (1998) measured PCN concentrations in human liver and adipose tissues [159]. In both tissues, congeners 33/34/37, 52/60, and 66/67 were the most abundant.

Temporal trends have been reported for PCN concentrations in Swedish human milk collected from 1972 to 1992 with concentrations declining nearly tenfold by 1992 from 3.1 ng g$^{-1}$ lw in 1972 [162, 163]. In general, PCN concentrations in human tissues appear to be decreasing following changes in production and use patterns. In breast milk of women from Los Angeles, USA, PCN concentrations ranged from 1.7 to 3 ng g$^{-1}$ lw [156].

# 7
## Conclusions and Recommendations for Further Research

Studies on the occurrence of PCNs in the Great Lakes region, while limited in scope, indicate that contamination by these compounds is as widespread as related compounds such as PCBs and PCDD/Fs. Atmospheric concentrations are greatest in urban areas, while sediment levels in the few areas examined to date are highest in the industrialized areas along the Detroit River, particularly Trenton Channel. PCNs have been measured in a range of aquatic biota in or near the Great Lakes, including benthic invertebrates, fish, and the eggs of cormorants and herring gulls. The fish with the highest concentrations were carp and walleye from the Detroit River and Lake Ontario lake trout. Homolog and congener profiles of PCN source and ambient environmental samples indicate that, although evaporative emissions from past use of technical Halowax mixtures or from PCB use are dominant, combustion sources do contribute and may be significant in particular areas of the Great Lakes.

Our knowledge and understanding of PCN occurrence, sources, and fate in the Great Lakes region are much less extensive than for PCBs and PCDD/Fs. This is probably a result of the attention attracted by the degree of contamination by PCBs and PCDD/Fs in the 1970s in the lakes, and their effects, but also because PCNs had largely been replaced by PCBs by that time. Certainly, challenges in PCN analysis also limited the extent of study. Once mass spectrometer-based methods became more routine in the 1980s and congener-specific methods were developed in the 1990s, interest in the occurrence and effects of PCNs increased. It is now apparent that PCNs significantly contribute to the burden of dioxin toxic equivalents (TEQ), particularly in fish from the Detroit River, but also in sediments from the Detroit River area and Lake Ontario. The physical-chemical properties and bioaccumulation behavior of PCNs have been found to be similar to PCBs.

Several aspects of the story of PCNs in the Great Lakes are yet to be told, and our understanding of the long-term persistence, fate, and effects of PCNs in the region would benefit from a more complete assessment of their sources, properties, and temporal and spatial trends. For example, there have been no retrospective studies of PCNs in sediment cores or archived matrices (sediments, fish, bird eggs, or humans) to date in the Great Lakes. Bioaccumulation and other impacts on the ecosystem should be examined further, including downstream levels and effects in organisms such as the St. Lawrence beluga whales, and the relative role of chemical cycling from historical sources compared to current inputs from combustion on observed concentrations in biota. TEQ contributions from PCNs should be compared to PCB and PCDD/F contributions in a range of fish species from across the Great Lakes to fully assess exposure and consider incorporating the findings into consumption guidelines. The environmental fate of PCNs remain unknown, particularly the extent and rates of degradation in various matrices. More detailed measurements of physical-chemical properties such as Henry's Law constants and octanol–water partition coefficients are needed for the application of fate models. Although production of PCNs no longer occurs within the basin, surveys to determine if "hot spots" of in-place contamination other than the Detroit River exist would be worthwhile. PCNs are proposed additions to the list of banned/restricted compounds under the POPs Protocol of the Convention on Long Range Transboundary Air Pollution sponsored by the United Nations Economic Commission for Europe. To this end, more information is required on combustion sources and on whether controls for PCDD/Fs are effective for PCNs.

## References


1. Laurent MA (1833) Ann Chim Phys Ser 2 52:275
2. Aylsworth JW (1909) US Patent 914:222
3. Aylsworth JW (1909) US Patent 914:223
4. Aylsworth JW (1914) US Patent 1:111–289
5. Brinkman UATh, Reymer HGM (1976) J Chromatogr 127:203
6. Falandysz J (1998) Environ Pollut 101:77
7. Jakobsson E, Asplund L (1999) Polychlorinated naphthalenes. In: Paasivirta J (ed) New types of persistent halogenated compounds. Springer, Heidelberg Berlin New York, p 97
8. Popp W, Norpoth K, Vahrenholz C, Hamm S, Balfanz E, Theisen J (1997) Am J Ind Med 32:413
9. Weistrand C, Lundén Å, Norén K (1992) Chemosphere 24:1197
10. Yamashita N, Taniyasu S, Hanari N, Horii Y, Falandysz J (2003) J Environ Sci Health A38:1745
11. Wiedmann T, Ballschmiter K (1993) Fres J Anal Chem 346:800
12. Hayward D (1998) Environ Res 76A:1
13. Crump-Weisner HJ, Feltz HR, Yates ML (1973) US Geol Surv J Res 1:603
14. Erickson MD, Michael LC, Zweidinger RA, Pellizzari ED (1978) J Assoc Off Anal Chem 61:1335
15. Erickson MD, Michael LC, Zweidinger RA, Pellizzari ED (1978) Environ Sci Technol 12:927
16. Ofstad EB, Lunde G, Martinsen K, Rygg B (1978) Sci Total Environ 10:219
17. Koeman JH, Van Velzen-Blad HCW, De Vries R, Vos JG (1973) J Reprod Fert, Suppl 19:353
18. Cooke M, Roberts DJ, Tillett ME (1980) Sci Total Environ 15:237
19. Harner T, Kylin H, Bidleman TF, Halsall C, Strachan WMJ, Barrie LA, Fellin P (1998) Environ Sci Technol 32:3257
20. Helm PA, Bidleman TF, Li HH, Fellin P (2004) Environ Sci Technol 38:5514
21. Herbert BMJ, Halsall CJ, Villa S, Fitzpatrick L, Jones KC, Lee RGM, Kallenborn R (2005) Sci Total Environ 342:145
22. Helm PA, Bidleman TF, Stern GA, Koczanski K (2002) Environ Pollut 119:69
23. Corsolini S, Kannan K, Imagawa T, Focardi S, Giesy JP (2002) Environ Sci Technol 36:3490
24. Muir D (2004) New contaminants in arctic biota. In: Smith S, Stow J, Carrillo F (eds) Synopsis of research conducted under the 2003–2004 Northern Contaminants Program, Indian and Northern Affairs, Ottawa, p 139
25. Hanberg A, Wærn F, Asplund L, Haglund E, Safe S (1990) Chemosphere 20:1161
26. Hanberg A, Ståhlberg M, Georgellis A, de Wit C, Ahlborg UG (1991) Pharmacol Toxicol 69:442

27. Blankenship AL, Kannan K, Villalobos SA, Villeneuve DL, Falandysz J, Imagawa T, Jakobsson E, Giesy JP (2000) Environ Sci Technol 34:3153
28. Villeneuve DL, Kannan K, Khim JS, Falandysz J, Nikiforov VA, Blankenship AL, Giesy JP (2000) Arch Environ Contam Toxicol 39:273
29. Järnberg U, Asplund L, de Wit C, Grafström A-K, Haglund P, Jansson B, Lexén K, Strandell M, Olsson M, Jonsson B (1993) Environ Sci Technol 27:1364
30. Falandysz J, Rappe C (1996) Environ Sci Technol 30:3362
31. Lundgren K, Tysklind M, Ishaq R, Broman D, van Bavel B (2002) Environ Sci Technol 36:5005
32. Hanari N, Kannan K, Horii Y, Taniyasu S, Yamashita N, Jude DJ, Berg MB (2004) Arch Environ Contam Toxicol 47:84
33. Crookes MJ, Howe PD (1993) Environmental hazard assessment: halogenated naphthalenes. TSD/13. Department of Environment, London
34. Kover FD (1975) Environmental hazard assessment report: chlorinated naphthalenes. EPA-560/8-75-001. US EPA, Office of Toxic Substances, Washington, DC
35. Beland FA, Geer RD (1973) J Chromatogr 84:59
36. Drinker CK, Warren MF, Bennett GA (1937) J Ind Hyg Toxicol 19:283
37. Greenberg L, Mayers MR, Smith AR (1939) J Ind Hyg Toxicol 21:29
38. Crow KD (1970) Trans St John Hosp Dermatol Soc 56:79
39. Kleinfeld M, Messite J, Swencicki R (1972) J Occup Med 14:377
40. Ward EM, Ruder AM, Suruda A, Smith AB, Fessler-Flesch CA, Zahm SH (1996) Am J Ind Med 30:225
41. Von Wedel H, Holla WA, Denton J (1943) Rubber Age 53:419
42. Olson C (1969) Bovine hyperkeratosis (X-disease, highly chlorinated naphthalene poisoning), historical review. In: Brandly CA, Cornelius CE (eds) Advances in veterinary sciences and comparative medicine. Academic, New York, p 101
43. Cornish HH, Block WD (1958) J Biol Chem 231:583
44. Pudelkiewicz WJ, Boucher RV, Callenbach EW, Miller RC (1959) Poult Sci 38:424
45. Campbell MA, Bandiera S, Robertson L, Parkinson A, Safe S (1983) Toxicology 26:193
46. Mantyla E, Ahotupa M (1993) Chemosphere 27:383
47. Omura M, Masuda Y, Hirata M, Tanaka A, Makita Y, Ogata R, Inoue N (2000) Environ Health Persp 108:539
48. Engwall M, Brunström B, Jakobsson E (1994) Arch Toxicol 68:37
49. Norrgren L, Andersson T, Bjork M (1993) Aquat Toxicol 26:307
50. Pesonen M, Teivainen P, Lundström J, Jakobsson E, Norrgren L (2000) Arch Environ Contam Toxicol 38:52
51. Holm G, Norrgren L, Andersson T, Thurén A (1993) Aquat Toxicol 27:33
52. Åkerblom N, Olsson K, Berg AH, Andersson PL, Tysklind M, Förlin L, Norrgren L (2000) Arch Environ Contam Toxicol 38:225
53. Villalobos SA, Papoulias DM, Meadows J, Blankenship AL, Pastva SD, Kannan K, Hinton DE, Tillitt DE, Giesy JP (2000) Environ Toxicol Chem 19:432
54. Laughlin RB Jr, Neff JM (1979) Mar Environ Res 2:275
55. Villeneuve DL, Khim JS, Kannan K, Giesy JP (2001) Aquat Toxicol 54:125
56. Behnisch PA, Hosoe K, Sakai S (2003) Environ Int 29:861
57. Lei YD, Wania F, Shiu W-Y (1999) J Chem Eng Data 44:577
58. Järnberg U, Asplund L, Jakobsson E (1994) J Chromatogr A 683:385
59. Harner T, Bidleman TF (1998) J Chem Eng Data 43:40
60. Chen J, Xue X, Schramm K-W, Quan X, Yang F, Kettrup A (2003) Comput Biol Chem 27:165

61. Staikova M, Wania F, Donaldson DJ (2004) Atmos Environ 38:213
62. Puzyn T, Falandysz J (2005) Atmos Environ 39:1439
63. Helm PA, Jantunen LM, Ridal JJ, Bidleman TF (2003) Environ Toxicol Chem 22:1937
64. Opperhuizen A (1987) Toxicol Environ Chem 15:249
65. Isnard P, Lambert S (1988) Chemosphere 14:1871
66. Lei YD, Wania F, Shiu WY, Boocock DGB (2000) J Chem Eng Data 45:738
67. Opperhuizen A, van der Velde EW, Gobas FAPC, Liem DAK, van der Steen JMD, Hutzinger O (1985) Chemosphere 14:1871
68. Law LM, Goerlitz DF (1974) Pestic Monit J 8:33
69. Armour JA, Burke JA (1971) J Assoc Off Anal Chem 54:175
70. Stalling DL, Smith LM, Petty JD (1979) Approaches to comprehension analyses of persistent halogenated environmental contaminants. In: Van Hall CE (ed) Measurement of organic pollutants in water and wastewater, ASTM STP 686 American Society for Testing and Materials, p 302
71. Hutzinger O, Safe S, Zitko V (1972) Int J Environ An Ch 2:95
72. Brinkman UATh, de Vries G, de Kok A, de Jonge AL (1978) J Chromatogr 152:97
73. Zimmerli B (1974) J Chromatogr 88:65
74. Cooke M, Nickless G, Prescott AM, Roberts DJ (1978) J Chromatogr 156:293
75. Jansson B, Asplund L, Olsson M (1984) Chemosphere 13:33
76. Harner T, Bidleman TF (1997) Atmos Environ 31:4009
77. Kannan K, Imagawa T, Blankenship AL, Giesy JP (1998) Environ Sci Technol 32:2507
78. Haglund P, Asplund L, Järnberg U, Jansson B (1990) J Chromatogr 507:389
79. Järnberg U, Asplund L, de Wit C, Egebäck A-L, Wideqvist U, Jakobsson E (1997) Arch Environ Contam Toxicol 32:232
80. Hanari N, Horii Y, Taniyasu S, Falandysz J, Bochentin I, Orlikowska A, Puzyn T, Yamashita N (2004) Pol J Environ Stud 13:139
81. Harner T, Kucklick JR (2002) Chemosphere 51:555
82. Hiedmann WA (1988) Chromatographia 25:8
83. Falandysz J, Kawano M, Ueda M, Matsuda M, Kannan K, Giesy JP, Wakimoto T (2000) J Environ Sci Health A35:281
84. Imagawa T, Yamashita N (1997) Chemosphere 35:1195
85. Williams DT, Kennedy B, LeBel GL (1993) Chemosphere 27:795
86. Helm PA, Jantunen LMM, Bidleman TF, Dorman FL (1999) J High Resol Chromatogr 22:639
87. Jakobsson E, Eriksson L, Bergman Å (1992) Acta Chem Scand 46:527
88. Horii Y, Kannan K, Petrick G, Gamo T, Falandysz J, Yamashita N (2005) Environ Sci Technol 39:4206
89. Järnberg UG, Asplund LT, Egebäck A-L, Jansson B, Unger M, Wideqvist U (1999) Environ Sci Technol 33:1
90. Klöpffer W, Haag F, Kohl E-G, Frank R (1988) Ecotox Environ Safe 15:298
91. Meylan WM, Howard PH (1993) Chemosphere 26:2293
92. Hites RA, Anderson PN, Hites (1996) Environ Sci Technol 30:1756
93. Keum Y-S, Li QX (2004) Bull Environ Contam Toxicol 72:999
94. Ruzo LO, Bunce NJ, Safe S, Hutzinger O (1975) Bull Environ Contam Toxicol 14:341
95. Ruzo LO, Bunce NJ, Safe S (1975) Can J Chemistry 53:688
96. Gulan MP, Bills DD, Putnam TB (1974) Bull Environ Contam Toxicol 11:438
97. Harner T, Bidleman TF (1998) Environ Sci Technol 32:1494
98. Finizio A, Mackay D, Bidleman T, Harner T (1997) Atmos Environ 31:2289
99. Falconer RL, Harner T (2000) Atmos Environ 34:4043
100. Helm PA, Bidleman TF (2005) Sci Total Environ 342:161

101. Hillery BR, Simcik MF, Basu I, Hoff RM, Strachan WMJ, Burniston D, Chan CH, Brice KA, Sweet CW, Hites RA (1998) Environ Sci Technol 32:2216
102. Ridal JJ, Kerman B, Durham L, Fox ME (1996) Environ Sci Technol 30:852
103. Zhang H, Eisenreich SJ, Franz TR, Baker JE, Offenberg JH (1999) Environ Sci Technol 33:2129
104. Green ML, DePinto JV, Sweet C, Hornbuckle KC (2000) Environ Sci Technol 34:1833
105. Barber JL, Thomas GO, Bailey R, Kerstiens G, Jones KC (2004) Environ Sci Technol 38:3892
106. Lee RGM, Burnett V, Harner T, Jones KC (2000) Environ Sci Technol 34:393
107. Harner T, Lee RGM, Jones KC (2000) Environ Sci Technol 34:3137
108. Jaward FM, Farrar NJ, Harner T, Sweetman AJ, Jones KC (2004) Environ Toxicol Chem 23:1355
109. Egebäck A-L, Wideqvist U, Järnberg U, Asplund L (2004) Environ Sci Technol 38:4913
110. Walsh GE, Ainsworth KA, Faas L (1977) Bull Environ Contam Toxicol 18:297
111. Green FA Jr, Neff JM (1977) Bull Environ Contam Toxicol 14:399
112. Kannan K, Nakata H, Stafford R, Masson GR, Tanabe S, Giesy JP (1998) Environ Sci Technol 32:1214
113. Burreau S, Axelman J, Broman D, Jakobsson E (1997) Environ Toxicol Chem 16:2508
114. Ruzo L, Jones D, Safe S, Hutzinger O (1976) J Agric Food Chem 24:581
115. Niimi AJ, Oliver BG (1988) Can J Fish Aquat Sci 45:222
116. Pellizzari ED (1982) Environ Sci Technol 16:781
117. Kauss PB (1991) Polychlorinated naphthalenes survey at Goodyear (Owen Sound Harbour), October 16–19, 1990. ISBN 0-7729-8747-5. Ontario Ministry of the Environment, Queen's Printer for Ontario, Toronto, ON
118. Furlong ET, Carter DS, Hites RA (1988) J Great Lakes Res 14:489
119. Kannan K, Kober JL, Kang Y-S, Masunaga S, Nakanishi J, Ostaszewski A, Giesy JP (2001) Environ Toxicol Chem 20:1878
120. Marvin C, Alaee M, Painter S, Charlton M, Kauss P, Kolic T, MacPherson K, Takeuchi D, Reiner E (2002) Chemosphere 49:111
121. Brack W, Kind T, Schrader S, Möder M, Schüürmann G (2003) Environ Pollut 121:81
122. Stevens JL, Northcott GL, Stern GA, Tomy GT, Jones KC (2003) Environ Sci Technol 37:462
123. Abad E, Caixach J, Rivera J (1999) Chemosphere 38:109
124. Benfenati E, Mariani G, Fanelli R, Zuccotti S (1991) Chemosphere 22:1045
125. Oehme M, Manø S, Mikalsen A (1987) Chemosphere 16:143
126. Helm PA, Bidleman TF (2003) Environ Sci Technol 37:1075
127. Imagawa T, Yamashita N (1994) Organohalogen Cmpds 19:215
128. Imagawa T, Lee CW (2001) Chemosphere 44:1511
129. Schneider M, Stieglitz L, Will R, Zwick G (1998) Chemosphere 37:2055
130. Aittola J-P, Paasivirta J, Vattulainen A, Sinkkonen S, Koistinen J, Tarhanen J (1996) Chemosphere 32:99
131. Theisen J, Maulshagen A, Fuchs J (1993) Chemosphere 26:881
132. Lee RGM, Coleman P, Jones JL, Jones KC, Lohmann R (2005) Environ Sci Technol 39:1436
133. Martí I, Ventura F (1997) J Chromatogr A 786:135
134. Meijer SN, Harner T, Helm PA, Halsall CJ, Johnston AE, Jones KC (2001) Environ Sci Technol 35:4205
135. Bowes GW, Mulvihill MJ, Simoneit BRT, Burlingame AL, Risebrough RW (1975) Nature 256:305

136. Haglund P, Jakobsson E, Asplund L, Athanasiadou M, Bergman Å (1993) J Chromatogr 634:79
137. Yamashita N, Kannan K, Imagawa T, Miyazaki A, Giesy JP (2000) Environ Sci Technol 34:4236
138. Albro PW, Parker CE (1978) J Chromatogr 169:161
139. Noma Y, Yamamoto T, Sakai S-I (2004) Environ Sci Technol 38:1675
140. Harner T, Shoeib M, Gouin T, Blanchard P (2005) Organohalogen Cmpds 67:674
141. Dörr G, Hippelein M, Hutzinger O (1996) Chemosphere 33:1563
142. Lee RGM, Thomas GO, Jones KC (2005) Environ Sci Technol 39(13):4729
143. Giesy JP, Jude DJ, Tillitt DE, Gale RW, Meadows JC, Zajieck JL, Peterman PH, Verbrugge DA, Sanderson JT, Schwartz TR, Tuchman ML (1997) Environ Toxicol Chem 16:713
144. Gomez-Belinchon JI, Grimalt JO, Albaigés J (1991) Water Res 25:577
145. Weston, Inc (1994) RCRA Facility Investigation – Phase I report, vol I. Elf Atochem North America, East Plant, Wyandotte, Michigan
146. Verbrugge DA, Othoudt RA, Grzyb KR, Hoke RA, Drake JB, Giesy JP, Anderson D (1991) Chemosphere 22:809
147. Kaminsky R, Kaiser KLE, Hites RA (1983) J Great Lakes Res 9:183
148. Gevao B, Harner T, Jones KC (2000) Environ Sci Technol 34:33
149. Kannan K, Yamashita N, Imagawa T, Decoen W, Khim JS, Day RM, Summer CL, Giesy JP (2000) Environ Sci Technol 34:566
150. Vannucchi C, Sivieri S, Ceccanti M (1978) Chemosphere 7:483
151. Falandysz J, Strandberg L, Bergqvist P-V, Kulp SE, Strandberg B, Rappe C (1996) Environ Sci Technol 30:3266
152. Falandysz J (2003) Food Addit Contam 20:995
153. Kannan K, Corsolini S, Imagawa T, Focardi S, Giesy JP (2002) Ambio 31:207
154. Kannan K, Hilscherova K, Imagawa T, Yamashita N, Williams LL, Giesy JP (2001) Environ Sci Technol 35:441
155. Domingo JL, Falcó G, Llobet JM, Casa C, Teixidó A, Müller L (2003) Environ Sci Technol 37:2332
156. Hayward DG, Charles JM, de Bettancourt CV, Stephens SE, Stephens RD, Papanek PJ, Lance LL, Ward C (1989) Chemosphere 18:455
157. Asplund L (1994) PhD thesis, Stockholm University
158. Takeshita R, Yoshida H (1979) Eisei Kagaku 25:24
159. Weistrand C, Norén K (1998) J Toxicol Env Heal A 53:293
160. Kawano M, Ueda M, Falandysz J, Matsuda M, Wakimoto T (2000) Organohalogen Cmpds 47:159
161. Witt K, Niessen KH (2000) J Pediatr Gastr Nutr 30:164
162. Norén K, Meironyté D (2000) Chemosphere 40:1111
163. Lundén Å, Norén K (1998) Arch Environ Contam Toxicol 34:414

Hdb Env Chem Vol. 5, Part N (2006): 307–353
DOI 10.1007/698_5_044
© Springer-Verlag Berlin Heidelberg 2005
Published online: 9 December 2005

# Polycyclic Aromatic Hydrocarbons in the Great Lakes

Matt F. Simcik[1] (✉) · John H. Offenberg[2]

[1]Division of Environmental Health Sciences, School of Public Health,
  University of Minnesota, MMC 807, 420 Delaware Street SE, Minneapolis, MN 55082,
  USA
  *msimcik@umn.edu*

[2]Department of Environmental Sciences, Rutgers, The State University of New Jersey,
  14 College Farm Road, New Brunswick, NJ 08901, USA
  *jho@envsci.rutgers.edu*



**Abstract** Polycyclic aromatic hydrocarbons (PAHs) are produced during the incomplete combustion of organic material. They can also be produced through natural, non-combustion processes, and may be present in uncombusted petroleum. Uncombusted petroleum can be a direct source to the waters of the Great Lakes, but combustion sources discharge PAHs into the coastal atmosphere. Atmospheric deposition of combustion re-lated PAHs seems to be the dominate source to the Great Lakes, except in nearshore areas where point sources can be significant. Once airborne, PAHs partition in the atmosphere between the gas and particle phases and can undergo long-range transport. During trans-port, PAHs can be degraded or modified by photochemical reactions. Both the original PAH species and their degradation products can be washed out of the atmosphere by wet and dry deposition, air–water exchange and air–terrestrial exchange. Once in an aquatic system, PAHs partition between the dissolved and particle phases. In general, PAHs are particle reactive and settle out in sediments. PAH contamination of Great Lakes sediments are higher in the nearshore regions where ports, harbors, and urban/industrial areas are the densest. In the open lake area, sediment concentrations are rather uniform, with Lake Superior having slightly less PAHs in its surficial sediments. That portion of the PAHs that does not partition to particles can bioaccumulate in the lipid reserves of organisms. PAHs accumulated in an organism may be metabolized to more toxic by-products or exert toxi-city in its original form. When combined with ultraviolet radiation this toxicity is greatly enhanced. In coastal areas where concentrations can be quite high, PAHs can be toxic to all forms of aquatic life during at least part of their life cycle. PAHs are expected to remain an ecological threat to the Great Lakes well into the future. The threat may even increase with the increasing combustion needed for the increasing population centers and greater transportation needs. Of particular concern is the short-term increase in PAH concentra-tions that can result from the dredging of ports and harbors where highly contaminated sediments have been buried.

**Keywords** Great Lakes · PAHs · Sediments · Air · Sources · Photo-enhanced toxicity

## Abbreviations

| | |
|---|---|
| $\phi$ | Fraction of contaminant associated with particles |
| $\phi_{p,f}$ | Fraction of contaminant associated with filter-retained particles |
| $\phi_{p,n,f}$ | Fractions of contaminant associated with nonfilter-retained particles |
| $\phi_T$ | Fractions of contaminant associated with all particles |
| $A$ | Operationally defined gas phase concentration |
| AOC | Area of concern |
| $a_{TSP}$ | Surface area of the TSP |
| $C_{air}$ | Total concentration in air |
| $C_d$ | Dissolved organic contaminant concentration in water |
| $C_g$ | Gaseous contaminant concentration in the atmosphere |
| $C_{p,air}$ | Concentration bound to airborne particles |
| $C_{p,rain}$ | Particle associated concentration in rainwater |
| $C_{rain}$ | Total concentration in rain |
| DOC | Dissolved organic carbon |
| ELA | Experimental lakes area |
| EROD | Ethoxyresorufin-O-deethylase |
| $F$ | Flux or operationally defined particle phase concentration |
| $F_{abs}$ | Absorption flux |
| *FFPI* | Fossil fuel pollution index |
| $f_{om}$ | Fraction of organic matter on the TSP |
| $F_{vol}$ | Volatilization flux |
| $F_{wet}$ | Wet deposition flux |
| $H$ | Henry's law constant |

| $H'$ | Dimensionless Henry's law constant |
|---|---|
| HOMO | Highest occupied molecular orbital |
| IADN | Integrated Atmospheric Deposition Network |
| $K_d$ | Water solid particles distribution coefficient |
| $K_{oc}$ | Organic carbon–water partition coefficient |
| $K_{OL}$ | Overall mass transfer coefficient |
| $K_{ow}$ | Octanol-water partition coefficient |
| LUMO | Lowest unoccupied molecular orbital |
| $MW_{OM}$ | Molecular weight of the organic matter phase |
| $N_S$ | Surface area concentration of adsorption sites |
| $P$ | Precipitation flux |
| $p_L^0$ | Sub-cooled liquid vapor pressure |
| PAHs | Polycyclic aromatic hydrocarbons |
| $Q_l$ | Enthalpy of desorption |
| $Q_v$ | Enthalpy of volatilization |
| $R$ | Molar gas constant |
| $SEM_w$ | Standard error of the weighted mean |
| $SPM$ | Concentration of suspended particulate matter |
| SQC | Sediment quality criteria |
| $T$ | Temperature |
| $TSP$ | Concentration of total suspended particulate matter |
| UCM | Unresolved complex mixture |
| $V_d$ | Deposition velocity |
| $VWM$ | Volume weighted mean |
| $W_g$ | Gas scavenging ratio |
| $W_p$ | Particle scavenging ratio |
| WQC | Water quality criteria |
| $W_T$ | Total scavenging ratio |
| $\zeta_{OM}$ | Activity coefficient of the absorbate in the organic matter |
| $\Delta H_{excess}$ | Free enthalpy of dissolution |
| $\Delta H_F$ | Enthalpy of fusion (melting) |
| $\Delta H_H$ | Enthalpy of the H |
| $\Delta H_{sol}$ | Enthalpy of dissolution |
| $\Delta H_{vap}$ | Enthalpy of vaporization |
| $\Delta H_{vap}$ | Enthalpy of vaporization |
| $\Delta S_{vap}$ | Entropy of vaporization |

# 1
## Introduction

Polycyclic aromatic hydrocarbons (PAHs) are organic compounds containing two or more aromatic rings. They are the product of incomplete combustion of organic matter, the product of diagenesis of organic compounds, and are present in uncombusted petroleum. Because they have both natural and anthropogenic sources, there have always been PAHs present in the environment. However, with the Industrial Revolution and its increase in combustion for heat, energy, and manufacturing, PAH concentrations in the environment have increased dramatically. PAHs are of environmental concern because they are persistent, bioaccumulative and toxic.

PAHs have been found all over the globe in all compartments of the environment. They are ubiquitous because the are persistent. Recalcitrance in PAHs may stem, in part, from the delocalized electrons in the planar pi orbitals of the aromatic structure. Their relatively high octanol-water partition coefficients, $K_{OW}$s, make them rather lipophilic. The lipophilicity of PAHs forces them from the dissolved phase to particles and also into lipid rich organisms, but they can be metabolized in higher organisms. However, these metabolites are often more toxic than their parent PAHs. When combined with other stressors, particularly ultraviolet radiation, PAHs can exert enhanced toxicity.

The US Environmental Protection Agency has identified 16 priority PAHs due to their known or suspected carcinogenicity. The structures of these 16 PAHs are presented in Fig. 1, but there are almost limitless possibilities to the number of PAHs that can be produced. In addition to the structures presented in Fig. 1, PAH can also have alkyl chains or heteroatoms within their rings. These compounds are most often produce by the combustion of organic material. In this chapter we will discuss the sources, transport, sinks and toxicity of PAHs and how each of these areas relates to the Great Lakes.

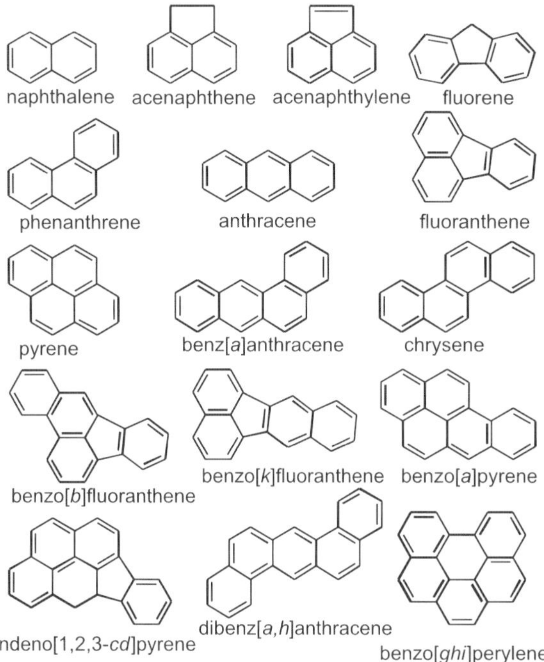


**Fig. 1** Structures of the 16 EPA priority PAHs

## 2
## Sources

There are many more PAHs than the EPA's 16 priority compounds shown in Fig. 1. These include other parent compounds as well as alkylated species. The large number of possible PAHs provides power to elucidate sources of PAHs to the Great Lakes. Because different sources produce different relative amounts of each PAH, the relative concentrations of PAHs in environmental media can be used to identify significant sources back out their contributions to environmental levels. Some compounds that are particularly helpful in the elucidation of sources, or may have both natural and anthropogenic sources are shown in Fig. 2.

### 2.1
### Pyrogenic and Petrogenic Sources

Combustion sources produce a wide array of PAHs from low to high molecular weight species. These combustion derived PAHs are often called pyrogenic PAHs, and are recognized by the large presence of higher molecular weight species (i.e., > 3 rings). Pyrogenic PAHs are formed during the combustion of large biomolecules present in fossil fuels and contemporary carbon sources. These biomolecules are pyrolized into small fragments that cool, polymerize, and aromatize in the exhaust of the combustion process. The combustion process can form the vast majority of PAHs found in the environment, but different combustion sources often produce a small subset of PAHs and in some cases there are unique individual compounds specific to certain sources.

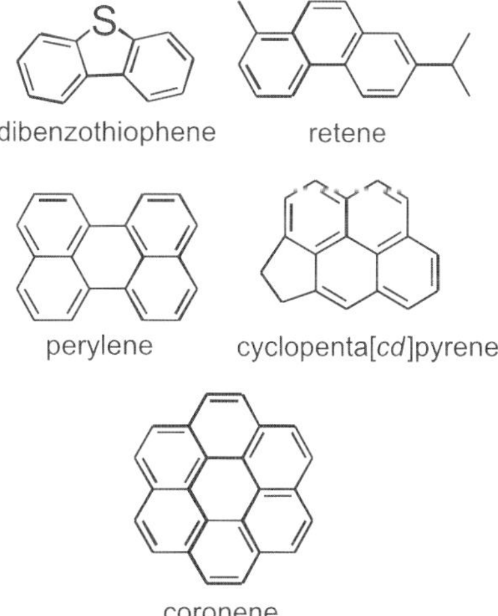


**Fig. 2** Structures of some source-relevant PAHs

For example, vehicular emissions can be identified by high concentrations of benzo[*ghi*]perylene and coronene [1–5], and a unique tracer of vehicle emissions may be cyclopenta[*cd*]pyrene [6–11]. Another example of a unique tracer is retene. Retene is a tracer for wood combustion, and although it can also be produced from non-combustion sources, this is usually minimal (see 2.1.1). Other sources are characterized by their relative contributions of several PAHs. Further differentiation between gasoline and diesel emissions can often be elucidated by the high contribution of benzo[*b*]fluoranthene, benzo[*k*]fluoranthene and indeno[*cd*]pyrene by diesel vehicles [2]. Coal combustion contributes relatively higher concentrations of alkylated PAHs than other combustion sources [12, 13]. A large source of energy for heating, cooking, and electrical generation in the Great Lakes Region [14], natural gas combustion is characterized by high relative amounts of benz[*a*]anthracene and chrysene [15]. Pyrogenic PAHs are a direct source to the atmosphere. Once in the atmosphere, these PAHs can be transported and deposited to the surface of the Great Lakes (see transport section).

Petrogenic PAHs, found in uncombusted petroleum, can be a source to the atmosphere through volatilization or directly to the Great Lakes through intentional or unintentional spills to the water. Unlike combustion sources, different sources of uncombusted petroleum cannot be reconciled by their PAH signatures. However, uncombusted petroleum can produce different PAH signatures than combustion sources. As a volatilization source of PAHs to the atmosphere, uncombusted petroleum naturally produces more of the lighter, less volatile PAHs such as naphthalene, acenaphthene, acenaphthylene. Uncombusted petroleum is also relatively high in alkylated PAHs [16, 17]. Boehm and Farrington [18] suggest that uncombusted, petroleum-derived PAHs can be determined by the relative abundance of parent and alkylated naphthalenes, dibenzothiophenes and phenanthrenes. The authors constructed a formula for estimating the contribution of uncombusted petroleum called the fossil fuel pollution index (FFPI) expressed as:

$$
\begin{aligned}
\text{FFPI} = \Bigg[ &\sum \text{naphthalenes}(C_0 - C_4) \\
&+ \sum \text{dibenzothiophenes}(C_0 - C_3) \\
&+ \frac{1}{2} \sum \text{phenanthrenes}(C_0 - C_1) \\
&+ \sum \text{phenanthrenes}(C_2 - C_4) \Bigg] \Big/ \sum \text{PAHs}
\end{aligned}
\tag{1}
$$

Boehm and Farrington [18] successfully applied this technique to marine sediments. However, lacking the analytical capabilities to resolve all of these alkylated species, other techniques must be employed. It has been determined that the most environmentally stable alkylated dibenzothiophenes from petroleum are those with a methyl in the 4 position [19], suggesting that 4methyldibenzothiophene could be a good tracer of uncombusted petroleum. Other techniques involve looking at the aliphatic hydrocarbons for any preference in odd chain length $n$-alkanes over even chain length $n$-alkanes or an unresolved complex mixture (UCM). Petroleum residues show no preference of odd or even chain length $n$-alkanes whereas plant waxes show a predominance of odd chain length $n$-alkanes [20]. An unresolved complex mixture is a large response underneath the n-alkane chromatogram of a gas chromatography mass spectrometry run and is indicative of either uncombusted petroleum or microbial activity [21]. The ratio of unresolved to resolved mass in the chromatogram is then used as another tool for source attribution [22]. High values of unresolved to resolved ratios ($> 3$) indicate petroleum sources where lower values ($< 4$) indicate microbial activity [21].

Several of the techniques and tracers mentioned above have been used to apportion sources of PAHs to the atmosphere and sediments of the Great Lakes. Simcik et al. [14] used a multiple linear regression/factor analysis model to apportion the PAHs in the Chicago, IL/southern Lake Michigan atmosphere. The factor analysis indicated individual or groups of individual PAHs that were co-correlated in the combined gas and particle phase concentrations. These PAHs were matched up with source signature information to assign a specific source with each factor and then tracer compounds of each factor were run on a multiple linear regression model to apportion the specific contributions of each source to the total PAH concentration in air. Results are summarized along with the 1994 energy use in Illinois in Fig. 3. The largest source of PAHs was concluded to be coal use followed by natural gas and a relatively small percentage from vehicle emissions. It is important to point out that air sampling was conducted near the large industrial complex of southern Lake Michigan where there is a lot of coal combustion for power generation and the steel industry.

This large influence of coal combustion on PAHs to the atmosphere of southern Lake Michigan is echoed in the PAH accumulation in the sediments of the lake. Simcik et al. [23] concluded that coal combustion particles from the southern end of the lake were responsible for the majority of PAH accumulation in the sediments of both the southern and northern basins. Evidence for this conclusion comes from similarities among PAH signatures in the sediments and air particles in Chicago, a lack of indication of uncombusted petroleum in the sediments from calculations of FFPI [23] and interpretation of aliphatic hydrocarbons [24]. The most convincing evidence, however, is the correlation of total PAH accumulation in sediments of the deep hole of the northern basin and coal use in Illinois over the same time period (Fig. 4). Taking a different approach to apportioning PAHs Christensen and Zhang [25] concluded that the decrease in PAH accumulation after about 1950 was a result of a switch from coal combustion to oil and natural gas for home heating, and that the influence of coal combustion and

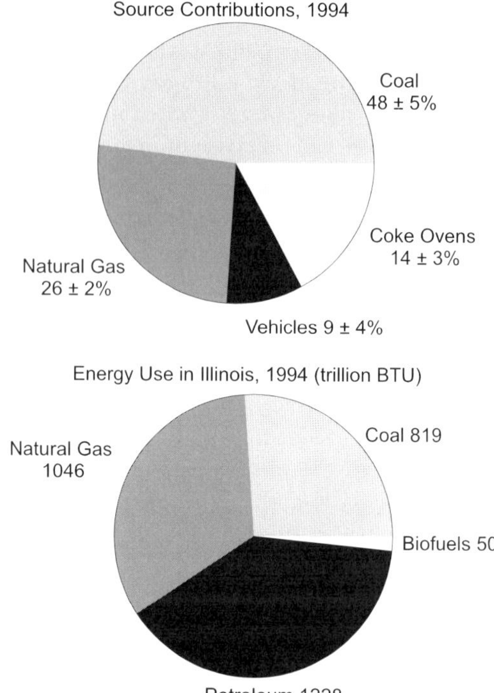


**Fig. 3** Source apportionment of PAHs in Chicago, IL/southern Lake Michigan atmosphere (reprinted with permission from [14])

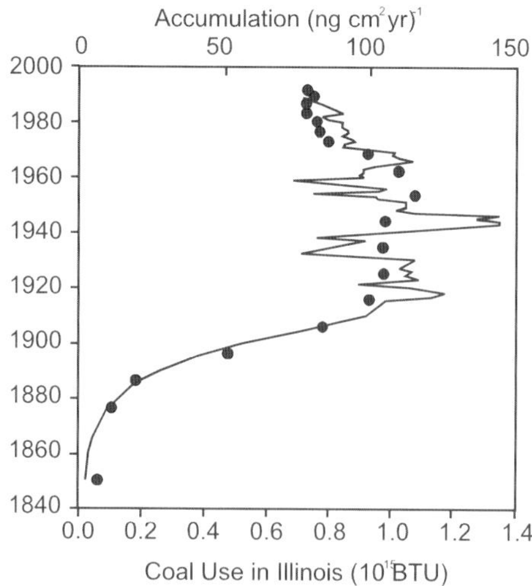


**Fig. 4** Accumulation of PAHs in sediments of Lake Michigan and coal use in Illinois (reprinted with permission from [23])

coke ovens decreases more slowly with increasing distance from the shore than does the contribution from petroleum sources. While neither Simcik et al. [23] nor Christensen and Zhang [25] saw a large influence of point sources to the open lake sediments, the nearshore areas of the Great Lakes can be impacted by local PAH sources.

There are several coastal areas within the Great Lakes that fall under regulatory scrutiny by both the Canadian and US governments because of PAH contamination. These sites have been given the designation of areas of concern (AOCs). Of the 46 AOCs around the Great Lakes, 15 specifically list PAH contamination as a reason of concern (often in addition to other contaminants). A list of the AOCs contaminated with PAHs is summarized in Table 1 below.

Because PAHs are often co-located with other contaminants, this list is by no means inclusive of the hot-spots of PAH contamination. Some other AOCs around the lakes and areas not listed as AOCs may contain significant PAH contamination. In fact, there are several Superfund sites around the Great Lakes that have been so listed because of PAH contamination. These areas of PAH contamination are not highly contaminated because of regional background sources of PAHs. In every case they have specific point sources of contamination. Studies of nearshore sediments by the Christensen group also show local sources of contamination in Lake Michigan. Rachadawong et al. [26] concluded that Lake Michigan sediments in the Milwaukee basin, near Milwaukee, WI, are contaminated with PAHs from a combination of coal

**Table 1** Great Lakes areas of concern identified as having a PAH problem

| AOC | Lake | State/Province |
| --- | --- | --- |
| Black R. | Erie | Ohio, USA |
| Buffalo R. | Erie | New York, USA |
| Detroit R. | Erie | Michigan, USA<br>Ontario, Canada |
| Grand Calumet R. | Michigan | Illinois, USA |
| Hamilton Harbor | Ontario | Ontario, Canada |
| Menominee R. | Michigan | Wisconsin, USA |
| Milwaukee Estuary | Michigan | Wisconsin, USA |
| Niagara R. | Ontario | New York, USA |
| Presque Isle Bay | Erie | Pennsylvania, USA |
| Sheboygan R. | Michigan | Wisconsin, USA |
| St. Clair R. | St. Clair | Michigan, USA<br>Ontario, Canada |
| St. Lawrence R. @ Messena | Ontario | New York, USA |
| St. Louis R. | Superior | Minnesota, USA |
| St. Mary's R. | Huron | Michigan, USA<br>Ontario, Canada |
| Thunder Bay | Superior | Ontario, Canada |

combustion and petroleum point sources. Closer to shore, within the Milwaukee Harbor, suggested sources are highway traffic and industrial discharges including uncombusted petroleum and coal piles [27]. In Green Bay, which is also a more nearshore environment than the open lake, Su et al. [28] concluded that coke, highway and wood burning were the dominant sources of PAHs to the sediments. Further up the Fox River, which flows into Green Bay the same sources were responsible for sedimentary PAH contamination [29].

Like coastal Lake Michigan, the Black River in Ohio is greatly influenced by local sources. Gu et al. [30] concluded that historic contamination of the sediments resulted from coke oven operation, but recent PAHs also receive significant inputs from highway dust and wood burning. Interestingly, the historic PAH contamination from coke ovens buried in the sediments of the Black River were released upon dredging and acted as a major source of PAH contamination to the river [30] and presumably to coastal Lake Erie. Fluvial inputs and nearshore atmospheric deposition to coastal areas of the Great Lakes has been shown by numerous studies for Lake Michigan [25–29, 31]. Similar conclusions have been made for Lake Erie. Smirnov et al. [32] measured PAHs in the sediments of Lake Erie and found much higher concentrations in cores near Detroit, MI, Cleveland, OH and Buffalo, NY than along the southern shore of the lake. In turn, the sediments along the southern shore were much higher than the sediments in the northern portion of the lake. Their conclusions were that the major source of PAHs to the sediments of Lake Erie are the three major cities and fluvial inputs. The atmospheric contribution from Buffalo was confirmed by Cortes et al. [33] using air concentrations and wind direction measured on the shore of Lake Erie 20 km southwest of Buffalo. The relative importance of Detroit and Cleveland to atmospheric deposition of organic contaminants has been implicated by Simcik [34] in which the author concludes that urban areas with a significantly higher population density than the average density around the lake will contribute an elevated atmospheric source of persistent organic pollutants including PAHs. Another suspected urban source of increased atmospheric deposition to Lake Erie from Simcik [34] is Toledo, OH. This is in close proximity to Detroit and Cleveland, and therefore my be indistinguishable from the other two. Simcik [34] also concluded urban areas of suspected increased atmospheric deposition for each of the other Great Lakes.

Lake Superior is expected to have increased atmospheric deposition from Duluth, MN and Thunder Bay, Ontario [34]. This increase from Duluth has been observed in both the atmosphere [35] and sediments [36] of western Lake Superior. Lake Michigan is suspected to have the greatest number of urban areas affecting PAH deposition including Green Bay, Milwaukee, Racine and Kenosha, WI and Chicago, IL and Gary, IN. Lake Huron is only expected to have an urban influence from the Saginaw Bay area. Lake Ontario is expected to have urban influences from Toronto and Hamilton, ON and Rochester, NY. The non-atmospheric influence of Hamilton has been observed by many investigators [37–42], which brings up an important point in that the non-atmospheric point sources along the coasts are often much greater than atmospheric deposition.

## 2.2
## Natural Non-Combustion Sources

There are several natural non-combustion sources of PAHs. A study in 1980 by Wakeham [43] concluded that phenanthrene could be created by the dehydrogenation of steroids, retene could be produced by the diagenesis of abietic acid, and alkyl chrysenes could form from the degradation of the pentacyclic triterpenes alpha- and beta-amyrin, which are components of higher plant waxes. In this section we will look at the natural non-combustion sources of retene and perylene and how these sources might impact the Great Lakes.

### 2.2.1
### Retene

Retene is a methyl isopropyl phenanthrene (Fig. 2). As mentioned above, Wakeham [43] was the first to address the natural non-combustion production of retene in relation to lake sediments. The starting material for retene is abietic acid, a diterpenoid found primarily in the resin of coniferous trees. When burned, the abietic acid forms retene. It can also degrade via one of two pathways to retene without combustion (Fig. 5). One pathway is through dehydroabietin and another intermediate to retene. The other proceeds through dehydroabietane, another intermediate to simonellite and finally to retene. These mechanisms can occur in both the atmosphere [44] and in aquatic systems. Therefore, where there is abietic acid, retene can follow. This becomes especially important in areas where there are high densities of conifers, and thus abietic acid.

One area where this has been found is the experimental lakes area (ELA) in northwestern Ontario. This area is located outside of the Great Lakes watershed, but has a similar geology type and forest to the northern Great Lakes basin. Jeremiason et al. [45] measured PAHs in two lakes within the ELA. Retene was the predominant PAH, by far, in all particulate media (suspended particulate matter, settling particles and sediments). The ELA is a large expanse of wilderness with no towns, industry or even camping, so combustion sources of PAHs to this region would most likely be forest fires. However, only two forest fires occurred in the watershed since industrialization [45]. The dominant forest type is spruce, a conifer, so both natural combustion and non-combustion sources can produce large amounts of retene. Since retene dominates all other PAHs including those produced by combustion, the mechanisms shown in Fig. 5 most likely dominate this system through watershed runoff into the lakes.

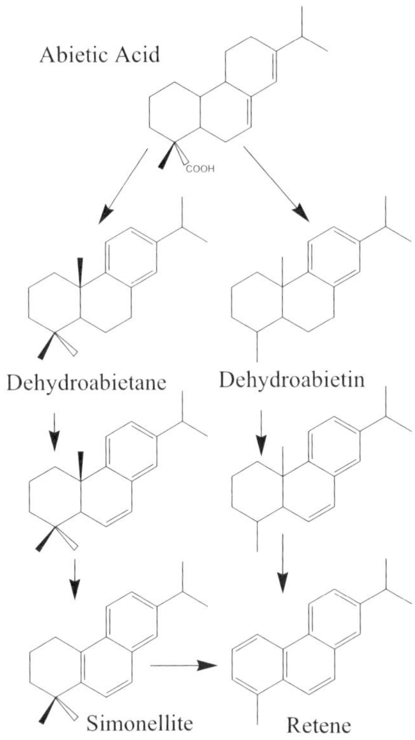

**Fig. 5** Proposed mechanism of natural non-combustion formation of retene

The Great Lakes are not as remote as the ELA, and not all lakes are dominated by coniferous forests, but there are still situations where natural retene may be important. While retene has been measured in sediments from Lakes Superior [36] and Michigan [23] and the atmospheres of Chicago, IL and southern Lake Michigan [31], it was relatively small compared to other PAHs. Areas of the Great Lakes that may be susceptible to natural retene are areas removed from anthropogenic combustion sources near coniferous forests. As it turns out, the areas with the highest densities of conifers also are the least developed. The northern half of the Great Lakes watershed is a separate ecoprovince from the southern half. The area from about the northern half of Lakes Michigan and Huron to Lake Superior is part of the Laurentian mixed forest ecoprovince [46], which is a mixture of hardwoods (deciduous) and conifers. The forest shifts from being dominated by hardwoods in the southern portion of the ecoprovince to spruce-fir in the northern portion of the ecoprovince. Given the dominance of watershed inputs of retene to the remote lakes of ELA, only the nearshore areas of the Great Lakes are expected to be susceptible to non-combustion sources of retene. Northern Lake Michigan is still relatively populated and contains more industry than Lake Superior and northern Lake Huron, especially in the Georgian Bay region. Therefore, areas of greatest potential for non-combustion retene are coastal wetland sediments around Lake Superior and northern Lake Huron.

## 2.2.2
## Perylene

Perylene has both combustion and non-combustion sources. Combustion sources are the same as have been described at the beginning of this section. The predominant non-combustion source of perylene appears to be diagenesis of organic matter in the sediments of lakes and marine systems. There is evidence that *in situ* production of perylene does occur in the sediments of the Great Lakes.

Wakeham et al. [43] were some of the first investigators to recognize the possibility of *in situ* production of perylene in lake sediments. They concluded that autochthonous organic matter was the precursor for perylene. Later in 1984, Louda and Baker [47] concluded from deep ocean sediments that the source of perylene was terrestrial pigments that contained a perylene base structure. These pigments would have a functional group attached at various positions on the perylene moiety (Fig. 2). That functional group was surmised to most likely be an isopropyl acetate side chain, but could also include a hydroxy aldehyde. The authors also concluded that the sediments must be anoxic for diagenesis to produce perylene. Then Venkatesan [48] in 1988 concluded that both terrestrial and aquatic organic matter could be precursors to natural *in situ* production of perylene. Venkatesan [48] also found that high perylene concentrations correlated with diatomaceous sediments and therefore concluded that diatoms were responsible for the diagenesis of organic matter producing perylene. Presumably because of the requirement of sediment anoxia, all of these investigators observed an increase in perylene accumulation and/or concentration with depth in the sediments that did not correlate with other PAHs, and in most cases predated the Industrial Revolution when the majority of combustion derived PAHs were formed. This has also been the case for select studies on Great Lakes sediments.

Simcik et al. [23] were the first to show perylene increasing with depth in Great Lakes sediments. One of the five cores sampled by Simcik et al. [23] showed this increase. That core was in the deep hole of the northern basin of Lake Michigan. Because only one core indicated natural sources of perylene, it is not expected to be a large source of perylene to Lake Michigan sediments. On the other hand, Silliman et al. [49] also observed increasing perylene accumulation in sediments for Lake Ontario. They sampled two cores from the eastern portion of the lake and noted that perylene accumulation was not correlated with other PAHs, terrestrial or aquatic organic matter. Therefore, they concluded that the sediments of Lake Ontario have natural *in situ* production of perylene resulting from the biotic or abiotic transformation of non-specific precursors in anoxic portions of the sediments. The perylene accumulation rates reported by Silliman et al. [49] were also much higher than those observed by Simcik et al. [23] in Lake Michigan (2 to 75 ng cm$^{-2}$ yr$^{-1}$ vs. 0 to 4 ng cm$^{-2}$ yr$^{-1}$), and lends further evidence that natural diagenesis is not an important source of perylene to Lake Michigan sediments, but may still be significant for Lake Ontario.

Natural production of perylene was also observed in Green Bay sediments. Silliman et al. [49] reported increasing perylene accumulation in one core sampled from the middle portion of the bay. They concluded that perylene there is a result of diagenesis of non-specific precursors that include both autochthonous and allochthonous organic matter. Furthermore, they conclude that perylene is produced microbially, but the microbes in Lake Michigan and Green Bay cannot compete in the upper portions of the sediments where there is a surplus of organic matter. Rather, they are active only in the deeper sediments where more reactive organic matter is depleted.

## 3
## Transport

Once produced PAHs can be transported through the atmosphere or the water column if directly discharged *via* uncombusted petroleum. In the air, PAHs partition between the gas and particle phases, can undergo photochemical and oxidation reactions, be washed out by precipitation and deposit to aquatic surfaces by both wet and dry deposition. Once in the aquatic system, PAHs partition between the dissolved and particulate phases, can undergo photochemical reactions and bioaccumulate in the lower trophic levels.

### 3.1
### Gas/Particle Partitioning

Because PAHs are often formed in close proximity to combustion particles in the atmosphere, an equilibrium between the gas and particle phases is established. In general, low molecular weight PAHs (naphthalene to fluorene) are found predominantly in the gas phase; mid-molecular weight PAHs (phenanthrene to chrysene) are split between the gas and particle phases; and high molecular weight PAHs (benzo[$b$]fluoranthene and higher) are predominantly associated with particles. This partitioning of PAHs between the gas and particle-bound phases has been parameterized by the partition coefficient, $K_p$ (m$^3$ mg$^{-1}$), according to following relationship [50, 51]:

$$K_p = \frac{F}{A \cdot \text{TSP}}, \tag{2}$$

where TSP is the concentration of total suspended particulate matter ($\mathrm{mg\,m^{-3}}$), $F$ and $A$ are the operationally defined particle and gas phase concentrations of the analyte of interest ($\mathrm{ng\,m^{-3}}$), respectively [50–52].

Measured partition coefficients, $K_P$, tend to be linearly correlated with the sub-cooled liquid vapor pressure ($p_L^0$) of the pure compound according to the form:

$$\log K_\mathrm{p} = m_\mathrm{r} p_L^0 + b_\mathrm{r} , \tag{3}$$

where the slope $m_r$ is often, though not necessarily, close to $-1$. Slopes of these regressions for PAHs have been found to range between $-0.61$ and $-1.04$ in a wide variety of urban and rural atmospheres [50, 53–58].

Correlation of $\log K_\mathrm{P}$ vs. $\log p_L^0$ have been explained as a combination of both adsorption to the surface of the particle and absorption into an organic surface layer [59]. More recently partitioning to the surface of soot carbon has been suggested as playing an important role in this gas/particle distribution [60]. While the relative importance of these mechanisms remains unclear, the overall behavior can be summarized by the following equation [59]:

$$K_\mathrm{p} = \frac{1}{p_L^0} \left[ \frac{N_S a_\mathrm{TSP} T e^{(Q_l - Q_v)/RT}}{2133} + \frac{f_\mathrm{OM} 101325 RT}{MW_\mathrm{OM} \zeta_\mathrm{OM} 10^6} \right] , \tag{4}$$

where $p_L^0$ is the sub-cooled liquid vapor pressure (Pa), $N_S$ is the surface area concentration of adsorption sites ($\mathrm{mol\,cm^{-2}}$), $a_\mathrm{TSP}$ is the surface area of the TSP ($\mathrm{cm^2\,mg^{-1}}$), $T$ is the temperature (K), $R$ is the molar gas constant ($8.314 \times 10^{-3}\,\mathrm{kJ\,K^{-1}\,mol^{-1}}$), $Q_1$ and $Q_v$ are enthalpies of desorption and volatilization, respectively ($\mathrm{kJ\,mol^{-1}}$), $f_\mathrm{om}$ is the fraction of organic matter on the TSP; $MW_\mathrm{OM}$ is the molecular weight of the organic matter phase ($\mathrm{g\,mol^{-1}}$), and $\zeta_\mathrm{OM}$ is the activity coefficient of the absorbate in the organic matter [58, 59]. The constants in Eq. 4 differ from those presented in Eq. 20 of Pankow [59] because of the use of the sub-cooled liquid vapor pressure in units of Pascals instead of Torr as described by Offenberg and Baker [61]. Additionally, the distribution of individual PAHs between the gas and particle phases was recently shown to depend upon not only the particle mass and carbon content, but the size of the particles [61]. The resulting size specific partition coefficient relates the partitioning of the gas phase compound to the particles in the size class of interest. Thus, the physical characteristics of the particles in each size class will determine the partitioning to that size class, as well as the relative importance of adsorptive and absorptive processes.

$$K_{P,i} = \frac{1}{p_L^0} \left[ \frac{N_{S,i} a_{\mathrm{PM}i} T e^{(Q_{l,i} - Q_v)/RT}}{2133} + \frac{f_{\mathrm{OM},i} 101325 RT}{MW_{\mathrm{OM},i} \zeta_{\mathrm{OM},i} 10^6} \right] . \tag{5}$$

With the terms $N_{S,i}$; $a_{PMi}$; $Q_{1,i}$; $f_{OM,i}$; $MW_{OM,i}$ and $\zeta_{OM,i}$ representing the analogous terms to those given in Eq. 4 for specific particle sizes. Note that the designation for the term $a_{TSP}$ from Eq. 4 was changed to $a_{PMi}$. Integrating across all particle sizes results in the overall bulk G/P partitioning behavior of the compound.

## 3.2
## Reactivity

PAHs are reactive in both the atmosphere and aquatic systems through photolysis and reaction with oxidizing species. In the gas phase PAHs react primarily by hydroxyl radical oxidation, nitrate radical or ozone, in that order [62]. Half-lives of various PAHs in the gas phase range from 0.9 hours to 1.6 days assuming a hydroxyl radical concentration of $1.9 \times 10^6$ molecules m$^{-3}$ (calculated from [62]). While there is no significant trend of molecular weight and reactivity, the higher molecular weight PAHs tend to be more reactive, but they are also more likely to be bound to particles in the atmosphere. Sorption to atmospheric particles has been shown to protect PAHs from atmospheric reactions [63–66]. This atmospheric reactivity can lead to variations in gas phase concentrations between daytime and nighttime hours. Simcik et al. [31] observed diurnal variations in gas phase PAHs in Chicago, IL and over southern Lake Michigan, and attributed the variation to gas phase hydroxyl radical reactions.

Once in the Great Lakes, photolysis reactions are predicted to result in relatively short half-lives [67], but the experiments to determine these photolysis rate constants were performed in pure water and some even involved the presence of photosensitizers. PAHs have been shown to be much more resistant to photo-degradation when sorbed to suspended particulate matter in natural waters [68]. Simcik et al. [23] found no evidence of photodegradation of PAHs that were atmospherically deposited to Lake Michigan, settled through the water column and collected in the sediments.

## 3.3
## Air-Water Exchange

The direction and magnitude of gas transfer of PAHs across the air–water interface can be calculated using a modified [69] two-layer resistance model. This model has been previously well described elsewhere [70] and is summarized here. The overall flux calculation is defined by

$$F = K_{OL} \left( C_d - \frac{C_g}{H'} \right), \tag{6}$$

where $F$ is the flux (ng m$^{-2}$ d$^{-1}$), $K_{OL}$ (m d$^{-1}$) is the overall mass transfer coefficient, and $(C_d - C_g/H')$ describes the concentration gradient (ng m$^{-3}$), The concentration gradient is calculated as $C_d$ (ng m$^{-3}$), the dissolved phase concentration of the compound in water, subtracted by $C_g$ (ng m$^{-3}$), the gas phase concentration of the compound in air, which is divided by the dimensionless Henry's Law Constant, $H'$. The $H'$ value is calculated as $H/RT$, where $R$ is the universal gas constant (8.314 Pa m$^3$ K$^{-1}$ mol$^{-1}$), $H$ is the temperature and salinity-corrected Henry's Law Constant (Pa m$^3$ mol$^{-1}$), and $T$ is the absolute temperature at the air–water interface (K). The overall mass transfer coefficient, $K_{OL}$, is calculated as the resistance to transfer across the water layer and the air layer and quantified as

$$\frac{1}{K_{OL}} = \frac{1}{k_a H'} + \frac{1}{k_w}, \tag{7}$$

Mass transfer coefficients ($k_w$ and $k_a$) have been empirically defined based on experimental studies using tracer gases [71–77] and converted to values for PAHs using differences in diffusivities. The magnitude of $K_{OL}$ for individual PAHs typically ranges from 0.05 to 0.7 m/d (e.g. [78]).

Wanninkhof and McGillis [79] recently established a cubic relationship for describing the effect of wind speed on $k_w$, an update of the relationships established by Liss and Merlivat [80]. The cubic relationship is a better predictor of field data for higher wind speed conditions ($> 6$ m/s) [79]. The details of calculating $k_w$ and $k_a$ are further discussed in Eisenreich et al. [81].

The Henry's Law Constants and $\Delta H_H$ values have been reported by Bamford et al. [82] for thirteen PAHs: 2-methylnaphthalene; 1-methylnaphthalene; acenaphthylene; acenaphthene; fluorene; phenanthrene; anthracene; 1-methylphenanthrene; fluoranthene; pyrene; benzo[$a$]fluorene; benzo[$a$]anthracene; and chrysene. The Henry's Law Constants ranged from $0.02 \pm 0.01$ Pa m$^3$ mol$^{-1}$ for chrysene at $4\,°$C to $73.3 \pm 20$ Pa m$^3$ mol$^{-1}$ for 2-methylnaphthalene at $31\,°$C. The $\Delta H_H$ values for other PAHs not investigated by Bamford et al. [82] can be calculated as the difference between the enthalpy of vaporization ($\Delta H_{vap}$) and the excess free enthalpy of dissolution ($\Delta H_{excess}$) of the compound [83]. The $\Delta H_{vap}$ is calculated from the boiling point and the entropy of vaporization ($\Delta S_{vap}$), which is calculated using the Kistiakowsky relationship [83]. The $\Delta H_{excess}$ is calculated from the enthalpy of dissolution ($\Delta H_{sol}$) by subtracting the enthalpy of fusion (melting) ($\Delta H_F$). For example, Gigliotti et al. [78] used the $\Delta H_{sol}$ measured for 12 PAHs [82] to develop a correlation between $\Delta H_{sol}$ and boiling point ($r^2 = 0.91$) which was then used to estimate $\Delta H_{sol}$ for the other PAHs. Using a different approach Simcik [84] correlated the Henry's Law Constants for PAHs to their sub-cooled liquid vapor pressure and obtained the following relationship:

$$\log H = 0.477 \log p_L^0 + 1.30 r^2 = 0.934, \tag{8}$$

which can also be used to estimate Henry's Law Constants of other PAHs, provided the sub-cooled liquid vapor pressures are known.

### 3.3.1
### Precipitation Scavenging

*Rain Scavenging:* Investigations of the total removal of semi-volatile organic contaminants from the atmosphere by precipitation lead to relating the total contaminant concentration in the surrounding air to that observed in precipitation [85, 86]:

$$W_{\mathrm{T}} = \frac{C_{\mathrm{rain}}}{C_{\mathrm{air}}} = W_{\mathrm{g}}\left(1 - \phi\right) + W_{\mathrm{p}}\phi\,, \tag{9}$$

where $W_{\mathrm{T}}$, $W_{\mathrm{p}}$, and $W_{\mathrm{g}}$ are total, particle, and gas scavenging ratios, respectively, $\phi$ is the fraction of contaminant in the air that is bound to particles and $C_{\mathrm{rain}}$ and $C_{\mathrm{air}}$ are the total concentrations of the compound of interest in rain and air, respectively. The particle scavenging ratio is defined as:

$$W_{\mathrm{p}} = \frac{C_{\mathrm{p,rain}}}{C_{\mathrm{p,air}}}\,, \tag{10}$$

where $C_{\mathrm{p,rain}}$ is the particle associated concentration in rainwater and $C_{\mathrm{p,air}}$ is the concentration of the compound of interest bound to airborne particles. Likewise, the gas scavenging ratio is defined as:

$$W_{\mathrm{g}} = \frac{C_{\mathrm{d}}}{C_{\mathrm{g}}} \tag{11}$$

where $C_{\mathrm{d}}$ is the dissolved organic contaminant concentration in rainwater and $C_{\mathrm{g}}$ is the gaseous contaminant concentration in the atmosphere. The fraction of contaminant bound to particles ($\phi$) is defined in the Gas/Particle Partioning section. This expression of the compound's gas/particle distribution is directly related to compound vapor pressure, temperature, and aerosol surface characteristics [87]. The portion of the contaminant in the gas phase will behave in accordance with Henry's law when distributing into the water droplets of precipitation. Thus, at equilibrium the gas scavenging ratio equals the dimensionless Henry's Law Constant ($RT/H$) where $R$ is the universal gas constant ($8.21 \times 10^5$ m$^3$ atm mol$^{-1}$ K$^{-1}$), $T$ is temperature (K), and $H$ is the Henry's Law Constant (atm m$^3$ mol$^{-1}$). Particle scavenging, however, is highly complex and is determined by both meteorology and size of both particles and rain droplets [88–90]. The relative importance of gas and particle scavenging on the total removal of contaminants from the atmosphere depends on the relative magnitudes of the terms $W_{\mathrm{p}}\phi$ and $W_{\mathrm{g}}(1 - \phi)$.

Poster and Baker [91] modified Eq. 8 to include the scavenging of submicron particles that are present in the filtrate of precipitation samples:

$$W_{\mathrm{T}} = \frac{C_{\mathrm{rain}}}{C_{\mathrm{air}}} = W_{\mathrm{p,f}}\phi_{\mathrm{p,f}} + W_{\mathrm{p,nf}}\phi_{\mathrm{p,nf}} + W_{\mathrm{g}}(1 - \phi_{\mathrm{T}}) \tag{12}$$

where $\phi_{p,f}$, $\phi_{p,n,f}$ and $\phi_T$ are the fractions of contaminant associated with filter-retained particles, nonfilter-retained particles, and all particles. This model relates the respective scavenging of operationally defined large and small particles from the atmosphere with the equilibrium partitioning of contaminants between the gas phase and the dissolved phase within falling raindrops. This relationship assumes no changes in particle size upon entrainment in the water droplet such as reduction in particle size due to partial dissolution of the aerosol or growth of the particle due to coagulation of small particles upon entrainment within the falling precipitation. Furthermore, this model assumes minimal importance of re-distribution once the contaminant has entered the drop while falling and before it is filtered in the rain sampling apparatus.

This model has been further expanded [92] to include a broader range of particle sizes, each exhibiting its own washout ratio, yet application of this expanded model is difficult due to the complexities in attributing particles of various sizes to their original atmospheric diameter and inefficiencies in separating small particles in rainwater. As such, the simpler model of Poster and Baker [93] is sufficiently specific and useful to examine the relative influences of operationally defined small and large particles on the total contaminant removal from the atmosphere by precipitation.

More recently Simcik [84] has shown that adsorption of PAHs to the surface of a rain drop may be much more important than previously thought and has modified Eq. 9 to include surface adsorption:

$$W_T = W_P\phi + \frac{(1-\phi)}{H'} + K_{ia}\frac{6000}{d_R}(1-\phi),\tag{13}$$

where the third term represents the additional washout of PAHs due to adsorption to the surface of the raindrop. $K_{ia}$ is the interfacial adsorption coefficient (m) defined as the ratio of the mass adsorbed per surface area of sorbent (i.e. raindrop) ($ng/m^2$) to the atmospheric vapor phase concentration ($ng/m^3$) and $d_R$ is the raindrop diameter. Pankow [94] inferred $K_{ia}$ values for PAHs by extrapolating gas-particle partitioning coefficients to 100% humidity on filter surfaces. Simcik [84] calculated $K_{ia}$ values for PAHs using solvation parameters and an equation from Roth et al. [95]. $K_{ia}$ values ranged from $10^{-3}$ to 100 m and correlated with the sub-cooled liquid vapor pressure:

$$\log K_{ia} = -0.827\log p_L^0 - 3.75 r^2 = 0.970.\tag{14}$$

The dependence of washout on raindrop diameter results from the fact that as diameter decreases, the surface to volume ratio increases and surface adsorption becomes proportionately more important. Note also that a rainfall will have a distribution of raindrop sizes, therefore Eq. 13 must be integrated over the entire raindrop size distribution:

$$W_{\mathrm{T}} = W_{\mathrm{P}}\phi + \frac{(1-\phi)}{H'} + K_{\mathrm{ia}}6000(1-\phi)\frac{\int\limits_{0}^{\infty} n(d_{\mathrm{R}})d_{\mathrm{R}}^2\,\partial d_{\mathrm{R}}}{\int\limits_{0}^{\infty} n(d_{\mathrm{R}})d_{\mathrm{R}}^3\,\partial d_{\mathrm{R}}}\,, \tag{15}$$

where $n(d_{\mathrm{R}})$ is the raindrop size distribution function. The raindrop size distribution is a function of rainfall rate. The reader is directed to the Simcik [84] paper for more in depth discussion of this subject. Ultimately, the author concluded that for PAHs and a typical rainfall rate, gas-phase washout of PAHs lower sub-cooled liquid vapor pressures less than $10 \times 10^{-5}$ Pa (i.e. five or more aromatic rings) is expected to be dominated by surface adsorption.

Combining Eq. 12, which included the washout of ultrafine particles and Eq. 15, which includes adsorption to the raindrop surface results in the following equation for total washout of semi-volatile organic compounds including PAHs:

$$W_{\mathrm{T}} = W_{\mathrm{p,f}}\phi_{\mathrm{p,f}} + W_{\mathrm{p,nf}}\phi_{\mathrm{p,nf}} + \frac{(1-\phi)}{H'} + K_{\mathrm{ia}}6000(1-\phi)\frac{\int\limits_{0}^{\infty} n(d_{\mathrm{R}})d_{\mathrm{R}}^2\,\partial d_{\mathrm{R}}}{\int\limits_{0}^{\infty} n(d_{\mathrm{R}})d_{\mathrm{R}}^3\,\partial d_{\mathrm{R}}}\,. \tag{16}$$

*Snow Scavenging.* Snow may be more efficient than rain at below-cloud scavenging of particles because of the larger size and surface area of snowflakes [96]. The particle scavenging efficiency of snow is related to crystalline shape with needles and columns less effective than stellar plates, dendrites, and snowflakes [97]. Snowflakes exhibit a "filtering effect" on atmospheric particles enroute to the surface due to their porosity, which allows air to pass through the falling solid. This process increases the ability of snowflakes to scavenge small particles (0.2–2 µm), which tend to follow the streamlines around a nonporous raindrop [98, 99]. Field experiments have demonstrated that below-cloud scavenging of particles by snow is about five times more efficient than by rain [100, 101].

Hydrometeors exiting the cloud base intercept additional particles and gases enroute to the surface. In precipitation occurring during unstable atmospheric conditions, below cloud scavenging is less important than in-cloud scavenging [100–103]. However, when stable atmospheric stratification causes PAHs to accumulate within the surface boundary layer, below-cloud scavenging may contribute significantly to wet deposition [104].

Differences in scavenging between rain and snow occur because of the dissimilarities in the physical state of the hydrometeor, liquid versus solid. Theoretically, a trace gas attaining equilibrium with a cloud droplet or raindrop is scavenged by dissolution according to Henry's law as seen in the rain scavenging section. For snow, this equation may not apply since only a thin liquid water film exists on the surface of ice crystals. At temperatures $< 1\,°C$ and in the absence of ions, the film maybe less than 5 nm thick, whereas impurities within the water layer may cause the film to thicken to 30 nm at $-15\,°C$ [105–107]. In this case, adsorption to the water surface may dominate due to a lack of bulk water for dissolution, and gas scavenging by snow may be better described as vapor sorption to a liquid interface [108, 109]:

$$W_{g,ads} = K_{ia} SAs \rho_{ice} \tag{17}$$

where $K_{ia}$ is the interfacial adsorption coefficient defined above, SAs is the specific surface area ($m^2\,g^{-1}$), and $F$ is density of ice ($0.917\,g\,cm^{-3}$). The specific surface area of snow is uncertain and depends on crystalline shape, but is thought to range from $< 0.1$ to $> 1.0\,m^2\,g^{-1}$ [109]. In rimed snow, both dissolution and adsorption mechanisms may apply with interfacial adsorption occurring to the surface water film and with Henry's Law dissolution into cloud droplets scavenged by snowflakes falling through the cloud. Because scavenged droplets freeze upon contact with the snowflake, Henry's law partitioning behavior may only be observed if there is negligible loss of contaminants during freezing of the droplets [94, 110–112].

The presence of organic surface films on snow crystals may also enhance gas adsorption. Such films are likely to be discontinuous, i.e., less than one monolayer thick [96]. About $200\,mg\,L^{-1}$ of surface-active organic matter is necessary for monolayer coverage of a 6 mm diameter platelike crystal [113]. Regardless, the presence of an organic film may potentially enhance adsorption of PAHs to snow. Yet, snow may be more efficient than rain at below-cloud scavenging of particles because of the larger size and surface area of snowflakes [99, 100, 102, 103, 114, 115].

## 3.4
## Deposition

### *Dry deposition*

Dry deposition flux, $F_{dry}$ ($ng\,m^{-2}\,d^{-1}$), can be calculated by multiplying the concentration of PAHs on atmospheric particles, $C_p$ ($ng\,m^{-3}$) by a particle deposition velocity, $V_d$ ($cm\,d^{-1}$):

$$F_{dry} = C_p V_d \tag{18}$$

Particle deposition velocities depend on a number of factors, including wind speed, atmospheric stability, relative humidity, particle characteristics (diameter, shape, and density), and receptor surface characteristics. Recent studies on dry particle deposition to surrogate surfaces and derived from atmospheric particle size distributions and micrometeorology suggest that a $V_d$ equal to about $0.5\,cm\,s^{-1}$ is applicable to urban/industrial regions [116–120].

### Wet deposition

Wet deposition flux, $F_{wet}$ (ng m$^{-2}$ d$^{-1}$) is calculated by multiplying the volume-weighted mean concentration of the PAH compound in rainwater, $VWM$ (ng L$^{-1}$), by the precipitation flux, $P$ (L m$^{-2}$ d$^{-1}$):

$$F = VWM\, P. \tag{19}$$

The uncertainty in the volume-weighted mean concentration is often expressed as the standard error of the weighted mean (SEM$_w$), calculated for each sampling location as an approximate ratio variance as expressed by Endlich et al. [121].

### Volatilization and absorptive air–water fluxes

The gross volatilization ($F_{vol}$) and gross absorption ($F_{abs}$) fluxes (ng m$^{-2}$ d$^{-1}$) are calculated as:

$$F_{vol} = K_{OL} C_{diss}, \tag{20}$$

$$F_{abs} = K_{OL} \frac{C_g}{H'}. \tag{21}$$

The net diffusive gas exchange flux is then calculated by subtracting the gross absorption flux from the gross volatilization flux. A positive flux indicates net volatilization out of the water column, and a negative flux indicates net absorption into the water column.

### Net atmospheric loading estimates

A comparison of the magnitudes of the dry particle depositional, the wet depositional, and the air-water diffusive gas fluxes is often useful in order to assess their relative importance to the total atmospheric loading to the water of the Great Lakes. Several researchers have attempted to synthesize the large number of measurements and calculations required to do so [122–125]. Many of these efforts were a result of the Integrated Atmospheric Deposition Program (IADN). IADN is a binational effort to estimate atmospheric deposition loadings to the Great Lakes. It has been monitoring the air of each of the Great Lakes since the early 1990s and is a combined effort of the US and Canadian governments.

The first report on loadings of contaminants including PAHs to all five Great Lakes from IADN was published in 1996 [124], and updated with two additional years of data for three of the lakes in 1998 [123]. In the 1996 paper Hoff et al. [124] reported estimates of wet and dry deposition for phenanthrene, pyrene, benzo[$k$]fluoranthene and benzo[$a$]pyrene seasonally and annually for each of the five Great Lakes and net gas exchange for Lakes Superior and Ontario for 1991–1992 and 1994. The authors concluded for Lakes Superior and Ontario that net gas exchange dominates the atmospheric loading of phenanthrene and pyrene, but wet and dry deposition together is comparable to net gas exchange for benzo[$k$]fluoranthene and benzo[$a$]pyrene. They also observed increases in wet and dry deposition from 1988 to 1994 for each all lakes except Lake Huron. In the 1998 paper, Hillery et al. [123] reported dry deposition estimates for each of the five Great Lakes and wet deposition and net gas exchange for all but Lakes Huron and Ontario for 1993 and 1994. Again, net gas exchange dominated the atmospheric loadings estimate for phenanthrene and pyrene, but wet and dry deposition dominated the loading estimates for benzo[$k$]fluoranthene and benzo[$a$]pyrene. Adding two more years of data, Buehler and Hites [122] report that atmospheric deposition loadings of PAHs for all five Great Lakes show no declining trend as do many banned industrial chemicals.

## 3.5
## Dissolved/Particle Partitioning

Similar to the dynamics of PAHs in the atmosphere, partitioning of PAHs between water and solid particles suspended in solution controls the transport characteristics of the individual PAHs. Partitioning is described by a distribution coefficient, $K_d$ (L kg$^{-1}$);

$$K_d = \frac{\left(C_P/\text{SPM}\right)}{C_d}, \tag{22}$$

where $C_p$ is the concentration in the operationally-defined particulate phase (ng L$^{-1}$), SPM is the concentration of suspended particulate matter (kg L$^{-1}$), and $C_d$ is the concentration of the compound in the operationally defined dissolved phase (ng L$^{-1}$). The partitioning of hydrophobic organic chemicals such as PAHs to aquatic particles is expected to increase as the hydrophobicity of the compound increases [126–129]. The organic carbon content (OC) of the solid phase is an important factor affecting partitioning in laboratory experiments so the distribution coefficient can be expressed in terms of the OC content of the sorbent as follows [127]:

$$K_{OC} = \frac{K_d}{f_{OC}}, \tag{23}$$

where $f_{OC}$ represents the fraction of the total solid that is organic carbon. For example, the organic carbon content of the particles in Green Bay [83] ranged from 11 to 37% with an average of 23%. The following relationship between the organic carbon normalized distribution coefficient ($K_{OC}$) and the octanol-water partition coefficient ($K_{OW}$) has been well established:

$$\log K_{OC} = a(\log K_{OW}) + b. \tag{24}$$

Laboratory studies using a variety of sorbents and HOCs show the slope of this relationship is approximately 1 [126–128]. The slope determined in field studies using PAH has also been shown to be near unity [130, 131]. Partitioning data generated by Eadie et al. [130] using field samples from southern Green Bay and radio-labeled HOC including benzo[*a*]pyrene and pyrene gave the following relationship:

$$\log K_{OC} = 0.90(\log K_{OW}) + 0.81 \, r = 0.94 \, . \tag{25}$$

## 3.6
## Bioaccumulation

Because PAHs are hydrophobic, they tend to accumulate in lipids of organisms that are unable to metabolize them. There are several partition coefficients that can describe the accumulation of PAHs in organisms. Direct partitioning of aqueous phase PAHs to an organism is described by the bioconcentration factor:

$$BCF = \frac{C_b}{C_d} \, , \tag{26}$$

where $C_b$ and $C_d$ (like units) are the biota and dissolved phase concentrations, respectively. Increases in concentration up a food chain is termed biomagnification and can be described by a biomagnification factor, $BMF$, or an assimilation efficiency (%):

$$AE = \frac{M_{PAH,b}}{M_{PAH,f}} \times 100 \, , \tag{27}$$

where $M_{PAH,b}$ and $M_{PAH,f}$ are the mass of PAH in biota and food, respectively. When a determination between bioconcentration and biomagnification cannot be made, or both occur within a food web the accumulation of PAHs in a trophic level is described by a bioaccumulation factor:

$$BAF = \frac{C_b}{C_d} \, , \tag{28}$$

where the only difference between $BCF$ and $BAF$ is that $BCF$ is used when no trophic transfer occurs. Another form of the bioaccumulation factor is used when describing particle reactive compounds such as PAHs and benthic organisms that reside in close proximity to sediments. The bio-sediment-accumulation factor, $BSAF$ is used to describe the increase or decrease in concentration from sediments to biota:

$$BSAF = \frac{C_b}{C_s} \, , \tag{29}$$

where $C_s$ is the sediment concentration.

Given the above definitions, it is important to note that PAHs do not accumulate in the higher trophic levels of the Great Lakes food webs to the extent that many other hydrophobic organic contaminants do. The reason for the lack of accumulation is the ability of higher organisms to metabolize PAHs. However, lower trophic organisms do accumulate PAHs to some degree. Much of the recent work on bioaccumulation of PAHs in lower trophic levels of the Great Lakes has been performed on zebra mussels (*Dreissena polymorpha*).

Zebra mussels are an invasive species that hitchhiked into the Great Lakes *via* ballast water of trans-Atlantic ships. First discovered in Lake St. Clair in 1988, zebra mussels are extremely efficient filter feeders ($\sim 1$ L day$^{-1}$ [132]) making them susceptible to bioaccumulation of PAHs. While zebra mussels filter large amounts of water they are selective in the ingestion of filtered particles, depurating unfavorable particles in pseudofeces [133]. Because of this selective ingestion, Gossiaux et al. [134] and Bruner et al. [135] observed greater bioaccumulation of PAHs in mussels for algae than for sediment particles. In both studies, radiolabeled PAHs were fed to zebra mussels. Zebra mussels in the Gossiaux et al. study assimilated $58 \pm 14\%$ of the pyrene and $45 \pm 6\%$ of the benz[*a*]pyrene from suspended sediment, but $92 \pm 4\%$ of the pyrene, $92 \pm 1\%$ of the benz[*a*]pyrene and $97 \pm 1\%$ of the chrysene from algae. The zebra mussels in the Bruner et al. study [135] assimilated just $53 \pm 6\%$ and $21 \pm 4\%$ of the benz[*a*]pyrene from algae and suspended sediment, respectively.

High lipid levels in zebra mussels also contribute to their susceptibility to bioaccumulation of hydrophobic organic contaminants including PAHs [132]. However, lipid content varies seasonally in zebra mussels and Bruner et al. [136] found that bioconcentration of hydrophobic PAHs such as benz[*a*]pyrene were higher in high lipid, pre-spawn mussels than in low lipid, post-spawn mussels. On the other hand, the *BCF* of a less hydrophobic PAH, pyrene, was not affected by lipid content. Lipid content varied from 4 to 20% of the mussel dry mass and *BCF*s ranged from $1.3 \times 10^4$ to $3.5 \times 10^4$ for pyrene and from $4.1 \times 10^4$ to $8.4 \times 10^4$ for benz[*a*]pyrene. Lipid normalized *BCF*s ranged from $8.4 \times 10^5$ to $1.9 \times 10^6$ and from $3.1 \times 10^6$ to $4.7 \times 10^6$ for pyrene and benz[*a*]pyrene, respectively.

In the field, Gewurtz et al. [137] observed *BSAF*s of several PAHs in dreissenids in western Lake Erie. *BSAF*s ranged from $10^{-1.6}$ to $10^{0.2}$ and were inversely proportional to $K_{OW}$. The authors concluded that this inverse relationship suggests metabolism of higher molecular weight PAHs. However, this is only one possible reason. As we have seen in the Dissolved/Particulate Partitioning section above, higher molecular weight PAHs may be more tightly bound to sediment particles making them less bioavailable. Since zebra mussels (a dreissenid) reject some particles in pseudofeces, they may not accumulate PAHs from the particle phase to a significant degree. By using the mussel and sediment concentrations from Gewurtz et al. [137] and equations 23-25 we can estimate the *BAF* for each PAH. The resulting *BAF*s are linearly correlated with $K_{OW}$ in a log-log plot (Fig. 6), which is statistically significant ($p < 0.001$). Therefore, it is not clear whether zebra mussels are able to metabolize PAHs.

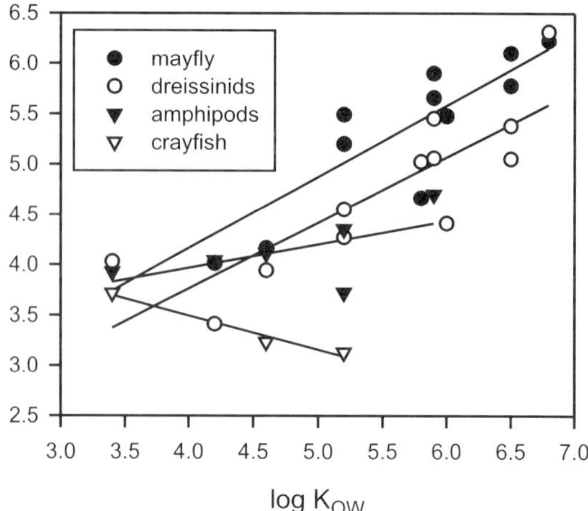


**Fig. 6** Log-log plot of estimated bioaccumulation factor vs. octanol–water partition coefficient of PAHs in Lake Erie (data from [137])

Gewurtz et al. [137] also measured PAHs in other organisms including mayfly larvae, amphipods and crayfish. Like the dreissenids, mayfly larvae also show a significantly ($p < 0.001$) positive relationship between log *BAF* and log $K_{OW}$ (Fig. 6), but the relationship for amphipods is not significant and the relationship for crayfish is significantly ($p < 0.005$) negative. These results suggest that amphipods are able to metabolize PAHs and confirms that higher trophic levels such as that represented by crayfish are able to metabolize PAHs.

Evidence for PAH exposure and metabolism in higher trophic levels has also been shown by measuring PAH metabolites in the bile of fish. While PAH metabolites were reported in all fish analyzed, significantly higher levels were present in PAH contaminated areas such as Hamilton Harbor, ONT and the Black River, OH [38] the Cuyahoga River, OH [138] and the Sheboygan River, WI [139]. Because higher trophic level organisms metabolize and depurate PAHs quickly, the metabolite concentrations are representative of only recent PAH exposure. Therefore, PAH *BAF*s are not meaningful in higher trophic levels of the Great Lakes.

# 4
# Sinks

PAHs have been measured in several compartments of the Great Lakes Ecosystem. These include almost all media, but most information is available for air, precipitation and sediments.

## 4.1
## Air

Atmospheric concentrations of polycyclic aromatic hydrocarbons have been measured since 1991 as part of the Integrated Atmospheric Deposition Network (e.g. [122]). Buehler et al. [35] report PAH concentrations from five US IADN sites (Table 2).

Detailed analysis of IADN results show intracies in the measured concentrations of PAHs. Particle phase PAH concentrations at Sturgeon Point were higher than at other IADN master stations, especially when winds blew from Buffalo, NY [33]. Likewise, elevated particle phase PAH concentrations were observed at Brule River, Wis, a small city 40 km southwest of Duluth, MN exhibited as compared with Eagle Harbor, MI [35]. Buehler and Hites [122] suggest that this indicates that even a small urban center can cause an increase in atmospheric concentrations.

In a field experiment directed toward understanding the role of the urban atmosphere in regional transport of PAHs, Simcik et al. [31], found that PAH concentrations over Lake Michigan increase by a factor of 12 when the wind is from the Chicago region. Simcik et al. [31] report vapor $\Sigma$-PAH concentrations that range from 27 to 430 ng/m$^3$ in Chicago in 1994–1995, 0.8 to 70 ng/m$^3$ over southern Lake Michigan and 4.1 to 55.1 ng/m$^3$ downwind at a rural location. For further comparison, Pirrone et al. [140], report an average of $150 \pm 103$ ng/m$^3$ at the same site in Chicago and $15.4 \pm 12.5$ ng/m$^3$ over-water in 1993, while and Cotham and Bidleman [55] reported a range of 75 to 1410 ng/m$^3$ at another site in Chicago during 1988.

As part of the IADN effort, Cortes et al. [33] found similar results, showing that Buffalo is a source of increased PAH concentrations to Sturgeon Point [33]. Conversely, when the winds are from across Lake Erie, the concentrations of PAHs are lowest, suggesting that it is not always appropriate to use shoreline measurements to estimate the concentrations in the atmosphere above the lake.

Cortes et al. [33] report that gas-phase concentrations of PAHs have been steadily decreasing at all sites since the inception of this extensive monitoring network. Interestingly, particle phase PAHs do not exhibit the same decreasing trend. All but one location (Sleeping Bear Dunes, MI) show no decrease in particle phase concentration over the same period from 1991 through 1999.

**Table 2** Average concentrations of PAHs during 1996–1998 (+ / –SE) at all US IADN stations as reported by Buehler et al. (2000)

|                 | Brule River     | Eagle Harbor    | Sleeping Bear Dunes | Sturgeon Point   | Chicago, IL |
| --------------- | --------------- | --------------- | ------------------- | ---------------- | ----------- |
| Particle phase  | $0.41 \pm 0.07$ | $0.16 \pm 0.2$  | $0.32 \pm 0.05$     | $1.2 \pm 0.12$   | $14 \pm 1.5$ |
| Gas phase       | $1.50 \pm 0.13$ | $1.00 \pm 0.08$ | $1.50 \pm 0.17$     | $5.50 \pm 0.41$  | $99 \pm 14$ |

Historically, concentrations of PAHs in the atmosphere have been infrequently measured. In five air samples collected over Lake Superior in August 1986 ranged in concentration from 2.5 to 6.3 ng m$^{-3}$ and average 3.9 ± 1.7 ng m$^{-3}$. Absolute and relative PAH concentrations in the gas phase were constant among the samples. The concentrations reported by Baker and Eisenreich [54] agreed well with those measured in 1983 over Isle Royale, but were lower than the 6.8 ng m$^{-3}$ measured over Lake Michigan in the mid 1970s [141]. Similarly, these concentrations were lower than those measured in air along the Niagara River, Ontario in 1982-83 where individual PAH species peaked at 40 ng m$^{-3}$ [142].

For comparison to other geographic regions, Tsai et al. [143], found monthly average Total PAH concentrations in the San Francisco bay area, which ranged from 8.0 to 37 ng m$^{-3}$, with 1 to 17% of the total PAH concentration bound to particles. Concentration measured at Concord, CA and Fremont/San Jose were generally higher than those found in the Central Bay near San Francisco. Measurements of PAHs in the St Lawrence Basin of Quebec from 1993 to 1996 found annual mean ΣPAH concentrations that ranged from 7.57 to 22.8 ng m$^{-3}$ at St Anicet and from 5.6 to 18.92 ng m$^{-3}$ at Villeroy Quebec.

Likewise, Offenberg and Baker [144] found PAH concentrations Total (gas + particulate) ΣPAH concentrations measured at all three sites ranged from 0.4 to 114 ng m$^{-3}$. Gas phase concentrations are dominated by phenanthrene, while the particle-bound phase concentrations are dominated by fluoranthene and pyrene. Gas phase ΣPAH concentrations range from 17 to 113 ng m$^{-3}$, accounting for ∼ 90% of the atmospheric PAH burden in Baltimore. Over-water gas phase concentrations range from 2.9 to 13.8 ng m$^{-3}$ and constitute ∼ 87% of the total measured at Hart-Miller Island. Rural gas phase ΣPAH concentrations range from 0.4 to 5.7 ng m$^{-3}$, comprising 89% of the total ΣPAH observed at Stillpond, MD. Dachs and Eisenreich [60] found atmospheric concentrations of PAHs to be generally two times higher in Baltimore than out over the Northern Chesapeake bay.

## 4.2
## Precipitation

Precipitation can take two forms in the Great Lakes. During the warmer months, precipitation falls in the form of rain, and during the winter months it often falls in the form of snow.

### 4.2.1
### Rain

Simcik et al. [148] reported concentrations of PAHs in rainwaters from around four of the five Great Lakes (except Huron). Monthly volume weighted

**Table 3** Comparisons for select gas and particulate PAH concentration data

Phen = phenanthrene (ng m$^{-3}$), pyr = pyrene (ng m$^{-3}$), BbkF = benzo[$b + k$]fluoranthene (ng m$^{-3}$), BaP = benzo[$a$]pyrene (ng m$^{-3}$) concentrations in ng m$^{-3}$ as mean ($\pm$ one standard deviation)

| Site | Ref. | phen (gas) | phen (part) | pyr (gas) | pyr (part) | BbkF (gas) | BbkF (part) | BaP (gas) | Bap (part) |
|---|---|---|---|---|---|---|---|---|---|
| Sturgeon Point | [33] | 3.60 (2.81) | 0.089 (0.064) | 0.36 (0.34) | 0.11 (0.08) | — | 0.259 (0.231) | — | 0.057 (0.046) |
| Sleeping Bear Dunes | [33] | 0.91 (0.79) | 0.028 (0.021) | 0.074 (0.074) | 0.032 (0.025) | — | 0.061 (0.051) | — | 0.015 (0.011) |
| Eagle Harbor | [33] | 0.65 (0.64) | 0.021 (0.027) | 0.075 (0.135) | 0.019 (0.023) | — | 0.033 (0.026) | — | 0.007 (0.005) |
| Eagle Harbor, MI | Hoff et al. 1994 | 0.86 (0.12) | 0.019 (0.069) | 0.19 (0.17) | 0.022 (0.016) | 0.019 (0.034) | 0.022 (0.016) | 0.0093 (0.023) | 0.011 (0.047) |
| Sturgeon Point, NY | Hoff et al. 1994 | 4.0 (0.068) | 0.0060 (0.077) | 0.51 (0.10) | 0.074 (0.071) | 0.019 (0.034) | 0.074 (0.071) | 0.013 (0.062) | 0.044 (0.076) |
| Chicago, IL | [31] | 64 (46) | 3.7 (7.4) | 9.0 (8.4) | 5.9 (11) | 0.29 (0.38) | 6.6 (2.4) | 0.080 (0.082) | 3.0 (5.9) |
| S. Lake Michigan | [31] | 9.9 (9.6) | 0.14 (0.15) | 1.6 (1.8) | 0.21 (0.17) | 0.12 (0.23) | 0.59 (0.74) | 0.014 (0.030) | 0.13 (0.14) |
| Baltimore, MD | [145] | 21 (19) | 0.33 (0.34) | 1.7 (1.2) | 0.35 (0.35) | 0.04 (0.05) | 0.41 (0.40) | — | 0.18 (0.20) |
| Northern Chesapeake Bay, MD | [145] | 2.3 (2.0) | 0.07 (0.08) | 0.32 (0.20) | 0.06 (0.06) | 0.002 (0.005) | 0.10 (0.12) | — | 0.04 (0.04) |
| New Brunswick, NJ | [146] | 8.9 (4.6) | 0.16 (0.16) | 0.69 (0.46) | 0.14 (0.12) | 0.012 (0.012) | 0.32 (0.30) | 0.037 (0.064) | 0.088 (0.096) |
| Sandy Hook, NJ | [146] | 4.8 (3.3) | 0.083 (0.052) | 0.41 (0.28) | 0.070 (0.052) | 0.0027 (0.0019) | 0.12 (0.12) | 0.0023 (0.00087) | 0.033 (0.035) |
| Haven Beach, VA | [147] | 2.9 (3.3) | 0.041 (0.031) | 1.2 (1.4) | 0.039 (0.033) | 0.0086 (0.013) | 0.11 (0.16) | 0.0044 (0.0062) | 0.032 (0.072) |
| Baltimore, MD | [60] | 13 (11) | 0.089 (0.034) | 2.1 (1.3) | 0.14 (0.070) | 0.0011 (0.0029) | 0.16 (0.071) | 0.00015 (0.00055) | 0.071 (0.041) |
| Chesapeake Bay | [60] | 5.6 (4.3) | 0.051 (0.057) | 0.55 (0.46) | 0.067 (0.14) | NDc | 0.085 (0.054) | NDc | 0.019 (0.015) |

mean concentrations ranged from 1 to nearly 100 ng L$^{-1}$. Site-specific average concentrations, ($\pm 1$ standard error) were $32 \pm 4$ ng L$^{-1}$, $56 \pm 5$ ng L$^{-1}$, $102 \pm 9$ ng L$^{-1}$, $147 \pm 37$ ng L$^{-1}$ at lakes Superior, Michigan, Erie and Ontario, respectively. The ensemble average was $85 \pm 20$ ng L$^{-1}$, and only the concentration measured at Lake Superior was significantly different from this value (99% confidence interval; [148]).

There has not been a measured decrease in rain concentrations of PAHs over the period from 1991 to 1998 [148], unlike the observed decreases in concentration of these same compounds in the gas phase. This is likely due to the dominance of particle scavenging on the observed precipitation concentrations. PAHs are present in both the gas and particle phases, and Cortes et al. [33] showed that only gas-phase PAH concentrations showed a decrease with time. Particle bound atmospheric PAH concentrations exhibited no such decrease, and precipitation does not as well. Simcik et al. [148] compared the profiles of individual compounds and found the distribution in precipitation closely resembles that associated with particles in the air ($r^2 = 0.925$) but not the gas phase ($r^2 = 0.085$). They went on to to explain that precipitation is an effective scavenger of particle phase PAHs, and so the lack of decrease in particle-bound PAHs in the air has lead to the same lack of decrease in PAH concentration in rain. Earlier measurements revealed concentrations of 109 to 459 ng L$^{-1}$ across all Great Lakes on the US Canada border [149], and support the conclusion that concentrations of PAHs in precipitation are not decreasing appreciably.

### 4.2.2
### Snow

Few measurements of PAH concentrations have been performed, and those that have show largely varying concentrations. Franz and Eisenreich [96] report total PAH concentrations ranging from 477 to 17 580 ng L$^{-1}$ in snowfall. Particle-associated PAHs dominate the observed concentrations, with 93 to 96% of the total PAHs measured in the particle bound phase. Strong similarities between chemical accumulations in the snowpack and concentrations in collected snowfall at Eagle Harbor support the hypothesis that dry deposition to the snowpack is a minor contributor to the observed snowpack concentrations. Measured snowpack concentrations of PAHs ranged from 35 to 3280 ng L$^{-1}$ [150]. Earlier measurements of PAH concentrations in snow on Isle Royal National Park [151], revealed geometric mean snow concentrations that were approximately 15 fold higher than rain concentrations measured at the same location during summer months. In all, Franz and Eisenreich [150] estimate tributary discharges from spring snowmelt into Lake Superior in 1992 were between 220 and 350 kg of PAHs.

## 4.3
## Sediments

Because PAHs are hydrophobic, they tend to associate with the sediments. When looking at sediment bound PAHs, one can use accumulation rates or concentrations (normalized to OC or dry weight) for the surface or throughout a core depth, or even inventories, which are aerial concentrations summed over the entire depth of the core. Because it is most often reported we will compare surface sediment concentrations (mass PAH per mass dry sediment) among many areas of the Great Lakes for $\Sigma$PAHs, the sum of the 16 priority PAHs. For many of the studies referenced, more or less than the 16 priority PAHs are listed or totaled. Where possible we have taken only the sum of the 16, in others where the majority of the 16 were present, or totals of greater than 16 where individuals were given, we assumed that the additional PAHs contributed by those compounds either included or excluded did not significantly affect the total. In comparing, we divided the sediments of the Great Lakes into two categories: open lake and coastal/riverine. The open lake concentrations of PAHs are relatively uniform within a lake and from lake to lake (Fig. 7). The greatest variability and highest concentrations of PAHs in Great Lakes sediments occurs in the coastal/riverine regions.

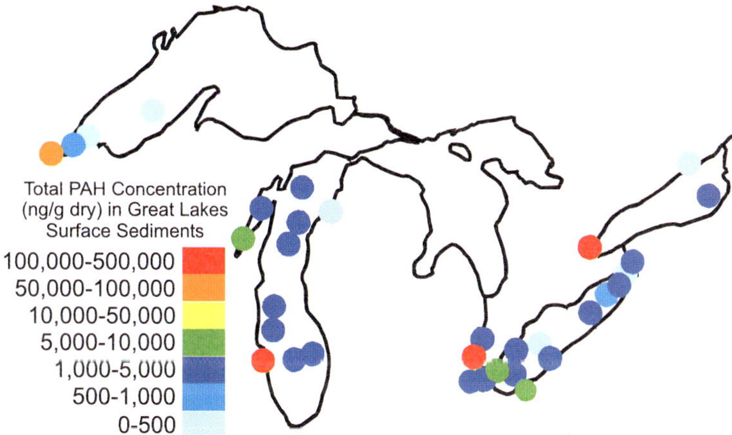


**Fig. 7** $\Sigma$PAH concentrations in surface sediments of the Great Lakes

## 4.3.1
## Open Lake

Only four of the five Great Lakes have had surficial sediment concentrations of $\Sigma$PAHs from the open areas of the lake since 1990 (no data is available for Lake Huron). Open lake sediment is defined as sediment from depositional zones typically deeper than the 100 m contour. The open lake surficial sediment concentrations are relatively uniform within each of the lakes. Lake Superior sediment $\Sigma$PAH concentrations ranged from 290 to 650 ng g$^{-1}$ dry weight [36]. The highest concentration (and accumulation rate) was observed for sites furthest west in the western arm of Lake Superior closer to Duluth, MN/Superior, WI. Lake Michigan sediment $\Sigma$PAH concentrations ranged from 1200 to 6500 ng g$^{-1}$ dry weight [23, 152], with no spatial trends observed in these concentrations. Silliman et al. [49] reported PAH concentrations from two cores of eastern Lake Ontario and the sum of four PAHs (perylene, benzo[$a$]pyrene, pyrene, fluoranthene and phenanthrene) equaled 1350 and 1550 ng g$^{-1}$ dry weight. The most sampled lake has been Erie. Smirnov et al. [32] reported $\Sigma$PAHs from 23 sampling points spanning the three main sedimentary basins of Lake Erie. Concentrations ranged from 200 to 5000 ng g$^{-1}$ dry weight [32]. The concentrations reported by Smirnov et al. [32] from the western end of Lake Erie near Middle Bass Island compare well with those reported by others [137, 153]. As mentioned in Sect. 2, the Lake Erie sediments showed some spatial trends in that concentrations of PAHs in sediments nearer cities were greater than those in along the southern portion of the lake, which were in turn greater than sediments along the northern portion of the lake.

While the range within each lake may span an order of magnitude, a spatial trend was only observed for Lake Erie. This trend may be a result of the larger number of sampling points measured in Lake Erie compared to the other lakes. In general, the lakes are similar in concentration to one another in the range of a few micrograms per gram dry weight $\Sigma$PAHs, although an argument could be made that Lake Superior sediments are less contaminated than the other three lakes for which PAH data is available.

### 4.3.2
### Nearshore

The sediments in nearshore areas of the Great Lakes are much more contaminated with PAHs than the open lake sediments. This is primarily because the coastal areas where one would sample contain the major ports where shipping and industry are concentrated. There are most certainly uncontaminated sediments in coastal areas, but for obvious reasons would not be sampled. The four highest reported sediment concentrations for $\Sigma$PAHs are Duluth/Superior Harbor on Lake Superior, Milwaukee Harbor on Lake Michigan, the Detroit River between Lake St. Clair and Lake Erie and Hamilton Harbor on Lake Ontario (Fig. 7). The highest concentration reported was for Hamilton Harbor at 580 $\mu$g g$^{-1}$, but that does not necessarily make it the most contaminated, as the range in concentrations were broad for all of the coastal areas.

Duluth/Superior and Milwaukee Harbors had wildly varying surficial sediment concentrations given the close proximity of samples, and they exhibited no spatial trends [27, 154]. Sediment concentrations of $\Sigma$PAHs in Duluth/Superior Harbor ranged from 3 to 54 $\mu$g g$^{-1}$. Sediment concentrations of $\Sigma$PAHs in Milwaukee Harbor ranged from 25 to 200 $\mu$g g$^{-1}$. Both sites are surrounded by industrial areas contributing point sources of contamination. This is evident from historical deposits of sediment in Milwaukee Harbor near a coke company that ranged from 380 to 1000 ng g$^{-1}$, and hot spots in Duluth/Superior Harbor near steel plants that also reached 1000 ng g$^{-1}$ [155]. The large spatial variability in these two sites and other sites like them may result from a multitude of individual point sources. However, even Green Bay where the presumed source of PAHs is the Fox River showed no spatial trend in surficial sediment concentrations. The range in $\Sigma$PAHs in Green Bay sediments ranged from 1600 to 5200 ng g$^{-1}$ [28, 156].

Sites that did show a spatial trend included Grand Traverse Bay, MI on Lake Michigan and the Detroit River. Grand Traverse Bay had relatively low sediment concentrations ranging from 200 to 500 ng g$^{-1}$. However, sediments contained the higher concentrations of $\Sigma$PAHs in the southern part of the bay closer to Traverse City, MI, the only major city near the bay. Detroit river sediments ranged from 24 to 200 $\mu$g g$^{-1}$ and the concentration decreased as one went downstream [157]. The PAH pattern also shifted downstream to a distribution that favored higher molecular weight species [157], most likely reflecting a loss of point source uncombusted petroleum. This decrease in concentration and increase in mean molecular weight of the PAHs was also described earlier by Furlong et al. [158]. This trend, however, was not observed by Metcalfe et al. [153] who reported a range of $\Sigma$PAHs (4 to 90 $\mu$g g$^{-1}$) that was highly spatially variable, but without a spatial trend. These latter concentrations are lower than previously observed and may not have adequately captured the source of PAHs to the Detroit River, or it may reflect a decrease in source as these samples were taken later than the other two studies. As was pointed out for Milwaukee Harbor, historically high concentrations of PAHs can become buried in the sediments once a source is removed. This was reported for the Black River, OH by Gu et al. [30], where current $\Sigma$PAH concentrations ($\sim$ 100 ng g$^{-1}$) are orders of magnitude less than deeper, older sediments ($\sim$ 200 $\mu$g g$^{-1}$).

# 5
# Toxicity

## 5.1
## Direct Toxicity

Many hydrophobic organic contaminants pose a threat to human health because they bioaccumulate to high concentrations in predatory fish that are eaten by people. From a human health standpoint, there is little concern of exposure to PAHs through a diet that includes fish from the Great Lakes. This is due to the lack of PAH bioaccumulation to higher trophic levels. However, PAHs can pose a threat to the aquatic organisms themselves by many direct toxicity mechanisms. These mechanisms include narcosis [159], increased ethoxyresorufin-$O$-deethylase (EROD) activity [38, 139], AhR induction [160], CYP1A induction [161], and mutagenesis [162].

At high enough concentrations, PAHs can induce narcosis in aquatic organisms [159]. Narcosis is a reversible anesthetic effect, therefore, removing exposure reverses the narcosis. Di Toro et al. [163] used a model involving lipid based LC50's to determine PAH water quality criteria (WQC) based on narcosis. Since the model used lipid based LC50s, species differences were expected to be eliminated. However, it does not take into account a species ability to metabolize PAHs, reversing the narcosis. The WQC reported by Di Toro et al. [163] range from $10^{-8.76}$ to $10 \times 10^{-5.46}$ mol L$^{-1}$ (0.5 to 500 ug L$^{-1}$) and is inversely proportional to $K_{OW}$ on a log-log basis (since higher $K_{OW}$ PAHs have a greater affinity for the lipids of an organism). None of the reported water concentrations in the Great Lakes is this high [164], but these reported values are for open water. Nearshore areas with high sediment concentrations may approach or exceed these levels.

In a companion paper Di Toro et al. [165] established sediment quality criteria (SQC) based on equilibrium partitioning discussed in the Dissolved/Particle Partioning section and the WQC from the first paper. SQC ranged from 5 to 7 μmol per gram organic carbon. Assuming 5% organic carbon in sediments, these values translate into mass concentrations of 33 to 94 μg per gram dry sediment for individual PAHs. Given the known additivity of narcotic chemicals and similar LC50's of the majority of PAHs reported by Di Toro et al. [165], any sediment with $\geq$ 30 μgg$^{-1}$ ΣPAH would have to be considered suspect for narcotic toxicity to aquatic organisms that cannot metabolize PAHs. From Fig. 7 we see that there are four areas around the Great Lakes that exceed this criterion. They are Duluth/Superior Harbor on Lake Superior, Milwaukee Harbor on Lake Michigan, the Detroit River emptying into Lake Erie and Hamilton Harbor on Lake Ontario.

Induction of hepatic ethoxyresorufin-$O$-deethylase is a defense mechanism of an organism against chemical exposure. It is induced to form more soluble metabolites than can be eliminated from the organism. However, PAH metabolites are often the toxic agents responsible for other effects. White suckers observed by Schrank et al. [139] showed a correlation of increased EROD activity and lower hematocrits with increased PAH metabolites in the bile and PAH contamination of sediments in the Sheboygan River, WI on Lake Michigan. Likewise, Arcand-Hoy et al. [38] observed increased EROD activity and increased biliary PAH metabolites in brown bullheads from the Black River, OH and Hamilton Harbor, Ontario.

PAHs have also been linked to tumors in fish from the Great Lakes. Baumann et al. [166] observed liver tumors in 22 to 39% of the adult brown bullheads sampled from the Black River, OH in the early 1980s. The high cancer rate corresponded to high PAH concentrations in the sediments. Later, when sediment PAHs dropped two orders of magnitude, the cancer rate was reduced by 75%. Fabacher et al. [167] was able to induce liver cancer in Medaka using sediment extracts of four industrialized rivers around the Great Lakes. While most of the PAH toxicity cited here is non-lethal, there is evidence that these toxicological effects can alter community structure. Lesko et al. [168] showed that at PAH contaminated sites in the Great Lakes, brown bullhead fecundity increased at the expense of other fish species. Likewise Canfield et al. [169] observed that the benthic invertebrate communities of PAH contaminated sites were dominated by two contaminant resistant taxa with oligochaetes and chironomids making up over 90% of the identified invertebrates by number.

## 5.2
## Photo-enhanced Toxicity

While PAHs alone are toxic, their toxicity is dramatically increased with the addition of ultraviolet (UV) light [170]. The increased toxicity can occur as a result of photomodification or photosensitization of PAHs. Photomodification can occur in the air, water or inside a cell. The resulting compound from photomodification is an oxidized form of the parent PAH (oxy-PAH). Oxy-PAHs can also be formed by reaction in the atmosphere with hydroxyl radicals as mentioned earlier. These oxy-PAHs have altered, physical, chemical and biological properties. Their vapor pressures are decreased and their solubilities are increased due to the unpaired electrons associated with the addition of oxygen in the form of a ketone or hydroxyl. The result is a decrease in the Henry's Law Constant. Therefore oxy-PAHs formed in the atmosphere would be driven into the aquatic system relative to the parent compounds. Being more soluble, they would also be more bioavailable than their parent compounds. In addition, to having a greater potential for exposure, oxy-PAHs are often more toxic than their parent compounds.

Both parent and photomodified PAHs can be photosensitized and exhibit phototoxicity. The mechanism for phototoxicity is presented in Fig. 8. A PAH or its photomodified analog can partition from the dissolved phase to a cell/organism in its ground (singlet) state. Once it has partitioned to the biota it may absorb UV light. If the energy of the light absorbed is sufficient, an electron from the highest occupied molecular orbital (HOMO) can be excited to the lowest unoccupied molecular orbital (LUMO). The excited electron, in the singlet state, can undergo thermal deactivation or fluorescence back to the ground state or it can undergo internal conversion to the excited triplet state. Once in the triplet state that electron can phosphoresce back to the ground state, cause photomodification or transfer its energy to oxygen. If energy is transferred, reactive oxygen species can be formed including singlet

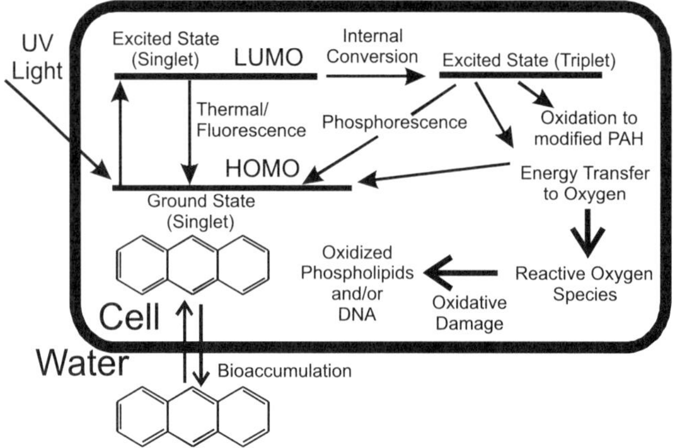


**Fig. 8** Mechanism of photo-enhanced toxicity of PAHs in aquatic organisms

oxygen [171]. These reactive oxygen species can cause oxidative damage to biological molecules in the cell including cell membranes [172–174] and those responsible for photosynthesis [37, 175–178]. It is also important to note that once an electronically excited PAH transfers its energy to create reactive oxygen species it returns to the ground state where it can undergo excitation numerous times.

Not all PAHs exhibit the same phototoxic effect. There are differences in the stability of electronically excited PAHs, the amount of energy available for transfer, and the absorption cross section of each PAH. The relationship of toxicity to the HOMO-LUMO gap in *Daphnia magna* is summarized in Fig. 9, modified from Mekenyan et al. 1994 [179] and Mekenyan et al. 1995 [180]. Line (a) indicates the relationship of toxicity with HOMO-LUMO gap based on the energy available for transfer to oxygen to create reactive oxygen species. The larger the gap the more energy is available. However, as the energy gap gets larger, more energy is needed to excite an electron to the next orbital, and therefore requires shorter wavelengths of light. If light intensity were equal at all wavelengths, the relationship would be approximated by line (a). However, neither incoming solar radiation nor radiation that penetrates natural waters is constant over all wavelengths. As wavelengths shorten, the intensity of light decreases over the range of energies plotted here, and so there is a counter acting force depicted by line (b). A third factor affecting toxicity is the stability of the PAH. The less stable a PAH is, the less chance it has to exert phototoxicity. Stability can be approximated by the HOMO-LUMO gap and is expressed as line (c). The resulting range of phototoxic PAHs have HOMO-LUMO gaps ranging approximately from 6.7 to 7.5 eV (shaded region of Fig. 9).

Rarely, if ever, are organisms exposed to a single PAH, so the question of phototoxicity from mixtures becomes relevant. Because the excitation of

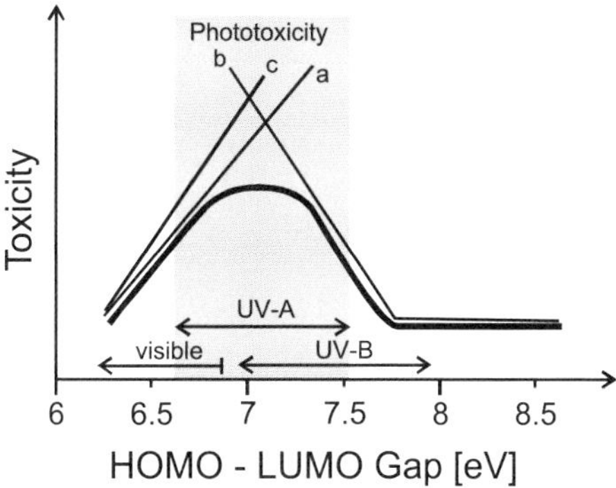



**Fig. 9** Relationship of photo-enhanced toxicity and HOMO-LUMO Gaps of PAHs; *bold line* indicates predicted toxicity based on available energy (a), light intensity (b) and stability of PAHs (c)

PAHs and transfer of energy to oxygen is abiotic and a function of the HOMO-LUMO gap, one would expect the addition of phototoxic PAHs would increase the toxicity in an additive manner. Erickson et al. [181] exposed Oligochaete (*Lumbriculus variegatus*) to binary mixtures of PAHs to investigate this very question. The toxicity of anthracene, fluoranthene and pyrene to *Lumbriculus* in single compound exposures and binary mixtures demonstrated that phototoxicity of PAHs could be explained with a concentration addition model, consistent with expectations. Because of this additivity in phototoxicity, it is important to measure all of the relevant PAHs within a system when considering phototoxicity. Alkylated PAHs often have similar HOMO-LUMO gaps as their parent species [182] and can therefore add to the phototoxicity of complex mixtures. Identification of relevant phototoxic PAHs becomes especially important for sites with uncombusted petroleum where alkyl species can account for large portions of the PAH concentration (see Sect. 2). Just such a case was found at a former steel mill site where PAH analysis underestimated the phototoxicity of sediments, presumably because of the lack of quantitation of alkyl species [183]. Below is a summary of phototoxic and non-photoxic parent PAHs and their calculated HOMO-LUMO gaps (Table 4).

Phototoxicity of PAHs may be important in the Great Lakes because of the confluence of sources and sensitive life stages of aquatic life. As noted in the Sediments section the highest PAH concentrations are found in the nearshore areas including rivers and harbors. These sites are also most likely to be influenced by point sources including uncombusted petroleum (see Sect. 2). Therefore the coastal Great Lakes have the highest potential for ecologically relevant contamination by phototoxic PAHs. Given the shallow water situations of coastal areas, any organisms residing in these areas may also be subject to higher UV exposure, because UV and visible light is attenuated

**Table 4** HOMO-LUMO gap and observed or expected phototoxicity of selected parent PAHs

| PAH | HOMO-[1] (eV) LUMO gap | Phototoxic?[2] |
| --- | --- | --- |
| Anthracene | 7.3 | Y |
| Benzo[a]pyrene | 6.8 | Y |
| Dibenzo[a, h]anthracene | 7.5 | Y |
| Fluoranthene | 7.7 | Y |
| Pyrene | 7.2 | Y |
| Benzo[a]anthracene | 7.4 | Y |
| Benzo[e]pyrene | 7.4 | Y |
| Benzo[a]fluorene | 7.8 | N |
| Benzo[b]fluorene | 8.0 | N |
| Acridine | 7.4 | Y |
| Benzo[k]fluoranthene | 7.4 | Y |
| Benzanthrone | 7.4 | Y |
| Phenanthrene | 8.2 | N |
| Fluorene | 8.5 | N |
| Carbazole | 8.2 | N |
| Triphenylene | 8.2 | N |
| Chrysene | 7.7 | Y |
| Perylene | 6.7 | Y |
| Benzo[b]anthracene | 6.5 | Y |
| Benzo[g, h, i]perylene | 7.0 | Y |
| Coronene | 6.9 | E |
| Dibenzo[a, j]anthracene | 7.0 | E |
| Benzo[b]chrysene | 6.5 | E |
| Benzo[b]triphenylene | 7.1 | E |
| Benzene | 8.0 | NE |
| Naphthalene | 10.1 | NE |
| Dibenzo[b, i]anthracene | 5.3 | NE |
| Benzo[a]chrysene | 7.1 | E |

[1] Calculated in Mekenyan et al. [179]

[2] Y = observed phototoxicity; N = no observed phototoxicity; E = expected phototoxicity; NE = no expected phototoxicity; to *Daphnia magna* [179]

in the water column. Therefore, if depth is the only consideration, shallower areas would have higher UV irradiance than deeper waters. However, there is often greater dissolved organic carbon (DOC) and higher suspended particulate matter (SPM) in coastal areas due to runoff of nutrients and soil erosion. These components can rapidly attenuate the UV in a water column. Therefore, some coastal areas of the Great Lakes may contain high PAHs, but not high UV or high UV but not high PAHs. It is the sites where PAHs and UV are both high where phototoxicity may be a concern to flora and fauna.

Coincidentally, the coastal areas also house most of the biological activity in the photic zone of the lakes. This results from the higher nutrient contents of the coastal areas contributing greater amount of plant material and spawning, hatching and development of many fish species in the rivers and wetlands of the Great Lakes [40, 184–188]. These early life stages of fish can be especially problematic for several reasons. The eggs of many species are laid close to the sediments where they can accumulate a high concentration of PAHs. Upon hatching the larval fish are translucent allowing for high UV doses. Some species will also seek out the air-water interface because the bending of light at the surface allows them to elude predation, further exposing them to UV. Presumably, larval fish have evolved the ability to avoid a lethal dose of UV, but if PAHs are added it is not clear whether they will avoid UV because of phototoxic stress or be subject to phototoxicity because of "UV only" avoidance. If larval fish do have the perception and avoidance of phototoxic stress, high PAH and UV doses may drive fish to shaded areas of a coastal area that may or may not be ideal habitat for development. This potential shrinking of habitat should also be considered an ecological effect of PAHs in the Great Lakes, but to date has not been quantified. Of potentially greater concern is the phototoxicity of PAHs to coastal plant life because they require light to grow, and therefore can't avoid UV and because photosynthesis occurs in the thylakoid membrane [175] where phototoxicity also can occur [172, 178].

Phototoxicity of PAHs in the Great Lakes has been recognized for several years. The Black River in Ohio, which flows into Lake Erie, is an AOC that specifically lists phototoxicity of PAHs being a concern during periods of low flow when SPM is low. More specific studies on phytoplankton have been performed by the group of Bruce Greenberg at the University of Windsor [37, 175–178, 189]. Marwood et al. [175] sampled Lake Erie phytoplankton, exposed them to 0.2 to 2 mg/L anthracene or 1,2-dihydroxyanthraquinone, an oxy PAH form through photomodification. The PAH exposed phytoplankton were then exposed either to darkness or 60 minutes of 50% sunlight. The sunlight and PAH exposed phytoplankton showed a decrease in chlorophyll fluorescence, a surrogate for photosynthetic efficiency, with increasing anthracene concentration. However, phytoplankton in the Great Lakes are not only in the coastal areas where concentrations can be high, they live throughout the open lake water where concentrations can be quite low. As a result, Marwood et al. [176] again sampled Lake Erie phytoplankton and exposed them to much lower concentrations of PAHs to determine if environmentally relevant concentrations of PAHs could produce the same effect. In this study the phytoplankton were exposed to 40–2000 µg/L of numerous individual PAHs. The authors concluded that anthracene, fluoranthene and phenanthrenequinone were the most photoxic of the PAHs studied (that is they showed the greater decrease in photosynthetic efficiency) resulting in EC50s of 314, 118 and 90 µg/L, respectively. They also concluded that toxicity was a function of both UV duration and PAH dose, therefore, at low concentrations like those observed in the Great Lakes long-term exposure to UV could cause inhibition of photosynthesis in phytoplankton. Other organisms that might have the ability to avoid UV, then presumably would need to be exposed in highly contaminated nearshore areas.

Davenport and Spacie [190] investigated phototoxicity of PAHs to *Daphnia magna* using sediments from southern Lake Michigan. Sediments from the Grand Calumet River, Indiana Harbor Canal and Waukegan Harbor were washed with clean water. The water was collected and used to expose *D. Magna* to PAH mixture at relevant concentrations. The *Daphnia* were exposed to natural sunlight and laboratory UV. Exposure to Grand Calumet River sediment water resulted in 100% mortality within 1 to 6 hours depending on the site within the river. Exposure to Indiana Harbor Canal sediments resulting in > 90% mortality within 4 hours. No mortality was observed with exposure to Waukegan Harbor sediments. Therefore, there are coastal sites within the Great Lakes with sufficient PAHs to cause ecologically relevant phototoxicity. However, no attempt was made to determine if the *in situ* UV exposure would be able to produce the same effects as those found in the field.

In order to better understand the wavelengths responsible for PAH phototoxicity from mixtures found in sediments around the Great Lakes, Diamond et al. [183] sampled amphipods (*Gammarus spp.*) from the Duluth/Superior Harbor, the St. Louis Rivers, which flows into the Harbor, both sites contaminated with point source PAHs and a non-contaminated reference site. These amphipods were exposed to UVA and UVB in separate experiments. The UVA exposed animals showed increased mortality where UVB exposed animals showed no effect. While amphipods are not likely to be exposed to UV in the field, their sensitivity to UV in the laboratory would indicate that organisms that spend part of their life cycle near the sediments and move higher in the water column later might be susceptible to phototoxicity.

## 6
## Lessons Learned for the Future

PAHs are compounds that will continue to be produced in the atmosphere of the Great Lakes, via combustion of organic matter. They will also remain a threat from direct discharge of uncombusted petroleum for as long as this remains a fuel source. They are expected to remain a concern for ecological health in the future as concentrations and loadings increase, urban areas exert increased atmospheric deposition to coastal areas, dredging re-introduces historically deposited PAHs and increased water clarity could increase the photo-enhanced toxicity of PAHs to aquatic organisms.

While sediments have shown a decrease in atmospheric deposition since ~ 1960 [23], atmospheric concentrations and loadings estimates show no decrease since the early 1990s [122]. Future concentrations and loadings depend on the changes in PAH sources over the next few decades. A major source of PAHs to the Great Lakes is coal combustion associated with power generation and steel production. Political and economic forces driving the future use of coal is currently unknown, but any increase in coal use is expected to increase PAHs in the atmosphere, water and sediments of the Great Lakes to a disproportionately greater extent than other sources of energy. This is evident from Fig. 3, where coal and coke produce much more PAHs per BTU than natural gas or petroleum. Another major source of PAHs is that emitted by vehicles. The number of vehicle miles travelled has continued to increase over the past few decades and is expected to continue to rise as the number of vehicle miles per household increases and the number of households increases (DOE statistics). Therefore, PAH concentrations and loadings are expected to increase in the future.

As urban areas get larger, their influence on the adjacent coastal atmospheres and on atmospheric deposition to adjacent coastal waters is expected to increase. Monitoring of these urban areas will be necessary to understand the loadings of PAHs to the Great Lakes. Simcik [34] suggests that urban sources not currently being monitored, but could be exerting increased atmospheric deposition include Duluth, Thunder Bay, Green Bay, Milwaukee, Racine, Kenosha, Saginaw Bay, Toronto, Hamilton and Rochester. As urban areas grow, this list is expected to grow.

Of more immediate concern are the stores of PAHs already buried in sediments of ports and harbors of the Great Lakes. As mentioned in the Sediments section, many of the ports around the Great Lakes experienced point sources of PAHs in their pasts that created hot spots. As the ports continue to deposit sediments, these PAHs become buried and are no longer bioavailable. But as the sediments build up, ports need to be dredged to allow ship traffic. These historically deposited sediments can then be resuspended during the dredging process making historic PAHs once again bioavailable. Just such an increase in PAHs due to dredging was observed in the Black River by Gu et al. [30]. This increase in PAHs was associated with a concurrent increase in hepatic tumors in brown bullheads from the Black River, but tumor incidence rapidly decreased later as PAHs were re-buried by depositing sediments [166]. Therefore, dredging in the Great Lakes poses a real but temporary threat to aquatic life. More permanent effects could result from increasing photo-enhanced toxicity of PAHs.

Since 1986 zebra mussels have been filtering the water of the Great Lakes with great efficiency. Their ability to filter water has caused increased clarity, which has been reported for Lake St. Clair, Lake Erie and Hamilton Harbor [134, 191, 192]. As the Great Lakes waters become clearer, especially coastal areas where zebra mussels tend to congregate, the UV dose to the water column increases. Therefore, the introduction of zebra mussels could increase photo-enhanced toxicity of PAHs to aquatic organisms. There is some evidence to suggest that this may be affecting fish community structure.

Since the 1980s the fish population of Lake Michigan has been in decline and the decline continued through the 1990s [193, 194]. Considerable effort has been expended to determine the reason for the decline of this popular sport and commercial fishery. A recent article by Fitzgerald et al. [195] suggests that mortality in Lake Michigan yellow perch occurs prior to the juvenile stage (i.e., larval stage). As mentioned in the Photo-Enhanced Toxicity section, larval fish may be especially vulnerable to photo-enhanced toxicity of PAHs. While no concrete causal relationship has been established, there is no doubt that increased water clarity is expected to increase photo-enhanced toxicity of PAHs, and may affect larval fish populations such as the yellow perch.

# References


1. Harrison RM, Smith DJT, Luhana L (1996) Environ Sci Tech 30:825
2. Li CK, Kamens RM (1993) Atmos Environ. Part A, General Topics p523
3. Miguel AH, Pereira PAP (1989) Aerosol Sci Tech 10:292
4. Pistikopoulos P, Masclet P, Mouvier G (1990) Atmos Environ 1189
5. Venkataraman C, Friedlander SK (1994) J Air Waste Manage Assoc 44:1103
6. Kristensson A, Johansson C, Westerholm R, Swietlicki E, Gidhagen L, Wideqvist U, Vesely V (2004) Atmos Environ 38:657
7. Sai P-J, Shih T-S, Chen H-L, Lee W-J, Lai C-H, Liou S-H (2004) Atmos Environ 38:333
8. Schauer JJ, Kleeman MJ, Cass GR, Simoneit BRT (2002) Environ Sci Tech 36:1169
9. Fraser MP, Cass GR, Simoneit BRT (1998) Environ Sci Tech 32:2051
10. Westerholm R, Christensen A, Rosen A (1996) Atmos Environ 30:3529
11. Westerholm R, Li H (1994) Environ Sci Tech 28:965
12. Lee ML, Prado GP, Howard JB, Hites RA (1977) Biomedical Mass Spectrom 4:182
13. Simo R, Grimalt JO, Albaiges J (1997) Environ Sci Tech 31:2697
14. Simcik MF, Eisenreich SJ, Lioy PJ (1999) Atmos Environ 33:5071
15. Rogge WF, Hildemann LM, Mazurek MA, Cass GR, Simoneit BRT (1993) Environ Sci Tech 27:2736
16. Grimmer G, Jacob J, Naujack K-W (1983) Anal Chem 314:29
17. Youngblood WW, Blumer M (1975) Geochim Cosmochim Acta 39:1303
18. Boehm PD, Farrington JW (1984) Environ Sci Tech 18:840
19. Chakhmakhchev A, Suzuki M, Takayama K (1997) Org Geochem 26:483
20. Simoneit BRT (1984) Atmos Environ 18:51
21. Simoneit BRT (1989) J Atmos Chem 8:251
22. Mazurek MA, Simoneit BRT (1984) In: Keith LH (ed) Identification and Analysis of Organic Pollutants in Air. Ann Anbor Sci/Butterworth Publishers, Woburn, MA
23. Simcik MF, Eisenreich SJ, Golden KA, Liu S-P, Lipiatou E, Swackhamer DL, Long DT (1996) Environ Sci Technol 30:3039
24. Simcik MF (1994) Polycyclic Aromatic Hydrocarbons in Lake Michigan Sediments. Master of Science, University of Minnesota
25. Christensen ER, Zhang X (1993) Environ Sci Tech 27:139
26. Rachdawong P, Christensen ER, Karls JF (1998) Water Res 32:2422
27. Li A, Razak IAA, Ni F, Gin MF, Christensen ER (1998) Water Air Soil Pollut 101:417
28. Su M-C, Christensen ER, Karls JF (1998) Environ Poll 99:411

29. Su M-C, Christensen ER, Karls JF, Kosuru S, Imamoglu I (2000) Environ Toxic Chem 19:1481
30. Gu S-H, Kralovec AC, Christensen ER, Van Camp RP (2003) Water Res 37:2149
31. Simcik MF, Zhang H, Eisenreich SJ, Franz TP (1997) Environ Sci Tech 31:2141
32. Smirnov A, Abrajano TA Jr, Smirnov A, Stark A (1998) Org Geochem 29:1813
33. Cortes DR, Basu I, Sweet CW, Hites RA (2000) Environ Sci Tech 34:356
34. Simcik MF (2005) Environ Monit Assess 100:201–216
35. Buehler SS, Basu I, Hites RA (2001) Environ Sci Tech 35:2417
36. Simcik MF, Jeremiason JD, Lipiatou E, Eisenreich SJ (2003) J Great Lakes Res 29:41
37. Lampi MA, Huang X-D, El-Alawi YS, McConkey BJ, Dixon DG, Greenberg BM (2000) ASTM Spec Tech Publ STP 1403:211
38. Arcand-Hoy LD, Metcalfe CD (1999) Environ Toxicol Chem 18:740
39. Fox ME, Khan RM, Thiessen PA (1996) Water Qual Res J Can 31:593
40. Leslie JK, Timmins CA (1992) J Great Lakes Res 18:700
41. Mayer T, Nagy E (1992) Water Pollut Res J Can 27:807
42. Metcalfe CD, Balch GC, Cairns VW, Fitzsimons JD, Dunn BP (1990) Sci Total Environ 94:125
43. Wakeham SG, Schaffner C, Giger W (1980) Geochm Cosmochm Acta 44:415
44. Simoneit BRT, Mazurek M (1982) Atmos Environ 16
45. Jeremiason JD, Eisenreich SJ, Paterson MJ (1999) Can J Fish Aquat Sci 56:650
46. Danz NP, Regal RR, Niemi GJ, Brady VJ, Hollenhorst T, Johnson LB, Host GE, Hanowski JM, Johnston CA, Brown T, Kingston J, Kelly JR (2004) Environ Monit Assess 102:41–65
47. Louda JW, Baker EW (1984) Geochim Cosmochim Acta 48:1043
48. Venkatesan MI (1988) Mar Chem 25:1
49. Silliman JE, Meyers PA, Eadie BJ (1998) Org Geochem 29:1737
50. Yamasaki H, Kuwate K, Miyamoto H (1982) Environ Sci Tech 16:189
51. Pankow JF (1987) Atmos Environ 21:2275
52. Pankow JF (1988) Atmos Environ 22:1405
53. Harner T, Bidleman TF (1998) J Chem Eng Data 43:40
54. Baker JE, Eisenreich SJ (1990) Environ Sci Tech 24:342
55. Cotham WE, Bidleman TF (1995) Environ Sci Tech 29:2782
56. Foreman WT, Bidleman TF (1990) Atmos Environ 24A:2405–2416
57. Ligocki MP, Pankow JF (1989) Environ Sci Tech 23:75
58. Simcik MF, Franz TP, Zhang H, Eisenreich SJ (1998) Environ Sci Tech 32:251
59. Pankow JF (1994) Atmos Environ 28:185
60. Dachs J, Eisenreich SJ (2000) Environ Sci Tech 34:3690
61. Offenberg JH, Baker JE (2002) Atmos Environ 36:1205
62. Finlayson-Pitts BJ, Pitts JN (2000) Chemistry of the Upper and Lower Atmosphere. Academic Press, San Diego
63. Behymer TD, Hites RA (1985) Environ Sci Tech 19:1004
64. Dunstan TDJ, Mauldin RF, Jinxian Z, Hipps AD, Wehry EL, Mamantov G (1989) Environ Sci Tech 23:303
65. Korfmacher WA, Natusch DFS, Taylor DR, Wehry EL, Mamantov G (1979) In: Jones PW, Leber P (eds) Polynuclear Aromatic Hydrocarbons. Ann Arbor Sci, Ann Arbor p 165
66. Yokley RA, Garrison AA, Wehry EL, Mamantov G (1986) Environ Sci Tech 20:86
67. Mackay D, Shiu WY, Ma KC (1992) Illustrated Handbook of Physical-Chemical Properties and Environ Fate for Organic Chemicals. Lewis Publishers, Chelsea, Michigan
68. Zepp RG, Schlotzhauer PF (1979) In: Jones PW, Leber P (eds) Polynuclear Aromatic Hydrocarbons. Ann Arbor Sci, Ann Arbor

69. Whitman WG (1923) Chem Metall Eng 29:146
70. Nelson ED, McConnell LL, Baker JE (1998) Environ Sci Tech 32:912
71. Wanninkhoff R (1985) Science 227:1224
72. Wanninkhof R, Ledwell J, Crusius J (1990) In: Wilhelms S, Gulliver J (eds) Air-Water Mass Transfer. Am Soc Civil Eng, New York, p 441
73. Watson AJ, Upstill-Goddard RC, Liss PS (1991) Nature (London, United Kingdom) 349:145
74. Wanninkhof R, Ledwell JR, Broecker WS, Hamilton M (1987) J Geophys Res (Oceans) 92:14567
75. Broecker W, Peng T (1984) In: Garcia G (ed) Gas Transfer at Water Surfaces. D. Reidel, Hingham, MA, USA p 479
76. Kanwisher J (1963) J Geophys Res 68:3921
77. Liss PS (1973) Deep-Sea Res and Oceanographic Abstracts 20:221
78. Gigliotti CL, Brunciak PA, Dachs J, Glenn Thomas Rt, Nelson ED, Totten LA, Eisenreich SJ (2002) Environ Toxic Chem 21:235
79. Wanninkhof R, McGillis WR (1999) Geophys Res Lett 26:1889
80. Liss PS, Merlivat L (1986) In: Baut-Menard P (ed) The Role of Air-Sea Exchange in Geochemical Cycling. Reidel Publishing Co, Dordrecht, Holland p 113
81. Eisenreich SJ, Hornbuckle KC, Achman DR (1997) In: Baker JE (ed) Atmospheric Deposition of Contaminants to the Great Lakes and Coastal Water. SETAC, Pensacola, FL, USA p 109
82. Bamford HA, Poster DL, Baker JE (1999) Environ Toxic Chem 18:1905
83. Achman DR, Hornbuckle KC, Eisenreich SJ (1993) Environ Sci Technol 27:75
84. Simcik MF (2004) Atmos Environ 38:491
85. Ligocki MP, Leuenberger C, Pankow JF (1985) Atmos Environ (1967–1989) 19:1609
86. Ligocki MP, Leuenberger C, Pankow JF (1985) Atmos Environ (1967–1989) 19:1619
87. Pankow JF (1994) Atmos Environ 28:189
88. Scott BC (1981) In: Eisenreich SJ (ed) Atmospheric Pollutants in Natural Waters, vol 3-21. Ann Arbor Science, Ann Arbor, MI
89. Slinn WGN (1983) In: Liss PS, Slinn WGN (eds) Air-Sea Exchange of Gases and Particles. Reidel, Dordrecht, The Netherlands p 299
90. Topol LE, Vijayakumar R, McKinley CM, Waldron TL (1986) J Air Pollut Control Assoc 36:393
91. Poster DL, Baker JE (1996) Environ Sci Tech 30:341
92. Offenberg JH, Baker JE (2002) Environ Sci Tech 36:3763
93. Poster DL, Baker JE (1996) Environ Sci Tech 30:349
94. Pankow JF (1997) Atmos Environ 31:927
95. Roth CM, Goss K-U, Schwarzenbach RP (2002) J Colloid Interface Sci 252:21
96. Franz TP, Eisenreich SJ (1998) Environ Sci Tech 32:1771
97. Miller NL, Wang PK (1991) Atmos Environ. Part A: General Topics 25A:2593
98. Mitra SK, Barth U, Pruppacher HR (1990) Atmos Environ. Part A: General Topics 24A:1247
99. Redkin YN (1973) In: Voloshchuk VM, Sedunov YS (eds) Hydrodynamics and Thermodynamics of Aerosols. Wiley, New York, p 226
100. Murakami M, Kimura T, Magono C, Kikuchi K (1983) J Meteorol Soc Jpn 61:346
101. Sparmacher H, Fuelber K, Bonka H (1993) Atmos Environ. Part A: General Topics 27A:605
102. Nicholson KW, Branson JR, Giess P (1991) Atmos Environ. Part A: General Topics 25A:771
103. Schumann T, Zinder B, Waldvogel A (1988) Atmos Environ (1967–1989) 22:1443

104. Zinder B, Schumann T, Waldvogel A (1988) Atmos Environ (1967–1989) 22:2741
105. Beaglehole D, Nason D (1980) Surface Sci 96:357
106. Kvlividze VI, Kiselev VF, Kurzaev AB, Ushakova LA (1974) Surface Sci 44:60
107. Valdez MP, Dawson GA, Bales RC (1989) J Geophysical Res (Atmospheres) 94:1095
108. Goss KU (1993) Environ Sci Tech 27:2826
109. Hoff JT, Wania F, Mackay D, Gillham R (1995) Environ Sci Tech 29:1982
110. Lamb D, Blumenstein R (1987) Atmos Environ (1967–1989) 21:1765
111. Hoekstra P, Miller RD (1967) J Colloid Interface Sci 25:166
112. Gross GW, Wu C-H, Bryant L, McKee C (1975) J Chem Phys 62:3085
113. Gill PS, Graedel TE, Weschler CJ (1983) Rev Geophys Space Phys 21:903
114. Leuenberger C, Czuczwa J, Heyerdahl E, Giger W (1988) Atmos Environ 22:695
115. Raynor GS, Hayes JV (1983) In: Pruppacher HR, Semonin RG, Slinn WGN (eds) Precip-
     itation Scavenging, Dry Deposition, and Resuspension. Elsevier, New York, p 249
116. Franz T, Eisenreich SI, Holsen TM (1998) Environ Sci Tech 32:3681
117. Caffrey PF, Suarez AE, Ondov JM, Han M (1996) J Aerosol Sci 27:S31
118. Caffrey PF, Ondov JM, Zufall MJ, Davidson CI (1998) Environ Sci Tech 32:1615
119. Yi S-M, Holsen TM, Noll KE (1997) Environ Sci Tech 31:272
120. Zufall MJ, Davidson CI, Caffrey PF, Ondov JM (1998) Environ Sci Tech 32:1623
121. Endlich RM, Eymon BP, Ferek RJ, Valdes AD, Maxwell C (1988) J Appl Meteorol
     27:1322–1333
122. Buehler SS, Hites RA (2002) Environ Sci Tech 36:354A
123. Hillery BR, Simcik MF, Basu I, Hoff RM, Strachan WMJ, Burniston D, Chan CH,
     Brice KA, Sweet CW, Hites RA (1998) Environ Sci Tech 32:2216
124. Hoff RM, Strachan WMJ, Sweet CW, Chan CH, Shackleton M, Bidleman TF, Brice
     KA, Burniston DA, Cussion S, Gatz DF, Harlin K, Schroeder WH (1996) Atmos Env-
     iron 30:3505
125. Mackay D, Bentzen E (1997) Atmos Environ 31:4045
126. Chiou CT, Porter PE, Schmedding DW (1983) Environ Sci Tech 17:227
127. Karickhoff SW, Brown DS, Scott TA (1979) Water Res 13:241
128. Schwarzenbach RP, Westall J (1981) Environ Sci Tech 15:1360
129. Voice TC, Rice CP, Weber WJ Jr (1983) Environ Sci Tech 17:513
130. Eadie BJ, Morehead NR, Val Klump J, Landrum PF (1992) J Great Lakes Res 18:91
131. Readman JW, Mantoura RFC, Rhead MM, Brown L (1982) Estuar Coastal Shelf Sci
     14:369
132. Fisher SW, Gossiaux DC, Bruner KA, Landrum PF (1993) In: Schloesser DW (ed)
     Zebra Mussels: Biology, Impacts and Controls. Lewis Publishers, Ann Arbor, p 465
133. Vanderploeg HA, Liebig JR, Carmichael WW, Agy MA, Johengen TH, Fahnenstiel
     GL, Nalepa TF (2001) Can J Fish Aquat Sci 58:1208
134. Gossiaux DC, Landrum PF, Fisher SW (1998) Chemosphere 36:3181
135. Bruner KA, Fisher SW, Landrum PF (1994) J Great Lakes Res 20:735
136. Bruner KA, Fisher SW, Landrum PF (1994) J Great Lakes Res 20:725
137. Gewurtz SB, Lazar R, Haffner GD (2000) Environ Toxicol Chem 19:2943
138. Yang X, Peterson DS, Baumann PC, Lin ELC (2003) J Great Lakes Res 29:116
139. Schrank CS, Cormier SM, Blazer VS (1997) J Great Lakes Res 23:119
140. Pirrone N, Keeler GJ, Holsen TM (1995) Environ Sci Tech 29:2123
141. Andren AW, Strand JW (1981) In: Eisenreich SJ (ed) Atmospheric Pollutants in Nat-
     ural Waters. Ann Arbor Sciences, Ann Arbor, p 459
142. Hoff RM, Chan KW (1987) Environ Sci Tech 21:556
143. Tsai W, Cohen Y, Sakugawa H, Kaplan IR (1991) Environ Sci Technol 25:2012–2023
144. Offenberg JH, Baker JE (1999) Environ Sci Tech 33:3324

145. Offenberg JH, Baker JE (1999) J Air Waste Manage Assoc 49:959
146. Gigliotti CL, Dachs J, Nelson ED, Brunciak PA, Eisenreich SJ (2000) Environ Sci Tech 34:3547
147. Baker JE, Leister DL, Clark CA, Church TM, Scudlark JR, Ondov JM, Dickhut RM, Cutter G (1997) In: Baker JE (ed) Atmosheric Deposition of Contaminants to the Great Lakes and Coastal Waters. SETAC, Pensacola, FL
148. Simcik MF, Hoff RM, Strachan WMJ, Sweet CW, Basu I, Hites RA (2000) Environ Sci Technol 34:361
149. Chan CH, Perkins LH (1989) J Great Lakes Res 15:465
150. Franz TP, Eisenreich SJ (2000) J Great Lakes Res 26:220
151. McVeety BD, Hites RA (1988) Atmos Environ 22:511
152. Zhang X, Christensen ER, Gin MF (1993) Estuaries 16:638
153. Metcalfe TL, Metcalfe CD, Bennett ER, Haffner GD (2000) J Great Lakes Res 26:55
154. Breneman D, Richards C, Lozano S (2000) J Great Lakes Res 26:287
155. Peterson GS, Axler RP, Lodge KB, Schuldt JA, Crane JL (2002) Environ Monitor Assess 78:111
156. Zhang X, Christensen ER, Yan LY (1993) J Great Lakes Res 19:429
157. Besser JM, Giesy JP, Kubitz JA, Verbrugge DA, Coon TG, Braselton WE (1996) J Great Lakes Res 22:683
158. Furlong ET, Carter DS, Hites RA (1988) J Great Lakes Res 14:489
159. Verhaar HJM, Van Leeuwen CJ, Hermens JLM (1992) Chemosphere 25:471
160. Marty GD, Short JW, Dambach DM, Willits NH, Heintz RA, Rice SD, Stegeman JJ, Hinton DE (1997) Can J Zool 75:989
161. Brinkworth L, Hodson P, Tabash S, Lee P (2003) J Toxicol Environ Health, Part A 66:627
162. Malins DC, McCain BB, Brown DW, Chan S-L, Myers MS, Landahl JT, Prohaska PG, Friedman AJ, Rhodes LD, Burrows DG, Gronlund WD, Hodgins HO (1984) Environ Sci Tech 18:705
163. Di Toro DM, McGrath JA, Hansen DJ (2000) Environ Toxicol Chem 19:1951
164. Offenberg JH, Baker JE (2000) J Great Lakes Res 26:196
165. Di Toro DM, McGrath JA (2000) Environ Toxicol Chem 19:1971
166. Baumann PC, Harshbarger JC (1998) Environ Monitor Assess 53:213
167. Fabacher DL, Besser JM, Schmitt CJ, Harshbarger JC, Peterman PH, Lebo JA (1991) Arch Environ Contam Toxicol 21:17
168. Lesko LT, Smith SB, Blouin MA (1996) J Great Lakes Res 22:830
169. Canfield TJ, Dwyer FJ, Fairchild JF, Haverland PS, Ingersoll CG, Kemble NE, Mount DR, La Point GW, Burton GA, Swift MC (1996) J Great Lakes Res 22:565
170. Larson RA, Berenbaum MR (1988) Environ Sci Technol 22:354
171. Foote CS (1968) Science 162:963
172. Newsted JL, Giesy JP (1987) Environ Toxicol Chem 6:445
173. Oris JT, Giesy JP Jr (1987) Chemosphere 16:1395
174. Hatch AC, Burton GA Jr (1999) Environ Pollut 106:157
175. Marwood CA, Smith REH, Solomon KR, Charlton MN, Greenberg BM (1999) Ecotoxicol Environ Safety 44:322
176. Marwood CA, Smith REH, Charlton MN, Solomon KR, Greenberg BM (2003) J Great Lakes Res 29:558
177. Huang X-D, Krylov SN, Ren L, McConkey BJ, Dixon DG, Greenberg BM (1997) Environ Toxicol Chem 16:2296
178. Krylov SN, Huang X-D, Zeiler LF, Dixon DG, Greenberg BM (1997) Environ Toxicol Chem 16:2283
179. Mekenyan OG, Ankley GT, Veith GD, Call DJ (1994) Chemosphere 28:567

180. Mekenyan OG, Ankley GT, Veith GD, Call DJ (1995) SAR and QSAR in Environ Res 4:139
181. Erickson RJ, Ankley GT, DeFoe DL, Kosian PA, Makynen EA (1999) Toxicol Appl Pharmacol 154:97
182. Veith GD, Mekenyan OG, Ankley GT, Call DJ (1995) Chemosphere 30:2129
183. Diamond SA, Milroy NJ, Mattson VR, Heinis LJ, Mount DR (2003) Environ Toxicol Chem 22:2752
184. Chubb SL, Liston CR (1986) J Great Lakes Res 12:332
185. Leslie JK, Kelso JRM (1977) Bull Environ Contam Toxicol 18:602
186. Leslie JK, Timmins CA (1995) Water Qual Res J Can 30:713
187. Höök TO, Eagan NM, Webb PW (2001) Wetlands 21:281
188. Petering RW, Johnson DL (1991) Wetlands 11:123
189. Greenberg A, Darack F, Harkov R, Lioy PJ, Daisey JM (1985) Atmos Environ 19:1325
190. Davenport R, Spacie A (1991) J Great Lakes Res 17:51
191. Charlton MN, Le Sage R (1996) Water Qual Res J Can 31:473
192. Holland RE, Johengen TH, Beeton AM (1995) Can J Fish Aquat Sci 52:1202
193. Francis JT, Robillard SR, Marsden JE (1996) Fisheries 21:18
194. Shroyer TS, McComish SM (1998) North Am J Fish Manage 18:19
195. Fitzgerald DG, Clapp DF, Belonger BJ (2004) J Great Lakes Res 30:227

Hdb Env Chem Vol. 5, Part N (2006): 355–390
DOI 10.1007/698_5_045
© Springer-Verlag Berlin Heidelberg 2005
Published online: 2 December 2005

# Brominated Flame Retardants in the Great Lakes

Ronald A. Hites

School of Public and Environmental Affairs, Indiana University, Bloomington, IN 47405, USA
*Hitesr@Indidna.edu*



**Abstract** Brominated flame retardants in the Great Lakes have not been as well studied as many of the polychlorinated pollutants, especially PCBs, but in the last 5–10 years there has been some significant progress. The ubiquity of these compounds in the sediment and fishes of the lakes has now been well established, and perhaps more alarmingly, it is now known that the concentrations of some of these compounds are actually increasing. This observation is particularly important given that the concentrations of most of the other persistent organic pollutants in the lakes are decreasing. Despite their production cessation in the mid-1970s, polybrominated biphenyls are still present in fishes and sediment from most of the lakes. In general, these PBB concentrations are decreasing slowly, if at all. Polybrominated diphenyl ethers are present in air, fishes, birds, and sediment from the lakes. In lake trout and in herring gull eggs, the PBDE concentrations have been doubling every 3–5 years; in sediment cores, the doubling time is $\sim 15$ years. Hexabromocyclododecanes (HBCD) are also present in fishes and sediment from the lakes, but at much lower levels compared to the PBDEs. Two novel flame retardants (1,2-*bis*(2,4,6-tribromophenoxy)ethane (TBE) and 1,2,3,4,5-pentabromoethylbenzene) have been found in the air and sediment of the lakes. Clearly, it would be good to monitor the concentrations of all of these compounds in (at least) the sediment and fishes of the lakes to determine long-term trends. One might want to focus especially on those BFRs that will continue to be in production; these include deca-BDE (congener 209), HBCD, and TBE.

**Keywords**  1,2,3,4,5-Pentabromoethylbenzene · 1,2-*Bis*(2,4,6-tribromophenoxy)ethane · Hexabromocyclododecanes · Polybrominated biphenyls · Polybrominated diphenyl ethers

**Abbreviations**

| | |
|---|---|
| BB-$x$ | PBB congener $x$ |
| BDE-$x$ | PBDE congener $x$ |
| BFR | brominated flame retardant(s) |
| HBCD | hexabromocyclododecane |
| $K_P$ | atmospheric vapor-particle partition coefficient |
| $N$ | number |
| $P_L^\circ$ | sub-cooled liquid vapor pressure |
| PBB | polybrominated biphenyl |
| PBDE | polybrominated diphenyl ether |
| PEB | 1,2,3,4,5-pentabromoethylbenzene |
| TBE | 1,2-bis(2,4,6-tribromophenoxy)ethane |

# 1
# Introduction

Brominated flame retardants (BFRs) are added to many consumer and commercial products to prevent these products from burning even if they have been exposed to a spark or a smoldering cigarette. BFRs are added to polyurethane foam that is used in furniture found in most homes and offices; to commercial fabrics used in, for example, auditorium seating; and to carpeting. BFRs certainly save lives by preventing large fires, but some of these chemicals have become environmentally ubiquitous.

BFRs work by releasing bromine free radicals when they are heated. These free radicals scavenge other free radicals that are part of the flame propagation process and thus reduce the rate of flame expansion and the extent of fire damage. Organobromine compounds are particularly well suited for this function because of the relatively low energy of the C – Br bond. When heated, these bonds break, and the relatively stable bromine free-radical is generated. The BFR's carbon skeleton is more or less irrelevant; it is just a convenient way to carry the bromine atoms. Thus, most BFRs have simple carbon skeletons to which several bromine atoms are bonded.

BFRs can be divided into two classes: those that react with the substrate and those that do not. The reactive BFRs include tris(2,3-dibromopropyl)-phosphate, which had been used on children's sleep-ware (pajamas, blankets, etc.). The non-reactive BFRs are simply mixed with or applied directly to the substrate. In general, the reactive BFRs tend to be relatively polar molecules, and these compounds seem to be generally immobile in the environment. The non-reactive BFRs tend to be non-polar, persistent compounds, and they seem to move throughout the environment and to enter humans' food supply. In fact, there is now good evidence that non-reactive BFRs are widely present in the Great Lakes. These compounds include the long-banned polybrominated biphenyls (PBBs), the recently restricted polybrominated diphenyl ethers (PBDEs), and the currently used decabromodiphenyl ether, hexabromocyclododecanes, and others. Thus, this chapter focuses only on the non-reactive BFRs in the Great Lakes. The emphasis is on the concentrations of these compounds in people, fishes, birds, air, and sediment from the Great Lakes.

# 2
# Polybrominated Biphenyls

## 2.1
## Background

The polybrominated biphenyls (PBBs) were a disaster for the State of Michigan. PBBs were accidentally mixed into dairy cow feed in mid-1973, and as a result, the milk of much of the state became contaminated. As Michiganders consumed this milk, they too became contaminated. The history of this accident makes an interesting lesson and is summarized here based on the account by Fries [1].

The Michigan Chemical Corp. operated a plant in St. Louis, Michigan, on an impounded section of the Pine River; see Fig. 1. This company manufactured several products from brine pumped from wells under the center of the lower peninsula of Michigan. The anionic components of this brine included bromide, which was converted to elemental bromine and used to brominate biphenyl to make PBBs. This mixture was marketed under the trade name of FireMaster BP-6, which was a brown waxy material, and FireMaster FF-1, which was white powder made by adding $\sim 2\%$ calcium silicate to FireMaster BP-6. The manufacturing of PBBs at this Michigan plant started in 1970 and ended in 1974, during which time about 2500–5000 metric tons had been produced. The congener composition of FireMaster was largely 2,2′,4,4′,5,5′-hexabromobiphenyl (also called BB-153 using the IUPAC numbering system), 2,2′,3,4,4′,5,5′-heptabromobiphenyl (BB-180), and 2,2′,3,4,4′,5′-hexabromobiphenyl (BB-138). These congeners were present in FireMaster at a ratio of about 10 : 3 : 1; see Table 1. The structures of these congeners are shown in Scheme 1.

Several other congeners were present in the commercial PBB product, but their total abundance was less than 20% that of BB-153; see Table 1. How-

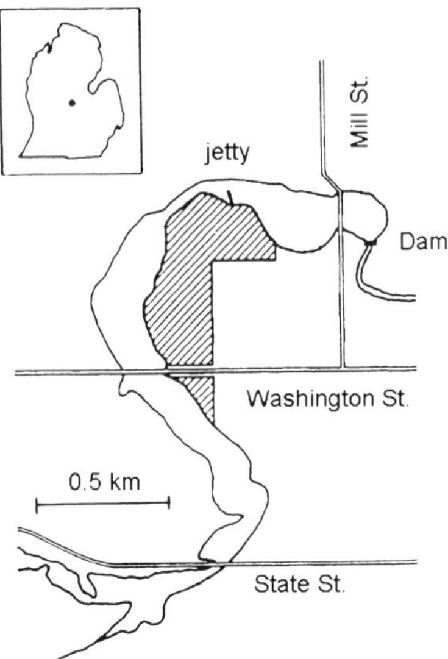



**Fig. 1** Map of the Pine River and its impoundment in the city of St. Louis, Michigan. The shaded area represents the site of the former Michigan Chemical plant that manufactured PBBs. Redrawn from Morris et al. [2]

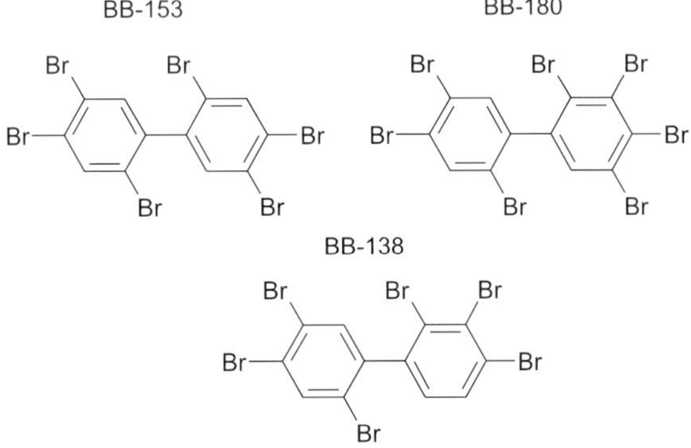



**Scheme 1** Structures of the major polybrominated biphenyl congeners present in the commercial FireMaster product

ever, given that the production of FireMaster was not a precise science, the exact congener composition varied from batch to batch, so the ratios given in Table 1 should be considered approximate.

This same plant also used the cationic components of the brine, which included alkali metals such as calcium and magnesium, to produce other useful products. Among these products was magnesium oxide, which was used as a nutritional supplement for dairy cows. This material had the trade name of NutriMaster and was also a white powder. Sometime in May 1973, there was apparently a shortage of the color-coded, printed paper bags in which these

**Table 1** Average composition (in % of the total) of FireMaster [1, 3, 4]

| Congener | Substitution | % of Total |
| --- | --- | --- |
| 101 | $2, 2', 4, 5, 5'$ | 2.8 |
| 118 | $2, 3', 4, 4', 5$ | 4.0 |
| 138 | $2, 2', 3, 4, 4', 5'$ | 7.9 |
| 149 | $2, 2', 3, 4', 5', 6$ | 1.3 |
| 153 | $2, 2', 4, 4', 5, 5'$ | 58 |
| 156 | $2, 3, 3', 4, 4', 5$ | 1.9 |
| 167 | $2, 3', 4, 4', 5, 5'$ | 4.5 |
| 170 | $2, 2', 3, 3', 4, 4', 5$ | 1.4 |
| 180 | $2, 2', 3, 4, 4', 5, 5'$ | 17 |
| 194 | $2, 2', 3, 3', 4, 4', 5, 5'$ | 1.7 |

two products were packaged, and as a result, some FireMaster was shipped to
a mill, which formulated dairy cow feed, in bags that were hand-labeled "NutriMaster." The exact details of this mistaken transaction will probably never
be known, but it is likely that something in the range of 100–300 kg of Fire-
Master was shipped to this feed mill, which was located in Climax, Michigan.
Unfortunately, the feed mill believed the labeling on the bags and mistakenly
added PBBs to the dairy cow feed as though it were NutriMaster.

By the late summer of 1973, the feed produced by this mill had been
shipped, both directly and through retailers, to dairy farms where this contaminated feed was consumed by cows. By late September 1973, it was clear
that the cows eating this material were not healthy. There was a precipitous
drop in their milk production, their hooves grew unnaturally, and they were
generally malnourished. PBBs were identified as the cause of these health
problems in April 1974 largely as a result of the persistence of a dairy farm
owner, Frederic L. Halbert. By the end of May, all dairy herds with a high
level of PBB contamination ($> 5$ ppm) were identified, quarantined, and eventually destroyed. During this period, some farm families had unknowingly
consumed milk from the contaminated cows and, in some cases, had even
eaten meat from the slaughtered cows. As a result, these dairy farmers and
their families were particularly contaminated. Eventually, the milk supply of
the entire state of Michigan (with the possible exception of the Upper Peninsula) had become contaminated with PBBs, and as a result, virtually everyone
in Michigan became contaminated to some extent.

Litigation and legislation ensued, the details of which are outside the scope
of this review. Suffice it to say that the production of FireMaster in Michigan
stopped on 20 November 1974 [5]. Michigan Chemical Corp. was purchased
by the Velsicol Chemical Corp., and the plant in St. Louis, Michigan was

closed in 1978. The plant was dismantled, and this site and the local county dump (the Gratiot County Landfill, located about 2.5 km to the southeast of the Michigan Chemical plant), which had been used during PBB production, were declared hazardous waste sites. The former plant site and the adjacent impoundment of the Pine River contained about 1 metric ton of PBBs [2], and the Gratiot County Landfill contained about 80 metric tons of PBBs [6]. Even after remediation during the 1982–1985 time period, these two sites are still highly contaminated with PBBs, the pesticide DDT, and chlorinated solvents. Remediation at these two sites and is continuing even today, 20 years later.

In addition to the contamination of the dairy industry and people of Michigan, PBBs also contaminated the environment and the Great Lakes. However, it is important to note that the duration of the input of PBBs into the environment itself was relatively short, starting in about 1970 and ending in about 1978. The following sections review the ambient levels of PBBs in people, fishes, birds, and sediment from the Great Lakes.

## 2.2
## PBBs in People

After the accidental input of PBBs into the food supply in Michigan, numerous studies were undertaken to quantify the level of contamination of the Michigan Chemical plant workers, the dairy farmers, and Michigan's general population. Thousands of adipose tissue and blood serum samples were analyzed, and these results are summarized in Table 2. In general, the measurements made prior to about 1990 reported only total PBB concentrations without giving congener-specific values, and the measurements made after 1990 often reported only BB-153 concentrations without giving the levels of the other PBB congeners. Because BB-153 is about 60% of the total PBBs in FireMaster, the older measurements will differ from the newer measurements by less than a factor of two; therefore, data of both vintages will be discussed here as though they were compatible with each other.

Because the PBB concentrations in people (see Table 2) are reported on a wet-weight basis, the adipose tissue samples show higher PBB concentrations compared to the serum samples. The workers at the Michigan Chemical plant showed the highest PBB levels, with the production workers showing concentrations $\sim$ 20-fold higher in their adipose tissue compared to the nonproduction workers. One wonders if the health of these people (especially the production workers) has been systematically studied over the last 30 years. The adipose tissue of the entire population of the sate of Michigan was contaminated to a level of 200 ng/g, but because of their relative isolation from the rest of the state, people from the Upper Peninsula were much less contaminated at 15 ng/g.

Serum from workers at the Michigan Chemical plant were the most contaminated, with levels of 100 ng/g for production workers and 6 ng/g for

**Table 2** PBB concentrations in human adipose tissue and serum (in ng/g wet weight), sequenced by decreasing PBB concentration[a]

| Type | Medium | Date | Conc. | N | Refs. |
|---|---|---|---|---|---|
| Mich. Chem. production workers | Adipose | 1976 | 47 000 | 7 | [7] |
| Mich. Chem. non-production workers | Adipose | 1976 | 2500 | 20 | [7] |
| Muskegon, MI region only | Adipose | 1978 | 500 | 84 | [8] |
| All Michigan adults | Adipose | 1978 | 200 | 821 | [8] |
| Detroit, MI region only | Adipose | 1978 | 160 | 255 | [8] |
| Upper Peninsula, MI only | Adipose | 1978 | 15 | 87 | [8] |
| Mich. Chem. production workers | Serum | 1976 | 110 | 10 | [7] |
| Farmers, quarantined farms | Serum | 1977 | 6.2 | 39 | [9] |
| Mich. Chem. non-production workers | Serum | 1976 | 6.1 | 45 | [7] |
| Consumers, non-quarantined farms | Serum | 1977 | 4.2 | 28 | [10] |
| All Michigan women | Serum | 1977 | 4.0 | 380 | [11] |
| Farmers, quarantined farms | Serum | 1977 | 3.9 | 283 | [10] |
| All Michigan farmers | Serum | 1977 | 2.6 | 524 | [12] |
| Consumers quarantined farms | Serum | 1977 | 2.2 | 40 | [10] |
| All Michigan women | Serum | 1975 | 2.0 | 1772 | [13] |
| All Michigan women | Serum | 1994 | 1.6 | 1170 | [13] |
| U.S. blood donors | Serum | 1988 | 1.8 | 12 | [14] |
| Muskegon MI region only | Serum | 1978 | 1.7 | 191 | [8] |
| Farmers, non-quarantined farms | Serum | 1977 | 1.6 | 23 | [9] |
| Farmers, non-quarantined farms | Serum | 1977 | 1.4 | 153 | [10] |
| All Michigan children | Serum | 1978 | 0.8 | 335 | [8] |
| Michigan fish consumers | Serum | 1993 | 0.7 | 14 | [15] |
| Swedish blood donors | Serum | 1997 | 0.6 | 20 | [14] |
| All Michigan adults | Serum | 1978 | 0.6 | 840 | [8] |
| Detroit region only | Serum | 1978 | 0.5 | 467 | [8] |
| Great Lakes fish consumers | Serum | 1993 | 0.4 | 30 | [15] |
| Ohio fish consumers | Serum | 1993 | 0.2 | 11 | [15] |
| Upper Peninsula MI only | Serum | 1978 | 0.2 | 232 | [8] |
| Wisconsin fish consumers | Serum | 1993 | 0.1 | 5 | [15] |
| Wisconsin farmers | Serum | 1977 | ND | 56 | [12] |

[a] The median concentration for the serum samples is 1.7 ng/g wet weight.

non-production workers. Clearly, it was not a good thing to be a production worker at this plant. There were many attempts to determine if PBB concentrations in serum from families from quarantined farms were higher than those from non-quarantined farms. Although there were some differences, these PBB concentrations were usually less than a factor of two different (4 versus 2 ng/g for the quarantined and non-quarantined farms, respectively), and it is not clear if this difference was statistically significant. The general population of Michigan had PBB levels of 1–2 ng/g. Women seemed to have somewhat higher levels than the general population and children somewhat less, but again it is not possible to say if these differences are statistically significant. It is clear, however, that all of the people from Michigan were more highly contaminated with PBBs than people from Wisconsin and Ohio, where concentrations were about a factor of 10 less.

The above concentrations were all measured in samples taken in 1976–1978, at the time of the Michigan Chemical accident. It is important to know if these concentrations have decreased in the intervening time. Three studies shed some light on this question: Sweeny et al. [13] reported on PBB concentrations in Michigan women sampled in both 1975 and 1994. Their data indicate that there is a small decrease (from 2.0 ng/g in 1975 to 1.6 ng/g in 1994), but the extent of this decrease depends on several factors such as the extent of original contamination. Sjödin et al. [14] measured PBB concentrations in blood that was donated in the U.S. in 1988 and found a median level of 1.8 ng/g, which is not much lower than the levels measured in the mid- to late 1970s in Michigan. Incidentally, Sjödin et al. [14] also measured PBBs in blood from Swedish blood donors collected in 1997 and found concentrations of 0.6 ng/g, indicating that PBB pollution is no longer just an issue in the U.S. Great Lakes region. Anderson et al. [15] measured PBBs in people who consumed fish from the Great Lakes. In samples taken in 1993, these authors reported PBB levels on the order of 0.3 ng/g, which are lower than the levels found in the general population of Michigan in the late 1970s, but again it is not possible to make statistically valid comparisons.

There is an additional problem in comparing PBB concentrations measured in different studies and, in some cases, measured 30 years apart: these differences could simply be due to different analytical methods; for example, the older work used packed gas chromatographic columns, and the more recent work used high-resolution capillary columns. Given these problems, it is only possible to say that the serum concentrations of PBBs in the general population of Michigan were about 1–2 ng/g wet weight in the late 1970s, and these levels have not changed much since then. This persistence of PBBs in people 30 years after the episode in Michigan suggests that the clearance rates of these compounds from people are slow.

## 2.3
## PBBs in Fishes

There have been periodic studies of PBB concentrations in fishes collected from the rivers downstream of the former plant site in St. Louis, Michigan and in the Great Lakes themselves. Initially the goal was to determine the geographical extent of contamination and to determine if the fish were fit for human consumption, but more recently, the goal was to determine the rate at which PBB concentrations are declining (if at all) in these fish.

Hesse and Powers [5] reported PBB concentrations in several species of small fish collected along the Pine River in 1974 and 1976. They could not detect PBBs in any fish collected upstream of the Pine River Impoundment; however, fish collected in and near this impoundment showed PBB levels of 13 000 ng/g lipid (the original data have been adjusted assuming 5% lipids relative to the wet weight of the fish). Other samples taken further downstream (at Magrudder Road, which is 20 km downstream of the former Michigan Chemical plant) showed PBB levels of 6000 ng/g lipid. Samples taken still further downstream of the plant (at Prairie Road, which is 50 km downstream of the plant) showed levels of 3000 ng/g lipid. The localized effect of the plant is obvious. The PBB concentrations found in this study were high and exceeded the FDA limit for the safe consumption of fish, which at that time was 300 ng/g lipid.

A more wide-ranging study was carried out on carp that were collected from the rivers in the Saginaw River watershed in the fall of 1983, well after PBB production ceased [16]. Carp from the Pine and Chippewa Rivers (see Fig. 2) had PBB concentrations of 6000–9000 ng/g lipid. Carp collected from four rivers downstream of the Pine and Chippewa Rivers (the Shiawassee, Tittabawassee, Flint, and Saginaw Rivers) had PBB levels of 20–150 ng/g lipid. Carp from the Saginaw Bay itself had PBB levels of 350 ng/g lipid. Similar to the results of Miller and Jude [17], PBBs could not be detected in fish from the AuSable River, which is a background site located about 50 km north of Saginaw Bay. These data indicate that PBBs were present at relatively high levels in fish collected in the contaminated region near St. Louis, Michigan, but these

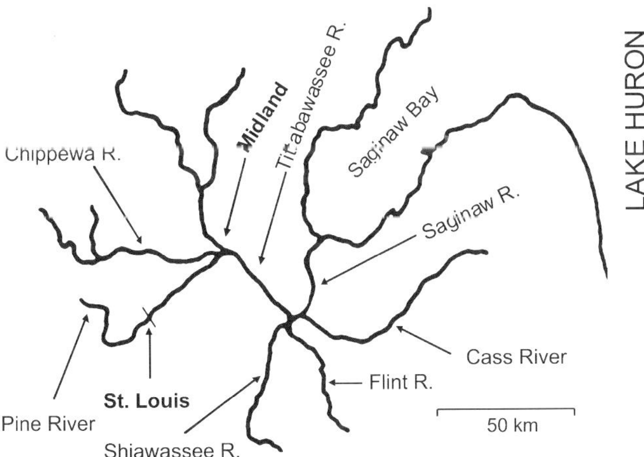


**Fig. 2** Map of rivers and the location of St. Louis, Michigan, the home of the former Michigan Chemical plant. Redrawn from Jaffe et al. [16]

concentrations decreased by $\sim$ 40-fold in fish collected below the confluence of the Pine and Chippewa Rivers with the Tittabawassee River. In addition, given that Hesse et al. [5] reported PBB levels in fishes collected from the Pine River in 1974–1976 of 3000–6000 ng/g lipid and that Jaffe et al. [16] reported PBB levels in fish collected from similar locations in 1983 of 6000–9000 ng/g lipid, it is likely that PBB levels had not changed much during the intervening 7–9 years.

In a later study, Luross [18] reported concentrations of PBB in about ten lake trout (a top predator) collected from four of the Great Lakes in 1997. The total PBB concentrations in these fish were as follows: Lake Superior, 1.5 ng/g lipid; Lake Erie, 1.5 ng/g lipid; Lake Ontario, 5.7 ng/g lipid; and Lake Huron, 15 ng/g lipid. Notice that the level in lake trout from the open waters of eastern Lake Huron in 1997 (15 ng/g lipid) is lower than the value measured by Jaffe et al. [16] for carp from the Saginaw Bay in 1985 (350 ng/g lipid). Presumably, this difference is due to the propinquity of the source to Saginaw Bay, to the differences in fish species sampled (carp vs. lake trout), or to the passage of 12 years, which might have allowed some clearing of PBBs from this ecosystem. Luross et al. [18] measured several PBB congeners and found that BB-101 was also present in these fish at a level of $\sim$ 20% relative to BB-153. Congeners BB-49 and BB-52 were also present but at much lower levels. Congener BB-101 was a minor component of the commercial FireMaster product (see Table 1), and thus its relatively high abundance in these lake trout samples is surprising. Is it possible that BB-153 is metabolized (or otherwise environmentally degraded) to BB-101 by the loss of a bromine atom from a *para* ring position?

To explore the detailed geographical distribution and temporal trends of flame retardants in the Great Lakes, lake trout from Lakes Superior, Michigan, Huron, and Ontario and walleye from Lake Erie, were collected during the period of 1980–2000 and analyzed for BB-153 (and other compounds) by Zhu and Hites [19]. These fishes were collected from all of the lakes biannually by the U.S. Geology Survey, composites of five individual fish were ground together, and the resulting puree was frozen and archived. Zhu and Hites [19] used fish collected in 1980, 1984, 1990, 1992, 1994, 1996, 1998, and 2000. All the samples from each lake were taken from the same locations each year, which were selected to be remote from urban areas. All of the fish sampled were about the same length, thus avoiding the effect of age. For 1980 and 1984, one composite from each lake was analyzed; for the other years, three composites from each lake were analyzed.

Zhu and Hites's [19] concentration data for BB-153 (see Fig. 3) indicate that concentrations of this compound have not changed significantly in fishes from the Great Lakes over the period of 1980–2000, with the exception of Lake Huron, where the concentrations have decreased with a half-life of 19 years. The average PBB concentrations over the time period of 1982–2000 are as follows: Lake Ontario, 24 ng/g lipid; Lake Huron, 19 ng/g lipid (post-1990 only);

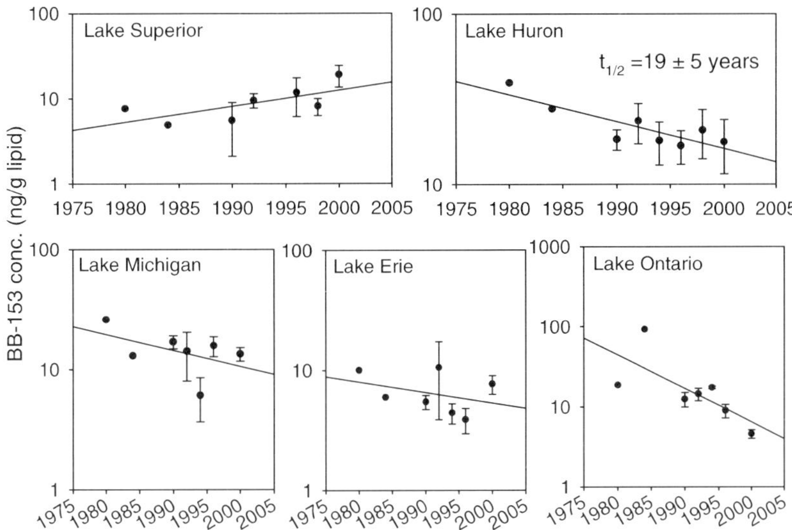

**Fig. 3** Temporal trend of BB-153 concentrations in fishes from the Great Lakes. The best-fit exponential curves are shown; the fitted lines have the following correlation coefficients ($r^2$): Superior, 0.481; Huron, 0.739; Michigan, 0.236; Erie, 0.128; and Ontario, 0.527. Only the regression for Lake Huron is statistically significant. From Zhu and Hites [19]

Lake Michigan, 15 ng/g lipid; Lake Superior, 9.5 ng/g lipid; and Lake Erie, 6.9 ng/g lipid. With the exception of Lake Huron, these concentrations are all 4–6 times higher than those measured by Luross [18] in lake trout collected in 1997. The reason for this difference is not known. Given the location of the PBB source, it is not surprising that the fish from Lake Huron had some of the highest PBB concentrations. It is, however, surprising that lake trout from Lake Ontario had relatively high levels of PBBs; perhaps there is another source of these compounds in this lake's watershed. The data in Fig. 3 indicate that the concentrations of PBBs in the fish from all five of the Great Lakes are decreasing only slowly, if at all, 30 years after the direct input of PBBs was stopped at the plant in St. Louis, Michigan.

## 2.4
## PBBs in Birds

There have been two studies of PBBs in the eggs of several species of waterfowl collected in 1977. All of these eggs were collected from nesting sites near Green Bay, which is part of Lake Michigan [20, 21]. Given that this sampling location was "upstream" of the major PBB source, PBBs were often not detected in these samples. The average PBB concentration found in all of the samples ($N = 33$) was 50 ng/g wet weight, or assuming the eggs averaged 15%

lipids by weight, 350 ng/g lipid. This concentration is similar to some of the concentrations found in the fish of Saginaw Bay, but higher than concentrations found in lake trout sampled from the open lake (see Fig. 3). Perhaps these fish-eating waterfowl are accumulating PBBs from their food. Unfortunately, there are no waterfowl samples from other locations around the lakes, so it is not possible to compare PBB levels in birds over the full geographical range of the lakes. Nevertheless, these data are indicators that PBBs are ubiquitous in Great Lakes, even "upstream" of the source.

## 2.5
## PBBs in Sediment and Soil

Like the fish, birds, and people in the vicinity of the PBB plant in St. Louis, Michigan, the sediment of the Pine River Impoundment and of the rivers leading to the Great Lakes are also contaminated with PBBs. Hesse and Powers [5] reported PBB concentrations in the Pine River Impoundment and in the Pine River itself as a function of distance downstream from the Michigan Chemical plant. Surficial sediment samples were collected in the summer of 1974, 1976, and 1977, but there were no systematic variations as a function of time, indicating the PBBs were not significantly degraded during these three years. The average PBB concentration in sediment from the Pine River Impoundment was 2500 ng/g dry weight. By 6 km downstream, the PBB concentration had dropped to 500 ng/g; by 10 km, it had dropped to 300 ng/g; and after 20 km, it had dropped to 150 ng/g. In 1989, sediment cores taken just west of Mill Street (see Fig. 1) from the Pine River Impoundment showed maximum PBB concentrations of 80 000–120 000 ng/g at sediment depths of 24–28 cm [2]. A sediment core taken just after Mill Street, showed a much lower PBB concentration of 2000 ng/g [2], a value compatible with the previous measurement of Hesse and Powers [5]. Unfortunately, the sedimentation rate in the Pine River Impoundment is not known; therefore, it is not possible to attach a date to the depths of 24–28 cm (see above) at this location. Nevertheless, it seems likely that the leakage of PBBs from the Michigan Chemical plant into this impoundment has been significant. Soil samples taken from the plant site itself showed PBB concentrations of 16 000 to 2 100 000 ng/g [4]. Although the exact locations from which these samples were taken were not given, it is noted that these are spectacularly high levels (up to 0.2%).

There has been one significant study of PBBs in sediment from the Great Lakes. Zhu and Hites [22] measured BB-153 in one sediment core from northern Lake Michigan and in one core from eastern Lake Erie. In both cases, the cores were dated (that is, core depths were correlated with particular years) and focusing factors were determined using the well-known lead-210 method. The concentrations (ng/g) of BB-153 are shown in Fig. 4 as a function of year of deposition for Lake Erie. At this location, BB-153 was found in the deepest layer, representing 1972; the concentrations reached a max-

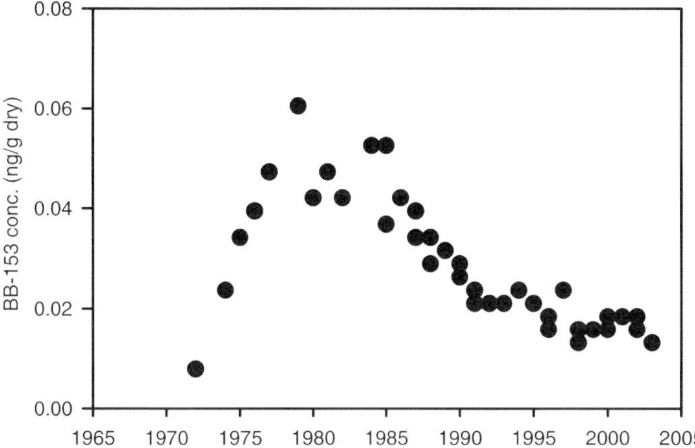

**Fig. 4** Concentrations of BB-153 as a function of deposition year in Lake Erie. From Zhu and Hites [22]

imum in $\sim$ 1980 and have decreased ever since. This profile was similar in Lake Michigan, where BB-153 was first detected in 1970, peaked in 1980, and decreased gradually afterwards.

Despite the ban on the production and use of PBBs in the U.S. in the mid-1970s, these sediment cores suggest that the peak concentration occurred about five years after the ban. This observation suggests that the time-dependent signal in sediment may lag behind the market's production and usage. The products that were manufactured before the ban were still in use, and PBB continued to be released from these products even after the ban. The history of PBB deposition revealed in these sediment cores suggests that the ban on these compounds was effective, given that the concentrations have decreased from their peak 20–25 years ago. However, the BB-153 concentrations in the surface layers of these cores are higher than at the bottom, which implies that PBBs are circulating in the Great Lakes environment even today. The inventories and fluxes for BB-153 are given in Table 3. The current loading rate of BB-153 into

**Table 3** BB-153 surface and maximum concentrations, inventories, burdens, surface fluxes, and load rates in sediment from Lakes Michigan and Erie. From Zhu and Hites [22]

| Site | Michigan | Erie |
| --- | --- | --- |
| Surface conc. (ng/g dry) | 0.052 | 0.018 |
| Maximum conc. (ng/g dry) | 0.10 | 0.060 |
| Inventory (ng/cm$^2$) | 0.038 | 0.087 |
| Burden (kg) | 22 | 22 |
| Surface Flux (ng·cm$^{-2}$yr$^{-1}$) | 0.0008 | 0.0017 |
| Load rate (kg/yr) | 0.46 | 0.43 |

Lake Michigan is estimated to be 0.46 kg/yr, and to Lake Erie it is estimated to be 0.43 kg/yr. The similarity of these two burdens suggests that atmospheric deposition is the source of BB-153 introduction to these two lakes.

Several studies have attempted to determine if PBBs degraded naturally in soil or sediment or if PBBs could be induced to degrade by the application of the optimum bacteria. The results of these studies are mixed: Hill et al. [4] reported that PBBs in a soil sample from the Michigan Chemical plant site with the highest level of PBB were degraded by photolytic debromination. In laboratory studies, Morris et al. [23] showed that anaerobic microorganisms from the Pine River have the capability to debrominate PBBs, and Bedard and vanDort [24] showed that anaerobic PCB-dechlorinating bacteria may also be able to debrominate PBBs. On the other hand, Jacobs et al. [25, 26] reported that PBBs were persistent after 52 weeks of incubation in soils and that photodegradation does not seem to be a significant pathway for the loss of PBBs from soil. Morris et al [2] reported that "limited *in situ* debromination of PBBs has occurred since 1970" based on the analysis of sediment from the Pine River. Despite the somewhat conflicting evidence of these focused degradation studies, it seems clear that PBBs would not be found in the environment 30 years after their release ended if they were easily degraded. It seems safe to conclude that PBBs are highly persistent compounds that will be present in the Great Lakes for many years to come.

# 3
# Polybrominated Diphenyl Ethers

## 3.1
## Background

With the demise of PBBs as a viable product, the brominated flame retardant industry turned to polybrominated diphenyl ethers (PBDEs) as a replacement. PBDEs have become a popular product; for example, furniture-grade polyurethane foam is now treated with 1–10% by weight of PBDEs to make this material safe for home use [27]. Not surprisingly, the use of PBDEs has increased over the years, and global annual sales are now $\sim 70\,000$ metric tons [27].

PBDEs are commercially available as three products: penta-, octa-, and deca-BDE. All of these products are mixtures of several congeners. The compositions of these three products, as measured in our laboratory at Indiana University, are given in Table 4. The penta-product contains primarily 2,2′,4,4′-tetrabromodiphenyl ether (BDE-47), 2,2′,4,4′,5-pentabromodiphenyl ether (BDE-99), and 2,2′,4,4′,6-pentabromodiphenyl ether (BDE-100) in a ratio of about 7:10:2. The octa-product contains primarily 2,2′,3,4,4′,5′,6-heptabromodiphenyl ether (BDE-183), 2,2′,3,3′,4,4′,5,6′-

**Table 4** Composition (in % of total) of the three main PBDE products produced by Great Lakes Chemical Co.

| Congener | Substitution | DE-71 ("penta-BDE") | DE-79 ("octa-BDE") | DE-83 ("deca-BDE") |
| --- | --- | --- | --- | --- |
| BDE-17 | 2,2′,4 | 0.2 | | |
| BDE-28 | 2,4,4′ | 0.2 | | |
| BDE-47 | 2,2′,4,4′ | 33 | | |
| BDE-49 | 2,2′,4,5′ | 0.5 | | |
| BDE-66 | 2,3′,4,4′ | 0.5 | | |
| BDE-85 | 2,2′,3,4,4′ | 2.5 | | |
| BDE-99 | 2,2′,4,4′,5 | 47 | | |
| BDE-100 | 2,2′,4,4′,6 | 8.3 | | |
| BDE-138 | 2,2′,3,4,4′,5′ | 0.3 | | |
| BDE-153 | 2,2′,4,4′,5,5′ | 3.9 | 4.3 | |
| BDE-154 | 2,2′,4,4′,5,6′ | 3.4 | 0.3 | |
| BDE-166 | 2,3,4,4′,5,6 | | 0.5 | |
| BDE-183 | 2,2′,3,4,4′,5′,6 | 0.1 | 25 | |
| BDE-190 | 2,3,3′,4,4′,5,6 | | 1.2 | |
| BDE-196 | 2,2′,3,3′,4,4′,5,6′ | | 7.8 | |
| BDE-197 | 2,2′,3,3′,4,4′,6,6′ | | 31 | |
| BDE-198/ 203 | 2,2′,3,3′,4,5,5′,6/ 2,2′,3,4,4′,5,5′,6 | | 4.5 | |
| BDE-206 | 2,2′,3,3′,4,4′,5,5′,6 | | 1.2 | 3.6 |
| BDE-207 | 2,2′,3,3′,4,4′,5,6,6′ | | 24 | 3.2 |
| BDE-208 | 2,2′,3,3′,4,5,5′,6,6′ | | | 0.2 |
| BDE-209 | 2,2′,3,3′,4,4′,5,5′,6,6′ | | 1.1 | 93 |

octabromodiphenyl ether (BDE-196), 2,2′,3,3′,4,4′,6,6′-octabromodiphenyl ether (BDE-197), and 2,2′,3,3′,4,4′,5,6,6′-nonabromo-diphenyl ether (BDE-207), in a ratio of about 8:3:10:8. The deca-product is almost entirely composed of decabromodiphenyl ether (BDE-209), but there are small but significant amounts of 2,2′,3,3′,4,4′,5,5′6-nonabromodiphenyl ether (BDE-206) and BDE-207. The structures of the major congeners found in the environment are given in Scheme 2.

Like most commercial chemical mixtures, the compositions of these products vary with manufacturer, with production batch, and with the year in which they were produced. Table 5 gives the global and North American market demands for these products in 1990, 1999, and 2001. Notice that > 95% of the penta-product is now used in the Americas.

Despite their societal benefits, PBDEs seem to be migrating from the products in which they are used and entering the environment and people. PBDEs are now ubiquitous; they can be found in air, water, fish, birds, marine mammals, and people, and in many cases, the concentrations of these compounds

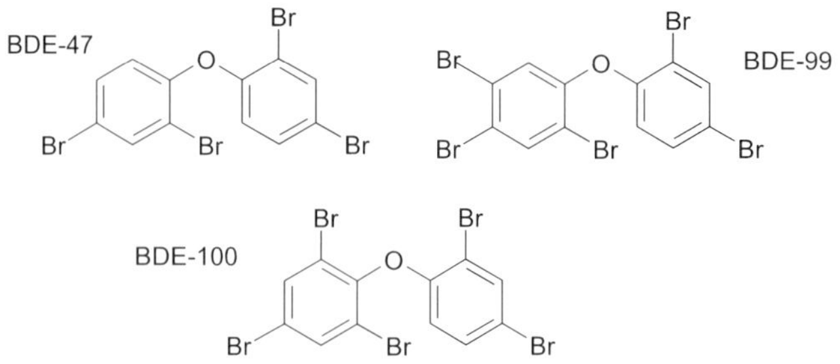

**Scheme 2** Structures of the major polybrominated diphenyl ether congeners present in the commercial products

**Table 5** Estimated global and North American demands for the major PBDE commercial products (in metric tons)

| Year | Type | Global demand | North American demand | % North American demand in global demand | Ratio of penta to octa |
|---|---|---|---|---|---|
| 1990[a] | Total | 40 000 | –[c] | –[c] | |
| | Deca | 30 000 | –[c] | –[c] | |
| | Octa | 6000 | –[c] | –[c] | |
| | Penta | 4000 | –[c] | –[c] | 0.67 |
| 1999[b] | Total | 67 125 | 33 965 | 51% | |
| | Deca | 54 800 | 24 300 | 44% | |
| | Octa | 3825 | 1375 | 36% | |
| | Penta | 8500 | 8290 | 98% | 2.2 |
| 2001[b] | Total | 67 390 | 33 100 | 49% | |
| | Deca | 56 100 | 24 500 | 44% | |
| | Octa | 3790 | 1500 | 40% | |
| | Penta | 7500 | 7100 | 95% | 2.0 |

[a] data from [28];
[b] data from [29];
[c] not available

are increasing over time [27]. For example, in human blood, milk, and tissues, total PBDE levels have increased exponentially by a factor of $\sim$ 100 during the last 30 years; this is a doubling time of $\sim$ 5 years. The environment and people from North America are much more contaminated with PBDEs compared to Europe [27]. Perhaps as a result of these observations, several governments have banned the use of the penta- and octa products, and the major U.S. manufacturer of these materials (Great Lakes Chemical) has voluntarily stopped producing these two products. The deca-BDE product is, however, still in use, and the flame retardant industry shows no signs of abandoning it [30].

## 3.2
## PBDEs in Fishes

There have been numerous measurements of PBDEs in fishes from the Great Lakes, and these data are summarized in Table 6. The concentrations range from 1000–3000 ng/g lipid for top predator fishes (such as trout and salmon) sampled recently down to 1–10 ng/g lipid for fish sampled in ∼ 1980. Taken as a whole, there is a strong relationship of concentration with time (see Fig. 5, note the logarithmic concentration scale), and this relationship holds

**Table 6** Total PBDE concentrations (ΣPBDE, in ng/g lipid) in fishes from the Great Lakes, sequenced by decreasing concentration

| Location | Species | Year | ΣPBDE | % BDE-47 | N | Refs. |
|---|---|---|---|---|---|---|
| Lake Michigan | steelhead | 1995 | 3000 | 57 | 6 | [31] |
| Lake Michigan | salmon | 1996 | 2440 | 65 | 21 | [32] |
| Traverse Bay, MI | salmon | 1997 | 2093 | 36 | 1 | [33] |
| Lake Michigan | lake trout | 2000 | 1395 | 59 | 3 | [19] |
| Lake Michigan | lake trout | 1996 | 1355 | 54 | 3 | [19] |
| Traverse Bay, MI | burbot | 1997 | 1110 | 50 | 1 | [33] |
| Lake Superior | lake trout | 2000 | 989 | 38 | 3 | [19] |
| Traverse Bay, MI | chub | 1997 | 958 | 48 | 9 | [33] |
| Lake Ontario | lake trout | 1998 | 945 | 50 | 10 | [34] |
| Traverse Bay, MI | lake trout | 1997 | 894 | 60 | 5 | [33] |
| Traverse Bay, MI | alewife | 1997 | 777 | 44 | 8 | [33] |
| Lake Ontario | lake trout | 1997 | 604 | 55 | 10 | [34] |
| Lake Erie | walleye | 2000 | 600 | 51 | 3 | [19] |
| Lake Ontario | lake trout | 1994 | 589 | 53 | 3 | [19] |
| Lake Ontario | lake trout | 2000 | 546 | 55 | 3 | [19] |
| Lake Ontario | lake trout | 1997 | 545 | 71 | | [35] |
| Lake Ontario | lake trout | 1996 | 543 | 55 | 3 | [19] |
| Traverse Bay, MI | smelt | 1997 | 478 | 64 | 4 | [33] |
| Lake Superior | lake trout | 1996 | 454 | 45 | 3 | [19] |
| Lake Ontario | lake trout | 1993 | 434 | 69 | 10 | [18] |
| Lake Huron | lake trout | 1998 | 433 | 57 | 3 | [19] |
| Lake Superior | lake trout | 1998 | 426 | 48 | 3 | [19] |
| Lake Michigan | lake trout | 1992 | 420 | 49 | 3 | [19] |
| Lake Superior | lake trout | 1997 | 392 | 62 | 10 | [34] |
| Lake Huron | lake trout | 2000 | 369 | 43 | 2 | [19] |
| Lake Huron | lake trout | 1996 | 301 | 57 | 3 | [19] |
| Lake Ontario | lake trout | 1992 | 297 | 50 | 3 | [19] |
| Traverse Bay, MI | whitefish | 1997 | 277 | 54 | 6 | [33] |
| Lake Erie | walleye | 1996 | 262 | 56 | 3 | [19] |

**Table 6** (continued)

| Location | Species | Year | ΣPBDE | % BDE-47 | $N$ | Refs. |
|---|---|---|---|---|---|---|
| Lake Huron | lake trout | 1997 | 247 | 69 | 10 | [34] |
| Lake Ontario | smelt | 1994 | 240 | 56 | 3 | [36] |
| Lake Huron | lake trout | 1997 | 237 | 46 | | [35] |
| Lake Erie | walleye | 1994 | 210 | 54 | 3 | [19] |
| Lake Michigan | lake trout | 1994 | 209 | 46 | 3 | [19] |
| Lake Huron | lake trout | 1994 | 195 | 51 | 3 | [19] |
| Lake Erie | walleye | 1992 | 181 | 52 | 3 | [19] |
| Lake Michigan | lake trout | 1990 | 176 | 45 | 3 | [19] |
| Lake Ontario | lake trout | 1988 | 171 | 50 | 10 | [34] |
| Lake Huron | lake trout | 1992 | 166 | 50 | 3 | [19] |
| Lake Superior | smelt | 1994 | 150 | 63 | 3 | [36] |
| Lake Ontario | lake trout | 1990 | 143 | 44 | 3 | [19] |
| Lake Superior | lake trout | 1997 | 135 | 62 | | [35] |
| Lake Huron | lake trout | 1990 | 119 | 50 | 3 | [19] |
| Lake Erie | lake trout | 1997 | 117 | 67 | 10 | [18] |
| Lake Ontario | lake trout | 1984 | 97 | 35 | 1 | [19] |
| Buffalo River | carp | 1991 | 91 | | 45 | [37] |
| Traverse Bay, MI | sculpin | 1997 | 86 | | 6 | [33] |
| Lake Michigan | lake trout | 1984 | 81 | 53 | 1 | [19] |
| Lake Erie | walleye | 1990 | 78 | 55 | 3 | [19] |
| Lake Superior | lake trout | 1992 | 76 | 50 | 3 | [19] |
| Lake Superior | lake trout | 1990 | 58 | 43 | 3 | [19] |
| Lake Huron | lake trout | 1984 | 33 | 33 | 1 | [19] |
| Lake Erie | walleye | 1984 | 20 | 40 | 1 | [19] |
| Lake Superior | lake trout | 1984 | 17 | 38 | 1 | [19] |
| Lake Michigan | lake trout | 1980 | 14 | 31 | 1 | [19] |
| Lake Huron | lake trout | 1980 | 11 | 32 | 1 | [19] |
| Lake Ontario | lake trout | 1980 | 10 | 34 | 1 | [19] |
| Lake Ontario | lake trout | 1983 | 8 | 50 | 10 | [34] |
| Lake Erie | walleye | 1980 | 7 | 24 | 1 | [19] |
| Lake Superior | lake trout | 1980 | 6 | 37 | 1 | [19] |
| Lake Ontario | lake trout | 1978 | 3 | 50 | 10 | [34] |

without regard to fish species or the location where the fishes were collected. This analysis shows that the concentrations of PBDEs have doubled in Great Lakes fishes every three years, which is fast.

To explore the geographical distribution and temporal trends of polybrominated diphenyl ethers (PBDEs) in the Great Lakes in more detail, lake trout from Lakes Superior, Michigan, Huron, and Ontario and walleye from Lake Erie, collected during the period of 1980–2000, were analyzed by Zhu and Hites [19]. The concentrations of fifteen PBDE congeners were de-

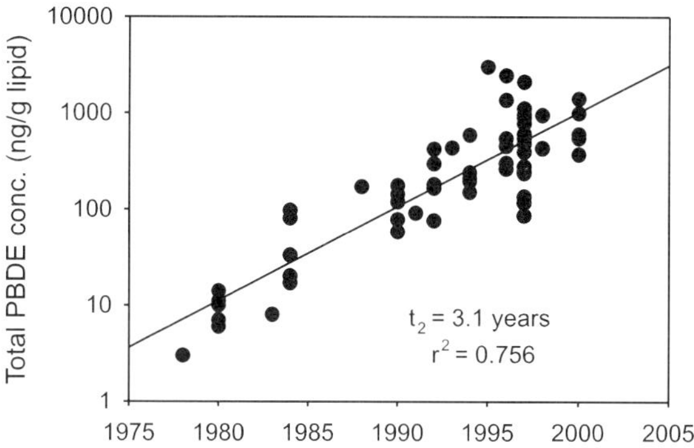


**Fig. 5** Total PBDE concentrations (in ng/g lipid) in fishes from the Great Lakes as a function of time; see Table 6

termined in each fish sample. In general, the total PBDE lipid-normalized concentrations measured in these Great Lakes fishes are on the order of 500–600 ng/g lipid for the years 1996–2000. These values are similar to PBDE concentrations in other fishes from various other locations in North America but are about five times higher than average PBDE concentrations measured in fishes from Europe [27]. This difference is smaller but in the same direction as the difference between PBDE concentrations in humans in North America and Europe; in people, North Americans have about 20 times more PBDE in their blood as do Europeans [27].

In the year 2000, the total PBDE concentrations (in ng/g lipid) in the five lakes (see Fig. 6) were as follows: Michigan, 1400; Superior, 990; Erie, 600; Ontario, 550; and Huron, 370. Lakes Michigan and Superior had the highest total PBDE concentrations in the year 2000, which is surprising given that Lake Superior is generally thought of as the most pristine of the Great Lakes. Perhaps the relatively high PBDE concentrations in Lake Superior are a result of the slow contaminant removal processes at work in the cold and deep waters of this lake. The finding that trout from Lakes Michigan and Ontario are contaminated with PBDE was reasonable, given the large human population residing on the shores of these lakes and given the intense industrial processes associated with these large urban regions.

The rate at which PBDE concentrations have increased in the environment and in humans has been of considerable interest. Temporal trend studies from Europe have indicated that PBDE levels in human milk increased markedly from 1972 to 1997, doubling every ~ 5 years [38]. Since 1997, the PBDE levels in human milk have decreased somewhat [39]. These recent ameliorations may be the result of changes in industrial practices in Europe. The European Commission, for example, has phased out the use of the commercial penta-BDE product because of concerns about its potentially adverse human health effects. Now, > 95% of the current global demand for the penta-BDE product

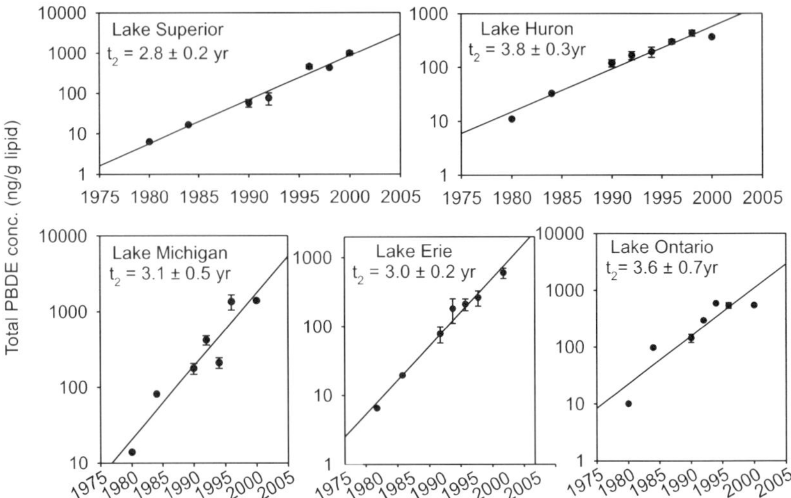

**Fig. 6** Temporal trend of ΣPBDE concentrations in fishes from the Great Lakes. The best fitted exponential curves are shown, the slopes of which were used to calculate the doubling time ($t_2 \pm 1$ standard error) for each lake. From Zhu and Hites [19]

is in North America; see Table 5. This suggests that the North American Great Lakes could be receiving a relatively high proportion of the less brominated homologues.

In their study, Zhu and Hites [19] found that the total PBDE levels in the fishes from all five of the Great Lakes increased exponentially as a function of time, doubling every 3–4 years; see Fig. 6. There are some indications that the total PBDE concentrations have leveled off in Lake Huron since 1998, in Lake Michigan since 1998, and especially in Lake Ontario since 1994. These temporal trend results are in general agreement with the findings by Luross et al. [34], who reported that PBDE concentrations in lake trout from Lake Ontario have increased 300-fold over the past 20 years, which is a doubling time of 2.4 years. Other studies have found that PBDE levels in biota and humans from North America have increased with doubling times of 4–6 years [27]. The differences in doubling times among all of these results are not statistically significant. Suffice it to say that the concentrations of PBDE in the Great Lakes environment have rapidly increased since 1980, but there are signs that these increases have slowed.

In the fish samples presented in Table 6, the congener patterns are dominated by BDE-47 (at an average of $52 \pm 13\%$ of the total PBDE concentration) followed by BDE-99, -100, -153, and -154. This congener pattern is more or less constant in these fish samples as it is for almost all environmental samples (including humans). Figure 7 shows the PBDE congener pattern for the penta-BDE product, for human blood and tissue samples, and for Great Lakes fishes. The penta-BDE product is dominated by BDE-47 and BDE-99; in this

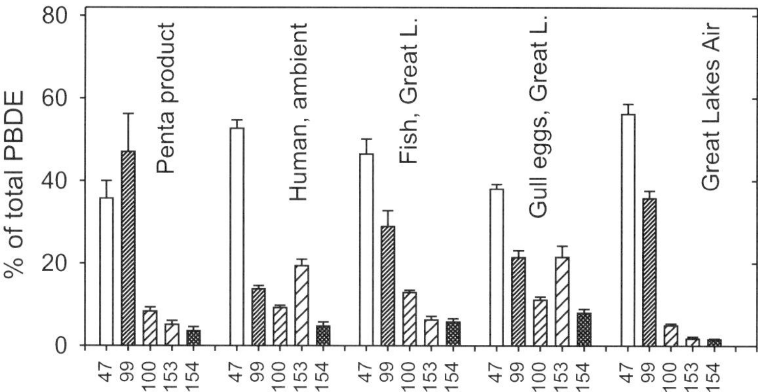



**Fig. 7** PBDE congener patterns for the commercial penta-BDE product, nonoccupationally exposed people, fishes from the Great Lakes, herring gull eggs from the Great Lakes, and air near the Great Lakes

case, BDE-99 is more abundant than BDE-47. The human samples are also dominated by BDE-47 and BDE-99, but in this case, BDE-47 is more abundant than BDE-99 and BDE-153 is more abundant than in the penta-BDE product. Presumably, this shift is due to the greater environmental mobility of BDE-47 (its vapor pressure is higher than the other congeners; see below) or the selective elimination of BDE-99 in some biota [40]. The congener pattern for the Great Lakes fishes is between that of the penta-BDE product and that found in humans. It is important, however, not to make too much out of these patterns. An extensive principal components analysis indicated that it is probably too early to say that a specific pattern of congeners can be used as a "marker" of a particular PBDE source or environmental process [27].

The average PBDE congener patterns measured by Zhu and Hites [19] in the archived lake trout samples were similar to those presented in Fig. 7, and these average patterns from all five lakes were similar to each other, even though their absolute concentrations were different. However, these congener patterns changed with time. The proportion of BDE-47 was usually smaller in 1984 than in 1996, and the opposite was true for BDE-153. To explore this observation systematically, Zhu and Hites [19] defined the following ratio:

$$R = \frac{[47] + [99] + [100]}{[153] + [154]} \tag{1}$$

where [x] refers to the concentration of congener x in any units. These ratios are plotted as a function of fish sampling year in Fig. 8. The correlations are all statistically significant ($P < 0.03$, on average).

Clearly, the congener patterns have changed systematically over time. Similar results were observed by Norstrom in herring eggs from the Great

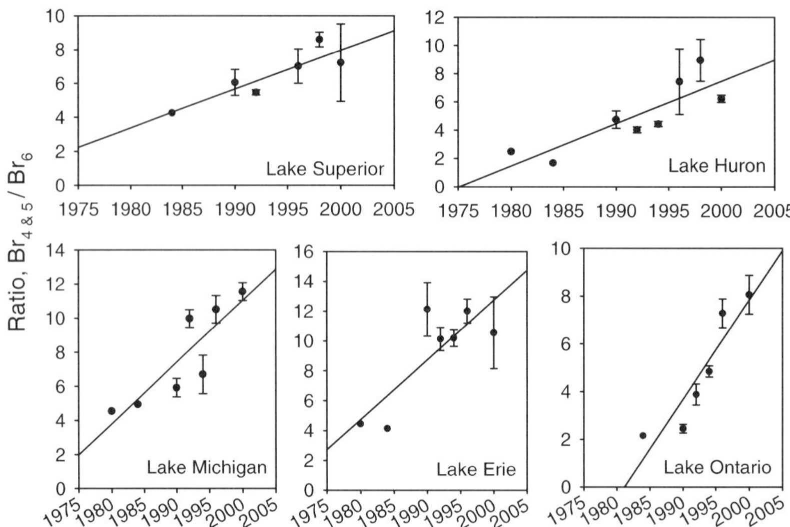

**Fig. 8** Temporal trends of the ratio of the sum of concentrations of BDE-47 plus -99 plus -100 to the sum of the concentrations of BDE-153 plus -154 in fishes from the Great Lakes. The fitted lines have the following correlation coefficients ($r^2$): Superior, 0.797; Huron, 0.714; Michigan, 0.757; Erie, 0.674; and Ontario, 0.855. From Zhu and Hites [19]

Lakes; see below. Differences in the production and use of the penta-BDE product versus the octa-BDE product may account for these differences. Table 5 lists the demand for the three major PBDE commercial products in the global and North American markets. In 1990, the ratio of the penta-BDE product to the octa-BDE product demand was $\sim 0.7$, but this ratio increased to $\sim 2$ in 1999 and 2001. In other words, more of the octa-BDE product, which is relatively high in BDE-153 and -154, was used in the 1980s relative to the penta-BDE product, which is high in BDE-47, but now the opposite is true. This shift toward more of the penta-BDE product in the marketplace is mirrored by an increase in the ratios ($R$—see Eq. 1) over the same time period. The increase in $R$ over the last 10 years has been about 50%, but the increase in the penta to octa ratio has been about 200%. This difference may be explained by the higher bio-availability of the less-brominated congeners compared with to the more-brominated ones.

## 3.3
## PBDEs in Birds

There has been one noteworthy study of PBDEs in herring gull (*Larus argentatus*) eggs collected from all five of the Great Lakes during the period from 1981 to 2000 published by Norstrom et al. [41]. These data are summarized in Fig. 9. It is clear from these data that PBDE concentrations have been in-

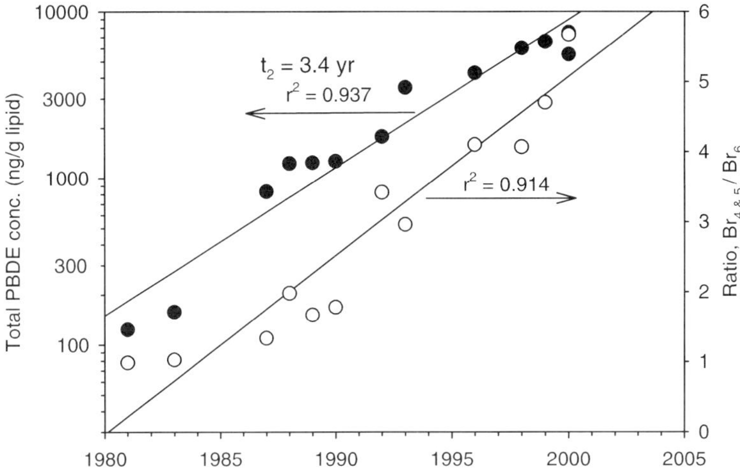



**Fig. 9** Total PBDE concentrations in herring gull eggs from the Great Lakes as a function of the year in which the samples were collected (*left axis*). Ratio of the sum of the concentrations of BDE-47, BDE-99, and BDE-100 divided by the sum of the concentrations of BDE-153 and BDE-154 (see Eq. 1) in herring gull eggs from the Great Lakes as a function of the year in which the samples were collected (*right axis*). Re-plotted from Norstrom et al. [41]

creasing rapidly in the gull eggs. The doubling time for PBDE in herring gull eggs from the Great Lakes is about 3.4 years, which is about the same rate at which PBDE concentrations have been increasing in fishes, the primary food of herring gulls. In general, both of these doubling times are about the same as those observed for people [27].

In the Great Lakes samples, the gull egg congener pattern also changes systematically with the year in which the sample was taken. In the 1981 samples, BDE-47 is $\sim 33\%$ of the total, but in 2000, it increased to 45% of the total; at the same time BDE-154 decreased from 16% in 1981 to 5% in 2000. These changes can be demonstrated by plotting the ratio of the less brominated congeners to the more brominated congeners ($R$—see Eq. 1) as a function of sampling year; see Fig. 9. The regression is excellent, perhaps indicating that the use or the composition of the various commercial PBDE products has changed systematically over the years. The average congener pattern for the Great Lakes gull egg samples is shown in Fig. 7; this pattern is similar to that of the human samples but dissimilar to that of the Great Lakes fishes—it is not clear why.

## 3.4
## PBDEs in Air

The only study of PBDEs in Great Lakes air is by Strandberg et al. [42], who measured the concentrations and spatial distribution of PBDE in air sampled at four sites near the Great Lakes in order to investigate possible sources of contamination. These authors used an urban sampling site on the south side of downtown Chicago, Illinois; two rural locations—one at Sleeping Bear Dunes, Michigan, on the northeast coast of Lake Michigan, and one at Sturgeon Point, New York, located near the shore of Lake Erie and about 30 km southwest of Buffalo; and a remote site at Eagle Harbor, Michigan, on the Keweenaw Peninsula near Lake Superior. Particle- and gas-phase concentrations were measured separately. It is known that the largest source of variability in the atmospheric concentrations of semi-volatile compounds is related to atmospheric temperature [43]; therefore, in order to minimize the variability of the data, Strandberg et al. [42] only used samples taken when the average air temperature was $20 \pm 3\,^{\circ}\mathrm{C}$ in the years 1997, 1998, and 1999.

The total PBDE concentrations (particle- plus gas-phases) are given in Table 7 as averages of the 12 samples collected at each site. PBDEs were found in all samples investigated, showing that these compounds are widely dispersed in Great Lakes air. The PBDE concentrations were relatively high in Chicago air, averaging about $50\,\mathrm{pg/m^3}$ and about $5$–$15\,\mathrm{pg/m^3}$ at the other locations. In fact, the PBDE concentrations in Chicago were significantly different from those measured at the other locations. The PBDE concentrations at the other three locations were not significantly different from each other, which suggests that (outside of Chicago) PBDEs are evenly distributed in Great Lakes air [42]. It was not possible from these data to determine a time-trend for PBDE concentrations near the Great Lakes.

Of the individual PBDE congeners, BDE-47, -99, -100, -153, and -154 were detected in all samples; BDE-209 was only detected at trace levels in the Chicago particle samples. Among the BDE congeners, BDE-47 accounted for 50–65% of the total mass of PBDEs, and BDE-99 accounted for another

**Table 7** Average concentrations (in $\mathrm{pg/m^3} \pm 1$ standard error, $N = 12$) of PBDEs in air samples collected near the Great Lakes 1997 to 1999[a]. From Strandberg et al. [42]

|  | Superior | Erie | Michigan | Chicago |
|---|---|---|---|---|
| BDE-47 | $2.9 \pm 0.8$ | $3.8 \pm 0.8$ | $8.4 \pm 2.1$ | $33 \pm 12$ |
| BDE-99 | $2.1 \pm 0.6$ | $2.8 \pm 0.6$ | $5.3 \pm 2.1$ | $16 \pm 4.8$ |
| BDE-100 | $0.29 \pm 0.05$ | $0.39 \pm 0.06$ | $0.80 \pm 0.18$ | $2.0 \pm 0.9$ |
| BDE-153 | $0.13 \pm 0.04$ | $0.19 \pm 0.07$ | $0.15 \pm 0.05$ | $0.41 \pm 0.17$ |
| BDE-154 | $0.09 \pm 0.03$ | $0.11 \pm 0.03$ | $0.25 \pm 0.07$ | $0.53 \pm 0.16$ |
| Total PBDEs | $5.5 \pm 1.1$ | $7.2 \pm 0.9$ | $15 \pm 5.1$ | $52 \pm 19$ |
| Ratio (see Eq. 1) | $25 \pm 5$ | $23 \pm 6$ | $39 \pm 24$ | $54 \pm 14$ |

[a] Congeners 190 and 209 were rarely present at limits of detection of 0.06 and $0.1\,\mathrm{pg/m^3}$, respectively, except for Chicago, where the concentration of BDE-209 averaged $0.3\,\mathrm{pg/m^3}$

35–40%. The other five congeners accounted for less than 10% of the total. This composition was the same at all of the sampling stations. This is also the most common congener pattern found in biological samples; see Fig. 7. Probably because of their relatively high volatilities, the atmosphere is high in BDE-47 and BDE-99 compared to the commercial penta-BDE product.

Semi-volatile organohalogen compounds, such as PBDEs, exist in the atmosphere in the gas-phase or associated with the particle-phase. The partitioning of compounds between these atmospheric phases is an important factor in their subsequent fate, transport, degradation, and human exposure assessment. Particle-to-gas partitioning is controlled largely by the physical properties of a compound, such as its vapor pressure and by the prevailing environmental conditions, such as the atmospheric temperature. As noted above, in the Strandberg et al. study, the samples were selected from days when the atmospheric temperature was $20 \pm 3\,^\circ$C [42]. At this temperature, the PBDEs were present in both the particle- and gas-phases, except for BDE-209, which was present only in the particle-phase.

The partitioning of semi-volatile organic compounds between the particle- and gas-phases is commonly described by a partition coefficient ($K_P$) [44, 45]:

$$K_P = \frac{(F/TSP)}{A} \tag{2}$$

where $F$ and $A$ are the particulate-associated and gas-phase concentrations, respectively (in pg/m$^3$), and $TSP$ is the total suspended particulate matter

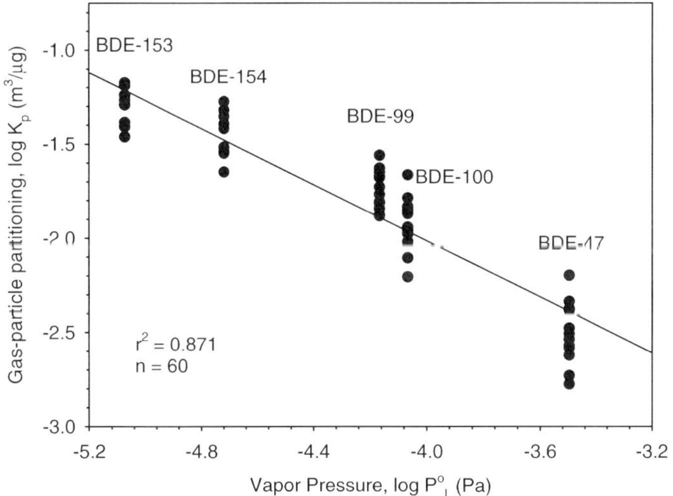


**Fig. 10** Comparison between the atmospheric particle- to gas-phase partition coefficients ($K_P$, in m$^3$/µg) for BDE-47, BDE-100, BDE-99, BDE-154, and BDE-153 (clusters from right to left) and their sub-cooled liquid vapor pressures ($P_L^o$, in Pascals) in the 1997–1999 Chicago air samples ($r^2 = 0.763$, $n = 60$). From Strandberg et al. [42]

level (in $\mu g/m^3$). At a given temperature, compound-dependent values of $K_P$ tend to be correlated with the sub-cooled liquid vapor pressure ($P_L^\circ$) [45]:

$$\log(K_P) \propto -\log(P_L^\circ) \tag{3}$$

As expected, Strandberg et al. observed large variations in $K_P$ for the various PBDE congeners [42]. For BDE-47, about 80% was in the gas-phase; for BDE-100 and BDE-99 about 55–65% was in the gas-phase; and for BDE-154 and BDE-153 only about 30% was in the gas-phase. Figure 10 shows the relationship between log ($K_P$) and log ($P_L^\circ$) for all Chicago samples for all three years. Clearly, there is a strong correlation between the volatility of PBDE and the ratio of their concentrations in the particle- to gas-phases. This observation suggests that it is important to measure both phases if one wants to get an accurate measurement of the concentrations of these compounds in the atmosphere.

## 3.5
### PBDEs in Sediment

There have been four studies of PBDEs in sediment from the Great Lakes; all have used dated sediment cores to determine the history of PBDE deposition. The studies by Song et al. [46–48] used 14 sediment cores from all of the lakes and that of Zhu and Hites [22] used one sediment core from northern Lake Michigan and one from eastern Lake Erie (see above). All cores were dated using the well-known lead-210 method.

The pattern of PBDE congeners in the sediment cores from all of the lakes was dominated by BDE-209, which makes up $\sim$ 95% of the total. BDE-47, BDE-99, BDE-206, and BDE-207 are also present but at relatively small concentrations. The high abundance of BDE-209 and the nonabrominated BDEs in sediment is probably the result of the high production of deca-BDE (25 000 tons were used in North America in 2001; see Table 5) and their high $K_{ow}$ values, which force them to partition to the sinking sediment particles.

BDE-209 is the dominant congener in most sediments, no matter whether they were from a river [49], an estuary [50], a lake [36], or a sea [51]. On the other hand, BDE-209 seems to be a minor PBDE congener in biota [27]; for example, BDE-209 was not found in any of the archived lake trout samples (see above). This observation suggests that, even though BDE-209 may be bioavailable, it may have a short half-life in biota.

For simplicity in the following discussion, $\Sigma$PBDE′ refers to the sum of the tri- through hepta-BDEs measured in this study. The surficial sediment PBDE concentrations in the lakes as measured by Zhu and Hites [22] and by Song et al. [46–48] are given in Table 8. The $\Sigma$PBDE′ and BDE-209 surficial concentrations in Lakes Michigan, Huron, and Erie are similar to one another; however, these concentrations in Lake Superior are lower than in these three lakes, and these concentrations in Lake Ontario are higher.

**Table 8** Average PBDE surface concentrations, doubling times, inventories, burdens, surface fluxes, and load rates in sediment from the Great Lakes. Data were from Zhu and Hites [22] and Song et al. [46–48]. The errors are $\pm 1$ standard error ($N = 2 - 4$); "NS" indicates that the concentrations did not change significantly with depth in this core

| Site | | Superior | Michigan | Huron |
|---|---|---|---|---|
| Surface conc. (ng/g dry) | $\Sigma$PBDE$'$ | $0.90 \pm 0.15$ | $2.9 \pm 0.5$ | $1.5 \pm 0.3$ |
| | BDE-209 | $8.9 \pm 2.4$ | $63 \pm 12$ | $29 \pm 4$ |
| Doubling time (yr) | $\Sigma$PBDE$'$ | $27 \pm 7$ | $15 \pm 2$ | $20 \pm 2$ |
| | BDE-209 | NS | $19 \pm 10$ | $10 \pm 2$ |
| Inventory (ng/cm$^2$) | $\Sigma$PBDE$'$ | $0.59 \pm 0.24$ | $2.2 \pm 0.7$ | $1.4 \pm 0.4$ |
| | BDE-209 | $4.8 \pm 0.7$ | $59 \pm 10$ | $16 \pm 5$ |
| Burden (tons) | $\Sigma$PBDE$'$ | $0.48 \pm 0.20$ | $1.3 \pm 0.4$ | $0.85 \pm 0.24$ |
| | BDE-209 | $3.9 \pm 0.6$ | $34 \pm 6$ | $9.5 \pm 2.9$ |
| Surface flux (ng$\cdot$cm$^{-2}$yr$^{-1}$) | $\Sigma$PBDE$'$ | $0.018 \pm 0.007$ | $0.057 \pm 0.017$ | $0.051 \pm 0.012$ |
| | BDE-209 | $0.14 \pm 0.01$ | $1.2 \pm 0.3$ | $0.97 \pm 0.22$ |
| Load rate (tons/yr) | $\Sigma$PBDE$'$ | $0.015 \pm 0.005$ | $0.033 \pm 0.010$ | $0.031 \pm 0.007$ |
| | BDE-209 | $0.11 \pm 0.01$ | $0.70 \pm 0.17$ | $0.58 \pm 0.13$ |

| Site | | Erie | Ontario |
|---|---|---|---|
| Surface conc. (ng/g dry) | $\Sigma$PBDE$'$ | $1.6 \pm 0.3$ | $5.6 \pm 0.7$ |
| | BDE-209 | $48 \pm 5$ | $230 \pm 15$ |
| Doubling time (yr) | $\Sigma$PBDE$'$ | $13 \pm 3$ | $13 \pm 4$ |
| | BDE-209 | $10 \pm 5$ | $13 \pm 6$ |
| Inventory (ng/cm$^2$) | $\Sigma$PBDE$'$ | $5.0 \pm 2.0$ | $3.3 \pm 0.9$ |
| | BDE-209 | $61 \pm 11$ | $110 \pm 30$ |
| Burden (tons) | $\Sigma$PBDE$'$ | $1.3 \pm 0.5$ | $0.63 \pm 0.18$ |
| | BDE-209 | $16 \pm 3$ | $22 \pm 5$ |
| Surface flux (ng$\cdot$cm$^{-2}$yr$^{-1}$) | $\Sigma$PBDE$'$ | $0.18 \pm 0.07$ | $0.17 \pm 0.03$ |
| | BDE-209 | $5.4 \pm 1.8$ | $7.0 \pm 0.4$ |
| Load rate (tons/yr) | $\Sigma$PBDE$'$ | $0.047 \pm 0.017$ | $0.033 \pm 0.005$ |
| | BDE-209 | $1.4 \pm 0.5$ | $1.3 \pm 0.1$ |

PBDEs have been studied in river, lake, and marine sediment samples all over the world, and the concentration of the $\Sigma$PBDE$'$ and BDE-209 are dependent on the sampling locations. In Europe, the highest BDE-209 concentration was found in the River Mersey, UK (at 1700 ng/g) [49]. At the other end of the scale, a sediment core taken from Drammenfjord in 1999 near Oslo, Norway [52] showed a total surficial PBDE concentration of 3.1 ng/g, which is lower than the total PBDE concentrations measured in all of the Great Lakes, even in Lake Superior. Incidentally, globally, the highest BDE-209 level (at 6000 ng/g) was found in estuarine sediment from the Kansai region of Japan [53], and in the United States, the highest level of BDE-209 was found in sediments of the Chesapeake Bay, where this concentration ranged from 9000 ng/g near a wastewater treatment plant to 2400 ng/g six kilometers downstream from that plant [54]. The concentrations of BDE-209 observed in surficial sediment from the Great Lakes (10–200 ng/g) are not particularly high in this context.

The PBDE concentrations in the core from Lake Erie are shown in Fig. 11 as a function of deposition year. Data from the other lakes are similar except the resolution with depth is much less. In this Lake Erie core, BDE-209 was first observed at core depths corresponding to 1979. This date is similar to the result reported by Zegers et al. [52], who found BDE-209 first appeared in 1978 in a sediment core from Drammenfjord, Norway. This "advent date" also coincides well with the increasing production of commercial deca-BDE beginning in the late 1970s [28]. Both the $\Sigma$PBDE$'$ and BDE-209 concentrations in this sediment core increased exponentially with time. This is consistent with the increasing demand for PBDEs in U.S. market over the last 30 years.

With the exception of Lake Superior, BDE-209 and $\Sigma$PBDE$'$ have doubling times of $14 \pm 5$ years; in Lake Superior, BDE-209 and $\Sigma$PBDE$'$ have much slower or insignificant doubling times (see Table 8). One possible reason for this difference may be the low population and industrial activity around the Lake Superior area as compared to the other lakes. This doubling time indicates that the input of PBDEs into the sediments of the Great Lakes has doubled every 14 years over that last 30–50 years. PBDEs have been studied as a function of depth in two dated sediment cores collected in Europe. In both cases, the doubling times were 8–11 years [52, 55]. These values are close to the BDE-209 and $\Sigma$PBDE$'$ results for the Great Lakes (except for Lake Superior). As outlined above, the doubling time of PBDEs in fish from the Great Lakes is about 3 years, which is much faster than observed in sediments, suggesting that the shorter doubling time of PBDEs in fish may be due to rela-

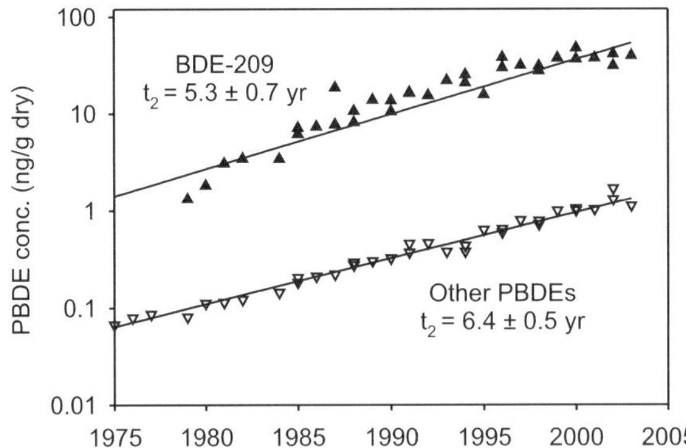


**Fig. 11** Concentrations of BDE-209 and other PBDEs as a function of deposition year in Lake Erie. Note the logarithmic scale for the concentrations. From Zhu and Hites [22]

tively high bioaccumulation factors of tetra- and penta-PBDEs in biota or to degradation of BDE-209 in the sediment or in the water column.

The PBDE inventories (in $ng/cm^2$) represent the total integrated mass of these compounds per unit area from onset to the present, and these results are given in Table 8. The focus-corrected $\Sigma PBDE'$ inventories were $0.6-5\ ng/cm^2$ in the lakes, but the inventories of BDE-209 were much higher, ranging from $5\ ng/cm^2$ in Lake Superior to $110\ ng/cm^2$ in Lake Ontario (see Table 8). These inventories suggest that the total burden of BDE-209 ranges from 4 metric tons in Lake Superior to 34 metric tons in Lake Michigan. Using the focus-corrected surficial sediment fluxes, the current load rate of BDE-209 ranges from 0.11 tons/yr in Lake Superior to 1.4 tons/yr in Lake Erie.

If the sites that have already been sampled in the Great Lakes are typical of the lakes as a whole, the sediment of the lakes has a burden on the order of 100 metric tons of PBDEs; of this, at least 90% is BDE-209, and at least a third of it seems to be in Lake Michigan. Although 100 tons may seem like a lot, it is a small fraction of the annual production of these compounds.

# 4
# Other Brominated Flame Retardants

Three other classes of BFRs have been found in the Great Lakes: the hexabromocyclododecanes (HBCDs), 1,2-*bis*(2,4,6-tribromophenoxy)-ethane (TBE), and pentabromoethylbenzene (PEB). All of these compounds are currently non-regulated alternatives for the PBDEs, and thus, the environmental concentrations of these compounds are likely to increase.

## 4.1
## 1,2,5,6,9,10-Hexabromocyclododecanes

Hexabromocyclododecanes are additive BFRs used in polystyrene. In 2001 the global market demand for HBCD was 16700 metric tons [29]. Of this amount, 2800 tons (17%) were used in North America, and 9500 tones (57%) were used in Europe [29]. The commercial product contains three isomers, among which the $\gamma$ isomer is the most predominant at 75–90% of the total. The $\alpha$ isomer is 10–13% of the total, and the $\beta$ isomer is $< 0.5-12\%$ of the total [56]. The structures of these compounds are given here, but because there has been some confusion about their exact three-dimensional structures, their Chemical Abstracts Registry numbers are also given. Incidentally, these compounds are chiral.

Tomy et al. [57] published a study on the biomagnification of HBCD in Lake Ontario's food web. Using liquid chromatographic techniques, they were able to differentiate the three isomers from one another. In general, the $\beta$

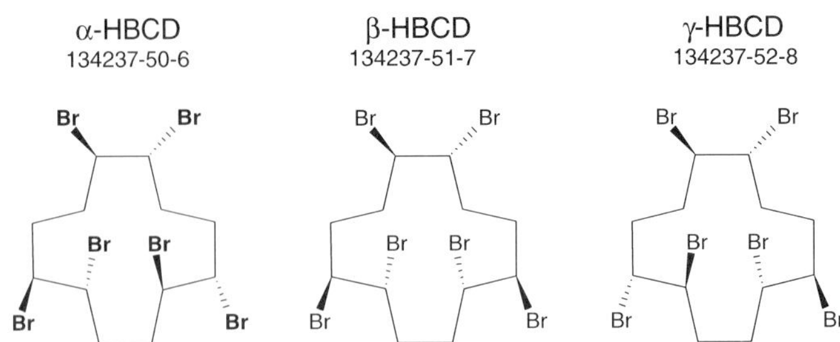

**Scheme 3** Structures of the hexabromocyclododecane isomers present in the commercial product; *Chemical Abstracts* registry numbers are included

isomer was not found in any of the samples, the $\alpha$ isomer was, on average, $\sim 84\%$ of the total HBCD load in the fish, and the $\gamma$ isomer was 16% of the total HBCD load. In the zooplankton samples, the $\alpha$ isomer was $\sim 72\%$ of the total HBCD load, and the $\gamma$ isomer was 28% of the total HBCD load. In both cases, this relative abundance of the $\alpha$ isomer was opposite to the commercial product in which the $\gamma$ isomer dominated. This result suggests some interconversion among the isomers as they are used in thermal setting polymers or in the environment after they escape from the materials in which they are used. It is also possible that the $\alpha$ isomer is preferentially metabolized [58]. The total HBCD concentrations (all in ng/g lipid weight) were, on average, 10 in lake trout, 21 in slimy sculpin, 20 in rainbow smelt, 4 in alewife, and 5 in zooplankton. Generally, there was a strong correlation between these HBCD concentrations and the trophic level of the organism as measured by the nitrogen-15 method [57]. These HBCD concentrations are about 100-fold lower than those of PBDEs measured in fishes collected recently from the lakes. HBCD has also been measured in suspended sediment from the Detroit River, where the maximum concentrations (in ng/g dry weight) by isomer were as follows: $\alpha$, 5; $\beta$, 2; and $\gamma$, 86 [59]. Unlike the biota data of Tomy et al. [57], these results show an isomeric distribution similar to the commercial product, suggesting that the isomer distribution is not changed in abiotic samples and that biota preferentially accumulate the $\alpha$ isomer.

## 4.2
### 1,2-*Bis*(2,4,6-tribromophenoxy)ethane

During the analysis of PBDE in the particle phase of atmosphere samples from various sites in the United States, Hoh et al. [60] noticed a significant, unknown, bromine-containing gas chromatographic (GC) peak in some samples. To identify this unknown substance, full-scan mass spectra were obtained in both electron-impact and electron-capture negative ionization modes. Based on this analysis, it was suspected that this GC peak might be 1,2-*bis*(2,4,6-tribromophenoxy)ethane (TBE). Identification was confirmed

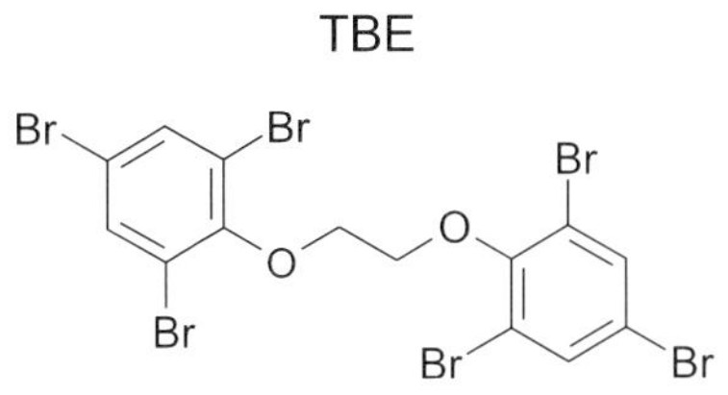

TBE

**Scheme 4** Structure of 1,2-*bis*(2,4,6-tribromophenoxy)ethane

by comparison of the GC retention times and of the full-scan mass spectra of the unknown to those of authentic TBE.

Hoh et al. [60] determined the TBE concentrations in four selected atmospheric samples from Chicago. The TBE concentrations ranged from 4 to $9\,pg/m^3$ in these samples. For comparison, the total PBDE concentrations in these same samples ranged from 31 to $67\,pg/m^3$, indicating that, at least in these samples, the TBE levels are on the order of 15% of the levels of the PBDEs, which is a significant amount.

The highest atmospheric TBE concentrations were detected in samples from a site in southern Arkansas. It is interesting to note that the only producer of TBE in the United States, Great Lakes Chemical, produces this compound at a facility located in El Dorado, Arkansas, which is 150 km west of this Arkansas sampling site. It is likely that the production of TBE will increase at this site given that Great Lakes Chemical has announced that they will market TBE (trade named FF-680) as an additive flame retardant to replace the discontinued octa-BDE product [60].

Hoh et al. [60] also found TBE in a sediment core from northern Lake Michigan. The levels of TBE in this sediment core were lower than those of BDE-209, but the concentration of TBE was almost 10 times higher than the total concentration of BDE-47, -99, and -100, which are the major compo-

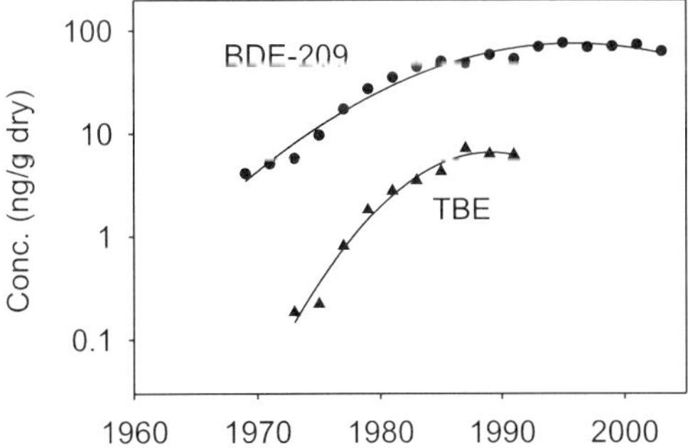


**Fig. 12** Concentrations of TBE and BDE-209 as a function of year in a northern Lake Michigan sediment core. Second-degree polynomials have been fitted to these data as visualization aids. From Hoh et al. [60]

nents of the penta-BDE commercial product. The concentrations of TBE and BDE-209 as a function of year of deposition in the core are shown in Fig. 12 (note the logarithmic scale). TBE first appeared in this core at a depth corresponding to 1973, which was four years later than when BDE-209 was first found in this core. Although the details are hazy, Great Lakes Chemical seems to have first produced TBE 25–30 years ago, a time compatible with the 1973 advent date found in this core [60]. The levels of TBE in this sediment core increased rapidly after 1973, with a doubling time of $\sim$ 2 years until 1985, after which time the TBE concentrations were relatively constant. This increase of TBE concentration was much faster than the rate of increase of BDE-209, the concentration of which doubled every $\sim$ 5 years between 1968 and 1982, with a relatively constant concentration after that time.

According to EPA Inventory Update Rule 2002, Great Lakes Chemical produced 4500–22 500 metric tons each year of TBE from 1986–1994, but the production decreased to 450–4500 metric tons per year after 1998 [60]. This diminution in production may be the cause of the relatively constant concentrations in the sediment core in sections dating after $\sim$ 1985. However, since Great Lakes Chemical plans to replace the octa-BDE product with TBE, there may be an increase in the production of TBE, which will show up as increased concentrations in surficial sediments in the future [60].

## 4.3
### 2,3,4,5,6-Pentabromoethylbenzene

Pentabromoethylbenzene was found by Hoh et al. [60] in gas- and particle-phase atmospheric samples collected in the summer of 2003 in Chicago. Its identity was verified by comparison of the unknown mass spectra and GC retention times with those of authentic material.

In one Chicago sample collected on 20 July 2003, the PEB concentration was 520 pg/m$^3$ in the gas phase and 29 pg/m$^3$ in the particle phase. In the same sample, the total PBDE (tri- through hexabrominated) concentration was 47 pg/m$^3$, and the BDE-209 concentration was 22 pg/m$^3$ [60]. PEB was present in most other Chicago atmospheric samples, but its abundance was low. The reason for the relatively high concentration of PEB in the Chicago atmosphere on this one summer's day is unclear.

Historically, PEB has been used as an additive flame retardant for thermoset polyester and thermoplastic resins during the 1970s and 1980s. In 1977, the production of PEB was 45–450 metric tons [60]. The production of PEB declined to 5–225 metric tons in 1986, and in 1988, there was no ongoing or intended production or processing of this substance [60]. Information on the current manufacturers or processors of PEB is not publicly available; in addition, information on the amount of PEB currently produced (if any) is confidential. However, PEB is listed as a low-production-volume chemical manufactured by Albemarle in France according to the European

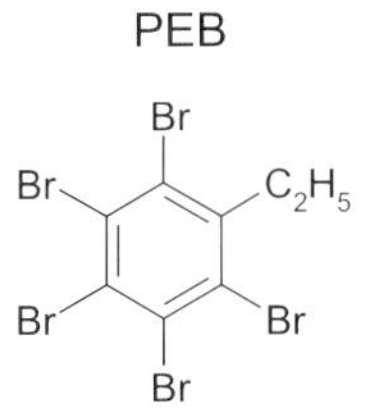

PEB

Scheme 5  Structure of 2,3,4,5,6-pentabromoethylbenzene

Chemical Substance Information System, where PEB production was listed at 10–1000 metric tons in Europe in 2002 [60]. There has been only one previous report about PEB in the environment [61]; in this case, the average PEB concentration was 30 pg/m$^3$ in the atmosphere of Chilton (United Kingdom) in 2001. This concentration was three times higher than the average total (tri-through heptabrominated) PBDE concentration at this site. The detection of small amounts of PEB in Chicago air samples could be due to residuals from the use of this compound 20 years ago or to the emission of fresh PEB used occasionally. Further measurements are required to investigate the environmental sources and fates of this chemical. PEB was not detected in sediment cores from northern Lake Michigan or from eastern Lake Erie.


# 5
# Summary and Conclusions

Brominated flame retardants in the Great Lakes have not been as well studied as many of the polychlorinated pollutants, especially PCBs, but in the last 5–10 years there has been some significant progress. The ubiquity of these compounds in the sediment and fishes of the lakes has been well established, and perhaps more alarmingly, it is now known that the concentrations of some of these compounds are actually increasing, doubling every 3–5 years. This observation is particularly important given that the concentrations of most of the other persistent organic pollutants in the lakes are decreasing. Table 9 summarizes the information at hand for the five main BFR classes discussed here.

Table 9 shows that there are reasonably good data available for the polybrominated diphenyl ethers (PBDEs), but there is much less information for the other BFRs. While the polybrominated biphenyls (PBBs) are no longer in production or use, it is clear that these compounds are persistent in the environment and in people; thus, it would be prudent to continue to monitor the concentrations of these compounds in (at least) the sediment and fishes of the lakes. The other BFRs (HBCD, TBE, and PEB) are not currently abundant in the lakes, but the likelihood of their increased use in the future suggests that they too should be monitored in the sediment and fishes of the lakes. Of course, the

**Table 9** Summary of available data for five brominated flame retardant classes found in the Great Lakes[a]

| | Air (pg/m$^3$) | Fishes (ng/g lipid) | Birds (ng/g lipid) | Sediment (ng/g dry weight) |
|---|---|---|---|---|
| PBBs | | $\gg$ 1000 upstream of Tittabawassee R.; 100–1000 downstream this river; $\ll$ 100 in open lakes<br><br>No significant decrease 1980–2000, except Lake Huron where $t_{1/2} = 19$ yr | 100–500 in eggs from Green Bay | 1000–100 000 near plant; 100–1000 after plant but upstream of Saginaw Bay; 0.01–0.1 in eastern Erie & 0.03–0.1 in northern Michigan |
| PBDEs | 5–15 near Lakes Michigan, Superior, & Erie; $\sim$ 50 in Chicago<br><br>Present in both the atmospheric gas and particle phases | 400–3000 in top predators in 2000; 1–10 in 1980<br><br>Lakes Michigan and Superior now highest; Lake Huron now lowest<br><br>Strong trend with time; $t_2 = 3$–4 years in all five lakes<br><br>Systematic changes in congener ratios with time | In herring gull eggs from all lakes: 5000–10 000 in 2000; 100–300 in 1980–1985<br><br>$t_2 = 3$ years<br><br>Systematic changes in congener ratios with time | Almost all is the deca congener<br><br>30–70 in Michigan, Huron, and Erie; 10 in Superior; 240 in Ontario<br><br>Burdens: 4 tons in Superior; 35 tons in Michigan; 10 tons in Huron; 17 tons in Erie; and 23 tons in Ontario<br><br>$t_2 = 14$ years in all lakes except Superior |
| HBCDs | | 4–20 in fishes & $\sim$ 4 in zooplankton from Lake Ontario<br><br>$\alpha$-isomer dominated | | < 90 in Detroit R. suspended solids<br><br>$\gamma$-isomer dominated |
| TBE | 4–9 in Chicago | | | 0.2–8 in Michigan |
| PEB | 500–600 in one Chicago sample | | | Not present in Lakes Michigan or Erie |

[a] $t_{1/2}$ is half-life and $t_2$ is doubling time.

PBDEs are not a closed issue either. Even though the penta and octa products are no longer produced, the deca product (BDE-209) is still being marketed aggressively, and the residues of the penta and octa products are still out there and are not going away soon. Given the increasing concentrations over time for these compounds, continued work on all of them is certainly warranted.

**Acknowledgements** Eunha Hoh and Lingyan Zhu made valuable contributions to some of the work presented here and to the review of this manuscript.

## References


1. Fries GF (1985) Critical Rev Toxicol 16:105
2. Morris PJ, Quensen JF, Tiedje JM, Boyd SA (1993) Environ Sci Technol 27:1580
3. Orti DL, Hill RH, Patterson DG, Needham LL, Kimbrough RD, Alley CC, Lee HCJ (1983) Arch Environ Contam Toxicol 12:603
4. Hill RH, Patterson DG, Orti DL, Holler JS, Needham LL, Sirmans SL, Liddle JA (1982) J Environ Sci Health B 17:19
5. Hesse JL, Powers RA (1978) Environ Health Perspect 23:19
6. Shah BP (1978) Environ Health Perspect 23:27
7. Wolff MS, Anderson HA, Camper F, Nikaido MN, Daum SM, Haymes N, Selikoff IJ (1979) J Environ Pathol Toxicol 2:1397
8. Wolff MS, Anderson HA, Selikoff IJ (1982) J Amer Med Assoc 247:2112
9. Wolff MS, Haymes N, Anderson HA, Selikoff IJ (1978) Environ Health Perspect 23:315
10. Lilis R, Anderson HA, Valciukas JA, Freedman S, Selikoff IJ (1978) Environ Health Perspect 23:105
11. Blanck HM, Marcus M, Hetrzberg V, Tolbert PE, Rubin C, Henderson AK, Zhang RH (2000) Environ Health Perspect 108:147
12. Wolff MS, Aubrey B, Camper F, Haymes N (1978) Environ Health Perspect 23:177
13. Sweeney AM, Symanski E, Burau KD, Kim YJ, Humphrey HEB, Smith MA (2001) Environ Res A 86:128
14. Sjödin A, Patterson DG, Bergman A (2001) Environ Sci Technol 35:3830
15. Anderson HA, Falk C, Hanrahan L, Olson J, Burse VW, Needham LL, Paschal D, Patterson DG, Hill RH (1998) Environ Health Perspect 106:279
16. Jaffe R, Stemmler EA, Eitzer BD, Hites RA (1985) J Great Lakes Res 11:156
17. Miller TJ, Jude DJ (1984) J Great Lakes Res 10:215
18. Luross JM, Alaee M, Sergeant DB, Cannon CM, Whittle DM, Solomon KR, Muir DCG (2002) Chemosphere 46:665
19. Zhu LY, Hites RA (2004) Environ Sci Technol 38:2779
20. Haseltine SD, Heinz GH, Reichel WL, Moore JF (1981) Pestic Monit J 15:90
21. Heinz GH, Erdman TC, Haseltine SD, Stafford C (1985) Environ Monit Assess 5:223
22. Zhu LY, Hites RA (2005) Environ Sci Technol 39:3488
23. Morris PJ, Quensen JF, Tiedje JM, Boyd SA (1992) Appl Environ Microbiol 58:3249
24. Bedard DL, vanDort HM (1998) Appl Environ Microbiol 64:940
25. Jacobs LW, Chou SF, Tiedje JM (1976) J Ag Food Chem 24:1198
26. Jacobs LW, Chou SF, Tiedje JM (1978) Environ Health Perspect 23:1
27. Hites RA (2004) Environ Sci Technol 38:945
28. World Health Organization (1994) Brominated diphenyl ethers, environmental health criteria no. 152. Geneva, Switzerland, p 34

29. Bromine science and environmental forum. http://www.bsef.com; accessed November 14, 2005
30. Hardy M (2005) Environ Sci Technol 39:377
31. Asplund L, Hornung M, Peterson RE, Turesson K, Bergman A (1999) Organohalogen Compd 40:351
32. Manchester-Neesvig JB, Valters K, Sonzogni WC (2001) Environ Sci Technol 35:1072
33. Stapleton HM, Baker JE (2003) Arch Environ Contam Toxicol 45:227
34. Luross JM, Alaee M, Sergeant DB, Whittle DM, Solomon KR (2000) Organohalogen Compd 47:73
35. Alaee M, Luross JM, Sergeant DB, Muir DCG, Whittle DM, Solomon KR (1999) Organohalogen Compd 40:347
36. Dodder NG, Strandberg B, Hites RA (2002) Environ Sci Technol 36:146
37. Loganathan BG, Kannan K, Watanabe I, Kawano M, Irvine K, Kumar S, Sikka HC (1995) Environ Sci Technol 29:1832
38. Meironyté D, Norén K, Bergman A (1999) J Toxicol Environ Health A 58:329
39. Darnerud PO, Aune M, Atuma S, Becker W, Bjerselius R, Cnattingius S, Glynn A (2002) Organohalogen Compd 58:233
40. Stapleton HM, Letcher RJ, Baker JE (2002) Organohalogen Compd 58:201
41. Norstrom RJ, Simon M, Moisey J, Wakeford B, Weseloh DV (2002) Environ Sci Technol 36:4783
42. Strandberg B, Dodder NG, Basu I, Hites RA (2001) Environ Sci Technol 35:1078
43. Hillery BR, Basu I, Sweet CW, Hites RA (1997) Environ Sci Technol 31:1811
44. Pankow JF (1994) Atmos Environ 28:185
45. Harner T, Bidleman TF (1998) Environ Sci Technol 32:1494
46. Song W, Ford JC, Li A, Mills WJ, Buckley DR, Rockne KJ (2004) Environ Sci Technol 38:3286
47. Song W, Li A, Ford JC, Sturchio NC, Rockne KJ, Buckley DR, Mills WJ (2005) Environ Sci Technol 39:3474
48. Song W, Ford JC, Li A, Sturchio NC, Rockne KJ, Buckley DR, Mills WJ (2005) Environ Sci Technol 39:5600
49. Sellström U, Kierkegaard A, de Wit C, Jansson B (1998) Environ Toxicol Chem 17:1065
50. Allchin C, de Boer J (2001) Organohalogen Compd 52:30
51. Christensen JH, Platz J (2001) J Environ Monitoring 3:543
52. Zegers BN, Lewis WE, Booij K, Smittenberg RH, Boer W, de Boer J, Boon JP (2003) Environ Sci Technol 37:3803
53. Watanabe I, Sakai SI (2003) Environ Int 29:665
54. Baker JE, Klosterhaus S, Liebert D, Stapleton HM (2004) Organohalogen Compd 66:3758
55. Nylund K, Asplund L, Jansson B, Jonsson P, Litzén K, Sellström U (1992) Chemosphere 24:1721
56. Peled M, Scharia R, Sondack D (1995) Ind Chem Library 7:92
57. Tomy GT, Budakowski W, Halldorson T, Whittle DM, Keir MJ, Marvin C, MacInnis G, Alaee M (2004) Environ Sci Technol 38:2298
58. Zegers BN, Mets A, van Bommel R, Minkenberg C, Hamers T, Kamstra JH, Pierce GJ, Boon JP (2005) Environ Sci Technol 39:2095
59. Marvin C, Alaee M, MacInnis G, Stern G, Reiner E, Kolic T, MacPherson K, Tomy G, Painter S, Muir DCG (2003) Organohalogen Compd 62:21
60. Hoh E, Zhu LY, Hites RA (2005) Environ Sci Technol 39:2472
61. Lee RGM, Thomas GO, Jones KC (2002) Organohalogen Compd 58:193

Hdb Env Chem Vol. 5, Part N (2006): 391–438
DOI 10.1007/698_5_046
© Springer-Verlag Berlin Heidelberg 2005
Published online: 20 December 2005

# Perfluorinated Compounds in the Great Lakes

J. P. Giesy[1,2] (✉) · S. A. Mabury[3] · J. W. Martin[4] · K. Kannan[5] · P. D. Jones[1] ·
J. L. Newsted[6] · K. Coady[7]

[1]Department of Zoology, National Food Safety and Toxicology Center,
 Center for Integrative Toxicology, Environmental Science and Policy Program,
 Michigan State University, East Lansing, MI 48824, USA
 *jgiesy@aol.com*

[2]Department of Biology and Chemistry, City University of Hong Kong,
 Kowloon, Hong Kong, SAR, China
 *jgiesy@aol.com*

[3]Department of Chemistry, University of Toronto, Toronto, ON M5S 3H6, Canada

[4]Department of Public Health Sciences, University of Alberta, Edmonton, AB T6E 2L3,
 Canada

[5]Wadsworth Center, New York State Department of Health
 and Department of Environmental Health and Sciences, School of Public Health,
 State University of New York at Albany, Empire State Plaza, PO Box 509,
 Albany, NY 12201-0509, USA

[6]Entrix, Inc., 4295 Okemos Rd., Suite 101, Okemos, MI 48864, USA

[7]Biology Department, Warner Southern College, Lake Wales, FL 33859, USA

**Abstract** Perfluoroalkyl acids (PFAs) are released to the environment via their manufacturing processes, their use in commercial products, or indirectly via oxidation of precursor molecules containing perfluoroalkyl chains. PFA precursors are diverse and include poly- and perfluorinated alcohols and perfluoroalkyl sulfonamide derivatives. Products in which PFAs and their precursors have been used include wetting agents, lubricants, stain resistant treatments, and fire-fighting foams. The PFAs in the environment comprise two general classes: perfluoroalkyl carboxylates such as $CF_3(CF_2)_yCO_2^-$ and perfluoroalkyl sulfonates, such as $CF_3(CF_2)_xSO_3^-$. The predominant PFA in biota samples from the Great Lakes is perfluorooctane sulfonate (PFOS), but a homologous series of perfluoroalkyl carboxylates, where $x = 6 - 13$, is also detected in most samples at lesser concentrations. The environmental behavior of most PFAs is not well studied, and our knowledge of the physicochemical properties of PFOS and PFOA is limited. Both compounds are persistent in the environment and are not expected to volatilize into the atmosphere to a significant extent, but have much greater water solubilities than similar chlorinated compounds. Concentrations of PFOS in surface waters are usually less than those of perfluorooctanoic acid (PFOA), but PFOS accumulates in aquatic organisms to a greater extent and appears to biomagnify in the food web of the Great Lakes region. PFAs and/or their precursors have been measured in air, surface waters, sediments, aquatic invertebrates, and in the tissues of fish, fish-eating water birds, mink, otter, and other wildlife from in and around the Great Lakes. Although the sources of PFAs to the Great Lakes are not well understood, fluorotelomer alcohols (FTOHs) and perfluorooctylsulfonamides degrade to perfluoroalkyl carboxylates and PFOS, respectively, in laboratory studies. Based on preliminary and incomplete information, current concentrations of PFOS in the Great Lakes environment do not seem to be sufficient to pose a significant risk to most aquatic organisms including fish. However, the margins of safety are less for mammals such as mink and birds and when the concentrations of all PFAs are considered together, current concentrations may pose risk to some sensitive species.

**Keywords** Perfluorinated acid · Precursors · Perfluorooctane sulfonamides · Fluorotelomer alcohols · Bioaccumulation · Food web · Surface waters · Toxicity · Risk Assessment · Fatty acids

**Abbreviations**

| | |
|---|---|
| AFFF | Aqueous film-forming foam |
| BAF | Bioaccumulation factor |
| BCF | Bioconcentration factor |
| CBR | Critical body residue |
| FOSA | Perfluorooctane sulfonamide |
| FTCA | Fluorotelomer acid |
| FTOH | Fluorotelomer alcohol |
| FTUCA | Unsaturated fluorotelomer acid |
| HPLC | High-pressure liquid chromatography |

$K_d$        Distribution coefficient in soil
$K_{oc}$      Organic carbon normalized adsorption coefficient
$K_{ow}$      Octanol-water partition coefficient
LC-MS    Liquid chromatography-mass spectrometry
LOAEL    Lowest observed adverse effect level
LOQ        Limit of quantitation
NOAEL    No observed effect
PFA         Perfluoroalkyl acid
PFCA      Perfluorinated carboxylates
PFFA      Perfluorinated fatty acid
PFHxS    Perfluorohexane sulfonate
PFOA      Perfluorooctanoic acid
PFOS      Perfluorooctane sulfonate
POSF      Perfluorooctanesulfonyl fluoride
SPE         Solid-phase extraction
TMF        Trophic magnification factor
TRV         Toxicity reference value

# 1
# Introduction

## 1.1
## Chemicals

Perfluoroalkyl acids (PFAs) are synthetic, perfluorinated, straight- or branched-chain organic acids characterized by a carboxylate or sulfonate moiety [1]. PFAs are manufactured directly but can also be formed through transformation of many precursor molecules containing a perfluoroalkyl moiety [2]. These include fluorotelomer alcohols (FTOHs) and perfluoroalkylsulfonamides (Table 1, Fig. 1), however, others likely exist. Collectively, this family of chemicals including PFAs and their polyfluorinated precursors will be referred to in this document as perfluoroalkyl substances (PFSs).

## 1.2
## Environmental Fate and Transport

## 1.2.1
## Sources and Manufacturing

Perfluoroalkyl substances are synthetic products that have been manufactured for over 50 years [1]. They have been used in many commercial products such as refrigerants, surfactants, polymers, pharmaceuticals, wetting agents, lubricants, adhesives, pesticides, corrosion-inhibitors, stain resistant treatments for leather, paper and clothing [1, 3, 4]. Other uses have included, aqueous film-forming foams (AFFF) for fire-fighting, mining and oil-well

**Table 1** PFAs and their precursor molecules

| Compound (synonyms) | CAS number | Molecular Formula | Molecular Weight |
|---|---|---|---|
| **PFAs** | | | |
| Perfluorobutane Sulfonate (C4, PFBS) | 29420-49-3 | $C_4F_9SO_3^-$ | 299 |
| Perfluorohexane Sulfonate (C6, PFHxS) | 432-50-7 | $C_6F_{13}SO_3^-$ | 399 |
| Perfluorooctane Sulfonate(PFOS) | 2795-39-5 | $C_8F_{17}SO_3^-$ | 499 |
| Perfluorooctane Sulfonic Acid | 1763-23-1 | $C_8F_{17}SO_3H$ | 500 |
| Pentadecafluorooctanoate (PFOA; C8) | — | $C_7F_{15}COO^-$ | 413 |
| Pentadecafluorooctanoic acid | 335-67-1 | $C_7F_{15}COOH$ | 414 |
| Tridecafluoroheptanoate (C7, PFHpA) | — | $C_6F_{13}COO^-$ | 363 |
| Tridecafluoroheptanoic acid | 375-85-9 | $C_6F_{13}COOH$ | 364 |
| Heptadecafluorononanoate (C9, PFNA) | — | $C_8F_{17}COO^-$ | 463 |
| Heptadecafluoronoic acid | 375-95-1 | $C_8F_{17}COOH$ | 464 |
| Nonadecafluorodecanoate (C10, PFDA) | — | $C_9F_{19}COO^-$ | 513 |
| Nonadecafluorodecanoic acid | 335-76-2 | $C_9F_{19}COOH$ | 514 |
| Perfluoroundecanoate (C11, PFUnA) | — | $C_{10}F_{21}COO^-$ | 563 |
| Perfluoroundecanoic acid | 2058-94-8 | $C_{10}F_{21}COOH$ | 564 |
| Perfluorododecanoate (C12, PFDoA) | — | $C_{11}F_{23}COO^-$ | 613 |
| Perfluorododecanoic acid | 307-55-1 | $C_{11}F_{23}COOH$ | 614 |
| Perfluorotridecanoate (C13, PFTrA) | — | $C_{12}F_{25}COO^-$ | 663 |
| Perfluorotetradecanoate (C14, PFTA) | — | $C_{13}F_{27}COO^-$ | 713 |
| Perfluorotetradecanoic acid | 376-06-7 | | 714 |
| Perfluoropentadecanoate (C15, PFPA) | — | $C_{14}F_{29}COO^-$ | 763 |
| **PFA-Precursors** | | | |
| Perfluorooctane Sulfonamide (FOSA) | 4151-50-2 | $C_8F_{17}SO_2NH_2$ | 499 |
| N-Methyl perfluorooctane sulfonamidoethanol (NMeFOSE) | 24448-09-7 | $C_8F_{17}SO_2N(CH_3)$ $C_2H_4OH$ | 557 |
| N-Ethyl perfluorooctane sulfonamidoethanol (NEtFOSE) | 1691-99-2 | $C_8F_{17}SO_2N(C_2H_5)$ $C_2H_4OH$ | 571 |
| N-Ethyl perfluorooctane sulfonamidoacetic acid (PFOSAA) | 2991-51-7 | $C_8F_{17}SO_2N(C_2H_5)$ $CH_2CO_2H$ | 585 |
| N-Ethyl perfluorooctane sulfonamide (NEtFOSA) | 4151-50-2 | $C_8F_{17}SO_2NH(C_2H_5)$ | |
| Perfluorooctane Sulfonylfluoride (POSF) | 307-35-7 | $C_8F_{17}SO_2F$ | 502 |
| Perfluorooctanesulfinate (PFOSulfinate) | — | | 483 |
| 6:2 Fluorotelomer Alcohol (6:2 FTOH) | — | $CF_3(CF_2)_5C_2H_4OH$ | 364 |
| 8:2 Fluorotelomer Alcohol (8:2 FTOH) | — | $CF_3(CF_2)_7C_2H_4OH$ | 464 |
| 10:2 Fluorotelomer Alcohol (10:2 FTOH) | — | $CF_3(CF_2)_9C_2H_4OH$ | 564 |

surfactants, acid-mist suppressants for metal plating and electronic-etching baths, alkaline cleaners, floor polishes, photographic film, denture cleaners, shampoos, and as an insecticide [5].

The historical global production volume of PFSs is unknown, but in 2000, 3M, the major manufacturer of perfluorooctanesulfonyl-based chemicals,

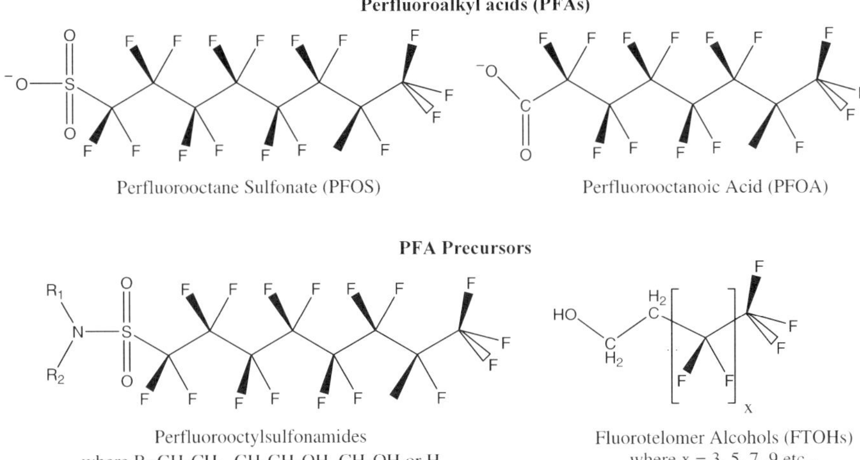



**Fig. 1** Example structures of perfluoroalkyl acids (PFAs) and some of their precursor molecules

produced 6.5 million lbs. [6]. Approximately 37% of this was used in surface treatment applications to provide soil, oil and water resistance to personal apparel and home furnishings. Approximately 42% of the production was used on paper products to provide grease, water, and oil resistance to plates, food containers, bags, and wraps [6]. In 1985, the US market for AFFF products containing perfluoroalkyl substances (i.e., 3 and 6% concentrates) was 6.8 million L [cited in 5].

The study of perfluoroalkyl substances is complicated by the fact that there are many fluorinated compounds produced, their production and usage volumes are largely unknown, and many of the functional chemistries remain the proprietary information of the chemical industry [1]. However, it can be generalized that there are two major processes used in the manufacturing of fluorinated surfactants: electrochemical fluorination and telomerization. Electrochemical fluorination involves reaction of a hydrocarbon feedstock with hydrofluoric acid in the presence of an electric current (Eq. 1). All hydrogen molecules of the hydrocarbon are replaced by fluorine and perfluorinated molecules result. Perfluorooctanesulfonyl fluoride (POSF) was historically the product of electrochemical fluorination, and this served as the basic building block for a series of perfluorooctanesulfonylamides that are used as surface protectors in carpets, leather, paper and packaging, fabric and upholstery and as surfactants.

$$2C_8H_{17}SO_2F + 34HF \rightarrow (4.5-7V) \rightarrow 2C_8F_{17}SO_2F + 17H_2 \qquad (1)$$

Electrochemical fluorination yields 30–40% straight-chain (normal) isomers and a mixture of byproducts of unknown and variable composition. The product of electrochemical fluorination is thus not a pure chemical but rather a mixture of isomers and homologues. The commercialized products derived from these materials are also a mixture of approximately 70% linear and 30% branched compounds, including impurities. With electrochemical fluorination, perfluorinated compounds with a continuous homologous series of even and odd number perfluorocarbons are generated based on the variable chain length in the starting material. Electrochemical fluorination was economically attractive because it is relatively inexpensive. However, much of the perfluorochemical production by this method ceased in 2000 when 3M, the major manufacturer, voluntarily phased out production of perfluorooctyl sulfonate-based chemistries by this method. In addition, the US Environmental Protection Agency instituted a Significant New Use Rule regarding production and import of PFOS and its precursors [6]. Today, electrochemical fluorination is only applied to the analogous production of perfluorobutyl-based substances.

An alternative production method to electrochemical fluorination is telomerization [7], a process by which molecules called telogens, consisting of two or more unsaturated fluorinated molecules called taxogens, are reacted to create a fluorinated telomer (Eq. 2). This process results primarily in production of linear molecules containing only even numbered carbon chain-lengths (Eq. 2), such as FTOHs. Today, telomerization is the most common synthetic route for manufacturing perfluoroalkyl substances containing straight (i.e.,normal) perfluoroalkyl chains of 6–14 carbons.

$$\text{YZ (Telogen)} + n\text{A (Taxogen)} \rightarrow \text{Y} - (\text{A})n - \text{Z (Telomer)} \tag{2}$$

### 1.2.2
### Physical/Chemical Properties

Because of the high-energy carbon – fluorine (C – F) bond and their highly oxidized state, PFAs are largely resistant to biotic and abiotic degradation mechanisms such as hydrolysis, photolysis, microbial degradation, and metabolism. Because of this, PFAs are environmentally persistent [1]. Due to their fluorinated tails and moderate to high surface activity, the phase-partitioning behavior of perfluoroalkyl substances is different than that of chlorinated hydrocarbons. Longer perfluoroalkyl chains are both oleophobic and hydrophobic and thus do not mix well with either water or oil [7]. This and the presence of an anionic functional group result in some perfluoroalkyl substances having surfactant properties which cause them to migrate to the interface of solutions due to the competing action of hydrophobic and hydrophilic moieties. It has been reported that PFOS actually forms a third immiscible layer between octanol, a surrogate for lipid, and water, confounding the measurement of octanol-water partition coefficients [8]. These properties are considerably different from those of chlorinated and brominated hydrocarbons, which are hydrophobic, but comparably lipophilic. The hydrophilic nature of some perfluoroalkyl substances is due to the presence of a charged moiety, such as carboxylates, sulfonates, or a quaternary ammonium, but may also be due to polar, yet neutral, moieties such as an alcohol group. PFAs can lessen the surface tension of water more than hydrocarbon-based surfactants and are thus more powerful wetting agents. In addition, fluorinated surfactants have greater chemical stability to acids, oxidizing agents, and alkalis compared to hydrocarbon-based surfactants [7].

Water solubility and vapor pressures of PFOS and PFOA are given in Table 2. These data were obtained from products that were not refined and as a result may contain more than one PFA such that these data may not be representative of the pure compounds, especially in environmental media. Due to the lack of accurate information on the physico-chemical properties, accurate prediction of the environmental fate and transport of most perfluoroalkyl substances has not yet been possible. The prediction of the distribution and ultimate fates of perfluoroalkyl substances is further complicated by their hydrophobic and lipophobic properties, such that the fugacity approach that has been useful in describing the environmental fates of organochlorines is less useful for describing the environmental fate of PFAs and their precursors. The bulk of the available physical and chemical information is for PFOS

**Table 2** Physical/chemical properties of select perfluorinated fatty acids

| Parameter | PFOS (Potassium salt) | Refs. | PFOA | Refs. |
|---|---|---|---|---|
| Melting point | $\geq 400\,^{\circ}\mathrm{C}$ | [60] | $45{-}50\,^{\circ}\mathrm{C}$ | [11] |
| Boiling point | Not calculable | [8] | $189{-}192\,^{\circ}\mathrm{C}$ at 736 mm Hg | [11] |
| Specific gravity[a] | $\sim 0.6$ (7–8) | [8] | — | |
| Vapor pressure | $3.31 \times 10^{-4}$ Pa @ 20 °C | [61] | $1.33 \times 10^{-5}$ Pa @ 25 °C (PFOA ammonium salt) $1.33 \times 10^{4}$ Pa @ 25 °C (PFOA) | [11] |
| Water solubility | | | | |
|   Pure water | 680 mg/L | [9] | 3.4 g/L | [11] |
|   Fresh water | 370 mg/L | [8] | — | |
|   Sea water | 12.4 mg/L | [62] | — | |
| Octanol solubility | 56 mg/L | [63] | — | |
| Log $K_{\mathrm{ow}}$[b] | $-1.08$ | [8] | — | |
| Henry's law constant[c] | $2.0 \times 10^{-6}$ | [64] | — | |

[a] pH values in parentheses
[b] Log $K_{\mathrm{ow}}$ calculated with solubility of PFOS in water and n-octanol
[c] Henry's Law constant calculated at 20 °C with the solubility for pure water

and PFOA. PFOS is moderately water soluble, non-volatile, and thermally stable. The potassium salt of PFOS has a reported mean solubility of 680 mg/L in pure water [9]. However, PFOS is a very strong acid, and in neutral water will fully dissociate to form the respective anion (conjugate base). Thus, the PFOS anion can form strong ion pairs with many cations, resulting in a salting-out effect in natural waters that contain high amounts of dissolved solids (Table 2). For instance, as salt content increases, solubility of PFOS decreases such that the solubility of PFOS in salt water is approximately 12.4 mg PFOS/L as compared to the 680 mg PFOS/L in pure water [10]. The reported water solubility of PFOA is 3.4 g/L, indicating that this compound is highly soluble in water [11]. Although no data exists, it can be generalized that as the length of the perfluorinated chain increases, perfluoroalkyl carboxylates become more hydrophobic and their water solubility diminishes.

PFOS has a reported mean solubility of 56.0 mg PFOS/L in pure octanol [12]. However, due to the surface-active properties of both PFOS and PFOA, when these compounds are added to octanol and water in a standard test system to measure the octanol-water partitioning coefficient ($K_{ow}$), three layers are formed. Thus, a $K_{ow}$ has not been directly measured for either PFOS or PFOA. The log $K_{ow}$ for PFOS has been estimated from its water and octanol solubility to be approximately – 1.08, but the utility of this estimate is questionable. Since the $K_{ow}$ of PFOS and PFOA can't be determined by standard methods, other more operationally defined or empirically measured physio-chemical properties, such as bioconcentration factors and soil adsorption coefficients can be used to describe the environmental fates of these compounds. It is not currently possible to estimate physical-chemical properties that are traditionally used to predict environmental fates in multi-media models by use of conventional quantitative structure activity relationship models (QSAR). In addition, the use of $K_{ow}$ may not be appropriate to predict these other properties since PFAs do not accumulate by partitioning into lipids. Enterohepatic recirculation, rather than accumulation in lipids, is partially responsible for the slow depuration of some PFAs (see Sect. 1.5.2) and PFOS binds to certain proteins in animals [13]. As a result, use of either water solubility or predicted $K_{ow}$ values may underestimate the accumulation of PFOS into organisms and other environmental media. Again, currently, the best methods to estimate the potential for bioaccumulation and biomagnification are empirically measured distribution constants.

Based on their vapor pressure and predicted Henry's law constant, neither PFOS (potassium salt) nor PFOA (ammonium salt) are expected to volatilize under prevalent environmental conditions (Table 2). PFOA has a relatively great vapor pressure, and would be expected to volatilize if present in the protonated form, however it is present in the environment as the conjugate base anion resulting in greatly diminished volatilization. The OECD has classified PFOS as a type 2, nonvolatile chemical that has a very low or possibly negligible volatility [8].

## 1.3
## Environmental Transport

Perfluoroalkyl substances are globally distributed and ubiquitous [2, 14–18] and are also detectable in water, air, and biota from the Great Lakes region [19], (Tables 11–17). Because of the relatively small vapor pressure and greater water solubility of PFA compounds such as PFOS and PFOA (Table 2), questions remain as to how these compounds have become globally distributed, particularly in remote regions of the world. While long-range transport in rivers or marine currents is still a feasible explanation, several researchers have hypothesized that neutral PFA precursors may be responsible for transporting perfluoroalkyl substances long-range in the atmosphere, and that through oxidation these may yield PFAs as degradation products. This contention is based on the knowledge that POSF and POSF-based products are ultimately degraded to PFOS, and evidence suggesting that PFOA could also be produced from POSF-based chemicals through abiotic pathways (Fig. 2a) [20].

Martin et al. [21] first presented evidence for volatile, neutral PFA precursors in the environment. A series of FTOHs and perfluorooctane sulfonamide derivatives were detected in the air (gas + particle phase) of southern Ontario (Table 17). These observations have been subsequently confirmed by other researchers [22–25]. The gas-phase atmospheric chemistries of perfluoroalkyl sulfonamides, such as $N$-EtFOSE or $N$-MeFOSE have not yet been reported. Thus, while it remains a hypothesis that these are partially responsible for the global dissemination of PFOS and PFOA, it is not yet possible to accurately assess their contribution to PFA concentrations in the Great Lakes or other remote locations. In contrast, the atmospheric chemistry of FTOHs has been studied extensively [26] and is briefly discussed here in Sect. 1.4.2 Atmospheric processes. It has been suggested that the atmospheric concentrations and lifetimes of FTOHs, and the measured profile of their perfluoroalkyl carboxylate degradation products, might explain the pattern of contamination of Arctic organisms with long-chain perfluoroalkyl carboxylates [17, 26]. The relative contribution of atmospheric degradation of FTOH to PFOA and other perfluoroalkyl carboxylates to the currently observed concentrations in the Great Lakes has not been assessed. However, the yield of PFCAs from FTOHs in the atmosphere is likely diminished around urban locations where NO concentrations are greater [26].

## 1.4
## Environmental Fate

Under environmental conditions, neither PFOS nor PFOA hydrolyze, photolyze, or biodegrade to any significant extent. Over the course of a 285-d field microcosm experiment, no degradation of PFOS was observed [27] confirm-

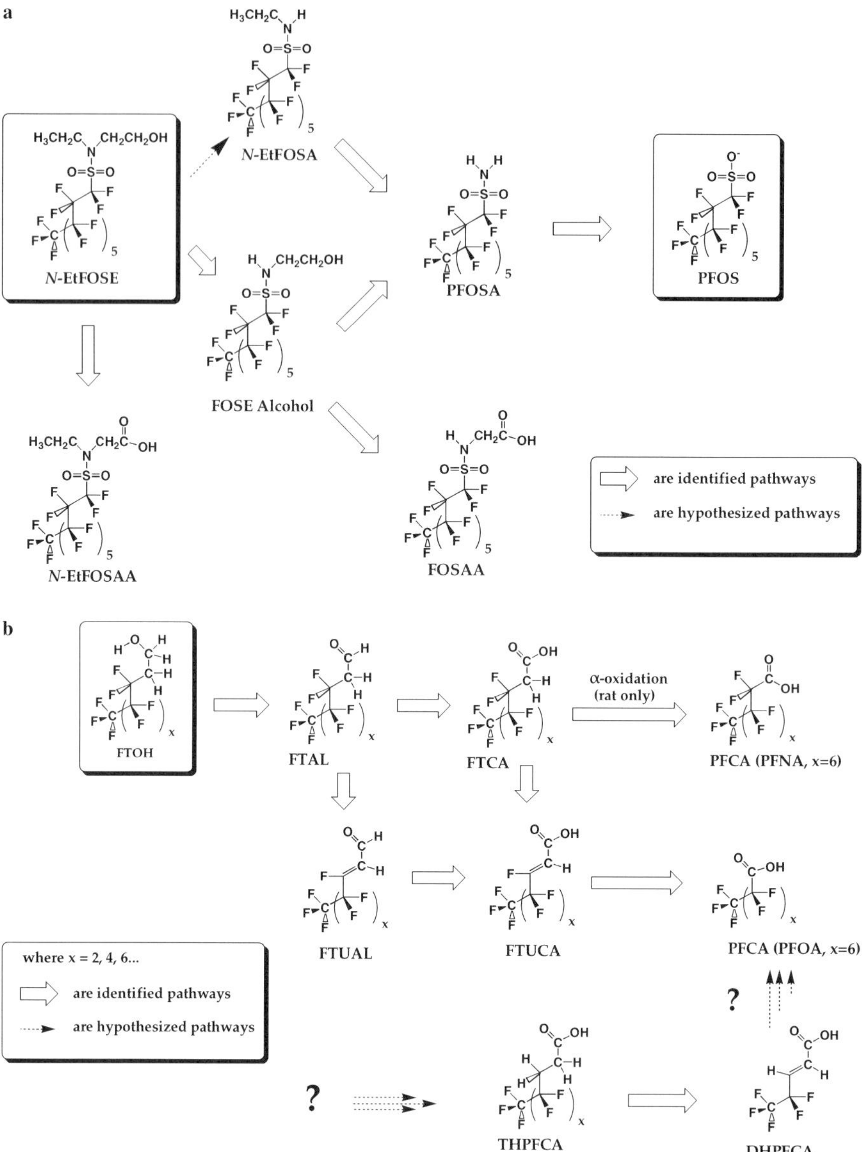



**Fig. 2 a** Overall metabolic scheme for the perfluoroalkylsulfonamide alcohols based on rat and fish studies. Presumably the conversion of the alcohol to the corresponding acetates (e.g., *N*-EtFOSAA and FOSAA) proceed via the aldehyde intermediate though this has yet to be identified. Secondary metabolites for the alcohols (e.g., glucoronide and sulfate) and the carboxylic acids have been omitted. **b** Overall metabolic scheme for fluorotelomer alcohols (FTOH) derived from published work from microbial and rat studies [52–54, 85]

ing that this compound undergoes little degradation in aquatic systems. Both PFOA and PFOS are considered to be persistent in the environment, and it is expected that longer-chained perfluoroalkyl carboxylates will behave similarly. Here we summarize some of the existing studies relating to photolysis, hydrolysis, biodegradation, and thermal stability for FTOHs, PFOS, and PFOA.

## 1.4.1
### Direct and Indirect Photolysis

It was previously noted that no evidence for direct photolysis for PFOS or PFOA has been observed experimentally. In aqueous solutions alone and in the presence of hydrogen peroxide ($H_2O_2$), iron oxide ($Fe_2O_3$) or humic material, PFOS has been observed to undergo some indirect photolysis [28] whereas PFOA did not undergo indirect photolysis [29]. Using an iron oxide photo-initiator matrix model, the indirect photolytic half-life for PFOS was estimated to be $\geq 3.7$ years at $25\,^{\circ}C$. The half-life of PFOA was estimated to be $> 349$ d.

More recently, studies have demonstrated that UV irradiation ($220-460$ nm) of aqueous PFOA solutions at room temperature in the presence of oxygen resulted in the formation of shorter-chain perfluoroalkyl carboxylates, $CO_2$, and $F^-$ [30]. The environmental significance for this photolytic pathway was not discussed, but it is not likely to be an important degradation route for PFOA in natural surface waters. PFOA absorbs UV mostly at wavelengths below 220 nm [30], and these energetic wavelengths are not present at the earth's surface.

## 1.4.2
### Atmospheric Processes

Since it is unlikely that the perfluorinated anions (e.g., PFOA or PFOS) would experience photolytic degradation under environmental conditions, most research has focused on identifying the mechanisms for the widespread dissemination of these compounds. Much attention has focused on the hypothesis that volatile perfluorinated precursors, released from industrial and consumer sources, are atmospherically degraded to the persistent perfluorinated anions. Specifically, it is proposed that airborne fluorotelomer alcohols (FTOHs) are precursors of the perfluorinated carboxylates (PFCAs) and that the perfluorinated sulfamidoethanols are precursors to PFOS [31]. The relevant physical properties of both the FTOHs and perfluorinated sulfamideothanols suggest significant partitioning to the atmosphere [22, 32] and both classes have been observed in the atmosphere across North America [21, 23]; see Sect. 1.3.

A collaborative effort between the groups of Mabury and Wallington investigated the atmospheric kinetics and reaction dynamics of the FTOHs.

FTOHs are linear fluorinated alcohols with the formula $C_nF_{2n+1}CH_2CH_2OH$ ($n = 2, 4, 6 \ldots$) and have proved to be sufficiently long-lived ($\sim 20$ days) in the atmosphere to support long-range transport to remote regions [33, 34]. Hydroxyl radicals initiate the transformation of FTOHs to yield, in essentially quantitative yields, the fluorotelomer aldehyde. The overall mechanism for the atmospheric processing of FTOHs is presented (Fig. 3), which represents a complement of smog-chamber investigations [34–39]. Critical to this mechanism was the discovery that reactions of an acylperoxy radical with HOO could yield the tetraoxide intermediate, that through one pathway yields ozone and the corresponding PFCA. Shorter-chain PFCAs were shown to be produced through a perfluorinated alcohol that was the product of a dismutation reaction between a perfluorinated peroxy radical and an acyl peroxy radical containing an alpha hydrogen, that yielded an acid fluoride which would quickly hydrolyze to the PFCA. The PFCAs themselves are not expected to react further in the atmosphere due to OH reactivity [40] but will be scavenged on the timescale of days through wet and dry deposition.

Overall, these reactions suggest potential atmospheric reaction pathways to produce PFCAs from FTOHs. Under environmental conditions, the pro-

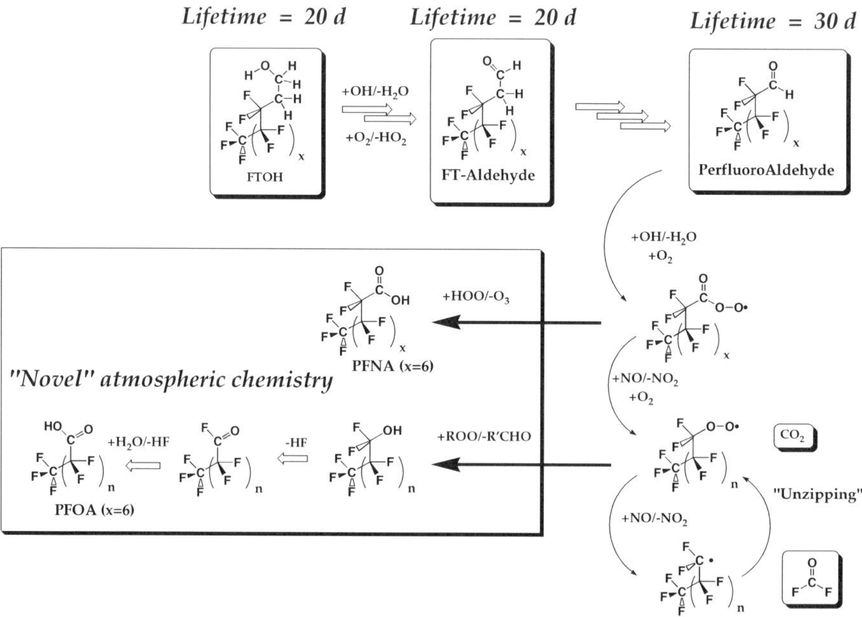


**Fig. 3** Overall scheme for the atmospheric processing of fluorotelomer alcohols (FTOHs) [26]

duction of PFCAs through either of these pathways requires air that is (relative to [ROO]) low in $NO_x$ since these PFCA pathways compete with the unzipping of the fluorochain through sequential loss of carbonyl fluoride. The balance of reactions is likely to be spatially important given the heavy influence urban areas have on $NO_x$ concentrations. Further modelling studies are required to assess the relative importance of whether this PFCA source will be significant for the input of PFCAs to the Great Lakes region. Measurement of PFCAs, PFOS, and related chemistries in precipitation in the region, would provide further support to the overall theory and the fluxes would provide a means to gauge the relative importance of these pathways.

Currently, evidence has been presented that supports the theory that FTOHs are the source of PFCAs to remote regions such as the Arctic [26, 34]. Critical to the veracity of this theory was the connection between the atmospheric mechanism showing production of the full suite of PFCAs, with the observation that Arctic biota showed a distinctive isomer pattern where the odd-chain length acid was higher in concentration than the next lowest PFCA [17]. For example, this could plausibly arise from roughly equivalent concentrations of PFOA and PFNA being delivered from the 8 : 2 FTOH to the environment [26] and then via food chain bioaccumulation where the larger PFCA would preferentially accumulate and result in higher concentrations. It is proposed that this theory will prove useful in elucidating the atmospheric fate of the polyfluorinated sulfamidethanols, though this research has yet to appear in the literature. An analogous set of reactions and pathways appears to be a reasonable supposition, in part, for the world-wide occurrence of PFOS. Of greater uncertainty is whether PFCAs, such as PFOA, will also be produced through OH mediated processing of these previously widely used long-chain perfluorinated surface active materials.

### 1.4.3
### Surface Water Processes

Although not expected to be found in surface waters in significant quantities due to very great Henry's Law constants [32], FTOHs were recently shown to yield PFCAs under indirect photolysis conditions [41]. FTOHs do not absorb actinic radiation but appear to react primarily with aqueous hydroxyl radicals to produce the fluorotelomer aldehyde, similarly to the corresponding pathway in the atmosphere. This aldehyde undergoes further oxidation to yield the fluorotelomer acid (FTCA) which can lose HF to yield the $\alpha, \beta$ unsaturated fluorotelomer acid (FTUCA), this latter pathway is unique to condensed phase systems. The unsaturated FTUCA provides the opportunity for OH addition reactions and is likely the critical step in the production of PFOA (starting with the 8 : 2 FTOH), which was the ultimate degradation product observed in irradiations of Lake Ontario water with an approximately 7% molar yield. Minor quantities of the nine carbon acid (PFNA) were observed and further experiments suggested they arose from OH reaction with the 8 : 2 FTCA.

### 1.4.4
### Hydrolysis

Under experimental conditions of $50\,^{\circ}$C and pH 1.5, 5, 7, 9, or 11, no hydrolytic loss of PFOS was observed in a 49-day study [42]. Based on mean values and precision measures, the half-life of PFOS was estimated as $\geq 41$ years at $25\,^{\circ}$C. However, it is important to note that this estimate was influenced by the analytical limit of quantitation and that no loss of PFOS was detected in the study. Likewise, under experimental conditions of $50\,^{\circ}$C and pH 1.5, 5, 7, 9, or 11, the hydrolytic loss of PFOA was observed for 109 days. Results showed that the degradation rate of PFOA was not dependent upon pH levels. The hydrolytic rate constant for PFOA at $50\,^{\circ}$C was determined to be 8.1E-5/d, and the minimum calculated half-life was estimated as 92 years [43].

### 1.4.5
### Biodegradation

Biodegradation studies where PFOS and PFOA were monitored analytically for loss of parent compound have been conducted using a variety of microbial sources and exposure regimes [20, 44–47]. In an 18-day biodegradation study with aerobic sludge, neither PFOS nor PFOA were measurably biodegraded [20]. In one study, no loss or biotransformation of PFOS was observed over a 20-week period with activated sludge under aerobic conditions nor were there any losses observed in a study conducted for a 56-day period with activated sludge under anaerobic conditions. The findings from these studies are supported by the results from a MITI-I test [48] that showed no biodegradation of PFOS after 28 days as measured by net oxygen demand, loss of total organic carbon, or loss of parent material.

The results of these studies differ from those observed by Schroder and co-workers that observed the disappearance of PFOS and PFOA from wastewater under anaerobic conditions [49, 50]. In these studies, either PFOS or PFOA were spiked into a bioreactor containing wastewater and aqueous PFOS concentrations were monitored by flow injection analysis-mass spectrometry (FIA-MS) and by LC-MS. Under aerobic conditions, no losses of PFOS or PFOA were observed in aqueous samples collected from the bioreactors. Under anaerobic conditions, PFOS concentrations decreased rapidly within 2 days to below the detection limit. PFOA also decreased but at a slower rate and was not detectable after 25 days. After 2 days, the PFOS concentrations in water decreased to less than the detection limit. However, no evidence of mineralization of either PFOS or PFOA was observed in that degradation products were not observed in the water samples nor were there any non-polar, volatile fluorinated compounds detected in the digester gases. In addition, there was no increase in aqueous fluoride ion concentrations that would be indicative of mineralization. While adsorption of either PFOA or PFOS to the glass walls of the reactor was observed, the authors did not account for potential losses to other component of the reactor system including suspended organic particles. As a result, PFOA and PFOS could have been adsorbed to organic material that was removed during the extraction and clean-up procedures. It is of interest that this phenomena only occurred under anaerobic conditions and indicates the need to better understand the fate of these chemical under differing environmental conditions.

Similar to atmospheric and photochemical processes, the bulk of metabolism research has focused on biological mechanisms and pathways for converting the polyfluorinated alcohols into the perfluorinated anions. Hagen, Belisle and colleagues, working in the early 1980s without the benefit of much of the advanced instrumentation utilized in the bulk of research described elsewhere in this chapter, published an important paper [51] that first reported the conversion of the 8 : 2 FTOH to PFOA in laboratory rats. The researchers suggested two of the likely intermediates in this pathway were the 8 : 2 FTCA and 8 : 2 FTUCA. The general outlines of this pathway were confirmed and extended in both a mixed microbial culture and activated sewage sludge [52]. Utilizing LC/MS/MS precluded the need for the derivatization required by the earlier researchers and provided MS/MS spectra that were confirmed to be identical to authentic standards. Further, consistent with the expectation that the alcohol would proceed through the aldehyde intermediate, the fluorotelomer aldehyde was observed in the headspace over the aerobic cultures. Additional experiments, dosing the cultures consecutively with FTCA and FTUCA, confirmed the pathway and specifically the progression from FTCA to FTUCA to PFOA. The conversion of the FTUCA to the PFOA was slow and reported to be thermodynamically unfavorable given the difficulty in oxidizing the $\beta$ carbon given its electrophilicity. The overall scheme suggested a process similar to $\beta$-oxidation, consistent with Hagen et al. [51].

A recent paper by Wang et al. [53] using $^{14}$C 8 : 2 FTOH and cultures derived from activated sludge also reported the identified acid products (e.g., FTCA, FTUCA, and PFOA), though not the expected aldehydes, and discovered a previously unobserved partially defluorinated product. This intermediate, given the putative molecular formula $C_{10}F_{15}H_4O_2$, was suggested to have the structure $CF_3(CF_2)_6CH_2CH_2COO^-$ from mass spectral evidence, and referred to as tetrahydropolyfluorinated decanoic acid (THPFCA); no standard is available but the evidence is persuasive. These researchers were skeptical for the relevance of the $\beta$-oxidation pathway and suggested the lack of hydrogen atoms in the reported intermediates (e.g., FTCA and FTUCA) obviates the intermediacy of the formal $\beta$-oxidation cycle and suggested the newly identified intermediate as a potential intermediate to the formation of PFOA. Even more recently Martin et al. [54], working with whole-rat and isolated-rat hepatocytes, observed all of the intermediates (starting with 8 : 2 FTOH) described above and discovered the $\alpha,\beta$-fluorotelomer aldehyde (FTUAL), which was likely formed through a loss of HF from the FTAL and was confirmed by mass spectral comparison to an authentic standard. Additional products were also suggested including the dihydro product of THPFCA with the formula $CF_3(CF_2)_6CH = CHCOO^-$, which would support the suggestion that the TH-PFCA could yield PFOA through a $\beta$-oxidation cycle [53]. Interestingly, Martin et al. also observed minor quantities of PFNA suggesting $\alpha$-oxidation is operable in at least some mammals; no evidence has been reported in any of the microbial studies for $\alpha$-oxidation. An overall metabolism scheme for the FTOHs, derived from these studies, is provided in Fig. 2a.

Similarly, the sulfamidethanols have been investigated for their metabolic conversion potential to the perfluorinated anion, PFOS. Xu et al. [55] used microsomes and liver slices from rats to investigate the biotransformation of EthylFOSE alcohol and specifically delineated the major metabolic pathways (Fig. 2a). Initial reaction involved $N$-dealkylation, either losing the ethyl or the ethanol moiety, to yield either $N$-EtFOSA or FOSE alcohol. Interestingly, $N$-EtFOSA was suggested to degrade quickly to PFOSA as it was not observed in incubations of the EthylFOSE alcohol. Overall the $N$-dealkylation appeared to be the limiting step in the overall production of PFOSA itself and is apparently mediated by a suite of P450 enzymes. Conversion of PFOSA to PFOS was only observed in rat liver slices, but not in microsomal or cytosolic fractions and occurred at a slow overall rate of biotransformation. Glucoronidation was shown to be an important secondary metabolic pathway where both $N$ and $O$ glucuronides were formed and was specifically suggested as a potential intermediate for the conversion of the sulfonamide PFOSA to PFOS. Further biotransformation reactions of PFOS have not been directly observed in this or any other published studies nor are they expected. An important side-reaction of the alcohols was oxidation to the corresponding carboxylates, which appear to be dead-end reactions. Much earlier work [56] showed biotransformation in rats of $N$-EtFOSA to PFOSA and PFOS while more recently these pathways have been confirmed in fish liver microsomes [57]. The general transformation pathways outlined in Fig. 2b may be general for a wide range of metabolic systems.

## 1.4.6
### Thermal Stability

Several studies suggest that PFOS would have relatively low thermal stability. This conclusion is based on the fact that carbon-sulfur $(C-S)$ bond energy is much weaker than carbon-carbon $(C-C)$ or carbon-fluorine $(C-F)$ bonds and as a result, would break first under incineration conditions [58]. This conclusion is supported by another study that indicated PFOS should be almost completely destroyed in an incineration system [59].

## 1.5
## Partitioning

### 1.5.1
### Adsorption/Desorption

Little information on environmental partitioning is available for most of the PFAs that have been measured in the environment. Thus, we will discuss the partitioning properties of PFOS and PFOA. PFOS -sorbs to soil, sediment, and sludge (Table 3) with an average distribution coefficient ($K_d$) greater than 10 L/kg and an organic carbon normalized adsorption coefficient ($K_{oc}$) greater than 700 L/kg [65]. However, it must be emphasized that normalizing PFOS concentrations to organic carbon may not be appropriate because PFOS does not readily partition into lipid or other forms of organic carbon such that using this approach may overestimate effect of organic carbon on the adsorption of this compound to soils and sediments. Based on these values, PFOS would not be considered qualitatively mobile as defined by OECD guidelines. Once adsorbed to these matrices, PFOS does not readily desorb even when extracted with an organic solvent. The average desorption coefficient ($K_{des}$) for soils was determined to be less than 10 L/kg. Adsorption and desorption equilibria in these matrices were achieved in less than 24 h with greater than 50% occurring after approximately 1 min of contact with the test adsorbents. As a result, PFOS exhibits little mobility in all test matrices and would not be expected to migrate any significant distance in

**Table 3** Adsorption and desorption of PFOS to soils, sediments and sludge[a]

| Soil type | Adsorption kinetics | | | Desorption kinetics | |
|---|---|---|---|---|---|
| | $K_d$ (L/kg) | $K_{oc}$ | $K_F^{\text{ads b}}$ (L/kg) | $K_{des}$ (L/kg) | $K_F^{\text{des b}}$ (L/kg) |
| Clay | 18.3 | 704 | 25.1 | 47.1 | 105 |
| Clay loam | 9.72 | 374 | 14.0 | 15.8 | 60.2 |
| Sandy loam | 35.3 | 1260 | 28.2 | 34.9 | 94 |
| River sediment | 7.42 | 571 | 8.7 | 10.0 | 44.6 |
| Domestic sludge | < 120 | NC[c] | 338 | < 237 | 3130 |

[a] Values of $K_d$, $K_{oc}$ and $K_{des}$ are averaged values
[b] Freundlich coefficient. The units for the isotherms assume that $n = 1$. More accurately the units are $[\mu g^{1-1/n} (L)^{1/n} kg^{-1}]$
[c] NC is not calculable

soils. The shape of the adsorption isotherm (H-type) indicates a very strong chemical/adsorption interaction. Since PFOS is a strong acid, it most likely forms strong bonds in soils, sediments and sludge via a chemisorption mechanism. While soil adsorption and desorption tests have not been conducted on PFOA, a study has been conducted with the ammonium salt of PFOA. The results of this study indicate that in Brill sandy loam soil, the ammonium salt of PFOA has high mobility with a $K_{oc}$ of 14 L/g and a $K_d$ of 0.21 L/g [66].

## 1.5.2
### Bioaccumulation

For many organic compounds that accumulate in lipids, bioaccumulation potential can be estimated from the $K_{ow}$. Since perfluoroalkyl compounds are inherently oleophobic, and do not partition preferentially to lipid rich tissues [67], the $K_{ow}$ is not a useful predictor of bioaccumulation and direct evidence derived from the laboratory or field is required. Additionally, tissue concentrations of perfluoroalkyl substances should never be lipid normalized. The dietary accumulation, bioconcentration, and biomagnification potential of various perfluoroalkyl carboxylates and sulfonates have been assessed in the laboratory and in the Great Lakes food web (Table 4).

The results of laboratory experiments [67, 68] have demonstrated that the bioaccumulation potential of PFAs increases with increasing perfluoroalkyl chain-length (Table 4). Bioconcentration factors (BCFs) range from < 1 to 23 000 for all PFAs but bioconcentration does not occur (i.e., BCF < 1) for perfluoroalkyl carboxylates shorter than PFOA, or for perfluoroalkyl sulfonates shorter than PFHxS. Above these size thresholds, log BCFs exceed zero and generally increase linearly as a function of increasing chain-length. Dietary accumulation factors follow a similar relative pattern, however none exceed 1.0. Given equal perfluoroalkyl chain-lengths, perfluoroalkyl sulfonates accumulate to a greater extent than carboxylates because of higher rates of uptake and slower rates of elimination. For bioconcentration, the relationship between log BCF and chain-length falls away from linearity between 6–14 carbons, presumably a result of diminished uptake across the gills due to size-limited diffusion. This deviation from linearity was not observed in dietary accumulation experiments.

Bioaccumulation occurs when the rate of uptake exceeds the rate of depuration (i.e., BAF > 1 if $k_{uptake} > k_{depuration}$). For PFAs, uptake from water across the gills and uptake from food across the gut are both very efficient. Aqueous uptake rate constants have been determined to be as great as 700 L/kg/d in trout, for perfluorododecanoate, corresponding to the highest rate of uptake of any known anionic surfactant [67] and representing a near maximal rate of branchial flux [69]. Assimilation efficiency from spiked fish food ranged from 59 to 130%, indicating efficient absorption of PFAs from ingested food. These assimilation efficiencies are greater than observed

**Table 4** Bioaccumulation parameters for PFAs in freshwater

| PFA | Organism | Laboratory exposure bioaccumulation parameters | | Field-based estimates of Bioaccumulation (organism) in Great Lakes | | |
| | | Dietary BAF | BCF (L/kg) | BAF | Lake Ontario pelagic trophic magnification factor[f] | Lake Ontario Lake trout BAF[e] |
|---|---|---|---|---|---|---|
| PFOS | Bluegill sunfish | | 1124[h] | | | 2.9 |
| | Rainbow trout | 0.32[a] | 1100[a] | | 5.9 | |
| | Leopard frog | 17.5–175[g] | | | | |
| | Carp | | 200–1500[h] | | | |
| | Various fishes | | | 160–7000[e] | | |
| | Shiner | | | 6300–125000[d] | | |
| | Bass and catfish | | | 830–26 000[k] | | |
| PFHxS | Rainbow trout | 0.14[a] | 9.6[b] | | | |
| PFOA | Rainbow trout | 0.038[a] | 4.0[b] | | 0.37 | 0.41 |
| | Fathead minnow | | 1.8[i] | | | |
| PFNA | Rainbow trout | 0.089[c] | 39[c] | | 1.0 | 2.3 |
| PFDA | Rainbow trout | 0.23[a] | 450[b] | | 3.7 | 2.7 |
| PFUnA | Rainbow trout | 0.28[a] | 2700[b] | | 4.7 | 3.4 |
| PFDoA | Rainbow trout | 0.43[a] | 18000[b] | | 1.0 | 1.6 |
| PFTrA | Rainbow trout | | | | 2.5 | 2.5 |
| PFTA | Rainbow trout | 1.0[a] | 23000[b] | | 1.0 | > 2.3 |
| NEtFOSE | Catfish and Bluegill sunfish | | ~ 400[j] | | | |

[a] Martin et al. [68]
[b] Martin et al. [67]
[c] PFNA data determined from linear regression equations [67, 78]
[d] BAF calculated from liver concentration compared to max and min water concentrations following a spill of AFFF in Etobicoke Creek (tributary to Lake Ontario)
[e] Calculated from whole-body concentrations of lake trout compared to diet composed of sculpin, alewife, and smelt [17]
[e] Field-based bioconcentration factor determined from average PFOS concentrations in fish muscle and mean water concentrations in the Great Lakes [25]
[f] Martin et al. [17]
[g] Ankley et al. [69]
[h] Apparent steady-state estimate OECD [8]
[i] Howell et al. [74]
[j] Elnabarawy [78]
[k] Giesy and Newsted [76]

for chlorinated contaminants such as polychlorinated biphenyls, where efficiencies in trout can range from 20 to 60% [70]. Rates of depuration were observed to be independent of exposure route and carcass depuration half-lives for juvenile rainbow trout (*Oncorhynchus mykiss*) ranged between 3 and 35 days for all PFAs examined in dietary and aqueous exposures [67, 68]. The reason that PFAs have relatively great BCF values, yet small dietary accumulation factors, is because the rate of uptake in a feeding experiment was limited by the rate of feeding. Under the experimental design used by Martin et al. [68] (e.g., 1.5% body weight/day), biomagnification cannot occur for compounds having a half-life of less than 46 days, even if their assimilation efficiency from food is equal to unity. The experimental feeding rate is representative of the actual feeding rate for this species in the Great Lakes.

Relatively great assimilation efficiencies, in addition to the measured tissue distribution of PFAs in fish, are supportive of the idea that enterohepatic recirculation may be an important factor limiting PFA depuration. The greatest concentrations of PFAs in tissues are measured in the blood > kidney > liver > gall bladder [68]. The tissue distribution of fish corresponds well with the observed tissue distribution in eagles (Table 16). Evidence of enterohepatic recirculation in rats has been demonstrated to affect the rate of elimination for PFAs [71].

There is largely good agreement between all PFOA and PFOS fish bioconcentration studies conducted in independent laboratories. For PFOA, Martin et al. [67] reported a BCF of 4.0 for juvenile rainbow trout, the Kurume Laboratory [72, 73] reported BCFs between 3.1 and 9.1 at two concentrations for carp, and Howell et al. [74] reported a BCF of 1.8. Similarly for PFOS, Martin et al. [67] reported a BCF of 1100 in rainbow trout, while the OECD [8] reported a BCF of 3614 for bluegill sunfish and 200–1500 for carp. This may be a fortuitous outcome because the kinetics of uptake were much slower in bluegill than in trout and the depuration rate was much slower in bluegill compared to trout (i.e., half life of 112 days versus 13 days, respectively). Reasons for the large difference in kinetics between species are not presently understood.

Reported field-derived bioaccumulation factors for PFAs have been highly variable and do not always agree with what is predicted from laboratory studies. Bioaccumulation factors calculated from liver and surface water PFOS concentrations ranged from 6300 to 125 000 in the common shiner (*Notropus cornutus*) collected from a tributary to Lake Ontario [75]; whereas the rainbow trout BCF based on liver concentrations was 5400 [67]. Thus the maximal BAF derived for shiners is 23-fold greater than expected based on laboratory study results. In a field study conducted in a reservoir in the Tennessee River near Decatur, Alabama, BCFs based on surface water PFOS concentrations and whole-body PFOS concentrations in catfish and largemouth bass ranged from 830 to 26 000 [76]. Thus, while there is relatively good agreement between the results derived from laboratory studies, BAFs estimated from field data vary greatly and in many cases exceeds values calculated in laboratory studies. Factors contributing to the large variation in field BAFs and BCFs include inter-species variation in accumulation from water and the importance of dietary accumulation may be greater than what has been predicted for juvenile trout in laboratory studies. In addition, temporal and spatial variability in surface water PFOS concentrations must be considered in evaluating field derived BAFs. For instance, in Moody et al. [75] BAFs were calculated from fish and surface water samples that were collected from Etobicoke Creek approximately 7 months after the spill had occurred. Inherent in these calculations was the assumption that the fish were at a steady state with water PFC concentrations. However, if the rates of dissipation and/or losses of PFCs were much greater than that for losses from fish, the use of these data would tend to overestimate the actual BAF(s). An additional consideration is that fish are also likely exposed to PFOS precursors that are more bioaccumulative than PFOS, and hence when metabolized to PFOS would result in an over prediction of the PFOS BAF. For example, Tomy et al. [57] showed that $N$-ethyl perfluorooctane sulfonamide ($N$-EtFOSA) could be metabolized to PFOS in trout. Similar suggestions were made in Moody et al. [75] where $^{19}$F NMR data indicated a large quantity of perfluorinated alkyl material in contaminated creek water suggesting a rationale for the large field BAF values for PFOS in Etobicoke Creek fish.

The US Interagency Testing Committee estimated BCFs for $N$-EtFOSE and $N$-MeFOSE using structure–activity models to be 5543 and 26000, respectively [77]. These values are likely in error because the rate of metabolism was not a consideration in this exercise. Although the data quality is suspect, the apparent bioconcentration factor for $N$-EtFOSE was approximately 400 in bluegill sunfish and channel catfish [78]. As far as we are aware, this is the only existing experimental bioaccumulation data for any PFA precursor.

Based on experimental dietary accumulation studies [68], significant biomagnification is not expected in the aquatic food web of the Great Lakes, however two recent reports suggest otherwise. Martin et al. [2] measured whole-body PFA concentrations in the benthic and pelagic food web at a location in Lake Ontario and related these to trophic level determined from stable isotopes of nitrogen. Species examined included invertebrates (*Diporeia* and *Mysis*), forage fish (sculpin, smelt, alewife), and a top predator, lake trout. Surprisingly, the lowest trophic level organism, *Diporeia*, had the highest concentrations of PFAs (Tables 14 and 15). This is unlike any other organohalogen contaminant examined in Lake Ontario and is strongly suggestive that the sediments, where *Diporeia* reside, are a source of PFAs to Great Lakes biota. Sculpin also had relatively great concentrations of most PFAs, presumably because their diet consists largely of *Diporeia*. In the mainly pelagic food web (i.e., omitting *Diporeia* and sculpin) most PFAs showed evidence of biomagnification. However, only PFOS, PFDA (C10), PFUnA (C11), and PFTrA (C12) had trophic magnification factors (TMFs) that were significantly greater than one (Table 4). PFOS had a TMF of 5.9, which was greater than all other PFAs, and is similar to the lipid-normalized TMF reported for PCBs in the same food web ($\Sigma$PCB TMF = 5.7). Interestingly, FOSA had a TMF that was significantly less than 1.0, indicating that is was most likely metabolized to PFOS as it moved up the food chain. This suggests that, to some extent, the TMF calculated for PFOS is an over prediction due to metabolism of precursors that may serve as additional sources to these organisms. PFOA also had a TMF significantly less than zero, indicating that it is easily eliminated by fish and is not biomagnified to any extent. It is unknown why all other PFAs showed evidence of biomagnification in lake trout (Table 4; BMFs ranged from 1.6 to 3.4), given that no PFA biomagnified in experimental work with juvenile rainbow trout. Two possibilities are hypothesized. First, fish in the Great Lakes may be exposed to PFA precursors that accumulate to a greater extent and are then metabolized to PFAs, or secondly, that larger fish have slower rates of depuration than the smaller fish used in experimental systems.

Biomagnification of PFOS, PFOA, PFHxS, and FOSA were assessed in the Great Lakes food web, including fish-eating birds [19]. Although the results are not entirely quantitative for long-chain carboxylates, (because feeding relationships were estimated, and whole bodies were not analyzed) some general trends were observed that agree with results presented in Martin et al. [68]. For example, although PFOA was detected in water samples, it could not be detected above the analytical LOQ in any aquatic organism, suggesting negligible biomagnification potential in the aquatic food web. Similarly, although FOSA was detected in low-trophic level organisms, such as round gobies and zebra mussels, it was not detected in high-trophic level fish, such as Chinook salmon. PFOS, however, showed clear evidence of biomagnification based on low concentrations that were measured in benthic algae, mussels, and amphipods ($< 2$–$4.3$ ng/g), greater concentrations in forage fish such as round gobies and small mouth bass ($2$–$41.3$ ng/g), and greater concentrations in chinook salmon ($100$ ng/g). The accumulation factors predicted based on measured concentration of PFOS observed in mink and birds, relative to their prey items were relatively great [19]. Mean mink and bald eagle liver concentrations were $18\,000$ ng/g, and $400$ ng/g, representing biomagnification factors on the order of 5–10. Additionally, PFOA, PFHxS, and FOSA were also detected in these predators, indicating that terrestrial animals have a greater accumulation potential for all PFAs than aquatic organisms.

## 2
## Identification and Quantification

Until recently, the identification and quantification of PFAs was limited largely by a lack of available instrumentation that was both sensitive and applicable to non-volatile anionic compounds [1, 18]. With the emergence of relatively inexpensive liquid chromatography instruments having electrospray ionization linked to mass spectrometry. This combination became the perfect tool with which to measure PFAs in complex environmental samples, and the first of such methods were published in 2001 [79]. Since that time, the number of scientific studies reporting on the environmental concentrations, behavior, or toxicology of PFAs has grown immensely, such that by the time this chapter was written it was difficult to keep the manuscript up-to-date with emerging reports. Historical measurements utilized cumbersome and non-specific methods, such as total organic fluorine, and thus were not suitable for environmental research. A historical perspective of PFA methods is provided briefly here, but is also available in more detail in Giesy and Kannan [1] and a more critical evaluation of modern methods is provided in Martin et al. [18].

Fluorine content of samples can be determined by nondestructive methods or by destruction of the organic matter by combustion or fusion. Nondestructive methods include neutron activation and X-ray fluorescence. These methods require specialized equipment that is not readily available. In addition, these techniques are not very sensitive and do not allow the identification or quantification of individual compounds. Destructive methods measure fluorine in organic compounds by converting organic fluorine to an inorganic fluoride via combustion. This process is termed mineralization. However, the carbon-fluorine bond is exceptionally strong and extremely vigorous conditions are required for quantitative mineralization. Mineralization techniques were used historically for the determination of total fluorine compounds in environmental and biological samples [80–83]. Total concentrations of organically bound fluorine in samples were determined by subtracting inorganic fluorine concentrations from total fluorine concentrations. Tests that measure methylene-blue-active substances have been used to detect the presence of anionic perfluorinated surfactants in environmental matrices, but this approach is non-specific [84].

Derivatization techniques coupled with gas chromatography followed by electron-capture detection (ECD) [49, 85] and mass spectrometric detection [5, 86] have been employed for the determination of perfluorinated surfactants. However, derivatization techniques in combination with gas chromatography/mass spectrometry have limited utility for the detection of perfluorinated surfactants in environmental samples. For example, PFOS is nonvolatile, and due to the excellent leaving group properties of the perfluoroalkanesulfonic group, its derivatives are unstable. Because most perfluorinated surfactants lack chromophores, they are not amenable to traditional absorbance or fluorescence techniques used in high-performance liquid chromatography (HPLC). Concentrations of perfluoroalkyl carboxylates in biological samples have been measured by HPLC with fluorescence detection after derivatization [87]. However, the multiple steps involved in this method, which include ion pairing and derivatization, limit its application to environmental samples.

Nuclear magnetic resonance ($^{19}$F NMR) has been used to determine concentrations of perfluorinated surfactants in biological samples. In the 1970s, NMR techniques were applied to analyze organic fluorinated compounds in human blood [49], but the techniques were not quantitative. Because of the poor sensitivity of NMR techniques, pre-concentration is generally required. The pre-concentration step not only concentrates the target compounds, but also potential interferences. This necessitates rigorous clean-up procedures. NMR techniques have been applied to measure selected fluorinated organic chemicals in contaminated water samples [75]. Recent developments of compound-specific methods for the analysis of PFAs using high-performance liquid chromatography-negative ion electrospray tandem mass spectrometry [79] have permitted the survey environmental distribution of fluorinated organic compounds in wildlife on a global scale [14–16, 88, 89]. Further developments in analytical methodology are needed to accommodate the range of PFAs in different types of biological and environmental matrices.

## 2.1
## Modern Instrumental Analyses

Current methodologies for the analysis of PFAs in water or tissue are predominantly based on high-performance liquid chromatography and negative electrospray ionization tandem (triple quadrupole) mass spectrometry, although some reports claim effective use of single-quadrupole instruments. In tandem mass spectrometry, identification is generally based on a combination of retention time and the pattern of product ions evolved from a given precursor ion (Table 5). The liquid chromatography system used for most of the analyses reported for biota in the Great Lakes was an HP Series 1100 chromatograph equipped with a BETASIL $C_{18}$ column ($2 \times 100$ mm, $5\,\mu$m particle diameter) or equivalent (Table 6). Ten microliters of extract is typically injected onto the reversed-phase column (e.g., Keystone BETASIL $C_{18}$) using a 2 mM ammonium acetate methanol/water mobile-phase gradient elution program. At a flow rate of between 200 and $300\,\mu$L/min, the gradient is increased to 100% methanol to elute PFAs.

A critical review of the inherent limitations of modern PFA analytical methods is available in Martin et al. [18], but some examples are discussed here. For quantitative determination, the HPLC system is often interfaced to a Micromass or Sciex tandem mass spectrometer operated in the negative ion electrospray mode. Instrumental parameters are optimized to transmit the [M – H] ion for all analytes (Table 5). When possible, multiple daughter ions are monitored, but quantitation is generally based on a single product ion (Table 5, Fig. 4). In the electrospray tandem mass spectrometry (ES MS/MS) system, the 499 Da $\rightarrow$ 80 Da transition can provide a stronger signal than

**Table 5** Example LC-ESI-MS conditions when using the Quattro Ultima (Micromass, Inc.) for analysis of selected PFAs

| Compound | Cone voltage (V) | Collision energy (eV) | Primary ion | Secondary ions | |
|---|---|---|---|---|---|
| PFOA | 35 | 10 | 412.6 | 368.6 | 168.8 |
| PFOS | 90 | 35 | 498.7 | 98.7 | 79.7 |
| PFHxS | 60 | 30 | 398.7 | 98.7 | 79.7 |

**Table 6** Example LC conditions when using the Agilent 1100 series Liquid Chromatograph

| | |
|---|---|
| Column | BETASIL $C_{18}$, $50 \times 2.1$ mm, 5 µm particle size |
| Pre column | XDB $C_8$, $2.1 \times 12.5$ mm, 5 µm |
| Mobile phase | MeOH/2 mM ammonium acetate |
| Flow rate | 0.3 mL/min |
| Injection volume | 10 µL |

the 499 Da $\rightarrow$ 99 Da transition of PFOS. In the analysis of tissue samples collected from some species of animals, an unidentified interferent is present for the 499 Da $\rightarrow$ 80 Da transition, thus making the widespread use of single-quadrupole instruments dubious. Although this interferent is rarely observed, to ensure complete selectivity quantitation should be based on the 499 Da $\rightarrow$ 99 Da transition. Due to the variety of matrices analyzed (with respect to both species and tissues), and due to evolving analytical methods, the limit of quantitation (LOQ) is often variable. Data quality assurance and quality control protocols should include matrix spike, surrogate spike, laboratory blank, continuing calibration verification, and standard addition quantification to verify that matrix ionization effects are minimal. The recent commercial availability of some stable isotopically labeled PFAs has greatly enhanced data quality through use as internal or surrogate standards. However, there is still a large demand for additional native and labeled standards (radiolabel and stable) [18].

Teflon or glass containers should be avoided in PFA analytical procedures; the former may cause analytical interferences, and the latter may bind the surfactants in an aqueous solution. The use of disposable polypropylene or plastic labware minimizes the possibility of sample contamination that can occur when glassware is reused. Any glassware used in the preparation of the reagents should be thoroughly rinsed with methanol prior to use. Recoveries of target analytes through the analytical procedure generally range from 60 to 130%, and are generally not corrected.

Currently, there are a number of limitations to instrumental analyses of PFAs, but some of the most severe are high levels of PFAs in blanks, lack

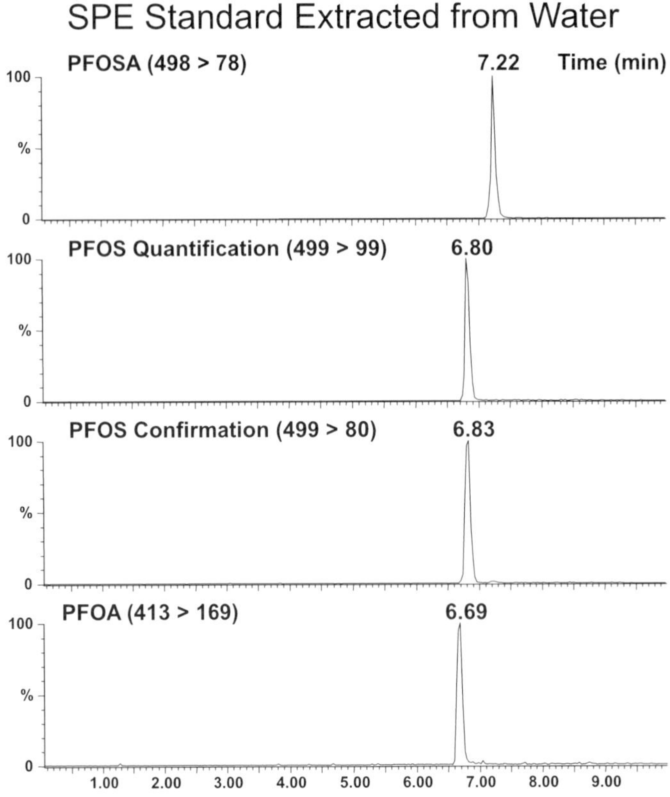


**Fig. 4** Chromatogram showing the retention times and abundances of selected transition ions for standards of PFOSA, PFOS and PFOS extracted from water by solid phase-extraction (SPE)

of standards, lack of isomer resolution in HPLC, and ionization suppression in electrospray [18]. The background comes from the fact that PFAs are in a number of laboratory appliances and result in contamination of not only samples, but laboratory reagents [90]. This contamination can come from everything from the septa used in autosampler vials to internal parts of the HPLC. We have found that to conduct analyses of PFAs with sufficiently low blanks to allow for reasonable LOQs, a number of internal parts on a standard LC/MS, must be replaced. The instrument in which the necessary changes are easiest is the 1100 series from Agilent Technologies. The necessary alterations are to substitute stainless steel or peek tubing instead of Teflon in all parts of the instrument including the inlet filter. In addition, the instrument should be operated without the degasser unit and inlet valves made of Teflon in the original configuration. Septa that contain Teflon or Viton contribute PFAs, including PFOA, PFOS, PFHxS and FOSA to samples. Polyethylene was found to be the best material for septa. Standards exist for relatively few perfluoroalkyl substances and these are often technical mixtures of varying isomer distributions.

## 2.2
## Water Analysis

Concentrations of PFAs in water can best be analyzed by use of solid-phase extraction followed by elution with methanol before analysis by high-performance liquid chromatography/electrospray negative ionization/tandem mass spectrometry [90–92, 103] (Figs. 4 and 5). Depending on the limits of quantification (LOQs) required, different volumes of water can be extracted. Volumes used have varied from as little as 50 mL in some rivers [76] to as much as 1 L of open ocean water [91]. Before extraction, a recovery spike is added to the water or a sub-sample of the water. Samples may need to be prescreened to determine an appropriate matrix spike level (typically 50–150% of sample concentration). Water is extracted by the use of Oasis HLB (hydrophobic-lipophilic-balanced) copolymer SPE cartridges (6 cc, 0.2 g, 30 μm particle size, part #: WAT106202, Waters, Inc.). The cartridges need to be cleaned and preconditioned before they can be used. This is accomplished by passing 100 mL methanol followed by 50 mL filtered type I water ($\sim$ 2 drop/s). It is important to not let the cartridges dry out during the conditioning. The sample is passed through the $C_{18}$ SPE cartridge and the eluate discarded (keep until the recovery has been calculated). The SPE cartridge is then washed with approximately 20 mL of 40% methanol in water and then the analytes of interest are eluted with approximately 30 mL of 100% methanol and the eluate collected. The target elution is then evaporated under nitrogen gas to 0.5 mL. The mass spectrometer used in these analyses was the Micromass Quattro Ultima equipped with MassLynx version 3.5 software. This general protocol results in good recoveries from both fresh [19] and marine waters [92, 93] (Table 7). The LOQ, based on a signal-to-noise ratio of 2× and a relative standard deviation of less than 30% was 15 ng/L

## SPE Methods For PFOS/PFOA in Water

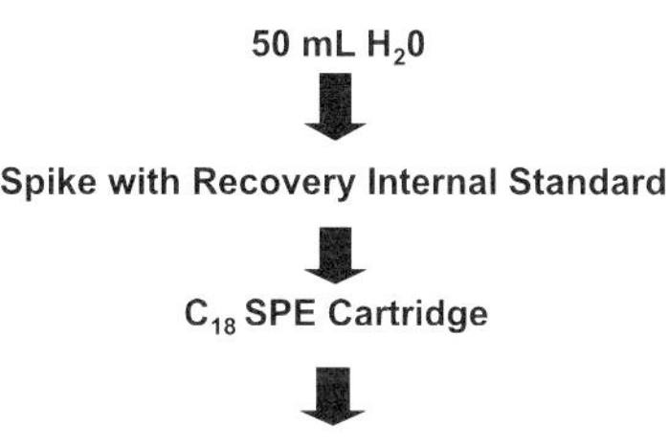


**Fig. 5** Simplified schematic of methods for extraction of PFAs from water

**Table 7** Mean % recoveries of PFOS and PFOA from water with solid-phase extraction

| PFOS | FOSA | PFOA |
| --- | --- | --- |
| (SD) [Range] | (SD) [Range] | (SD) [Range] |
| 103 (13.4) | 122 (31) | 125 (27) |
| [77–146] | [67–151] | [94–147] |

(ppt). While this method works well for PFOS and PFOA, the recoveries of the shorter-chain PFAs are poor, and it is suggested that these compounds be quantified by use of an Oasis mixed-mode Weak Anion-eXchange (WAX) and reversed-phase sorbent. The WAX sorbent retains and releases strong acids (e.g., sulfonates). These solid-phase adsorbents are an extension of the Oasis sorbent family with novel attributes for recovery of strongly acidic compounds that are otherwise difficult to recover. Recoveries should be in the range of 90% or greater.

## 2.3
## Biota Analysis

One of the first reliable methods for determining the concentrations of PFAs in tissues was the ion pairing method of Hansen et al. [79] (Fig. 6). Concentrations of PFOS in tissues, such as liver and blood plasma have been measured using high-performance liquid chromatography (HPLC) with electrospray mass spectrometry [1]. One half milliliter of serum, 5 µL of an internal standard (e.g., tetrahydroperflurooctane sulfonic acid), 1 mL of 0.5 M tetrabutyl

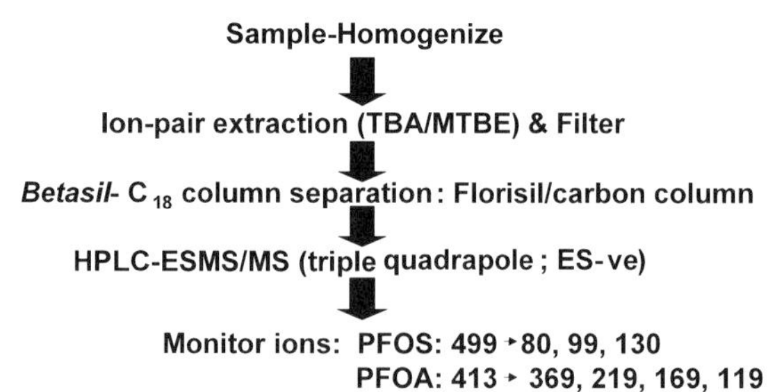



**Fig. 6** Simplified schematic of methods for ion-pairing (IP) methods of extraction of PFAs from tissue [103]

ammonium hydrogen sulfate (TBA) solution (adjusted to pH 10), and 2 mL of 0.25 M sodium carbonate buffer are added to a 15-mL polypropylene tube for extraction. After thorough mixing, 5 mL of methyl-*tert*-butyl ether (MTBE) is added to the solution, and the mixture is shaken for 20 min. The organic and aqueous layers are separated by centrifugation, and an exact volume of MTBE (4 mL) is then removed from the solution. The aqueous mixture is then rinsed with MTBE and separated twice; all rinses are then combined in a second polypropylene tube. The solvent is evaporated under a stream of nitrogen before being reconstituted to 0.5–1 mL with methanol. The extract is then vortex mixed for 30 s and passed through a 0.2 μm nylon mesh filter into an autosampler vial. For the extraction of liver samples, a liver homogenate of 1 g of liver to 5 mL of Milli-Q water is prepared. A 1.0 mL aliquant of the homogenate is added to a polypropylene tube, and the sample is extracted according to the procedure described above.

An alternative method utilizing solid-phase extraction (SPE) has more recently been developed to extract PFAs from biota. (Figs. 7 and 8). To compare the methodologies, several biological matrices from the Great Lakes were spiked with PFOS, PFOA, or PFHxS at concentrations 2- to 3-fold that in the original sample, and then extracted by either the ion-pairing method (IP) or the solid-phase extraction method (SPE) (Tables 8 and 9). The limits of quantification (LOQs) are less, and the recoveries are better with the SPE method than those obtained with the older ion-pairing method (Table 10). In addition, the SPE cleanup methods removed possible interferences that were a limitation with the earlier method, resulting in very clean extracts from which it is relatively easy to quantify PFAs.

Because PFAs can bind tightly to protein, a new method was developed by applying a combination of alkaline digestion and solid-phase extraction to improve the extraction of various short-chained and long-chained per-

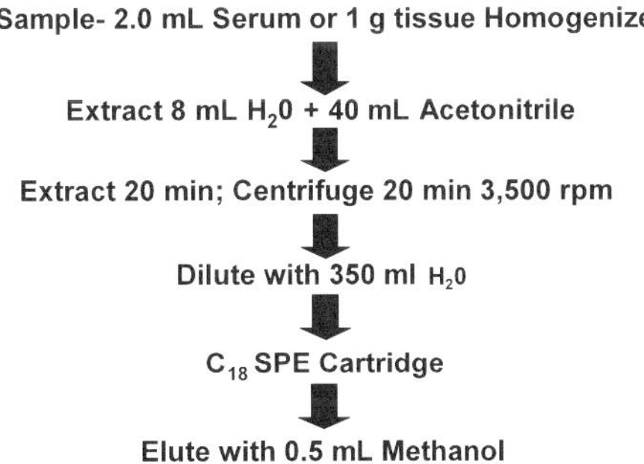



**Fig. 7** Simplified schematic of methods for solid-phase extraction (SPE) of PFAs from tissues

## Solid Phase Extraction of Biological Samples

### Tissue sample preparation
- Homogenize the sample with dry ice in a blender until the sample is homogenous.
- After homogenization, the sample is transferred to 50 ml polypropylene tube and then is loosely capped.

### Tissue sample extraction
- Aliquot 5g sample is transferred to a 50ml polypropylene centrifuge tube.
- Return the unused portions to the freezer after extraction amounts have been removed. The volume removed is recorded on the worksheet.
- Spike blanks, samples and standards, ready for extraction with surrogate standard. Spike each calibration standard matrix with appropriate amount of standard for the calibration curve standards and each QC sample.  Vortex the standard curve samples and QC samples for approximately 5 sec.
- To each sample and standard, add 30 ml of acetonitrile, cap and vortex or shake by hand by approximately 15 sec.
- Place all samples on the shaker at an appropriate speed (300 rpm) for 30 minutes to adequately mix.
- Remove from the shaker and centrifuge at 2500 rpm for 15 min to adequately pellet the precipitate.
- Remove samples from the centrifuge and decant the supernatant into a new 50ml tube, taking care not to introduce any of the matrix solids into the solution.
- Cap and mix by inverting several times.

### Clean-up procedures
- Transfer the supernatant through a florisil/carbon column. Florisil/carbon column is prepared by adding 3 g carbon (Supelclean Envi. carb, 120/400; from Supelco, 50 g bottles are available) and 3 g florisil (JT baker, 60-100 mesh).  Florisil is activated at 675 deg C prior to use. Use PP columns, not glass columns.
- Collect the ACN in a PP tube. Add another 15 ml of ACN to sample residue and shake, centrifuge and elute through the above florisil/carbon column. Combine the 2 ACN eluates. Then elute with 5 ml methanol through the column. Combine ACN and methanol (total 50 ml) and blow it down to dryness under a gentle stream of nitrogen. Reconstitute the solvent with 1ml methanol and filter it through the nylon filter, prior to instrumental analysis.
- The ACN extract is extracted by the use of Oasis® HLB (Hydrophobic-Lipophilic-Balanced) copolymer SPE cartridges (10g, 35mL; Sep Pak Vac 35cc, 10g, cartridges part #, Waters). The sample is passed through the $tC_{18}$ SPE cartridge and the eluate discarded (keep until the recovery has been calculated).
- The cartridges need to be cleaned and preconditioned before they can be used.  This is accomplished by passing 100mL methanol followed by 50mL filtered type I water (~2drop/sec).  It is important to not let the cartridges dry out during the conditioning.
- The SPE cartridge is then washed with approximately 20mL of 40% methanol in water and then the PFFAs of interest are eluted with approximately 30mL of 100% methanol and the eluate collected.  The target elution is then evaporated under nitrogen gas to 0.5mL.

**Fig. 8** Detailed methods for solid-phase extraction of PFAs from tissues

**Table 8** Mean recovery of perfluoroalkyl substances from Great Lakes tissues by two independent techniques

| Method | Mean recoveries (%) (SD) | | | |
| --- | --- | --- | --- | --- |
| | PFOS | FOSA | PFOA | PFHxS |
| SPE | 112.6 (28) | 28 (7.5) | 177 (181) | 102 (39) |
| Ion Pairing | 114.8 (56) | 98 (14) | 84 (37) | 2518 (5667) |

**Table 9** Limits of quantification for PFAs based on ion-pairing and solid-phase extraction methods

| Method | LOQ (ng/g, ww (ppb)) | | | |
| --- | --- | --- | --- | --- |
| | PFOS | FOSA | PFOA | PFHS |
| SPE | 0.2–2.0 | 0.2–2.0 | 0.2–5.0 | 0.2–5.0 |
| Ion Pairing | 19 | 4 | 7.4 | 7.5 |

**Table 10** Recoveries (%) and relative standard deviation (RSD) of multiple samples of tissues from the Great Lakes

| Matrix | PFOS | | PFOA | | PFHxS | |
| --- | --- | --- | --- | --- | --- | --- |
| | Recovery (%) (± RSD) | | Recovery (%) (± RSD) | | Recovery (%) (± RSD) | |
| | SPE | Ion Pairing | SPE | Ion Pairing | SPE | Ion Pairing |
| Amphipod | 125 (3) | < LOQ | < LOQ | < LOQ | 155 (12) | 774 (53) |
| Benthic algae | 68 (3) | < LOQ | 76 (43) | < LOQ | 99 (1) | 322 (35) |
| Crayfish muscle | 133 (35) | < LOQ | 432 (143) | < LOQ | 66 (4) | 304 (5) |
| L.M. Bass Muscle | 136 (26) | 53 | < LOQ | < LOQ | < LOQ | 175 (3) |
| Mink live | 154 (5) | 204 (2) | 22 (7) | 120 (3) | 53 (3) | 176 (0) |
| Pseudo feces | 110 (33) | < LOQ | < LOQ | < LOQ | 150 (10) | < LOQ |
| Round goby muscle | 76 (19) | 113 (3) | < LOQ | 47 (3) | 65 (18) | 184 (13) |
| S.M. Bass muscle | < LOQ | 89 (1) | < LOQ | < LOQ | < LOQ | 232 (8) |
| Zebra muscle | 99 (87) | < LOQ | < LOQ | < LOQ | 125 (22) | 655 (120) |

L.M. = Large Mouth Bass, S.M. = Small mouth bass

fluorosulfonates and PFAs from biological matrices [93]. These techniques produced adequate recoveries and reporting limits with small quantities of PFAs and small quantities of tissues.

# 3
# Environmental Monitoring

Recently, a monitoring study was conducted to evaluate the magnitude and extent of PFA distribution in the environment as well as bioaccumulation and biomagnification potential of several perfluoroalkyl substances in wildlife of the Great Lakes [14–16, 19, 88, 89, 94]. Among the compounds monitored, PFOS was commonly present in the tissues of wildlife, FOSA and PFOA were detected in the tissues of a few species, while PFHxS was rarely detected. While PFOS is a metabolic product of various sulfonated PFAs, other compounds such as, FOSA, PFOA, and PFHxS are intermediates in the production of other perfluorinated compounds. FOSA and PFOA are also products used in various applications.

## 3.1
## Concentrations in the North American Great Lakes

Relatively few measurements of PFAs have been made in abiotic and biotic compartments of the North American Great Lakes. Some results have been reported previously [1, 16, 19, 89, 94, 95] and here we report some additional information.

### 3.1.1
### Water

In the Great Lakes region, surface water samples were analyzed for PFAs in Lakes Ontario, Erie, and their tributaries, and various Michigan rivers (Table 11). Most reports are for PFOS and PFOA, whereas PFHxS and FOSA are often below method detection limits and no data exists for longer-chain perfluoroalkyl carboxylates. It can be generalized that the concentrations of PFOA slightly exceed the concentrations of PFOS at all background sites.

In the largest water sampling campaign, PFOS was detected in 89% of Michigan water samples [96]. Background concentrations were found to be between 2 and 5 ng/L for PFOS and between < 8 and 16 ng/L for PFOA. The maximum concentrations measured were 29 ng PFOS/L and 36 ng PFOA/L, both in the waters of southwestern Michigan where there are several paper mills that may be sources of these higher concentrations. These concentrations of PFOA and PFOS are similar to those reported for the Raisin and St. Clair Rivers [19], and for the control Etobicoke Creek site [75], upstream of an AFFF spill that resulted in PFOS and PFOA concentrations as high as 2.2 mg/L and 11 ug/L, respectively. PFOS could not be detected at any sites unaffected by the spill.

These findings, however, are in contrast to the PFOS and PFOA data presented in Boulanger et al. [97] for the open waters of Lakes Ontario and Erie

**Table 11** Mean concentrations or range of perfluorinated compounds (ng/L) in water from the Great Lakes and their tributaries (standard error is in parentheses)

| Sample | Location | Notes | N | PFOS [b] | PFHxS | FOSA | PFOA | NEtFOSAA | PFOS-sulfinate | FOSAA |
|---|---|---|---|---|---|---|---|---|---|---|
| Water[a] | Saginaw Bay and Michigan rivers | | 44 (9 sites) | 1.8–16.1 | | | < 8 -- 21.6 | | | |
| Water[b] | Lake Ontario | | 8 | 54.3 | | 1.0 | 42.5 | 5.7 | 0.6 | 0.9 |
| Water[b] | Lake Erie | | 8 | 31.2 | | 0.9 | 35.6 | 7.5 | 5.0 | 0.1 |
| Water[c] | Etobicoke Creek | AFFF spil. site | | 160-2.2E+6 | n.d.-134 | | n.d.-11E+3 | | | |
| Water[c] | Etobicoke Creek | Control site | | n.d. | n.d. | | 8–33 | | | |
| Water[d] | Raisin River | Tributary to L. Erie | | 3.5 | < 1 | < 10 | 14.7 | | | |
| Water[d] | St. Clair River | | | 2.6 | < 1 | < 10 | 4.4 | | | |

[a] Sinclair et al. [96]
[b] Boulanger et al. [97]
[c] Moody et al. [75]
[d] Kannan et al. [19]

(Table 11). Here, mean concentrations are generally greater than the maximum concentrations observed in the aforementioned tributaries. However, this study utilized single quadrapole MS for quantification and thus may have over predicted PFOS concentrations in water. More data are required for comparison; however, this is suggestive that PFA precursors may degrade in the larger lakes, releasing an additional burden of PFOS and PFOA to the Great Lakes in addition to direct PFA inputs from tributaries. This is supported by the detection of various PFOS precursors in water from the Great Lakes [97].

### 3.1.2
### Biota

Concentrations of PFAs have been determined in a range of biota from the North American Great Lakes (Tables 12–16 and Fig. 9). Concentrations of PFAs in the biota of the Great Lakes were similar to those from other urbanized and industrialized areas of North America and Europe [16, 94, 98–100]. However, they are less than those in wildlife in more remote locations such as the open ocean or Arctic and Antarctic [17, 88, 89, 95, 101].

### 3.1.3
### Air

Airborne concentrations of perfluoroalkyl substances are limited largely to neutral chemicals containing a perfluoroalkyl moiety, however, one study reported detection of PFOS in 4 out of 8 samples of airborne particulates collected above Lake Ontario (mean $= 6.4\,\mathrm{pg/m^3}$) [25]. While this finding has yet to be replicated by other investigators, it does suggest that PFOS may be

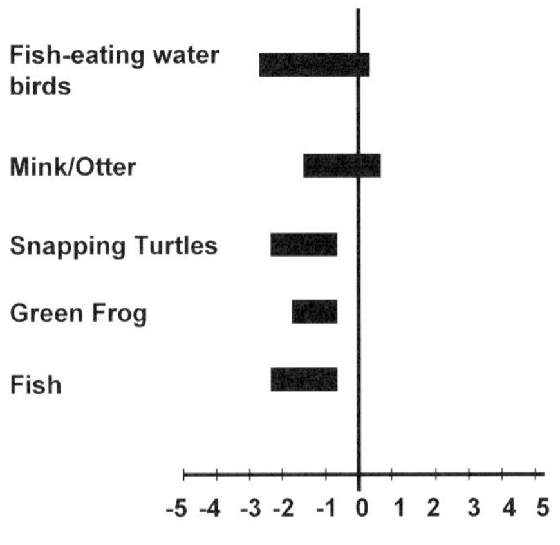


**Fig. 9** Ranges of PFOS concentrations (mg/kg, ww) in Laurentian Great Lakes biota

**Table 12** Mean concentrations of perfluorinated compounds ($\mu$g/kg or $\mu$g/L) in reptiles and amphibians from the Great Lakes[a]

| Species | Tissue | N | Location | PFOS[b] | PFHxS | FOSA | PFOA |
|---|---|---|---|---|---|---|---|
| Snapping turtle | Blood | 5 | Lake St. Clair MI, USA | 73.7 (31.8) | 1.3 (0.00) | 4.14 (2.84) | 3.1 (0.00) |
| Green frog | Liver | 4 | Southwest MI, USA | 97.3 (62.6) | 6.8 (0.00) | 18.8 (0.00) | 71.9 (0.00) |
| Green frog | Whole | 5 | Swans Creek and Calkins Dam MI, USA | 18.2 (0.84) | 34.2 (0.00) | 18.8 (0.00) | 18.0 (0.00) |

[a] Data taken from [16] and unpublished results [102]
[b] Standard error in parentheses

**Table 13** Mean concentrations of perfluorinated compounds ($\mu$g/kg or $\mu$g/L) in mink and otter from the Great Lakes[a]

| Species | Tissue | N | Location | PFOS[b] | PFHxS | FOSA | PFOA |
|---|---|---|---|---|---|---|---|
| Mink | Liver | 3 | Lab-Fed Saginaw Bay Fish, MI, USA | 3.38E+03 (1.43E+03) | 1.10 (0.10) | 37.5 (0.00) | 35.9 (0.00) |
| River Otter | Blood | 1 | Lake Superior, Great Lakes, USA | 39.0 (0.00) | 1.00 (0.00) | 112 (0.00) | 35.9 (0.00) |

[a] Data taken from [95] and unpublished results [102]
[b] Standard error in parentheses

a degradation product of some airborne precursor molecule, or perhaps that PFOS can be mobilized from terrestrial settings on wind-blown particles (e.g., soil). There are only two reports of airborne FTOH concentrations [21, 23] (Table 17) and there is a need for more data from independent laboratories. Only three independent laboratories have reported concentrations of airborne neutral perfluorooctylsulfonamides, and the methods utilized by each for sample collection, extraction, and analyses have been variable. Martin et al. [21] and Stock et al. [23] utilized a system containing a quartz-fiber filter for collection of particulates, upstream of a sandwiched combination of polyurethane foam and XAD-2 resin utilized for collection of the gas-phase compounds. This method has a collection efficiency for 87–120% of gas-phase analytes as determined in breakthrough tests. In contrast, Boulanger et al. [25] used a glass-fiber filter in combination with XAD-2 (i.e., no polyurethane foam), while Shoeib et al. [24] used glass-fiber filters in combination with polyurethane foam (i.e., no XAD-2). Shoeib et al. demonstrated that between 0 and $\sim$ 33% of MeFOSE could breakthrough in their system by using a backup

**Table 14** Mean concentrations of perfluorinated compounds ($\mu$g/kg or $\mu$g/L) in fish from the Great Lakes (standard error in parentheses)

| Species | Tissue | N | Location | PFOS[b] | PFHxS | FOSA | PFOA | $\Sigma$PFCAs[c] |
|---|---|---|---|---|---|---|---|---|
| Chinook salmon[a] | Liver | 6 | Great Lakes, MI, USA | 108 (24.9) | < 17.1 (0.00) | < 71.9 (0.00) | < 18.8 (0.00) | |
| Lake[a] whitefish | Liver | 5 | Great Lakes, MI, USA | 66.8 (8.73) | < 17.1 (0.00) | < 71.9 (0.00) | < 18.8 (0.00) | |
| Brown trout[a] | Liver | 10 | Great Lakes, MI, USA | 18.3 (0.86) | < 17.1 (0.00) | < 71.9 (0.00) | < 18.8 (0.00) | |
| Lake[a] whitefish | Egg | 2 | Great Lakes, MI, USA | 263 (118) | < 34.2 (0.00) | < 18.0 (0.00) | < 18.8 (0.00) | |
| Brown trout[a] | Egg | 3 | Great Lakes, MI, USA | 64.0 (7.77) | < 34.2 (0.00) | < 18.0 (0.00) | < 18.8 (0.00) | |
| Chinook[a] salmon | Muscle | 6 | Great Lakes, MI, USA | 90.5 (26.5) | < 34.2 (0.00) | < 35.9 (0.00) | < 18.8 (0.00) | |
| Lake[a] whitefish | Muscle | 5 | Great Lakes, MI, USA | 132 (15.6) | < 34.2 (0.00) | < 35.9 (0.00) | < 18.8 (0.00) | |
| Brown trout[a] | Muscle | 10 | Great Lakes, MI, USA | 10.9 (3.90) | < 34.2 (0.00) | < 35.9 (0.00) | < 18.8 (0.00) | |
| Carp[a] | Muscle | 10 | Saginaw Bay, MI, USA | 124 (25.1) | < 34.2 (0.00) | < 35.9 (0.00) | < 18.8 (0.00) | |
| Carp[a] | Whole | 4 | Saginaw Bay, MI, USA | 22.5 (2.92) | < 34.2 (0.00) | < 35.9 (0.00) | < 18.8 (0.00) | |
| Largemouth[b] Bass | Muscle | 2 | Saginaw River Raisin River | 26.3 (22.9) | < 1.00 (0.00) | 3.82 (2.82) | < 2.00 (0.00) | |
| Round goby[b] | Muscle | 14 | St. Clair River Raisin River Calumet River Lake Superior | 15.8 (5.16) | < 2.00 (0.00) | 2.56 (0.59) | < 0.20 (0.00) | |
| Smallmouth Bass[b] | Muscle | 14 | St. Clair River Raisin River Calumet River | 9.35 (2.87) | < 1.00 (0.00) | 2.01 (0.41) | < 2.00 (0.00) | |
| Lake trout[c] | Whole | 7 | Lake Ontario (2001) | 170 (64) | | 16 (9.6) | 1.0 (0.1) | 30 |
| Alewife[c] | Whole | 12 | Lake Ontario | 46 (15) | | 4.0 (3.2) | 2.0 (1.1) | 9.2 |
| Smelt[c] | Whole | 30 | Lake Ontario | 110 (55) | | 72 (23) | 2.0 (1.1) | 30 |
| Slimy sculpin[c] | Whole | 15 | Lake Ontario | 450 (98) | | 150 (17) | 44 (12) | 180 |

[a] Data taken from [16] and from unpublished results [102]

[b] Data taken from 3M Report entitled "Quantitative Analysis of Fluorochemicals in Environmental Samples Obtained from Michigan State University (MSU)" Dated April 15, 2003. 3M Laboratory LIMS No. E01-1419

[c] $\Sigma$PFCAs is the sum of perfluoroalkyl carboxylates (C8 – C15) [2]

**Table 15** Mean concentrations of perfluorinated compounds ($\mu$g/kg) in invertebrates and benthic algae from the Laurentian Great Lakes (standard error in parentheses)

| Species | Tissue | N | Location | PFOS | PFHxS | FOSA | PFOA | ΣPFCAs[c] |
|---|---|---|---|---|---|---|---|---|
| Amphipod[a] | — | 3 | St. Clair River<br>Raisin River<br>Calumet Harbor | 2.29<br>(0.29) | 1.00<br>(0.00) | 2.00<br>(0.00) | 5.00<br>(4.75E-8) | |
| Benthic algae[a] | — | 3 | St. Clair River<br>Raisin River<br>Calumet Harbor | 2.71<br>(0.20) | 2.00<br>(0.00) | 1.00<br>(0.00) | 0.20<br>(0.00) | |
| Pseudofeces[a] | — | 3 | St. Clair River<br>Raisin River<br>Calumet Harbor | 2.00<br>(0.00) | 1.00<br>(0.0) | 2.00<br>(0.00) | 5.00<br>(4.75E-8) | |
| Crayfish[a] | Muscle | 3 | St. Clair River<br>Raisin River | 3.47<br>(0.56) | 2.00<br>(0.00) | 1.34<br>(0.18) | 0.20<br>(0.00) | |
| Zebra Mussel[a] | Muscle | 7 | St. Clair River<br>Raisin River<br>Calumet River | 2.16<br>(0.16) | 1.00<br>(0.00) | 4.41<br>(2.01) | 5.00<br>(0.00) | |
| Mysis[b] | Whole | 3 | Lake Ontario | 13<br>(1.8) | | 130<br>(4.2) | 2.5<br>(0.4) | 12 |
| Diporeia[b] | Whole | 3 | Lake Ontario | 280<br>(33) | | 180<br>(26) | 90<br>(8.5) | 260 |

[a] Kannan et al. [19]
[b] Martin et al. [2]
[c] ΣPFCAs is the sum of perfluoroalkyl carboxylates (C8 – C15)

foam plug, while no collection efficiency data was provided in Boulanger et al. [25]. Data are summarized in Table 17, and in general the concentrations of perfluorooctylsulfonamides reported in Martin et al. [21] and Stock et al. [23] exceed concentrations reported by Shoeib, which in turn exceed concentrations reported by Boulanger et al. [25]. This may be an accurate reflection of the concentrations at the different sites and in different years, however, more data and better quality control are required among laboratories to ensure that these trends are not artifacts of different sampling strategies. A general conclusion that can be made for FTOHs and perfluorooctylsulfonamides is that concentrations are highest in air around populated regions and seem to decrease at semi-rural sites and over-water in the Great Lakes.

# 4
# Hazard Identification

Perfluoroalkyl substances were historically considered to be metabolically inert and non-toxic [104]. Furthermore, since these compounds were often

**Table 16** Mean concentrations of perfluorinated compounds (µg/kg or µg/L) in birds from the Great Lakes[a]

| Species | Tissue | N | Location | PFOS[b] | FOSA | PFOA | PFHxS |
|---|---|---|---|---|---|---|---|
| Bald eagle | Blood | 26 | MN, WI and MI, USA | 343 (126) | 12.5 (4.38) | 29.9 (0.00) | 1.35 (0.19) |
| Cormorant | Blood | 13 | Lake Huron and Lake Superior | 169 (33.8) | 16.2 (4.53) | 32.7 (1.77) | 1.14 (0.00) |
| Herring gull | Blood | 4 | Little Charity Is., Lake Huron | 219 (91.7) | 6.26 (0.00) | 29.9 (0.00) | 1.46 (0.32) |
| Great horned owl | Egg | 3 | MI, USA | | < 18.8 (0.00) | < 18.8 (0.00) | < 18.8 (0.00) |
| Tree swallow | Egg | 5 | MI, USA | 67.6 (8.70) | < 18.8 (0.00) | < 18.8 (0.00) | < 18.8 (0.00) |
| Cormorant | Egg | 6 | MI, USA | 1.26E+03 (188) | < 18.8 (0.00) | < 18.8 (0.00) | < 18.8 (0.00) |
| Herring gull | Egg | 2 | MI, USA | 220 (135) | < 18.8 (0.00) | < 18.8 (0.00) | < 18.8 (0.00) |
| Caspian tern | Egg | 5 | MI, USA | 2.21E+03 (468) | < 18.8 (0.00) | < 18.8 (0.00) | < 18.8 (0.00) |
| Ring billed Gull | Egg | 3 | Sulfur Is., Great Lakes | 7.83 (34.3) | < 75.0 (0.00) | 191 (5.51) | 6.83 (0.00) |
| Bald eagle | Liver | 6 | MI, USA | 394 (275) | 62.5 (7.91) | 21.9 (3.10) | 13.8 (3.17) |
| Bald eagle | Kidney | 4 | MI, USA | 508 (673) | < 75.0 (0.00) | < 3.74 (0.00) | < 18.8 (0.00) |
| Bald eagle | Gall Bladder | 1 | MI, USA | 1490 | < 75.0 (0.00) | < 3.74 (0.00) | < 18.8 (0.00) |
| Bald eagle | Muscle | 6 | MI, USA | 35.3 (16.8) | < 75.0 (0.00) | < 3.74 (0.00) | < 18.8 (0.00) |
| Bald eagle | Testes | 1 | MI, USA | 183 | < 75.0 (0.00) | < 3.74 (0.00) | < 18.8 (0.00) |
| Bald eagle | Ovary | 1 | MI, USA | 68 | < 75.0 (0.00) | < 3.74 (0.00) | < 18.8 (0.00) |

[a] Data taken from [88]; Kannan et al. [89], Kannan et al. [19]; and from unpublished results (Final Report to 3M entitled "Perfluorooctane sulfonate and Related Fluorochemicals in Fish-eating Water Birds"). By Giesy and Kannan (1999)
[b] Standard error in parentheses

inserted into polymeric materials of very great molecular weight, they were not considered to be mobile in the environment or bioavailable. However, PFAs have been found to be both mobile in the environment and biologically active [1]. PFAs can bind to structural and functional proteins [13] and cause changes in gene expression [105, 106] and changes in biological membranes [107, 108]. In biota, PFAs can cause peroxisomal proliferation, increased activity of lipid and xenobiotic metabolizing enzymes, and alter-

**Table 17** Mean concentrations or range (pg/m$^3$) of neutral perfluoroalkyl substances in North American air samples

| Site | Site Description | 6:2 FTOH | 8:2 FTOH | 10:2 FTOH | NEt -FOSE | NMe -FOSE | NEt -FOSA | Refs. |
|---|---|---|---|---|---|---|---|---|
| Toronto, ON | Urban | 87 | 55 | 29 | 205 | 101 | 14 | [21] |
| | | 80 | 83 | | 84 | 35 | | [23][a] |
| Long-Point, ON | Semi-rural (north shore of L. Erie) | 29 | 32 | 17 | 76 | 35 | n.d. | [21] |
| | | 25 | n.d. | n.d. | 13 | 28 | 10 | [23][a] |
| Griffin, GA | | 20 | 130 | < 10 | 50 | 359 | 10 | [23][a] |
| Reno, NV | | 40 | 37 | < 10 | 199 | 30 | 65 | [23][a] |
| Winnipeg, MB | | n.d. | 10 | n.d. | n.d. | 20 | n.d. | [23][a] |
| Cleves, OH | | 65 | 65 | n.d. | n.d. | 20 | 45 | [23][a] |
| Lake Ontario | Over-water on boat | | | | n.d. – 0.6 | | n.d. – 1.5 | [25] |
| Lake Erie | Over-water on boat | | | | n.d. – 1.0 | | n.d. – 2.2 | [25] |
| Semi-Urban | Unkown location | | | | 9.8, 8.5 | 32, 16 | | [24] |

[a] Values estimated from vertical bar graphs

ations in other important biochemical processes in exposed organisms [3, 109–114]. In wildlife, PFAs accumulate primarily in the blood and in liver tissue [14]. Therefore, while the major target organ for PFAs is presumed to be the liver, other organs such as the pancreas, testis, and kidney could also be affected [115]. PFOS appears to be the ultimate degradation product of a number of commercially used perfluoroalkyl sulfonamide compounds and the concentrations of PFOS found in wildlife are greater than other perfluorinated compounds [1, 88, 89]. In addition, PFOS is the PFA for which the greatest amount of toxicological information is currently available where acute and chronic studies have been conducted in rats, mice, rabbits, and monkeys [111, 116–122].

A large body of ecotoxicological information generated over a period of more than 20 years, exists for various salts of PFOS [8, 64]. However, until recently, definitive information was not available on chemical purity and validated analytical methodology did not exist to measure exposure concentrations in many of the early studies. Therefore, data generated prior to 1998 is somewhat unreliable relative to actual substance(s) used in the test, as well as the fact that exposure concentrations were not measured as part of these studies. Based on these data, a PFOS concentration in water thought to be protective of aquatic organisms including fish and aquatic invertebrates of 1.2 μg PFOS/L has been proposed [123, 128] (Fig. 10). This protective value was based on both acute and chronic toxicity tests and using US EPA methodologies

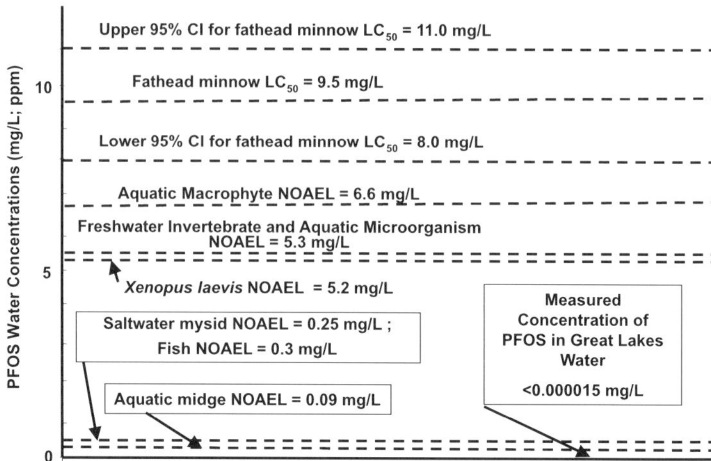



**Fig. 10** Concentrations of PFOS in Laurentian Great Lakes waters, compared to values for the protection of aquatic organisms [123, 128]

developed for the Great Lakes under the Great Lakes Initiative (GLI) [124]. The GLI document provides specific procedures and methodology for utilizing environmental fate and toxicity data to derive water quality values that are protective of aquatic organisms, upper trophic level wildlife, and humans in the Great Lakes region. While this approach was considered appropriate for the Great Lakes, it was also selected because it is one of the most comprehensive assessment strategies currently being used by regulatory agencies.

PFOS concentrations in tissues thought to be protective of aquatic organisms have also been derived based on laboratory results. The critical body residue (CBR) hypothesis provides a framework for analyzing aquatic toxicity data in terms of its mode of action and the accumulation of a chemical into tissues [125, 126]. The key assumption of the hypothesis is that adverse effects are elicited when the molar concentration of a chemical in an organism's tissues exceeds a critical threshold. Implicit in this hypothesis is the assumption that a chemical is accumulated in tissues via a partitioning process and it has reached a steady state within the test period. Thus, the CBR is a time-independent measure of effect for organisms exposed to the chemical. A critical body residue level for PFOS in fish was estimated from a bluegill bioconcentration study where significant mortality occurred at the greatest dose [127]. In this study, bluegill were exposed to 0.1 and 1.0 mg PFOS/L for up to 62 days followed by a depuration period. However, at 1.0 ppm, mortality was noted by day 12 with 100% mortality being observed by day 35. Using the lower 95% confidence of the NOAEL based on mortality noted in this study, the critical body residue PFOS concentration was determined to be 87 mg PFOS/kg, wet weight. Based on this analysis, concentrations in aquatic organisms in the Great Lakes would not be expected to cause adverse effects (Fig. 11). A complete literature review of the studies on which the calculations were based as well as a complete description of the assumptions made and calculations can be found in Beach et al. [123]. Based on these values, concentrations of PFOS in waters of the Great Lakes do not exceed concentrations that would cause adverse effects.

Less information is available on the toxicity of PFOS to fish-eating organisms, but recently studies were conducted with model bird species the mallard and the bobwhite quail [129] (Newsted JL, Coady KK, Beach SA, Gallagher S, Giesy JP (2005), personal communication). Adult mallards and bobwhite quail were fed dietary concentrations of 0, 10, 50, or 150 mg/kg PFOS for up to 21 weeks. Adult health, body and liver weight, feed consumption, gross morphology and histology of organs, as well as egg production, hatchability, embryo viability, and offspring health were examined. Concentrations of PFOS were measured in the blood, liver, and eggs of both bird species. Several weeks into the study, bobwhite quail and mallards at the 50 and 150 mg PFOS/kg exposure levels experienced a host of adverse effects, including high rates of mortality. Exposure to 10 mg PFOS/kg did not result in overt signs of toxicity in either the bobwhite quail or the mallard. The lowest observable adverse effect level (LOAEL) for northern bobwhite quail was determined to be 10 mg PFOS/kg based on greater liver weights in female quail, smaller testes size in male quail, and decreased survivorship in 14-d-old quail offspring. A no observable adverse effect level (NOAEL) of 10 mg PFOS/kg in the diet was determined for female mallards, while a LOAEL of 10 mg PFOS/kg was determined for male mallards based on decreased testis size. Concentrations of PFOS measured in the liver and blood were greater in male birds compared

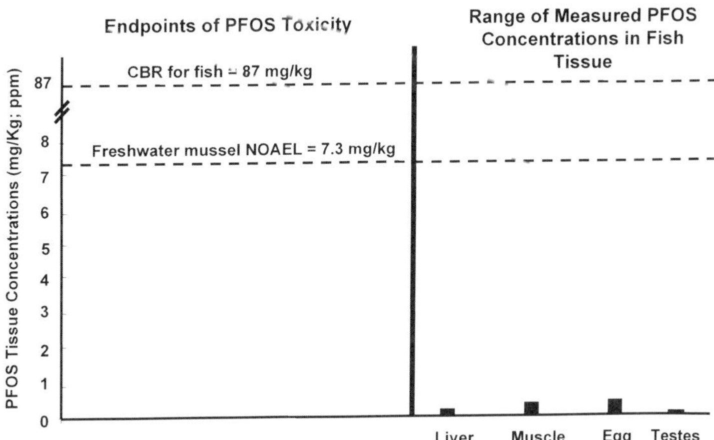


**Fig. 11** Concentrations of PFOS in tissues of fish from the Laurentian Great Lakes, compared to tissue concentrations for the protection of fish [123]

to female birds, and PFOS in the eggs was mostly concentrated in the very low density lipoprotein fraction of the egg yolk. Based on this information, it is not possible to derive a very accurate threshold for effects, based on tissue concentrations that could be compared to the concentrations of PFOS measured in the tissues of birds of the a Great Lakes. Toxic reference values (TRV) were estimated utilizing USEPA GLI and European Commission methods for assigning uncertainty factors (Table 18) [130]. Based on the uncertainty factors assigned by use of the US EPA GLI methodology, the dietary TRV was 0.021 mg PFOS/kg bw/d while the TRV based for eggs was 1.7 µg PFOS/ml yolk. TRVs for serum and liver for the protection of avian species were 1.0 µg PFOS/ml and 0.6 µg PFOS/g ww, respectively. However, these values were based on a subtle change in body weight gain of chicks and rates of normal testicular regression in adult males, which seemed to be accelerated by exposure to PFOS. It is unlikely that ecologically relevant, population-level effects on survival, growth or reproduction would be expected to occur at these doses. A likely threshold for the onset of significant population level effects would be closer to 6.0 mg PFOS/kg in the diet. At this point, the only conclusion that can be drawn from the available information is that the current concentrations of PFOS in the tissues of birds are within a factor 10 of the threshold for likely ecologically relevant effects (Fig. 12). It is not possible, without additional information to determine whether current concentrations of PFOS are causing any ecologically relevant effects on birds, but that the nearness to the likely threshold is reason for considerable concern.

The production and use of PFOS has been greatly curtailed since 2000. Thus, if concentrations of PFOS in the environment decline, the hazard posed to Great Lakes biota should also decrease and not reach critical concentrations. However, there are some uncertainties in this simple hazard assess-

**Table 18** PFOS TRV values for a generic Trophic Level IV predator based on dietary, liver, and serum toxic doses[a]

|  | Male | | | Female | | |
|---|---|---|---|---|---|---|
|  | LOAEL | TRV[b] | PNEC[c] | LOAEL | TRV[b] | PNEC[c] |
| ADI (mg PFOS/kg/d)[d] | 0.77 | 0.021 | 0.013 | 0.77 | 0.021 | 0.013 |
| Liver (µg PFOS/g, ww) | 88 | 2.4 | 1.5 | 4.9 | 0.14 | 0.08 |
| Serum (µg PFOS/ml) | 141 | 3.9 | 2.4 | 8.7 | 0.24 | 0.15 |
| Egg yolk (µg PFOS/ml) |  |  |  | 62 | 1.7 | 1.0 |

[a] LOAEL values based on bobwhite quail definitive study
[b] TRV estimated with total uncertainty factor derived for a generic Level IV predator by the USEPA GLI protocol [112]
[c] PNEC estimated with total assessment factor of 60
[d] ADI = Average daily intake (mg PFOS/kg body weight per day); estimates were based on pen averages

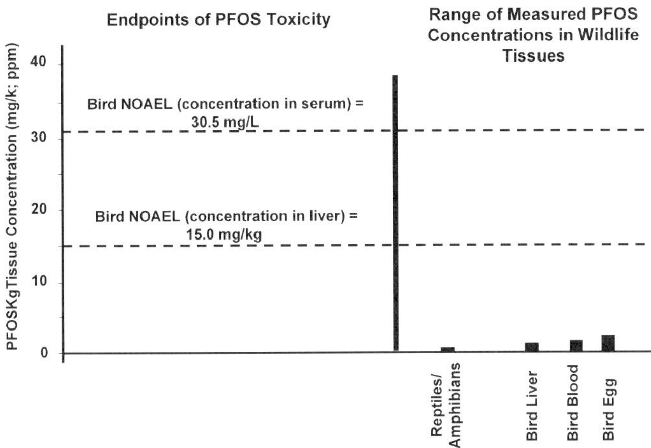


**Fig. 12** Concentrations of PFOS in the tissues of wildlife from the Laurentian Great Lakes, compared to tissue concentrations for the protection of wildlife [123]

ment. First of all, the environmental burden of PFOS precursors has not been well characterized or quantified. It is known that perfluorooctylsulfonamides are present in the air around the Great Lakes and in the water column, and it is expected that a fraction of this will eventually degrade to PFAs, including PFOS. Because the source of these compounds is unknown, it cannot be concluded that a manufacturing phase-out will necessarily lead to a swift reduction in the environmental burden of PFOS. The other uncertainty is how to consider the risk posed by combinations of PFAs. PFOS occurs in greater concentrations than other PFAs, but relative toxicities have not been defined, nor has it been deemed appropriate that a toxic equivalency approach to risk assessment is valid for these substances. To make a collective and comprehensive assessment of the hazard posed by PFAs, the collection of additional exposure information for all of these compounds and the development of additional toxicity information is required. With this information in hand, a more refined risk assessment would be possible.

**Acknowledgements** The authors wish to thank the 3M Company for supporting much of the research on which we report here, making information available, and supporting the preparation of this manuscript through a grant to Michigan State University. S.A.M. acknowledges financial support of the Natural Sciences and Engineering Research Council through a Strategic Grant.

# References


1. Giesy JP, Kannan K (2002) Environ Sci Technol 36:146A–152A
2. Martin JW, Whittle DM, Muir DCG, Mabury SA (2004) Environ Sci Technol 38:5379

3. Sohlenius AK, Andersson K, Bergstrand A, Spydevold O, De Pierre JW (1994) Biochem Biophys 1213:63
4. Key BD, Howell RD, Criddle CS (1997) Environ Sci Technol 31:2445
5. Moody CA, Field JA (2000) Environ Sci Technol 34:3864
6. US Environmental Protection Agency (2000) Perfluorooctyl Sulfonates: Proposed Significant New Use Rule. United States Environmental Protection Agency, Federal Register 65, No. 202, 62319–62333
7. Kissa E (2001) Fluorinated Surfactants and Repellents, 2nd Edn, Revised and Expanded. Marcel Dekker, New York
8. Organization for Economic Cooperation and Development (2002) Hazard assessment of perfluorooctane sulfonate (PFOS) and its salts. ENV/JM/RD (2002) 17/Final. 21 November, 2002. Paris. p. 362
9. 3M Company (2001) Solubility of PFOS in water. 3M Company, 3M Environmental Laboratory, Laboratory Project Number E00–1716, EPA Docket AR226-1030a025
10. 3M Company (2001a) Kow (Solubility in Water, Natural Seawater, An Aqueous Solution of 3.5% Sodium Chloride, and n-Octanol with Subsequent Calculation of the n-Octanol Water Partition Coefficient (Kow) of PFOS for each of the Aqueous Matrices). USEPA Docket No. AR226–1030a031
11. US Environmental Protection Agency (2002) Revised draft: hazard assessment of perfluorooctanoic acid and its salts. Office of Pollution Prevention and Toxics, Risk Assessment Division. EPA Docket no. AR226–1136
12. 3M Company (2001b) Environmental Laboratory Project Number EOO-17 16. Solubility in Octanol—Shake Flask Method. 3M Company, St Paul, MN. US EPA AR 226–0052
13. Jones PD, Hu W-Y, DeCoen W, Newsted JL, Giesy JP (2003) Environ Toxicol Chem 22:2639
14. Giesy JP Kannan K (2001) Accumulation of Perfluorooctane Sulfonate and Related Fluorochemicals in Fish Tissue. Final Report submitted to 3M on June 20, 2001. USEPA Docket AR226–1030a156
15. Giesy JP Kannan K (2001) Perfluorooctane Sulfonate and Related Fluorochemicals in Fish-eating Water Birds. Final Report Submitted to 3M on June 20, 2001. USEPA Docket AR226–1030a159
16. Giesy JP, Kannan K (2001) Environ Sci Technol 35:1339
17. Martin JW, Smithwick MM, Braune BM, Hoekstra PF, Muir DCG, Mabury SA (2004) Environ Sci Technol 38:373
18. Martin JW, Kannan K, Berger U, deVoogt P, Field J, Giesy JP, Harner T, Muir DCG, Scott BF, Kaiser M, Järnberg U, Jones KC, Mabury SA, Schroeder H, Simcik M, Sottani C, vanBavel B, Kärrman A, Lindström G, van Leeuwen S (2004) Environ Sci Technol 38:248A
19. Kannan K, Tao L, Sinclair E, Pastva SD, Jude DJ, Giesy JP (2005) Arch Environ Contam Toxicol 48:559
20. Lange CC (2001) The 18-day aerobic biodegradation study of perfluorooctanesulfonyl-based chemistries. Pace Analytical Services, Inc. Report No. E01–0415, EPA Docket No. AR-226-1030a038
21. Martin JW, Muir DCG, Moody CA, Ellis DA, Kwan WC, Solomon KR, Mabury SA (2002) Anal Chem 74:584
22. Stock N, Ellis D, Deleebeeck L, Muir DCG, Mabury SA (2004) Environ Sci Technol 38:1693
23. Stock NL, Lau FK, Ellis DA, Martin JW, Muir DCG, Mabury SA (2004) Environ Sci Technol 38:991

24. Shoeib M, Harner T, Ikonomou M, Kannan K (2004) Environ Sci Technol 38:1313
25. Boulanger B, Peck AM, Schnoor JL, Hornbuckle KC (2005) Environ Sci Technol 39:74
26. Ellis DA, Martin JW, De Silva AO, Mabury SA, Hurley MD, Sulbaek-Andersen MP, Wallington TJ (2004) Environ Sci Technol 38:3316
27. Boudreau TM, Wilson CJ, Cheong WJ, Sibley PK, Mabury SA, Muir DC, Solomon KR (2003) Environ Toxicol Chem 22:2739
28. Hatfield T (2001) Screening studies on the aqueous photolytic degradation of potassium perfluorooctane sulfonate (PFOS). 3M Environmental Laboratory, St. Paul, MN, EPA Docket AR226–1030a041
29. Hatfield T (2001) Screening studies on the aqueous photolytic degradation of perfluorooctanoic acid (PFOA). 3M Environmental Laboratory, St. Paul, MN, EPA Docket AR226–1030a091
30. Hori H, Hayakawa E, Einaga H, Kutsuna Koike K, Ibusuki T, Arakawa R (2004) Environ Sci Technol 38:6118
31. Renner R (2001) Environ Sci Technol 35:154A
32. Lei YD, Wania F, Mathers D, Mabury SA (2004) J Chem Eng Data 49:1013
33. Ellis DA, Martin JW, Mabury SA, Hurley MD, Sulbaek-Andersen MP, Wallington TJ (2003) Environ Sci Technol 37:3816
34. Hurley MD, Sulbaek Andersen MP, Wallington TJ, Ellis DA, Martin JW, Mabury SA (2004b) J Phys Chem A 108:1973
35. Sulbaek Andersen MP, Hurley MD, Wallington TJ, Ball JC, Ellis DA, Martin JW, Mabury SA (2003a) Chem Phys Lett 381:14
36. Sulbaek Andersen MP, Hurley MD, Wallington TJ, Martin JW, Ellis DA, Mabury SA, Nielsen OJ (2003b) Chem Phys Lett 379(1–2):28
37. Hurley MD, Sulbaek Andersen MP, Wallington TJ, Ellis DA, Martin JW, Mabury SA (2003) J Phys Chem A 108:615
38. Sulbaek Andersen MP, Stendby C, Nielsen OJ, Hurley MD, Ball JC, Wallington TJ, Ellis DA, Martin JW, Mabury SA (2004a) J Phys Chem A 108:6325
39. Sulbaek Andersen MP, Nielsen OJ, Hurley MD, Ball JC, Wallington TJ, Ellis DA, Martin JW, Mabury SA (2004b) J Phys Chem A 108:5189
40. Hurley MD, Ball JC, Sulbaek Andersen MP, Wallington TJ, Ellis DA, Martin JW, Mabury SA (2004a) J Phys Chem A 108:5635
41. Gauthier S, Mabury SA (2005) Environ Chem Toxicol 24:1837–1846
42. Hatfield T (2001) Hydrolysis reactions of perfluorooctane sulfonate (PFOS). 3M Environmental Laboratory, St. Paul, MN, EPA Docket AR226-1030a039
43. Hatfield T (2001) Hydrolysis reactions of perfluorooctanoic acid (PFOA). 3M Environmental Laboratory, St. Paul, MN, EPA Docket AR226-1030a090
44. Lange CC (2001b) The 35-d aerobic biodegradation study of PFOS. Pace Analytical Services, Minneapolis, MN. 3M Project Number E01-0444. EPA Docket AR226-1030a040
45. Gledhill WE, Markley BJ (2000a) Microbial metabolism (biodegradation) studies of perfluorooctane sulfonate (PFOS). I. Activated sludge/sediment. Springborn Laboratories, Inc., Wareham, MA. EPA Docket AR226-1030a034
46. Gledhill WE, Markley BJ (2000b) Microbial metabolism (biodegradation) studies of perfluorooctane sulfonate (PFOS). II. Aerobic soil biodegradation. Springborn Laboratories, Inc., Wareham, MA. EPA Docket AR226-1030a035
47. Gledhill WE, Markley BJ (2000c) Microbial metabolism (biodegradation) studies of perfluorooctane sulfonate (PFOS). III. Anaerobic sludge biodegradation. Springborn Laboratories, Inc., Wareham, MA. EPA Docket AR226-1030a036
48. Kurume Laboratory (2002) Final report, Biodegradation test of salt (Na, K, Li) of perfluoroalkyl (C = 4 – 12) sulfonic acid, test substance number K-1520 (test

number 21520). Kurume Laboratory, Chemicals Evaluation and Research Institute, Japan

49. Schroder HF (2003) J Chromatgr A 1020:131
50. Meesters RJW, Schroder HF (2004) Water Sci Technol 50:235
51. Hagen DF, Belisle J, Johnson JD, Venkateswarlu V (1981) Anal Biochem 118:336
52. Dinglasan MJ, Ye Y, Edwards E, Mabury SA (2004) Environ Sci Technol 38:2857
53. Wang N, Szostek B, Folsom PW, Sulecki LM, Capka V, Buck RC, Berti WR, Gannon JT (2005) Environ Sci Technol 39:531
54. Martin JW, Mabury SA, O'Brien PJ (2005) Chem Biol Inter 155:165–180
55. Xu L, Krenitsky DM, Seacat AM, Butenhoff JL, Anders MW (2004) Chem Res Toxicol 17:767
56. Arrandale RG, Stewart JT, Manning R, Vitayavirasuk BJ (1989) J Agric Food Chem 37:1130
57. Tomy GT, Tittlemier SA, Palace VP, Budakowski WR, Braekevelt E, Brinkworth L, Friesen K (2004) Environ Sci Technol 38:758
58. Dixon DA (2001) Fluorochemical Decomposition Process. Theory, Modeling, and Simulation. W.R. Wiley Environmental Molecular Sciences Lab. Pacific Northwest National Lab. Richmond WA
59. Yamada T, Taylor PH (2003) Laboratory-scale thermal degradation of perfluorooctanyl sulfonate and related substances. Environmental Sciences and Engineering Group, University of Dayton, Research Institute, Dayton, OH
60. Jacobs RL, Nixon WB (1999) Determination of the melting point/melting range of perfluorooctance sulfonate. Wildlife International Ltd. Project No. 45C-106. USEPA Docket No. AR226-0045
61. van Hoven RL, Stenzl JI, Nixon WB (1999). Determination of the vapor pressure of perfluorooctane sulfonated (PFOS) using the spinning rotor gauge method. Wildlife International Ltd., Project No. 454C-105. USEPA docket No. AR226-0048
62. 3M Company (2001) Solubility of PFOS in natural seawater and an aqueous solution of 3.5% of sodium chloride. 3M Company, 3M Environmental Laboratory, Laboratory Project Number E00–1716, EPA Docket AR226-1030a026
63. 3M Company (2001) Solubility of PFOS in octanol. 3M Company, 3M Environmental Laboratory, Laboratory Project Number E00–1716, EPA Docket AR226-1030a027.
64. 3M Company (2003) Environmental and Health Assessment of Perfluoroocatane sulfonate and its salts. US EPA Docket No. AR226-1486
65. 3M Company (2001d) Soil adsorption/desorption study of potassium perfluorooctane sulfonate (PFOS). 3M Company, 3M Environmental Laboratory, Laboratory Project Number E00–1311, EPA Docket AR226-1030a030
66. Welsh SK (1978) Technical report summary-adsorption of FC-95 and FC-143 on soil. 3M Environmental Laboratory. Project No. 9907612633, EPA Docket No. AR 226-0489.
67. Martin JW, Mabury SA, Solomon KR, Muir DCG (2003a) Environ Toxicol Chem 22:196
68. Martin JW, Mabury SA, Solomon KR, Muir DCG (2003b) Environ Toxicol Chem 22:189
69. Ankley GT, Kuehl DW, Kahl MD, Jensen KM, Butterworth BC, Nichols JW (2004) Environ Toxicol Chem 23:2745
70. Fisk AT, Norstrom RJ, Cymbalisty CD, Muir DCG (1998) Environ Toxicol Chem 17:951
71. Johnson JD, Gibson JS, Ober RE (1984) Fund Applied Toxicol 4:972
72. Kurume Laboratory (2001a) Final report, Bioconcentration test of 2-perfluoroalkyl (C = 4 – 14) ethanol [This test was performed using 2-(perfluorooctyl)-ethanol (test substance number K-1518)] in carp. Kurume Laboratory, Chemicals Evaluation and Research Institute, Japan. EPA Docket AR226-1276

73. Kurume Laboratory (2001b) Bioaccumulation test of Perfluoroalkylcarboxylic acid (C = 7 – 13) [This test is performed using perfluorooctanoic acid (test substance number K-1519)] in carp. Kurume Laboratory, Chemicals Evaluation and Research Institute, Japan. December 18

74. Howell RD, Johnson JD, Drake JB, Youngblom RD (1995) Assessment of the bioaccumulative properties of ammonium perfluorooctanoate: static. 3M Technical Report. May 31. EPA AR 226-0496

75. Moody CA, Martin JW, Kwan WC, Muir DCG, Mabury SA (2001) Environ Sci Technol 36:545

76. Giesy JP, Newsted JL (2001) Selected Fluorochemicals in the Decatur Alabama Area. 3M Report, Project No. 178041. EPA Docket AR226-1030a161

77. Environment Canada (2004) Environmental Screening Assessment Report on Perfluorooctane Sulfonate, its Salts and its Precursors that Contain the C8F17SO2 or C8F17SO3 Moiety (Draft for Public Comments)

78. Elnabarawy MT (1977) Bioconcentration of FM 3422 in Bluegill Sunfish and in Channel Catfish. EPA Docket AR226-0350

79. Hansen KJ, Clemen LA, Ellefson ME, Johnson HO (2001) Environ Sci Technol 35:766

80. Sweetser PB (1965) Anal Chem 28:1766

81. Kissa E (1983) Anal Chem 55:1445

82. Kissa E (1986) Determination of organofluorine in air. Environ Sci Technol 20:1254

83. Taves DR (1968) Nature 217:1050

84. Levine AD, Libelo EL, Bugna, Shelley GT, Mayfield H, Stauffer TB (1997) Sci Total Environ 208:179

85. Belisle J, Hagen DJ (1980) Anal Biochem 101:369

86. Moody CA, Field JA (1999) Environ Sci Technol 33:2800

87. Ohya T, Kudo N, Suzuki E, Kawashima Y (1998) J Chrom 720:1

88. Kannan K, Koistinen J, Beckmen K, Evans T, Gorzelany JF, Hansen KJ, Jones PD, Helle E, Nyman M, Giesy JP (2001) Environ Sci Technol 35:1593

89. Kannan K, Franson JC, Bowerman WW, Hansen KJ, Jones PD, Giesy JP (2001) Environ Sci Technol 35:3065

90. Taniyasu S, Yamashita N, Pettrick G, Kannan K, Gamo T, Lam P, Giesy JP (2003b) Perfluorooctane Sulfonate (PFOS) and Related Compounds in Open Ocean Water. 6th Annual Meeting of Japan Society of Endocrine Disruptors Research, December, 2–3, 2003

91. Taniyasu S, Kannan K, Horii Y, Hanari N, Yamashita N (2003a) Environ Sci Technol 37:2634

92. Yamashita N, Kannan K, Taniyasu S, Horri Y, Okazawa T, Petrick G, Gamo T (2004) Environ Sci Technol 38:5528

93. So MK, Taniyasu S, Yamashita N, Giesy JP, Zheng J, Fang Z, Im SJ, Lam PKS (2004) Environ Sci Technol 38:4056

94. Giesy J P, Kannan K, Jones PD (2001d) The Scientific World 1:627

95. Kannan K, Newsted JL, Halbrook RS, Giesy JP (2002b) Environ Sci Technol 36:2566

96. Sinclair E, Taniyasu S, Yamashita N, Kannan K (2004) Organohalogen Compounds 66:4069

97. Boulanger B, Vargo J, Schnoor JL, Hornbuckle KC (2004) Environ Sci Technol 38:4064

98. Kannan K, Hansen KJ, Wade TL, Giesy JP (2002a) Arch Environ Contam Toxicol 42:313

99. Kannan K, Choi J-W, Iseki N, Senthikumar KK, Kim DH, Masunaga S, Giesy JP (2002d) Chemosphere 49:225

100. Van De Vijver KI, Hoff PT, Van Dongen W, Esmans EL, Blust R, De Coen WM (2003) Environ Toxicol Chem 22:2037

101. Kannan K, Corsolini S, Falandysz J, Oehme G, S. Focardi S, Giesy JP (2002c) Environ Sci Technol 36:3210
102. Giesy JP, Kannan K (1999) Sulfonated fluorochemicals in aquatic mammals, turtles, and cormorants. Quarterly Report Submitted to 3M
103. Hansen KJ, Johnson HO, Eldridge JS, Butenhoff JL, Dick LA (2002) Environ Sci Technol 36:1681
104. Sargent J, Seffl R (1970) Fed Proc 29:1699
105. Hu W-Y, Jones PD, Giesy JP (2005) Environ Toxicol Pharmacol 19:57
106. Hu W-Y, Jones PD, DeCoen W, Giesy JP (2005) Environ Toxicol Pharmacol 19:153
107. Hu W-Y, Jones PD, DeCoen W, King L, Fraker P, Newsted JL, Giesy JP (2002) Comp Biochem Physiol Part C 135:77
108. Hu W-Y, Jones PD, Upham BL, Trosko JE, Lau C, Giesy JP (2002) Toxicol Sci 68:429
109. Obourn JD, Frame SR, Bell RH, Longnecker DS, Elliott GS, Cook J (1997) Toxicol Appl Pharmacol 145:425
110. Berthiaume J, Wallace KB (2002) Toxicol Lett 129:23
111. Luebker DJ, Hansen KJ, Bass NM, Butenhoff JL, Seacat AM (2002) Toxicol 176:175
112. Sohlenius AK, Andersson K, DePierre JW (1993) Pharmacol Toxicol 72:90
113. Starkov AA, Wallace KB (2002) Toxicol Sci 66:244
114. Shipley JM, Hurst CH, Tanaka SS, DeRoos FL, Butenhoff JL, Seacat AM, Waxman DJ (2004) Toxicol Sci 80:151
115. Olsen CT, Anderson ME (1983) Toxicol Appl Pharmacol 70:362
116. Case MT, York RG, Christian MS (2001) Int J Toxicol 20:101
117. Lau C, Thibodeaux JR, Hanson RG, Rogers JM, Grey BE, Stanton ME et al. (2003) Toxicol Sci 74:382
118. Lau C, Butenhoff JL, Rogers JM (2004) Toxicol Appl Pharmacol 15:231
119. Seacat AM, Thomford PJ, Hansen KJ, Olsen GW, Case MT, Butenhoff JL (2002) Tox Sci 68:249
120. Seacat AM, Thomford PJ, Hansen KJ, Clemen LA, Eldridge SR, Elcombe CR, Butenhoff JL (2003) Toxicol 183:117
121. Thibodeaux JR, Hanson RG, Rogers JM, Grey BE, Barbee BD, Richards JH, Butenhoff JL, Stevenson LA, Lau C (2003) Repro Develop Toxicol 74:369
122. Christian MS, Hyberman AM, York RG (1999) Combined oral (gavage) fertility, developmental and perinatal/postnatal reproduction toxicity of PFOS in rats. Argus Res. Laboratory, Inc., Horsham, PA. USEPA Docket No. 8EHQ-0200–00374
123. Beach SA, Newsted JL, Coady K, Giesy JP (2005) Rev Environ Contam Toxicol 186:133–174
124. U.S. Environmental Protection Agency (1995) Final Water Quality Guidance for the Great Lakes System: Final Rule. Federal Register 60:15366-15425. USEPA, Washington, DC, USA
125. McCarty LS, Mackay D (1993) Environ Sci Technol 27:1719
126. Di Toro DM, McGarth JA, Hansen DJ (2000) Environ Toxicol Chem 19:1951
127. Drottar KR, Van Hoven RL, Krueger HO (2001) Perfluorooctane sulfonate, Potassium salt (PFOS): A flow-through bioconcentration test with the bluegill (Lepomis macrochirus). Wildlife International, Ltd. Project Number 454A-134. EPA Docket AR226-1030a042
128. MacDonaald M, Sibley PK, Warne A, Stock N, Mabury SA, Solomon KR (2004) Environ Toxicol Chem 23:2116
129. Newsted JL, Beach S, Gallagher S, Giesy JP (2005a) Arch Environ Contamn Toxicol, in press
130. Newsted JL, Jones PD, Coady K, Giesy JP (2005b) Environ Sci Technol 39:9357–9362

# Subject Index

Printing: Krips bv, Meppel
Binding: Stürtz, Würzburg

DATE DUE

Printed
In USA